U.S. Marine Corps Physical Readiness Training for Combat

Plus

Marine Operations on Urbanized Terrain

Department of Defense

Pentagon
Publishing

UNITED STATES MARINE CORPS
Marine Corps Combat Development Command
Quantico, Virginia 22134-5001

29 January 1988

FLEET MARINE FORCE REFERENCE PUBLICATION............0-1B

MARINE PHYSICAL READINESS TRAINING FOR COMBAT

1. PURPOSE

Fleet Marine Force Reference Publication (FMFRP) 0-1Marine Physical Readiness Training for Combat is intended for use by all Marines. It provides the information and references necessary to establish and conduct physical conditioning programs to prepare Marines for the physical demands of combat.

2. APPLICABILITY

Although not directive, this handbook is intended for use as a reference by all Marines in developing unit physical readiness training for combat. This FMFRP does not establish policy for the conduct of physical fitness training. Refer to Marine Corps Order 6300.3_ for applicable Marine Corps policy on physical fitness. It may also be used as a basis for instruction conducted by unit leaders or as a reference for instruction or training.

3. RECOMMENDATIONS

To improve this manual, the MAGTF Warfighting Center seeks comments and recommendations from the user. Officers and noncommissioned officers with constructive comments are encouraged to respond. Lack of response will be considered as concurrence. The User Suggestion Form which follows can be detached and forwarded to - -

Commanding General
Marine Corps Combat Development Command (WF 06)
Quantico, Virginia 22134-5001

4. DISTRIBUTION

Initial distribution of this FMFRP 0-1B is shown below. Addressees on this list should ensure that appropriate subordinate organizations are provided copies.

JOHN I. HOPKINS
Brigadier General, U.S. Marine Corps
Assistant Chief of Staff, Warfighting
Marine Corps Combat Development Command
Quantico, Virginia

DISTRIBUTION: TDJ

From:

To: Commanding General, Marine Corps Combat Development Command (WF 06), Quantico, Virginia 22134-5001

Subj: FMFRP 0-1B, MARINE PHYSICAL READINESS TRAINING FOR COMBAT, RECOMMENDATIONS CONCERNING

1. In accordance with the Promulgation Page to FMFRP 0-1B, which invites individuals to submit suggestions concerning this FMFRP directly to the above addressee, the following unclassified recommendation is forwarded:

Page________Article/Para No.________Line No.________Figure/Table No.________

Nature of Change:_____Add_____Delete_____Change_____Correct

2. Proposed New Verbatim Text: (Verbatim, double spaced; continue on additional pages as necessary.)

3. Justification/Source: (Need not be double spaced.)

NOTE: Only one recommendation per page.

PREFACE

MARINE PHYSICAL READINESS TRAINING FOR COMBAT

Chapter 1
PHYSICAL READINESS LEADERSHIP
Section I. Role of Physical Fitness in Combat

1101. PURPOSE OF PHYSICAL FITNESS TRAINING

Physical fitness training in the Marine Corps has one purpose: to prepare Marines to physically withstand the rigors of combat. All other goals of physical fitness training are subordinate to and must support attainment of this goal. The idea that only infantry or reconnaissance units and their attachments normally face physically demanding combat is wrong. This error must not influence the priority commanders of combat support, combat service support, aviation, and headquarters units give to physical fitness for combat. Physical fitness for combat has a high priority for all Marines. A sound, effective unit program, requiring limited time and material, offers a greater payoff in combat than many more expensive and time-consuming training programs. This handbook describes the concept and provides the ingredients of a program for commanders to use in carrying out one of their most serious responsibilities: ensuring that their Marines are physically ready for combat.

1102. SCOPE OF THE MANUAL

This handbook provides guidance for all leaders, trainers, and planners of physical training programs. It describes unit physical fitness for combat training in the following chapters:

a. Physical Readiness Leadership. Chapter 1 provides guidance to leaders in the conduct of physical combat readiness training.

b. Physical Readiness Training Programs. Chapter 2 discusses how to structure programs to meet physical readiness goals in a variety of combat training situations.

c. Physical Conditioning Activities. Chapter 3 discusses the primary physical conditioning activities which commanders may use to attain readiness for combat goals. Sections in this chapter address foot marches under load, strength-building activities, and activities which build confidence and the aggressive spirit needed in combat.

d. Combat Water Survival. As an amphibious force, Marines cannot ignore the importance of combat water survival training. Chapter 4 describes a program for commanders to conduct this training.

e. Competitive Conditioning Activities. Chapter 5 describes competitive activities which stress the relationship to combat actions, featuring team-building types of competition.

f. Evaluation of Performance During Training. Chapter 6 describes tests which measure individual and unit physical fitness for combat.

g. The Human Body. Chapter 7 covers the structure and functioning of the body.

1103. COMMANDER'S ROLE

Major General Commandant John A. Lejeune in the 1921 edition of Marine Corps Manual expressed our philosophy of leadership. Among other things, General Lejeune stated that "it will be necessary for officers ... to endeavor to enlist the interest of their men in building up and maintaining their bodies in the finest physical condition." It is significant that the Commandant prefaced these writings in 1921 with the statement that his thoughts were greatly influenced by the experience of World War I. We have never varied as a Corps from this belief that one of a commander's most serious responsibilities is to have Marines physically prepared for combat. Senior commanders and their staffs carry out this role primarily by providing command emphasis, including planning, support, and supervision. Commanders at the battalion/squadron and company/battery level execute a program of physical fitness for combat by close supervision and, most importantly, leadership by example.

1104. HISTORY OF MILITARY PHYSICAL READINESS

Every war has revealed our military physical deficiencies during the initial periods of mobilization. This realization followed the Civil War and has recurred regularly with each national emergency.

a. Training programs in each war were geared to the physical need of the era. Success was dependent upon the amount of time available during training to physically prepare Marines for battle conditions. Frequently, casualties in initial engagements were attributed to the inability of our Marines to physically withstand the rigors of combat over rugged terrain and under unfavorable climatic conditions. With adequate preparation, Marines have always handled the test of battle.

b. During World War II, the first physical conditioning doctrine that could be scientifically justified by testing was introduced. As the war progressed, this program was effective in the physical conditioning of millions of Marines for combat.

c. Postwar periods have traditionally been a time of consolidation. Unfortunately, some leaders considered the conditioning phase of training to be a wartime tool. With such a philosophy prevalent between wars, physical readiness was relegated to secondary importance resulting in a serious lowering of combat effectiveness. The initial commitment of Marines in Korea dramatically displayed this failure to recognize the extreme physical demands of warfare. Call-up of large numbers of reserve Marines and pressing them into the war within weeks after call-up mainly caused the loss in effectiveness.

d. Costly lessons learned from our military experiences over a period of years and the course of several wars led to an increasing interest in the physical conditioning of the individual Marine. No longer can we afford to emphasize physical fitness during wartime and de-emphasize it during peacetime. It is evident that, in spite of increased mechanization and modern weapons, physical readiness retains a vital place in the life of each individual Marine and in every unit within the Marine Corps regardless of mission.

e. Commanders are well aware of the need for rugged and well conditioned Marines. The daily demands of housekeeping, maintenance, support, training, operations, and other time-consuming tasks make it necessary for commanders to set aside time for frequent, regular, and vigorous exercise periods.

1105. PHYSICAL DEMANDS OF COMBAT

There are three primary elements of effective physical fitness for combat: lower body strength and stamina; upper body strength and stamina; and a competitive, combative spirit. Cardiovascular functioning is not treated as a separate element since it is a necessary foundation for the other three. Exercises and activities designed to improve Marines' condition in the three primary elements will foster good cardiovascular and respiratory condition.

a. Lower Body Strength and Stamina. Some common demands that combat places on lower body strength and endurance are marching long distances under load and functioning effectively at the destination; moving quickly and evasively under fire; and carrying wounded Marines to safety.

b. Upper Body Strength and Stamina. Some common demands that combat places on upper body strength and stamina are rapidly emplacing crew-served weapons; handling large-caliber ammunition for extended periods; climbing walls, cliffs, and other high obstacles; and performing field maintenance on air craft or heavy machinery.

c. Competitive, Combative Spirit. While part of this critical aspect of combat readiness is mental and emotional, robust physical condition and a training program which promotes physical aggressiveness greatly enhance a competitive and combative spirit. Activities which force Marines to overcome natural physical fear, which directly leads to fatigue, are particularly valuable.

1106. PHYSICAL EFFECTS OF COMBAT STRESS

It has been thoroughly documented that the added uncertainty and stress of combat have a major physical effect on Marines as well as the acknowledged psychological effect. In other words, in combat, fear equals fatigue. In training, we cannot easily reproduce this combat stress which reduces the effectiveness of individuals and units. However, we can produce fatigue and psychological doubt in training situations by developing a challenging physical training program which stresses the participants. This compels Marines to exercise their ability to continue to function under difficult and trying conditions. Such activities as long foot marches under load and difficult confidence courses train Marines to overcome their own fear and fatigue. Rugged competitive activities such as martial arts and pugil stick fighting prepare Marines psychologically to overcome an opponent as well as their own fear and fatigue. This sort of physically demanding training builds self-confidence and unit morale. The adage that "the unit that sweats more in peacetime will bleed less in battle" certainly applies to its physical training program. Athletic coaches have long followed this approach by such practices as having teams shoot foul shots after wind sprints in order to strengthen their mental conditioning and toughness. Although pushing Marines to their limit is necessary, commanders often set lower standards initially than ultimately desired. The program should then call for steadily increasing the difficulty. A good guide for a starting point is a standard that will be physically and psychologically demanding for no less than three-fourths of the unit. Expressed another way, no more than one-fourth of the unit should be working at less than their limit. The program should progress quickly, augmented, if needed, by remedial training for those unable to meet the unit standards, until the training is challenging for all but the most exceptional Marines in the command. Commanders must never forget that Marines expect to be challenged and thrive on being pushed to their limit.

1107. LIMITS OF PHYSICAL READINESS

A well-conditioned Marine unit has a significant advantage in combat, but this does not mean that physical conditioning can substitute for technical knowledge or good planning. For instance, units must, as discussed in chapter 3, pursue a continuing and challenging program of marches under loads. This does not disprove the axiom that fighting men, if required to carry over one-third of body weight, will generally become ineffective in battle. Leaders and planners must decide what is essential to be carried and must use an effective logistic distribution system, such as unit trains, to make remaining items available when they are needed. In a similar vein, the valid requirement to foster a competitive and combative spirit in Marines does not excuse leaders from the responsibility to practice tactics which will not needlessly risk lives. Physical readiness for combat is a vital goal, but it is only one part of a unit's overall readiness for the lest of combat.

Section II. Fundamentals of Physical Fitness

1201. FITNESS FOR MARINES

Total combat readiness includes technical proficiency and mental, emotional, and physical fitness. If any one of these attributes is lacking, combat effectiveness suffers proportionately. Without technical fitness, Marines lack the knowledge and skill to fight; without mental and emotional fitness, they lack the incentive and will to fight; and without physical fitness, they lack the physical ability and confidence to fight. Physical fitness in a Marine means a whole and healthy body, the capacity for skillful and sustained performance, the ability to recover from exertion rapidly, the desire to complete a designated task, and the confidence to face any eventuality.

1202. COMPONENTS OF PHYSICAL FITNESS

A sound body, free of disease and defect, does not in itself constitute physical fitness. Before an untrained Marine can be considered physically fit for combat, the following traits must be developed:

a. Strength. Every Marine needs enough strength to perform the heaviest task encountered in routine and emergency activities. The basic areas where strength is required are in the arm and shoulder girdle, abdomen, back, and legs. Muscles increase in size, strength, and firmness with regular and strenuous exercises. Without work, they grow flabby and weak.

b. Endurance. Each Marine needs sustaining power to maintain the maximum ability without undue fatigue. There are two types of endurance:

(1) Muscular Endurance. The Marine needs muscular endurance to fight the enemy under the most tiring combat conditions. Muscular endurance is the ability to perform continuous work over long periods of time. Endurance depends on the bloodstream's ability to deliver large amounts of oxygen and nutrition to the muscle masses and then carry away the waste products quickly.

(2) Cardiovascular-Respiratory Endurance. The development of cardiovascular-respiratory endurance ("wind") is necessary to maintain muscular endurance. Cardiovascular-respiratory endurance depends on the efficiency of the blood vessels, lungs, and heart. The maximum effort a Marine can exert over a period of time is limited by the capacity to absorb oxygen and expel carbon dioxide. The average Marine's cardiovascular-respiratory capacity can be greatly increased by exercise.

c. Agility. A Marine must be able to change direction quickly and as faultlessly as possible. The ability to react instantly and to maintain orientation during rapid changes of body position is important to survival. This important ability may be developed by conditioning exercises that require varied and rapid changes of body position on the ground and in the air, such as obstacle courses.

d. Coordination. Coordination is the ability to move all parts of the body in a smooth, efficient, and concerted effort (commonly called timing). A well-coordinated individual does not make useless movements. An individual who moves with precision and accuracy saves energy. Coordination is best developed by practicing diversified muscular activities and skills affecting all body parts.

1203. TYPES OF EXERCISE

Basically, there are two forms of exercise: isotonic and isometric. Both forms are contained in the activities found in this manual.

a. Isotonic. Isotonic exercises are those in which the expenditure of energy is regulated and released during concentrated efforts. The regulated expenditure of energy is controlled by both the mode of exercise and the individual's effort. This type of effort is common to most exercise and sports. To develop endurance, coordination, and flexibility, isotonic exercise must be used. Strength can also be developed or increased through this type of exercise.

b. Isometric. Isometric exercises are those in which maximum effort is applied and held until the engaged muscle energy is depleted during a single contraction effort. The individual exerts full force against an immovable object for a relatively short period of time (6 to 10 seconds) and then repeats the contraction several times with a short period of rest between each contraction. This type of exercise develops only strength; therefore, it has limited application.

1204. PRINCIPLES OF PHYSICAL CONDITIONING

Men and women vary in their physical makeup. Physiological (body) function and reaction also vary in proportion to heavy demands placed upon the body. To attain the maximum benefit without sacrificing Marines' welfare, fitness must grow with a careful program of conditioning. To allow for adjustments in body functioning as the conditioning program progresses and to ensure attainment of objectives, the following principles of physical conditioning must be applied:

a. Overload. As strength and endurance increase, the physical load must be increased until the desired level of fitness is reached.

b. Progression. In beginning stages, the load must be moderate. Gradual progression from this low state of fitness to a higher state is possible through application of a progressive program.

c. Balance. An effective program utilizes various types of activities and provides for the concurrent development of strength, endurance, and coordination as well as basic physical skills.

d. Variety. Some programs fail because the routine becomes boring. The most successful programs always include conditioning activities, competitive events, and military physical skill development.

e. Regularity. There is no easy or occasional way to develop physical fitness. Regularity of exercise is a must, with daily exercise preferred.

1205. THREE STAGES OF PHYSICAL CONDITIONING

Unconditioned or poorly conditioned Marines pass through the following stages in reaching the desired state of physical condition.

a. Toughening Stage. This stage is approximately 2 weeks long and is usually characterized by muscular stiffness and soreness followed by recovery.

b. Slow Improvement Stage. This stage is approximately 6 to 10 weeks long and is characterized by slow and steady improvement until the desired level, or a high level, of fitness is attained.

c. Sustaining Stage. This stage goes on indefinitely in order to maintain the level of conditioning achieved by passage through the previous stages.

1206. EFFECTS OF CLIMATIC CONDITIONS

Temperature, both atmospheric and body, affects the physical performance of Marines. Proper maintenance of body temperature through warm up exercise, proper dress in cold weather, and removal or adjustment of clothing in hot weather is necessary for effective performance and health. Climatic factors to be considered are the following:

a. Exercise In High Temperatures and High Humidity. Marines can endure strenuous physical activity in extremely hot temperatures if they are given an opportunity to become acclimated and if they take enough salt and water. It is essential to continue physical training programs in hot climates. Marines can better withstand high temperatures when they are well-conditioned. High humidity combined with high temperatures presents a serious danger. These conditions prevent the natural cooling of the body by the evaporation of perspiration. Training schedules should conform to the provisions of the base commander in respect to wet bulb conditions. Those who conduct training under these conditions should monitor weight loss and be careful to make fluid replacement adjustments.

b. Exercise at High Altitudes. Certain problems are encountered in conditioning Marines stationed in high altitudes because the heart undergoes greater exertion during exercise. It is particularly important that only light exercise be given initially at such altitudes. A Marine's body gradually adjusts to high altitudes within a few weeks. After this adjustment, progressively greater amounts of exercise are possible.

c. Exercise in Arctic Regions. Military duty in the arctic is so arduous that a high level of physical conditioning is essential. Because of the difficulties of carrying on physical conditioning exercises in extreme cold, Marines should be conditioned to the highest level possible before they arrive. A sustaining program will then maintain that level. When exercising in cold weather, Marines should be required to remove excess clothing to prevent them from becoming damp with perspiration.

1207. WARMING UP AND COOLING OFF

It is a fundamental physiological principle that Marines should warm up gradually before taking strenuous exercise. Such action speeds up the circulation to prepare the body to take an overload and helps to prevent injury to muscles and joints after exercising. Marines should be kept mildly active, walking, stretching, or performing some other mild muscular activity until breathing and temperature have returned to normal. Marines should never be allowed to cool off too rapidly; in cool or cold weather, they should put on additional clothing during the cooling-off period.

1208. PHYSICAL ACTIVITY AS AGE INCREASES

Combat makes severe physical demands on Marines. All Marines, regardless of age, must be physically ready to meet these demands. There is no physiological reason to cease exercise or exertion with age. Increased age usually brings increased responsibility which, in many instances, leads to a routine that can become almost devoid of physical activity. The key to fitness with increased age is to continue exercising at a reasonable level and to include exercise of a vigorous type in the daily routine. Older persons who have not regularly maintained a reasonable state of physical fitness require more time than younger persons to become fit. Such individuals usually require a longer period of time to recover from physical effort than younger Marines. If general health is good, evidence shows that older Marines can develop and maintain a degree of fitness which permits vigorous activity and proper performance of duties. It is both the individual's and the commander's responsibility that all Marines continue a daily sustaining exercise program.

Section III. Goals of Physical Readiness Training

1301. NECESSITY FOR PHYSICAL READINESS TRAINING

A very important objective of training is attainment and maintenance of operational readiness. Marines must be physically ready for operations at any time and under all conditions of climate and environment. A combination of training to develop proficiency in physical skills and conditioning to improve strength and endurance results in physical readiness for combat. The degree of physical fitness required of Marines can be acquired through physical exercises in a normal well-coordinated and closely monitored training routine. The performance of purely military duties, such as drills and marching, is not enough to build all the desired areas of fitness. Few recruits are physically fit for the arduous duties ahead of them. The softening influences of our mechanized civilization and the difficulties of conditioning Marines make physical fitness more important than ever before. If Marines are to be fully developed to and maintained at the desired standard of physical fitness, a well-conceived plan of physical readiness training must be a basic part of every training program. Marines cannot be adequately prepared in any other way for the hard work and arduous demands required on the battlefield.

1302. OBJECTIVE OF PHYSICAL READINESS TRAINING

The overall objective of the physical readiness training program is to develop individuals and units who are physically able and ready to perform their duty assignments or missions during training and in combat. To attain the objective of physical readiness, exercise activities must develop the following areas:

a. Strength and Endurance. Developing adequate strength to perform required duties and adequate endurance to sustain activity over a long period of time.

b. Muscle Tone. Developing muscle tone adequate to maintain proper posture and reasonable weight control.

c. Skills. Developing military physical skills which are essential to personal safety and effective combat performance. As skill is developed, agility and coordination are attained. The essential skills are-

(1) **Marching Under Load.** Marching with individual and unit weapons and equipment.

(2) **Running.** Distance and sprint running on roads and cross-country.

(3) **Jumping.** Broad jumping and vertical jumping downward from a height.

(4) **Dodging.** Changing body direction rapidly while running.

(5) **Climbing and Traversing.** Vertical climbing of ropes, poles, walls, and cargo nets. Traversing horizontal objects such as ropes, pipes, and ladders.

(6) **Crawling.** High crawling and low crawling for speed and stealth.

(7) **Throwing.** Propelling objects, such as grenades, for a distance with accuracy.

(8) **Vaulting.** Surmounting low objects, such as fences and barriers, by use of hand assists.

(9) **Carrying.** Carrying objects and employing man-carries.

(10) **Balancing.** Maintaining proper body balance on narrow walkways and at heights above normal.

(11) **Falling.** Contacting the ground from standing, running, and jumping postures.

(12) Surviving in Water (Or Other Specialized Situations) Using water survival techniques.

d. Character Traits. Jnstilling character traits which help accomplish military missions to include-

(1) Confidence. Confidence develops through achieving progressively more difficult tasks as physical ability grows.

(2) Aggressiveness. Participation in combative activities and contests develops desire and willingness to overcome an opponent.

(3) Reaction Under Pressure. Competitive contests and game situations train Marines to think and to act quickly while under pressure.

(4) Teamwork. Teamwork develops through competitive events in which a number of Marines must coordinate their efforts to accomplish a physical task.

1303. BENEFITS OF EXERCISE

The benefits of exercise are not always understood. Some of the more important results of exercise are listed below:

a. Improved Muscle Tone. Muscular tone improves and, at the same time, muscular strength and endurance are built up.

b. Cardiovascular-respiratory Endurance. Cardiovascular-respiratory endurance, or wind, improves through a process of opening up dormant lung capacity to absorb greater amounts of oxygen.

c. Circulation. Circulation of the blood speeds up and extends to a greater portion of the body as exercise forces the blood to service all parts of the body. The efficiency and effectiveness of the heart, lungs, and blood vessels improve.

d. Flexibility. A wider range of muscular movement is possible and rapidity in physical skills grows.

e. Elimination of Body Waste. Bending and twisting the body and the general speedup of body processes caused by exercise regulate and help eliminate body wastes.

f. Tension. Working off excess nervous energy and relief from daily worries and cares relieve tension. Participation in exercises leaves little time for worry.

g. Sleep. Sleep improves because muscles are healthfully tired after about of exercise. A by-product of sound sleep is relief of tension.

h. Obesity Control.. Control of obesity (fat) is made possible by using up excessive amounts of fat-producing food elements.

i. Injury Susceptibility. Susceptibility to injury is reduced through exercise. Muscles, tendons, and joints are strengthened. Injuries such as hernia, back strain, and joint sprains are less likely to occur if muscles are maintained in proper tone.

Section IV. Leadership Roles

1401. PSYCHOLOGICAL LEADERSHIP

The full development of a Marine's resources is not all physical. To be effective in developing physical readiness, leaders must realize that mind and attitude are also important to success. The more important psychological considerations are to—

a. Promote Understanding of the Value of Physical Readiness. A desire to be physically ready should be created in all Marines. Motivation is increased and Marines take greater interest in their individual physical fitness if they understand the value and benefits of vigorous exercise. When Marines realize their efforts are an investment in their own personal welfare, it should not be difficult to obtain their cooperation. Marines should understand the objectives, the benefits, and the value of each type of exercise activity in their program. They should also understand the relation of physical readiness to survival in combat.

b. Maintain a Positive Approach. Physical readiness training for combat is strenuous and demanding. It is a responsibility of leadership to create an atmosphere where all desire to participate fully. This attitude should be fostered. A negative approach must not be identified with physical readiness training, even with those having difficulty. Only in unusual cases should fear of punishment be the motivating factor behind good performance. For those few who cannot keep up or attempt to malinger, an effective remedial program is essential.

c. Seek Cooperation and Develop Morale. In a program placing maximum physical stress upon individuals, it is necessary to gain their cooperation. Favorable reaction is enhanced by proper planning and organization, challenging requirements, use of competition, and application of a progressive program resulting in physical fitness. As physical fitness grows, morale also grows.

1402. COMMAND AND SUPERVISORY FUNCTIONS

a. Command Functions. Commanders should take the following actions to support physical readiness training:

(1) Lead by personal example.

(2) Instill command interest and indicate to subordinate personnel the importance of this training to the welfare of the organization.

(3) Allot sufficient time for the achievement of objectives and monitor the use of such allotted time. The substitution of other training or routine duties for scheduled physical readiness training is unsound and unwise.

(4) Assign and properly utilize qualified personnel to supervise and conduct physical readiness training. If leaders are not competent, take action to ensure they become competent quickly. Reassign those who do not meet this standard to other duties.

(5) Make necessary facilities and funds available to support a program to develop physical readiness within all personnel.

(6) Measure the physical fitness of individuals and units in order to evaluate progress and to determine if the program is successful.

b. Supervisory Functions. Leaders responsible for planning, conducting, and supervising physical readiness training should take the following actions:

(1) Prepare physical readiness training schedules which apply the principles of physical conditioning and which aim for a particular type of program plan.

(2) Provide for wide participation of as many Marines as possible. All Marines, regardless of position or age, will benefit from regular exercise. In some instances, special efforts are necessary to overcome obstacles to regular and frequent training. Special effort is also necessary to ensure remedial conditioning. Such conditioning should occur for those who are physically substandard and after extended absence due to leave, sickness, injury, and travel.

(3) Prevent waste or unwise use of time allotted for physical readiness training. Time-wasters include unprepared instructors; assignment of one instructor to a group larger than a platoon; progression which does not keep pace with the physical development of the Marines; extreme formality; inadequate equipment or facilities which require waiting turns to exercise; and lengthy rest periods between exercises which interfere with the application of overload.

(4) Ensure that the program contains vigorous physical activity. Such activity places progressively greater demands upon the body during each exercise session and also over the duration of the training program. To be of benefit, exercise must tire the muscles and cause the heart to increase its rate of beat.

(5) Set an overall objective for each physical fitness program.Observe the training as necessary to ensure that the established objectives are being achieved.

(6) Observe physical readiness training to insure the use of a positive approach. To implement a positive attitude, small-unit leaders and instructors should personally set the example; have an understanding, fair, and sympathetic attitude; recognize individual differences; and motivate Marines toward their best effort.

(7) Guide and inform small-unit leaders and instructors concerning approved techniques, directives, and literature. As necessary, arrange for local training of instructors to include clinics, conferences, schools, and demonstrations.

(8) Determine the effectiveness of physical readiness training by personal participation in and observation of training, analysis of field inspection reports, and analysis of individual physical fitness test scores. Scores may be combined to reflect the fitness of the unit.

1403. SMALL-UNIT LEADERS AND INSTRUCTORS

a. Responsibility. The instruction and conduct of physical readiness activities are the function of company/battery commanders, platoonleaders and persons assigned as instructors. Experience has proved the effectiveness of physical fitness development when conducted in company-and platoon-size units under direct control of the leader with overall supervision by the parent-unit commander. For example, all the platoons of a company may exercise at the same time under the general supervision of the company commander, with each platoon conducting the assignment separately and under its own leadership.

b. Leader's Assignment. A small-unit leader or an instructor in a school or training activity is assigned to a combat unit or to a support unit. In this assignment, the leader is responsible for all training to include physical readiness training. In a different situation, the noncommissioned officer (NCO) or officer may beassigned as a full-time physical readiness instructor. This contrasts to a unit leader assignment where only part of the time is devoted to such training. In either case, the leader will hold an important and vital position for the physical fitness of Marines.

c. Leader's Training. Leaders may come to the assignment either fully or partially trained. It may be their first responsibility for the development of physical fitness. If they have had previous training through experience, make certain that their information is supplemented with study of this manual. If they have had professional training in physical education during civilian life, but no military experience, they should also use this manual to learn the methods used by the Marine Corps. A new leader should take advantage of various ways to learn including attendance at leader training courses, self-study, practice, and discussion with more experienced leaders.

d.Leader's Objective. As a physical readiness training instructor, the leader has two general objectives. The first is to motivate Marines to want to be physically fit. The second is to conduct a program that will develop a high degree of physical fitnss. Motivated Marines will react enthusiastically to such a program. It aids greatly in achieving local program objectives.

e. Leader's Personal Fitnes. A unit leader who must instruct and demonstrate physical activities must be in physical condition to do the job without undue physical stress.The leader should be able to do those things that must be demonstrated. The leader's strength, endurance, posture, and skill should set the example. This does not mean that the leader must excel, as other Marines do not expect championship performance. However, they do expect, and deserve, a creditable showing of fitness for the job.

f. Leader's Knowledge. The leader must have three types of knowledge to properly administer physical readiness training. Theyare- -

(1) Knowledge of Marines. The leader must understand Marines, know how to lead and motivate them, understand how they learn, and apply this knowledge wisely in the day-to-day training situation.

(2) Understanding of Body Functioning. A more intelligent exercise program results from understanding and applying the principles which govern physical conditioning of the body. The leader with such knowledge can better prescribe, adjust, and regulate exercise types, amounts, and progression to attain fitness.

(3) Understanding Exercise Activities. The leader needs to understand the contribution each type of physical activity makes to physical fitness, and how to use each activity to develop fitness. Skill to demonstrate and lead the various activities is a necessary part of technique and is invaluable to the instructor or small-unit leader.

Chapter 2
PHYSICAL READINESS TRAINING PROGRAMS
Section I. Development of a Program

2101.GUIDANCE FOR PLANNERS

This chapter instructs planners on physical readiness training procedures. It contains program planning guidance including factors to consider when developing programs; steps in assembling a program; definitions of activity packages and systems of exercise; selection of activity packages; and selection of systems used in implementing those packages.

2102. MARINE CORPS MISSION

The mission of the Marine Corps is to seize and defend advanced naval bases and to perform other missions as the President of the United States may direct. Traditionally, to perform this mission, Marines have been projected into the area of operations by Navy vessels and have attacked hostile beaches over the shore in landing craft, assault amphibian vehicles, and helicopters. Recently, performance of this task has been somewhat complicated by the introduction of the concepts of maritime prepositioning and the air-landed Marine expeditionary-brigade. Under these concepts, Marines, landed at airfields in foreign countries, will join their equipment at commercial ports. The distance between these airfields and ports is often 10 or more miles. In time of international crisis, it can be expected that transportation assets will not be adequate. Marines from all elements of a Marine Air-Ground Task Force may have to march from the airfield to the site of their equipment. All Marines must be physically conditioned to rigors of conducting foot marches with individual weapons and equipment.

2103. EVALUATION OF FITNESS

The Marine Corps mission is such that all Marines must constantly and consistently achieve a high level of fitness that prepares them for the demands of combat. While Marines have many different MOS's, all Marines must be prepared for the demands of marching under load and performing basic infantry tasks, such as rear security and patrolling. While the physical fitness test remains our universal measure of individual fitness, it is the commander's responsibility to observe and evaluate the unit's ability to perform effectively in combat. Often the commander's most effective evaluation tool is weekly physical training sessions. The commander should participate in these sessions and see that they are fully integrated into the unit physical conditioning program.

2104. INTEGRATION OF TRAINING

Training time must be used efficiently and wisely. Every opportunity to integrate physical training into other training activities should be seized. For example, foot marches under load can be integrated into the normal daily activities as a way to move from one training or work site to another. This integration reflects realism in training and should be used to maintain overall proficiency. To accomplish this requires imagination and enthusiasm on the part of the commander. The area of operations in which the training is conducted determines what can be integrated and how. If possible, training should also consider the terrain and climate of the area in which the unit will subsequently conduct training or operations. Training should include familiarization with special equipment and the application of specialized techniques to tactical principles. Activities such as gun drills and command post emplacement drills also have physical conditioning value, particularly if supervised and done competitively or measured against time standards.

2105. ACTIVITIES AT THE UNIT LEVEL

Physical readiness training is a command responsibility and is generally conducted in Marine units at the company and platoon levels. Battalion/squadron commanders must constantly supervise, making sure that all Marines are physically prepared for combat.

a. Physical readiness training should usually be conducted weekly by each battalion/squadron-level unit, using events designed for all personnel (e.g., marches under load, battalion/squadron runs, military field days, etc.).These events enable the commanders to demonstrate personal leadership, and observe and evaluate the physical condition and combat physical readiness of their Marines. This exercise period should be designed to elevate the unit's morale and emphasize unit identity through the wearing of uniforms and the carrying of battalion and company guidons. The battalion/squadron commander should make every effort to be a visible participant in the unit's physical readiness program on a weekly basis. These activities could be conditioning marches under load or a weekly calisthenics/unit run.The suggested times for these activities are early on Monday morning or late on Friday afternoon after the training week has been completed. The uniform may be physical training gear orutilities and boots.

b. Commanders should be cautious of long runs in boots and utilities. This is not to suggest that units should not run in utilities, but to remind commanders that boots are designed for marching, not running. If commanders choose boots and utilities, they should select off-road routes which will provide a cushioning effect.

c. Units should be made well aware of the commander's intent before the unit falls out for exercise. If the exercise period is conducted early in the morning, it need not necessitate an extensive shower/clean-up period afterward, so long as all Marines from the commander on down remain dirty. Commanders should inform their Marines not to fall out for these sessions in clean, pressed uniforms.

d. The physical readiness training of headquarters units is often difficult to manage. Often physical training is left for individuals or sections to develop and manage on their own. This approach can be effective, but if not supervised, it can result in a "paper" program. Headquarters unit commanders should endeavor to conduct weekly integrated training which will enable them to evaluate the physical readiness of their personnel. Often this training has to be scheduled before or after normal working hours to maximize participation of the entire unit.

2106. STEPS IN PLANNING

To implement workable and effective programs (as directed by CMC ALMAR 261/87) the planner must-

a. Determine the Type of Program Needed. Marine units are inherently different in organization and mission. The physical readiness program must be tailored to the mission and to the current physical condition of most unit personnel. Programs to meet this need are of the following types:

(1) **Developmental Programs.** Marines in a beginning or poor state of physical readiness need a program which will develop strength, endurance, physical skills, and character traits which are beneficial to successful accomplishment of military missions. Such programs should be applied progressively to rise gradually to a peak of fitness and skill.

(2) Maintenance Programs. Once Marines reach the sustaining stage of conditioning, their goal is then to maintain this level by participation in a maintenance program.

(3) Remedial Programs. The term "remedial" is usually applied to those individuals or groups who possess substandard physical fitness. For example, a remedial physical conditioning program could be applied to persons who are overweight, who fail to reach physical fitness test standards, or who have missed extended periods of conditioning due to illness, injury, extended hospitalization, or other absence. Saturday training sessions are designed as remedial sessions for all individuals who either miss the daily training or have failed to meet adequate standards. Note that personnel in light duty or no-duty status should be expectedto attend all training sessions, observing, supervising, or assisting as needed.

b. Determine the Time Required. The amount of time for training operations varies considerably. However, every unit can find time to conduct physical readiness training. Frequent (i.e.,daily) physical training of short duration-30 to 60 minutes-is preferred over occasional longer periods. Other demands for training time are so urgent that every minute of time allotted for physical readinesstraining should be used. Determine the time required per week anddivide it into daily blocks.

c. Organize for Various Group Size. It is essential to stress exercise rather than formality. Marines must complete the program where they are- -on the training field, in the motor pool, on the range, next to the classroom, in the office area, in the shop, aboard ship, or elsewhere. Although desirable, it is not always possible to assemble company-size units for physical training. Platoon-size groups are appropriate for the proper conduct of physical conditioning activities. Certain situations may require exercise programs for section- or squad-size units. This manual outlines programs for all situations and types of organizations.

d. Allow for Weather and Exercise Area. In programing and scheduling, the climate and terrain often govern the selection of activities.

(1) Weather changes cause differences in temperature, rainfall, wind chill, and snow. These changes should be anticipated as they dictate the type of program. Alternate plans should be part of the schedule. Nevertheless, normal weather changes and rain should not drive training inside.

(2) Local terrain and available exercise areas may also influence the selection of activities and the type of program which it is possible to support. Some activities can be completed in nearly any area.

e. Plan for Seasonal Change. As most physical readiness training is conducted outdoors, it is necessary to recognize seasonal change. A program should be divided into fall, winter, spring, and summer parts. Seasonal change also causes change in light. For example, an early morning program started in the summer will have ideal lightconditions, yet in the fall or winter, darkness will occur at that same hour and interfere with the conduct of the program. Develop programs in seasonal blocks, and make provisions for anticipated changes.

f. Consider Needed Facilities. An excellent program can be conducted with practically no facilities since there are exercises which require no equipment. However, a better program can be developed when supported by certain facilities and items of equipment. Proper command support, plus ingenuity, will solve this problem. (Items of equipment, when necessary to support the recommended exercises, are included in the chapters on exercise activities.)

g. Specify Appropriate Uniform. The uniform worn for exercising depends upon the season of the year, the state of the weather, and local regulations. Whenever possible, Marines should be dressed alike. Undershirts are preferred as the upper garment when the weather permits. A uniform that restricts the free movement of the body should not be worn when exercising.

h. Consider Availability of Instructors. Leaders who can lead and direct the scheduled activity must be available. Organizational units should train junior officers and noncommissioned officers down to squad or section leaders to instruct and lead the various activities.

i. Select Activity and System. With the type of program needed and the objectives in mind, select an activity package or an integrated training package for each day's scheduled physical training. At this same time, the system to be used in employment of the selected activity or activities must be determined. This selection affects equipment, areas, instructors, transportation, and other support requirements.

j. Secure Command Participation and Support. Prepare and brief the commander to assure full understanding of the objectives and administration of the program. The full participation and support of the commander will greatly improve the level of success.

k. Supervise Execution. Determine needs, publish the program, and supervise its execution as a necessary part of developing the unit's training schedule.

2107. SELECTION OF ACTIVITIES AND SYSTEMS

a. Exercise Activities. The planner must determine exercise activities which will be appropriate to include in the program. An exercise activity is a single means of exercise usually identified by the name applied; for example, running, log exercises, and obstacle course. Many exercise activi ties can become part of activity packages.

b. Activity Packages. Many of the physical activities described in this manual are arranged in prescribed sequences and are known as activity packages. An activity package is a number of exercises of the same type, assembled as a group or a set, and arranged in a specific sequence. Exercise packages are organized in such manner that not more than 15 minutes will be required to complete the execution of any package. Each type of activity is explained in laterchapters. The number of available drills, tables, or circuits; the manner of organization; and the contribution each makes to the total program are covered. Full understanding of this information will greatly assist in developing effective programs. Various designations are used to identify exercise packages; for example, conditioning exercises when arranged in a set order are known as drills, and other packaged activities are designated as tables or circuits. The following are activity packages:

(1) Activity packages for groups:

o Conditioning Drill Two
o Conditioning Drill Three
o Rifle Drill
o Log Drill
o Grass Drill
o Running Tables
o Guerrilla Tables
o Circuit-Interval Table
o Combatives Tables
o Relay Tables

(2) Strength Circuits:

o Fixed Circuit
o Movable Circuit
o Simplified Circuit (Circuit Interval Table)

(3) Activity packages for individuals:

o The 6-12 Plan
o Weight Training
o Isometric Exercise

c. Advantages of Using Packages. These of exercise packages simplifies scheduling and conducting of exercise and results in the following benefits:

(1) Schedule development is simplified as the planner assembles packages which will satisfy the training objective. There is no need to deal with selection of individual activities or to be concerned about the amount of time to be expended on each.

(2) Any 15-minute period, and in some cases less time, can be scheduled or used to perform an activity.

(3) If longer periods of time are available or if the objective demands, several packages can be assembled to provide a more complete period of activity.

(4) Marines are assured a balanced set of exercises or activities as each package is carefully arranged to reach all muscle groups.

(5) The instructor can concentrate on the conduct of a vigorous workout as the type and duration of the activity have already been determined.

d. Nonpackaged Activities. Several types of activity are not packaged. Activities in this category are conditioning marches, unit runs, obstacle courses, combat water survival swimming, teamcontests, and team sports. These activities can be scheduled in combination with packaged activities, or they may be scheduled separately. Often they require a longer period of time. Most nonpackaged activities require a 50-minute period to satisfactorily complete their objective. For conditioning marches, a few hours are required. The benefits of these activities should not be overlooked as some desirable objectives cannot be attained without them.

e. Systems of Exercise. Several methods or systems of organizing exercise and activity packages can be used. Each system is based upon a specific organization as follows:

(1) Single Activity System. The unit leader immediately assumes command of the unit at the beginning of the exercise period. The leader moves the unit to a predesignated exercise site at double time, forms the unit in a circle around the leader, groundsclothing and equipment as appropriate, and quickly moves into the exercise routine. There is usually no time to teach; therefore,the Marines must know the activity to be used. At the conclusion of a 5- to 15-minute period, the unit leader returns the unit to the instructional area at double time and releases the unit for the next scheduled activity.

(2) Progressive Activity System. All Marines (company or platoon) complete activities in the same order during the period. For example, Drill One is followed by dual combatives, and finally a 1-mile run. This system is usually progressive from a warm up activity, such as Drill One, to an activity which contributes in a major way to one of the objectives, such as aggressiveness development through combatives or cardiovascular development, such as running.

(3) Rotating Activity System. Set up the same number of activities or stations as there are platoons in the company. Each platoon rotates through each station in turn. With four platoons in a 50-minute period, about 10 minutes can be devoted to each station. With three platoons, approximately 15 minutes may be spent at each station. Activities must be of a type that can be covered in the time allotted. For example, with three platoons, Station 1 could be Conditioning Drill One; Station 2, running; and Station 3, a team contest.

(4) Circuit System. Set up a number of stations to provide various types of exercise equipment and/or items of apparatus. The idea is to keep all Marines busy and exercising vigorously for a short period of time at each station. The fixed strength circuits and the movable strength circuits are examples of this system. Station changes must be rapid, and the exercise must be started quickly after each change. Since the objective is to exercise at top speed, the motivation comes from frequent change of activity by moving to another station. Rotation by station groups continues until all Marines have covered all stations.

(5) Interval System. This system stresses the development of strength and endurance. It involves heavy work for a given distance within a specified time, alternated with lighter work and recovery, but never stopping during the workout. This procedure is repeated, and the intensity is increased gradually as physical condition improves, but always with adequate recovery. The important concept is stress, recover, stress, recover, and so on. This system is often applied through running, but other activities of a continuous nature may also be used. An example is the Circuit-Interval Table.

f. Steps in Selecting Activities and Systems. The planner should follow these steps.

(1) Evaluate the needs of one unit with emphasis upon unit mission, objective, and time available.

(2) For each day, select an activity package or a combination of activities which will contribute to the objective.

(3) Then determine the system to be used in implementing the selected activities. There are several possible choices and much opportunity for flexibility in program development. For example, the weekly program may contain various exercise packages, systems, and time periods. Additional guidance is contained in following chapters.

2108. SAMPLE WEEKLY PROGRAMS

The variety of weekly programs given shows available alternatives. Programs and schedules other than those illustrated may be assembled to provide schedules to fit any situation. Many other scheduling combinations are possible through use of the packaged activities, varied time periods, and different types of assigned duty.

a. Sample A. Figure 2-1 illustrates a 60-minute time allotment in which the single activity system is used with activity packages that vary from day to day. In this program, 5 hours of training are scheduled for the week. This program needs some type of integrated unit training to supplement it each week. A bimonthly conditioning march of 10 miles in 3 hours is suggested to round out this program. The Saturday session is designed for remedial physical training.

NOTE: A physical training activity must be conducted at a minimum of once every 72 hours. After 72 hours, muscles begin to atrophy.

Figure 2-1 is available in the illustration section at the end of the book

b. Sample B. A more comprehensive weekly program (fig. 2-2) illustrates the scheduling of various-length periods, a variety of activities, and the use of three systems during the week. In this schedule, 3.5 hours are included for the week. The following should be noted concerning this schedule:

Figure 2-2 is available in the illustration section at the end of the book

(1) On Monday, a 60-minute period is available for physical readiness training. Three 20-minute packages are scheduled. Each platoon, under its own leadership, will progress through each scheduled activity in turn.

(2) On Tuesday, Wednesday, and Thursday, only 30-minute periods are available. Here each platoon leader supervises each platoon in a single activity system.

(3) On Friday, the company commander directs training to exhibit personal leadership example and supervision. In this example, only one strength circuit and one cross-country course are available, and each will accommodate only one platoon. For this reason, Conditioning Drill One station is included. A platoon rotates to each station within the 60-minute period.

(4) The Saturday session once again is designed for remedial physical training.
(5) Bimonthly, a 10-mile/3-hour march under load will be necessary to round out this program.
NOTE: Cardiovascular/respiratory exercise to improve endurance must be of long enough duration (20 minutes or more) to induce and maintain oxygen debt.

c. Sample C. A third sample schedule (fig. 2-3) illustrates the use of the single activity system for 5.25 hours of training. This schedule includes sustaining-type activities for a unit that has passed through both the toughening and slow improvement stages of conditioning. The use of competitive activities is featured to holdinterest and provide self-motivation. A bimonthly 10-mile/3-hour march under load will be necessary to round out this program. The Saturday session once again is designed forremedial physical training.

Figure 2-3 is available in the illustration section at the end of the book

NOTE: Thc body will respond to hard work or physical training without soreness and stiffness provided the work or physical training is conducted regularly.
NOTE: Excrciscs of short duration increase strength and bulk but do little to effect endurance.
NOTE: Thc times shown in the charts arc participation times.

Section II. Remedial Physical Training

2201. NEED FOR ATTENTION

Remedial physical training is a process by which physically substandard individuals are conditioned to meet standard levels of performance. To achieve a full degree of operational readiness, it is necessary to bring all Marines up to the prescribed standard. Experience has demonstrated that some Marines have difficulty due to a poor state of fitness, obesity, or lack of motivation. Attention to these individual deficiencies will help improve unit combat readiness.

2202. IDENTIFICATION OF INDIVIDUALS

The company or battery commander identifies persons who cannot achieve the prescribed level of proficiency through the physical fitness test or as a result of observation during training. The commander notes particular weakness of body parts as indicated by failure of certain test events. These Marines are then placed in a special remedial program either at platoon, company/battery, or battalion/squadron level. Extra time is devoted to overcoming the weakness. These extra periods of conditioning may be during or after duty hours, as determined by the commanders. Other persons need reconditioning as result of hospitalization. Restoring physical fitness to damaged areas of the body is accomplished through progressively graded physical activities under professional supervision, not by the physical fitness trainers.

a. Types of Deficiencies. Physical deficiencies which can be corrected by exercise fall into several categories.

(1) Lack of Strength in One or More Body Parts. The major muscle areas concerned are the arms and shoulder girdle, back, abdomen, and legs.

(2) Lack of Overall Endurance. Usually, there is deficient muscular and cardiovascular endurance.

(3) Deficiency in Coordination and Agility. In these cases, physical skill is not developed to a satisfactory degree in activities such as crawling, running, jumping, climbing, traversing, vaulting, pushing, pulling, lifting, and carrying.

(4) Overweight or Underweight. Either condition may interfere with physical fitness and mission accomplishment. Lack of exercise is not always the cause. The cause may be malfunction of normal physiological functions or it may be poor health habits such as over- or under-eating, lack of adequate rest, or overconsumption of alcohol.

(5) Lack of Motivation. Not all persons are motivated to attain or maintain **a** desirable state of fitness. Some Marines do not understand the importance of physical fitness, some find proper exercise too difficult, and others find it inconvenient.

b. Causes of Deficiencies. There are several causes for Marines being deficient in physical fitness:

(1) Absence of exercise.

(2) Exercise which fails to develop all muscle groups and components of fitness.

(3) Exercise which is not vigorous enough or which lacks progression.
(4) Injury or illness which depletes fitness.
(5) Inadequate amounts of sleep or rest.

c. Methods of Detection. The commander has several means by which to detect physical deficiencies:

(1)

from those who pass. Make an analysis to determine the cause of failure as revealed by test scores.

(2) Observe Marines as they perform physical tasks-

o Marines who have difficulty during training or in physical types of work,

o Marines who have become obese and therefore experience difficulty.

(3) Be alert to those Marines who are often on sick call, returnees from hospitalization, or newly assigned.

2203. ADMINISTRATION OF REMEDIAL ACTION

a. Group. Attitude Marines who are singled out as being physically deficient are self-conscious and are not always convinced they need extra help. Within a deficient group, motivation may be low. These Marines must be convinced that a remedial program which is tailored to their needs will help them and will eliminate their deficiencies. Encouragement is often needed and desirable.

b. Leadership. The leader of this group must study individuals and know their deficiencies. The unit leader should counsel Marines individually, observe them closely as they progress through the remedial program, maintain records, and adjust the program as required.

c. Measurement.. Whenever improvement in performance is noted, Marines should be measured by the physical fitness test either individually or by group. When testing reveals an individual to be satisfactory, that individual should be released from the remedial group. There may be exceptions to this policy in the case of Marines who are overweight or in the execution of an exercise program prescribed for some special purpose.

d. Organization of a Remedial Group

(1) A remedial group is usually a small group of Marines within a company or battalion. In some situations, it may be a larger group numbering 50 to 75 Marines. In the case of a larger group, divide the Marines initially into subgroups according to ability. Prescribe exercise loads commensurate with their ability. General conditioning is sufficient in the early part of the program to qualify Marines who are on the borderline. Qualification will reduce the size of the group, permitting more individual and specialized attention to those who remain.

(2) As the program progresses, regroup individuals who have like deficiencies in order that they can concentrate on their weakness. For example—

(a) One group may be weak in the arms and shoulders as revealed by failure in the chin up event. This group, with an assistant instruction charge, can work on pull-ups, rope climb, pushups, rifle or logexercises, horizontal ladder, and similar types of exercise.

(b) Another group may be weak in general endurance as revealed by the3-mile run. This group could profit from participation in conditioning drills, running, grass drill, and strength circuit.

(3) It may be necessary to form some groups to overcome weaknesses in skills, such as an inability to throw, to quickly change direction while running, to crawl rapidly, or to carry a load. Lack of coordination or lack of practice may cause these deficiencies. In this instance, instructors must provide an opportunity to practice and correct poor form and other errors as they are noted.

Chapter 3
PHYSICAL CONDITIONING ACTIVITIES
Section I. Marching Under Load

3101. GENERAL

Few physical fitness activities are as directly related to readiness for combat as foot marches under load. In addition to the obvious physical conditioning and unit cohesiveness benefits, marches under load prepare most Marines for numerous foreseeable tasks in combat. The idea that only infantry and reconnaissance units actually require training marches to prepare them for combat is clearly a mistake. In addition to the very real possibility of Marines from all types of units being used to fill combat-depleted infantry units, it is likely in maritime prepositioning force operations that aviation, combat service support, and command element personnel will have to march with weapons and equipment, from the arrival airfield to the equipment marshalling site. In American coastal cities, the harbor is nearly always 10 or more miles from the airport. It is difficult to imagine any different condition in a lesser developed country .A further combat-related benefit is that foot marching under load is the most effective way to develop leg strength in the context of a unit training program. In his classic study of troop performance in World War II, Men Against Fire, S.L.A. Marshall noted that the greatest single weakness of replacements of all occupational fields was leg strength. Truck drivers must push trucks which are stuck in mud, and headquarters communicators must climb hills to emplace antennas. The demands on the lower body in combat apply to all Marines, and foot marching is one activity which can help build these muscles while being conveniently integrated with the rest of the training program.

3102.TRAINING GOALS

The standard for success of a foot march is very simple to measure: did the unit arrive at the destination at the prescribed time with Marines in condition and required equipment present to accomplish the mission?A progressive program can increase a unit's readiness. It can instill pride by increasing distances and rates of march and by selecting routes over increasingly challenging terrain as the Marines become better conditioned. However, the ability to execute the mission at the conclusion of the march must remain the standard of success.

3103.MOVEMENT PLANNING

The basic considerations in planning a foot march are the mission, tactical situation, terrain and weather, and the units to participate. The success of the march will depend largely upon the thoroughness with which it is planned. A successful march is characterized by adherence to prescribed routes and time schedules, the efficient employment of the means available, and the ability of the unit to accomplish its assigned mission upon arrival at the destination.

a. Movement Orders. Movement planning culminates in the preparation and issuance of an operation order prepared in the standard, five-paragraph format. Necessary annexes are attached to furnish detailed information required for the movement. Written movement orders are rarely prepared at company level. A discussion of the annexes is contained in this chapter.

b.March Planning. March planning, as discussed here, is the planning conducted at battalion level. March planning may be organized into the following steps:

(1) **Preparation and Issuance of the Warning Order.** In order to afford subordinate units the maximum possible time to prepare for a pending move, a warning order containing all available information about the march is issued. The amount of planning time available will determine the time of issuance and the content of the warning order.

(2) Estimate of the Situation. In his estimate, the commander considers the mission, terrain, weather, time, and space factors, available routes, available transportation for the movement of equipment and/or shuttling of Marines, enemy capabilities, disposition of own forces, physical condition and training of Marines, and courses of action available to the command.

(3) Organization and Dispatch of a Reconnaissance. Party very march plan is based on as thorough and complete a ground reconnaissance as time and the situation will permit. Map and aerial reconnaissance are valuable in formulating a plan, but are not a substitute for ground reconnaissance. Route reconnaissance is accomplished by a reconnaissance party which usually consists of a reconnaissance element, an engineer element from the attached or supporting engineer unit, and a traffic control element. Unit standing operating procedures generally establish the basic composition of the reconnaissance party. It is modified as necessary to meet the requirements of a particular march. A recommended method of reporting information obtained by the route reconnaissance party is shown in figure 3-1.The minimum information required from the reconnaissance party is—

(a) Available routes and conditions.(Routes may be specified by higher headquarters.)

(b) Recommended rate of march.

(c) Selection of start point and release point, or confirmation of the suitability of start point and release point previously selected by map reconnaissance.

(d) Confirmation of location of the assembly or bivouac area.

(e) Location of critical points on the route.

(f) Distance between critical points on the route and total distance from start point to release point.

Figure 3-1 is available in the illustration section at the end of the book

(g) Location of obstacles and estimation of necessary Marines and equip- ment needed to repair and maintain routes.

(h) Number of guides required and their location on the route.

(4) Development of Detailed Movement Plan. Some of the elements included in a detailed movement plan are-

(a) Organizing the Column. To facilitate control and scheduling, units will be organized into serials and march units and given an order of march. In determining the order of march, the march planner must consider the enemy situation and the desirable order of arrival of the units at the destination. Where dispersion is required, a unit may be organized into two or more columns, each assigned a different route.

(b) Using Reconnaissance Information. Results of the route reconnaissance will be used to select the route(s); determine the start point, critical points along the route(s), and the release point; and select the rate of march.

(c) Determining March Computations. March unit pass time is based on the strength, formation, and rate of march. The pass time of the marching columns, plus necessary time distance computations, will be used to determine the completion time of the march.

(d) Drafting of Road Movement Table. Using the completed march computations, a draft road movement tabic is compiled.

(e) Checking the Plan. Using the draft road movement table and a road movement graph, the movement plan is checked to ensure that it conforms to the directive of the higher headquarters and the battalion commander's instructions.

(5) Preparation and Issuance of the Road Movement Order. After the plan has been checked and approved by the commander, an operation order is prepared and issued. The operation order may be in written form or issued orally, and is accompanied by a road movement table, overlay and/or strip map, and appropriate administrative details.

(6) Road Movement Table. A road movement table, prepared as an annex to the operation order, provides serial commanders with arrival and clearance times at critical points along the route of march. It also provides the column commander with information as to the proposed location of elements of the column at various times.

(7) Overlay/Strip Map. An overlay serves the normal purpose and should show, as a minimum, the present location of units, route of march, critical points, and the new location of units at the destination. A strip map is a schematic diagram of the route of march and shows landmarks and critical points with the distances between them. A strip map may be issued as an annex to the road movement order, in addition to or in lieu of an overlay.

c. March Computations. Prior to issuance of the operation order, the S-3 must verify time and space computations as they provide him with the necessary data for the preparation of a road movement table.

(1) Time-Distance. Time distance (TD) is determined by dividing the distance to be traveled (D) by the rate of inarch (R):

TD = D/RWhere TD = hours D = distance in kilometers R = kilometers per hour

(2) Length of ColumnThe length of column (LC) is used to determine the pass time of a column. The sum total of the following two parts determines the length of column. (See pars, (a) and (b)).The two parts are the space occupied by the Marines alone (including the distance between Marines) and the sum of the distances between the units of the column (column gap).

(a) The length of column of Marines alone is determined by multiplying the number of Marines by the appropriate factor selected from the table below. The length of column does not include distances between units. LC Marines = No. of Marines x factor.

(b) The total distance in meters between units is determined as follows:

Step One. Determine the number of gaps between serials (total serials minus one).

Step Two. Multiply the number of serial gaps from Step One by the length (in meters) between respective units.

Step Three. Determine the number of gaps between march units (total march units minus one, minus the number of serial distances).

Step Four. Multiply the number of march unit gaps from Step Three by the length (in meters) between respective units.

Step Five. Add the totals from Step Two and Step Four to get the total meters for the column gap.

EXAMPLE:A battalion foot column is organized into 12 platoon-sized march units and 3 company-sized serials. REQUIRED: total column gap distances when there are 100 meters between serials and 50 meters between march units.

(3) Pass-Time.Pass-time (PT) is the time a unit takes to pass a specified point. For foot columns, the pass-time is determined by applying the following formula:PT (minutes) = LC x FACTOR (for appropriate rate of march).

PASS-TIME FACTORS - FOOT TROOPS

.0150 for 4.0 km/h

.0187 for 3.2 km/h

.0250 for 2.4 km/h

.0375 for 1.6 km/h

EXAMPLE: Determine the pass-time of a unit whose length of column is 1,500meters and is marching at a rate of 4 km/h.PT (min) = 1,500 x .0150 (the factor for 4.0 km/h) = 22.5 mm.

(4) Completion Time. Completion time is the time of day that a march will be completed. Completion time is determined by using the following formula: Completion Time = SP time + TD + PT + Scheduled Halts (other than normal hourly halts).

EXAMPLE: A column's starting point (SP) time is 0700.The time-distance is6 hours and 40 minutes. Pass-time of the column is 30 minutes. A 35-minutelunch halt has been scheduled. What is the completion time of the march?Employing the 24-hour clock system, the formula can be applied as follows for simplified addition of the times:

HrMin SP time................0700

TD....................0640

PT....................0030

Lunch halt..............0035

Completion time.........13 105or 1445

The march will be completed at1445.

(5) Experience Tables. Based on previous movements made by a unit, data is accumulated to facilitate march planning. Such data includes approximate pass-times lor various elements of the battalion. The S-3 can utilize these data rather than computing them each time a march is scheduled. Such experience tables serve to reduce the time required to complete the computation phase of march planning. Matter appropriate to the unit standing operating procedure should be integrated therein.

d. Road Movement Graph

(1) A road movement graph is a time-distance diagram used in planning, preparing, or checking road movement tables, and for controlling marches. The graph provides a visual representation of a march plan so that conflicts and discrepancies may be prevented in the planning stage, before congestion occurs on the route. It is not usually issued as a part of the order. Road movement graphs may be applied to small units, to movements of a single column, or to a large organization scheduling separate elements, moving by various means, with different rates of march, over one or more routes.

(2) To construct a road movement graph, use the following steps: (See fig. 3-2.)

Figure 3-2 is available in the illustration section at the end of the book

Step One. Determine the number of hours available for the march. Designate the lower left corner of a sheet of graph paper as the SP time or the earliest even hour before the march is to begin. Select a scale and plot the hours available in sequence from the left on the horizontal axis.

Step Two. Determine the distance to be moved in kilometers. Indicate the SP at the lower left corner of the graph sheet and, using an appropriate scale, plot the number of kilometers on the vertical scale from the SP to the release point. Indicate the location of critical points on the vertical scale.

Step Three. At the proper distance from the SP, draw a horizontal line indicating the location of the RP. Indicate by a vertical line the hour when the movement must be completed. Plot lines representing route restriction, if any, at the proper distances and times on the graph.

Step Four. Determine the pass-times of foot and motor elements of the column.

Step Five. Starting at the SP at the specified hour, plot the movement of the head of the leading element. If the rate of march is 3 kilometers per hour, the lead element will have moved 1.5 kilometers in 30 minutes, 3 kilometers in 1 hour, etc. Plot the trace of the lead man to the RP. The last Marine will cross the SP in pass-times after the lead man. Measure this time on the graph and plot the trace of the last Marine of the column. The lines describing the head and tail of a march serial are parallel. Indicate the time subsequent serials reach the SP and plot the trace of the head and tail. Check to see that the plan complies with all restrictions and orders. If there are any violations or conflicts, the plan may be made to conform by changing the order of march, the starting time of the serials, the planned rate of march, or the organization of the column.

e. Conversion to the Metric System. The adoption of the metric system for expressing linear distances requires a simple means of conversion.

(1) To accomplish this, use the following tables:

o Multiply inches x 2.54 to obtain centimeters. Multiply centimeters x .39 to obtain inches,

o Multiply yards x .91 to obtain meters. Multiply meters x 1.1 to obtain yards,

o Multiply miles x 1.6 (or 8/5) to obtain kilometers. Multiply kilometers x .62 (or 5/8) to obtain miles.

(2) Linear distances can be accurately measured on maps which have been converted to the metric system by using a straightedge graduated in centimeters (cm).The centimeter scale can be used in lieu of the graphic scale when the scale of the map is shown, as follows:

Map Scale:	One cm Equals:
1:500,000	5,000 meters
1:250,000	2,500 meters
1:100,000	1,000 meters
1:50,000	500 meters
1:25,000	250 meters

3104. INDIVIDUAL LOAD

Since the backbone of Marine ground combat forces is the infantry, tactical mobility is largely dependent on the marching ability of the individual Marine. The load that each Marine must carry is the greatest impediment to mobility. In general, the commander must make every effort to reduce the individual load of his Marines to the absolute minimum. While the load has the greatest effect on Marines marching on foot, the value of carrying the minimum essential supplies and equipment applies for all Marine forces.

a. Fighting Load. The fighting load consists of items of clothing, equipment, weapons, and ammunition that are carried by, and essential to, the effectiveness of the combat Marine and the accomplishment of the immediate mission. The fighting load should not weigh more than 40 pounds. The commander must ensure that his Marines' loads- -and his own- -be stripped to the bare minimum. The addition of small, light-weight items in excess of the limit should not be tolerated, since cumulatively, these items will take a toll in energy. Every extra pound a Marine must carry decreases his combat effectiveness.

b. Existence Load. The existence load consists of items other than those in the fighting load that are required to sustain or protect the combat Marine, that may be necessary for increased personal and environmental protection, and that are not normally carried by the individual. The existence load is normally carried by the Marine's unit. Marching, but not engaged in combat, a Marine may be expected to carry a maximum of 50 pounds of supplies and equipment.

c. Considerations

(1) The primary consideration is not how much a Marine can carry, but how much he can carry without impaired combat effectiveness-moral or physical. The combat strength of a unit is not counted simply in numbers of Marines, but in the numbers of willing and physically able Marines.

(2) It is generally better to risk temporary inconvenience from lack of health and comfort items than to exhaust Marines due to overloading.

(3) It is a fundamental truth that men become physically exhausted more quickly when under the stress of combat. Marines must be required to carry less into battle than they are conditioned to carry in training. Marines should be conditioned for carrying weight, but should be equipped in combat for fleetness of foot.

(4) A common mistake is to base the fighting load on the gear and supplies necessary to meet every contingency. The commander cannot reasonably expect the Marines to carry enough gear for every possible eventuality. The items to be carried must be based on reasonable expectations.
(5) It is the commander's responsibility to procure the transport to carry any additional gear. As a rule of thumb, a rifle company, or unit of similar size, requires one 5-ton truck and trailer in support to carry additional gear. In cold weather, or other conditions in which the necessary personal gear increases, this requirement will increase.
(6) The commander must ensure that the supply system provides, on a dependable and timely basis, the balance of essential supplies and equipment not carried by the unit. Marines must feel confident that they will be supported with the necessary supplies and equipment.
(7) In training, commanders must implant in their Marines the pride in operating under austere conditions. To effectively demonstrate the need for this spartan mentality, the commander must personally set the standard. Marines should be trained in field craft, foraging techniques, and the use of caches and field expedients. Maximum use should be made of captured stores.

3105. MARCH TECHNIQUES AND PROCEDURES

a. Organization. A command executing a march is organized into march units, march serials, and march columns, as necessary for control.
(1) March Units. A march unit is a unit of command which moves and halts at the command of a single commander. The march unit normally corresponds to one of the smaller Marine units such as a platoon or company.
(2) March Serials. A march serial consists of one or more march units organized under the senior officer and given a numerical or alphabetical designation to facilitate control. A serial is usually a battalion, but may be a company if the battalion is marching alone.
(3) March Columns. A march column is composed of elements of a command moving over the same route. It may be composed of one or more serials. To facilitate control, a column commander is designated. A column is normally a regiment or larger unit, but may be a battalion if marching alone.

b. Order of March. The order of march is determined by the mission, terrain, and the commander's desired order of commitment of units into action.

c. Control and Coordination Measures. The commander establishes initial control of the march by designating control measures in his march order. The most frequently used measures include:
o Start point and release point.
o Time at which head or tail of column passes the start point.
o Rate of march.
o Order of march.
o Route of march.
o Location of command post.
o March communications.

d. The Pacesetter. The pace-setter is an experienced individual carrying the same load as the majority of the Marines and marching from 4 to 10 meters at the head of the column. (See fig. 3-3.)The pace-setter's primary duty is to maintain the rate of march ordered by the column commander.

(l) Setting the Pace. The pacesetter does this by establishing the pace (length of step) and cadence (steps per minute) to obtain the prescribed rate of march. He should be of medium height so normal strides will be taken. Overstriding or understriding tends to tire leg muscles quickly and affects the combat efficiency of marching Marines. The officer marching at the head of the column supervises the pacesetter to ensure that he takes normal strides and maintains a uniform cadence.

Figure 3-3 is available in the illustration section at the end of the book

(2) Reducing Accordion Effect. Accordion effect occurs when the column alternately lengthens and shortens, causing the rear elements of the column to double time to maintain distance. It is caused by changes in the rate of march by the lead elements, usually after passing a slowing obstacle. To reduce this effect as much as possible, lead elements should slow their rate of march for a sufficient time after crossing obstacles to allow the rear elements to maintain the prescribed distance without running. The rear elements can signal the head when the last man has cleared the obstacle and the rate can be resumed. Since some accordion effect is nearly inevitable, the order of march should be periodically rotated to prevent units in the rear from becoming physically exhausted at the conclusion of the march.

e. Length, Rate and Pace

(1) **Length of March.** The length of march varies depending on terrain and weather, enemy situation, and physical and mental condition of Marines. The normal length of march for a 24-hour period is from 20 to 32 kilometers (12 to 20 miles) marching from 5 to 8 hours at a rate of 4 kilometers (2.5 miles) per hour (km/hmi/h).A march in excess of 32 kilometers in a 24-hour period is considered a forced march. Well-trained units that have been progressively trained will be able to achieve a unit goal of 60 miles in a 3-day period.

(2) Rate of March. The same factors of terrain, weather, load to be carried, and condition of Marines affect the rate of march. The commander considers these and selects a rate which will place his unit at the destination in the shortest possible time in condition to accomplish the mission. Unit standing operating procedure should state the standard rate of march for that unit over normal terrain for both road and cross-country movement and in both day and night. The commander then modifies this rate if the situation requires. Normally prescribed standards are as follows:

Figure is available in the illustration section at the end of the book

(3) Pace and Cadence. The normal pace is 30 inches. A pace of 30 inches and a cadence of 106 steps per minute result in a speed of 4.8 kilometers per hour or 3 miles per hour and a rate of 4 kilometers per hour or 2.5 miles per hour if a 10-minute rest halt per hour is taken. Since the pace of each man may vary, the cadence may have to be adjusted to maintain the prescribed rate. Slope of ground and surface footing affect pace. A normal cadence is relatively simple to maintain on level or gently rolling terrain on firm footing. Mud, sand, loose gravel, and hills all greatly affect pace.

f. Halts. Halts during a day's march are taken at regular intervals to rest personnel and adjust loads .Halts are specified by standing operating procedures or march order. Such factors as midday heat or enemy action may force the adoption of long midday halts or night marches. At long halts, each unit disperses to a previously selected location near the route of march.

(1) Normal Time and Frequency. Under normal conditions, a 15-minute halt is made after the first 45 minutes of marching. Following the first halt, a 10-minute halt is made after each 50 minutes of marching. Variations of this schedule are made when a scheduled halt time occurs when passing through a built-up area or when cover and concealment are required by the tactical situation and none is available. Observation posts may be established, if required for security of the unit during halts.

(2) Actions at Halts. All units in the column should be halted at the same time. At the halt signal, Marines should move to the side of the road, staying within the immediate vicinity of their unit. Marines should remove or loosen gear and sit or lie with feet elevated. Commanders inspect Marines and equipment, and corpsmen administer medical treatment, as required.

g. Communications. The four primary means of communications used in controlling foot marches are radio, visual, sound, and messenger.

(1) Radio. Radios are normally used for communications between platoons and higher headquarters in the march column. Radio transmissions should be held to the absolute minimum necessary for control and should be at minimum power required.

(2) Visual.. Visual communications may include flashlights, luminous markers, panel sets, pyrotechnics, and hand and arm signals. When contact with the enemy is not expected, the loss of light discipline from some of these signals may be a lesser security risk than the risk of disclosure from radio transmissions.

(3) Sound. Sound communications include voice and such signaling devices as horns and whistles. Some of their uses include assembling Marines at the conclusions of halts and providing warnings of air or chemical attacks.

(4) Messengers. Messengers are particularly effective during periods of radio silence and reduced visibility. Messages should be simple and brief to preclude mistakes by the messenger.

h. Security. Both active and passive measures are used against attack by enemy aircraft and indirect fire weapons during movement. Active measures include the use of air sentries and organic and attached weapons in accordance with the unit air defense standing operating procedure. Passive measures include the use of concealed routes and assembly areas, night marches, and extended interval between elements of the column. With imagination and planning, all of these measures can be practiced in training without detracting from the physical conditioning benefit of the march. In fact, such drills as rapidly taking cover during simulated attacks and then quickly resuming the march can prove physically challenging.

i. Reconnaissance. Reconnaissance determines in advance critical points along the march route such as bridges, fords, and obstacles in order that congestion or delay may be held to a minimum and local security provided. Paragraph 3103.C describes organization of the reconnaissance party.

3106.TACTICAL MARCHES

These marches are movements of Marines and equipment not in direct ground contact with the enemy but expecting early ground contact either en route or upon arrival at the objective. Some characteristics of this type of movement include security elements to the front, flanks, and, if necessary, to the rear of the march column. Additional considerations include the selection of concealed areas on defendable terrain for start and release points and rest stops. In training, infantry units of regimental size or smaller will normally derive the greatest benefit from tactical marches. The security requirements usually dictate a slower rate of march and, therefore, reduce the physical conditioning value for other types of units which would seldom move in combat with the expectation of early ground contact with the enemy. Two key points should be remembered regarding tactical marches in a training program.

a. Security. Although security requirements often slow the rate of march, it is a valid and important training objective for infantry units to work toward reducing march times while still practicing sound tactical security. March standing operating procedures, good land navigation skills, and simple but complete march orders contribute as much to this goal as physical conditioning. The idea that proceeding slowly enhances security is seldom true. Keeping the enemy off balance by rapid movement frequently is the best guarantee of security.

b. Training. In peacetime training, it is critical that support personnel who normally are located with infantry units in combat, participate completely with infantry units in forced march training. Not only must they not slow the infantry rate of march because of inferior conditioning, but they must be thoroughly familiar with the infantry unit's standing operating procedure for security on the march. Personnel from within the ground combat element such as artillery forward observer teams and combat engineer units, can be easily identified for this training and should be habitually associated with the same infantry units. Personnel from other elements, such as low-altitude air defense teams and radio battalion detachments, are more difficult to identify and earmark for support of the same infantry unit. Their normal mode of employment is in vehicles and their units frequently provide general support vice direct support. This does not, however, lessen their requirement to be prepared to provide support to dismounted infantry over terrain or in a tactical situation not conducive to vehicles. Such training needs to take place, and planning and cooperation among the elements of the MAGTF can overcome the obstacles.

3107.ADMINISTRATIVE MARCHES

These marches are movements of Marines and equipment made when no enemy interference is expected except from aircraft or indirect fire. During these marches, units are administratively grouped for ease of control and speed of movement.The term administrative foot marches should not be taken to imply that tactical considerations are disregarded.There are simply less stringent security measures used than when ground contact with the enemy is a distinct possibility.The normal formation is the route column with one file moving on either side of the road and with negligible security. Air sentries are used and dispersion between individuals and units is practiced. Although roads usually provide the fastest route, there may be reasons to conduct these marches cross-country (particularly as part of a training program). Administrative marches are the type of foot marches most practical for training programs for non-infantry units. These marches may also play a key role in the physical conditioning program of infantry units, because of the greater speeds possible due to lessened security requirements.

3108. FORCED MARCHES

A forced march is a march which covers a greater distance than 32 kilometer (20 miles) in a 24-hour period. Normally the extra required distance should be achieved by increasing the number of hours marched in a day beyond the norm (8 hours) rather than increasing the rate of march beyond the norm (4 kilometers per hour/2.5 miles per hour). However, there will be occasions when the situation or mission demands an increase in the rate of march.

a. Maximum Recommended Distances. The maximum recommended distances for forced marches are-56 kilometers (35 miles) in 24 hours; 96 kilometers (60 miles) in 48 hours; or 128 kilometers (80 miles) in 72 hours.

b. Sample Time Breakdown. A sample time breakdown for a forced march of 52 kilometers, beginning at first light is as follows:

Figure is available in the illustration section at the end of the book

3109. NIGHT MARCHES

Night marches are characterized by closed formations, more difficult control and reconnaissance, and a slower rate of march than day marches.

a. Control. Control is increased by reducing the distance between individuals and units, and by using connecting files to maintain contact between platoons and companies. Connecting files are normally constituted from personnel from the rear most march unit.

b. Safety on Roads. Because of reduced visibility, night marches on roads used by vehicles require attention to safety procedures to prevent accidents. Commanders should not think these safety measures apply only to peacetime training. In combat, nearly all vehicles operate with blackout lights at night, making visibility even more difficult than in garrison driving. If consistent with the tactical situation, the following measures can reduce the possibility of marchers being struck by vehicles:

(1) Use off-road trails and routes as much as possible.

(2) Guards to the front and rear of march units should be marked with strips of luminous tape and carry red-filtered flashlights. These measures are generally consistent with light discipline, while still affording reasonable warning to approaching drivers.

Section II. Conditioning Drills (One, Two, and Three)

3201. GENERAL

a. Description and Objective. Conditioning drills are calisthenic exercises. Each drill contains seven exercises organized and numbered in a set pattern. Each drill takes 15 minutes to complete. The objective of conditioning drills is to exercise all major muscle areas in order to develop strength, endurance, coordination, and flexibility.

b. Area and Equipment. Any level area is satisfactory for conduct of the drills. Drills One and Two contain ground exercises. If ground conditions are unsatisfactory, Drill Three should be used as it contains no ground positions. Usually, no equipment is required;however, if the group exceeds a platoon in size, an instructor's stand is necessary.

c. Formation. The extended rectangular formation is used in this drill. (See app. A.)

d. Starting Level and Progression. The starting level is six repetitions of each exercise. An increase of one repetition for each three periods of exercise in which the drill is performed is an acceptable rate of progression. This rate is continued until 12 repetitions can be completed. To maintain, continue the drill at 12 repetitions. To progress, move to 6 repetitions at a more difficult drill which exercises the same muscle group. Progression can also be gained by moving from Drill One to Drill Two, as Drill Two is more demanding.

e. Starting Positions. Starting positions vary with the exercise and are explained in each exercise.

f. Leadership. A principal instructor demonstrates and leads the drills. The instructor must be familiar with leadership techniques peculiar to conditioning drills to include the exercises, commands, counting cadence, cumulative count, formation, method of teaching the exercises, and utilization of assistant leaders.

g. Use With Other Programs. Conditioning Drills One, Two, and Three reach all major muscles of the body. They are easy to learn and to perform, and they are simple to administer and supervise. These features, coupled with the short time required for completion, the fact that no equipment is necessary, and adaptability to most areas of execution, make these drills possible in any programs.

3202. CONDITIONING DRILL ONE

This conditioning drill is similar to the calisthenics drill commonly known throughout the Marine Corps as the "Daily Seven."The "Daily Seven" may be used as a substitute for Conditioning Drill One.

a. Exercise l:High Jumper

(1) Starting Position. Feet separated shoulder width, knees flexed, body bent forward at the waist, arms aligned with the trunk and hips, elbows locked, palms facing, fingers extended and joined, head and eyes to the front.(See fig. 3-4, A.)(Elbows remain locked throughout the exercise.)

Figure 3-4 is available in the illustration section at the end of the book.

(2) Cadence. Moderate.

(3) Movement. A four-count exercise: at the count of—

(a) ONE -- Take a slight jump into the air, swinging the arms forward and up to shoulder level.

(b) TWO -- Take a slight jump into the air and swing the arms downward and back, returning to the starting position.

(c) THREE -- Take a vigorous leap into the air, swinging the arms forward and up to an overhead position, momentarily looking skyward, on returning to the ground the knees are flexed, head and eyes return to the front.

(d) FOUR--Repeat the action of count two.

b. Exercise 2: Bend and Reach

(1) Starting Position. Feet spread more than shoulder width, arms overhead, elbows locked, palms facing, fingers extended and joined, head and eyes to the front.(See fig. 3-4, B.)

(2) Cadence. Moderate.

(3) Movement. A four-count exercise: at the count of—

(a) ONE-Bend at the knees and waist, swing the arms straight downward and reach between the legs. Touch the ground as far to the rear as possible and look to the rear. (Elbows remain locked throughout the exercise).

(b) TWO-- Recover sharply to the starting position.

(c) THREE-- Repeat the action of count ONE.

(d) FOUR-- Repeat the action of count TWO.

c. Exercise 3: Pushup

(1) Starting Position. Front leaning rest position. To assume this position there is a silent one-two count:on the silent count of one, assume the squatting position, heels together, elbows locked inside the knees, hands flat on the ground directly beneath the shoulders; on the silent count of two, thrust the legs to the rear, toes and heels together, body straight from head to heels.(Seefig. 3-4, C.)

(2) Cadence. Moderate.

(3) Movement. A four-count exercise: at the count of—

(a) ONE -- Flex the elbows lowering the body until the thick portion of chest touches the ground.

(b) TWO -- Raise the body until elbows are straight and locked.

(c) THREE-- Repeat the action of count ONE.

(d) FOUR -- Repeat the action of count TWO. (On returning to position of attention, the silent one-two count is used in reverse).

d. Exercise 4: Trunk Twister

(1) Starting Position. Feet are spread more than shoulder width apart, fingers placed behind neck, thumbs pointing downward, elbows back. (See fig. 3-4, D.)(Elbows remain well back throughout the exercise).

(2) Cadence. Slow.

(3) Movement. A four-count exercise: at the count of-

(a) ONE-- Keeping the knees locked and the back straight, bend forward at the waist sharply, with a slight recovery.

(b) TWO-- Twist the trunk to the left vigorously at the waist, keeping the elbow back. The left elbow is higher than the right.

(c) THREE-- Twist vigorously to the right, so the left elbow comes under the right.

(d) FOUR-- Straighten sharply to the starting position. NOTE: Do not attempt to touch the elbows to the knees on counts two and three.

e. Exercise 5: Squat Bender

(1) Starting Position. Feet are spread less than shoulder width apart, hands on hips, thumbs in small of back, elbows back. (See fig. 3-4, E.)

(2) Cadence. Moderate.

(3) Movement. A four-count exercise: at the count of—

(a) ONE-- Assume the squatting position, maintain balance on the balls of the feet, with trunk erect thrust arms forward to shoulder level, elbows locked, palms down.

(b) TWO-- Recover to starting position. Elbows are well back.

(c) THREE-- Keeping the knees locked, bend forward at the waist, touching the ground in front of the toes.

(d) FOUR-- Vigorously recover to the starting position.

f. Exercise 6: Body Twist

(1) Starting Position. On the back, arms extended sideward on the ground, palms down. The legs are raised to a near vertical position, feet together, knees locked.(See fig. 3-4, F.)

(2) Cadence. Slow-fast.

(3) Movement. A four-count exercise: at the count of-

(a) ONE-- Lower legs slowly to the left until they touch the ground near the left hand, keeping the knees straight and shoulders on the ground.

(b) TWO-- Recover the starting position by quickly raising the legs, keep knees straight and feet together.

(c) THREE-- Repeat movement of count ONE, except the movement is to the right side.

(d) FOUR-- Recover sharply to the starting position.

g. Exercise 7: Stationary Run

(1) Starting Position. Position of attention. (See fig, 3-4, G.)

(2) Cadence. Fast.

(3) Movement.

(a) At the command of execution, start running in place at double-time cadence, lifting the left foot first time cadence. Follow the instructor as he counts two repetitions of cadence;

e.g., 1, 2, 3, 4--1, 2, 3, 4. The instructor then gives informal commands such as FOLLOW ME, running on the toes and balls of the feet, keeping the back straight, speeding up the cadence to a sprint, raising the knees high, leaning forward at the waist, and pumping the arms vigorously.

(b) To halt the exercise, the instructor will count two repetitions of cadence as the left foot strikes the ground: 1, 2, 3, 4 - 1 , 2, 3, HALT.

NOTE: When counting cadence, the instructor counts only as the left foot strikes the ground. The duration of the exercise is approximately 1 1/2 minutes.

3203. CONDITIONING DRILL TWO

a. Exercise 1: Jumping Jack

(1) Starting Position. Feet separated more than shoulder width, arms overhead.(See fig.3-5, A.)

(2) Cadence. Moderate.

(3) Movement. A four-count exercise: at the count of~

(a) ONE-- Jump to position with the feet together and assume the squatting position, swinging the arms sideward and downward, placing the hands palms down on the ground, elbows locked inside the knees.

(b) TWO-- Recover to the starting position by jumping to the side straddle and swinging the arms sideward overhead.

(c) THREE-- Repeat the action of count ONE.

(d) FOUR-- Recover to the starting position.

b. Exercise 2: Turn and Bend

(1) Starting Position. Side straddle, arms overhead.(See fig. 3-5, B.)

(2) Cadence. Moderate.

(3) Movement. A four-count exercise: at the count of-

(a) ONE-- Turn the trunk to the left and bend forward over the left thigh, attempting to touch the fingertips to the ground outside the left foot. Keep the left knee straight. On successive repetitions attempt to touch farther and farther to the side.

(b) TWO-- Recover to the starting position.

(c) THREE-- Turn the trunk to the right and bend forward over the right thigh, trying to touch the hands to the ground outside the right foot. Keep the right knee straight.

Figure 3-5 is available in the illustration section at the end of the book.

(d) FOUR-- Recover to the starting position.

c. Exercise 3: Eight-Count Pushup

(1) Starting Position. Position of attention. (See fig. 3-5, C.)

(2) Cadence. Moderate.

(3) Movement. An eight-count exercise: at the count of—

(a) ONE-- Assume the squatting position, palms on the ground directly beneath the shoulders, elbows locked inside the knees.

(b) TWO-- Thrust the legs to the rear assuming the front leaning rest position.

(c) **THREE**-- Flex the elbows until the thick portion of the chest touches the ground.

(d) **FOUR**-- Raise the body on a straight plane until the elbows are locked.

(e) **FIVE**-- Repeat the action of count THREE.

(f) **SIX**-- Repeat the action of count FOUR.

(g) **SEVEN**-- Recover to the squatting position as in count ONE (elbows locked inside the knees).

(h) **EIGHT**—Return sharply to the position of attention.

d. Exercise 4: Turn and Bounce

(1) **Starting Position.** Feet spread more than shoulder width apart, arms sideward at shoulder level, palms up. (See fig. 3-5, D.)

(2) **Cadence.** Slow.

(3) **Movement**. An eight-count exercise: at the count of—

(a) **ONE**-- Turn sharply to the left as far as possible, then recover slightly.

(b) **TWO**-- Again turn to the left as far as possible and recover as in ONE.

(c) **THREE**-- Repeat the action of count TWO.

(d) **FOUR**-- Recover sharply to the starting position.

(e) **FIVE**—Turn sharply to the right as far as possible, then recover slightly.

(f) **SIX**—Again turn to the right as far as possible and recover as in FIVE.

(g) **SEVEN**-Repeat the action of count SIX.

(h) **EIGHT**-Return to the starting position.

NOTE: The head and hips remain to the front throughout the exercise and the knees and elbows are locked at all times.

e. Exercise 5: Squat Stretch

(1) **Starting Position**. Attention. (See fig. 3-5, E.)

(2) **Cadence**. Moderate.

(3) **Movement**. A four-count exercise: at the count of—

(a) **ONE**-- Squat, placing the hands on the ground about 12 inches in front of the feet.

(b) **TWO**— Keeping the fingertips on the ground, straighten the knees completely and raise the hips.

(c) **THREE**-- Recover to position ONE.

(d) **FOUR**-- Recover to the starting position.

f. Exercise 6: Leg Circular

(1) **Starting Position.** On the back, arms stretched sideward, palms down, feet raised foot from ground, knees straight. (See fig 3-5,F.)

(2) **Cadence.** Slow.

(3) **Movement** A four-count exercise: at the count of—

(a) **ONE**-- Swing the legs as far as possible to the left, keeping the knees straight and the legs together.

(b) **TWO**-- Swing the extended legs over head with the thighs as close as possible to the trunk.

(c) **THREE**-- Swing the legs as far as possible to the right.

(d) **FOUR**-- Recover to the starting position.

g. Exercise 7: Stationary Run

(1) Starting Position. Position of attention. (See fig. 3-5, G.)

(2) Cadence. Fast.

(3) Movement

(a) At the command of execution, start running in place at double-time cadence, lifting the left foot first. Follow the instructor as he counts two repetitions of cadence; for example: 1, 2, 3, 4--1, 2, 3, 4. The instructor then gives informal commands such as FOLLOW ME. Running on the toes and balls of the feet, keeping the back straight, speeding up the cadence to a sprint, raising the knees high, leaning forward at the waist, and pumping the arm vigorously.

(b) To halt the exercise, the instructor will count two repetitions of cadence as the foot strikes the ground: 1, 2, 3, 4-1,2, 3, HALT.

NOTE: When counting cadence, the instructor counts only as the left foot strikes the ground. The duration of the exercise is approximately 1 1/2 minutes.

3204. CONDITIONING DRILL THREE

a. Exercise 1: Side Straddle Hop

(1) Starting Position. Position of attention. (See fig. 3-6, A.)

(2) Cadence. Moderate.

(3) Movement. A four-count exercise: at count of-

(a) ONE-- Take a slight jump into the air, moving the legs sideward (more than shoulder width apart); at the same time, swing the arms overhead (to an overhead position) clapping the palms together.

Figure 3-6 is available in the illustration section at the end of the book.

(b) TWO-- Take a slight jump into the air, swing the arms sideward and downward returning to the starting position.

(c) THREE-- Repeat the action of count ONE.

(d) FOUR-- Repeat the action of count TWO.

b. Exercise 2: Back Bender

(1) Starting Position. Standing, feet 12 inches apart, fingers placed behind the head. (See fig. 3-6, B.)

(2) Cadence. Slow.

(3) Movement. A four-count exercise: at the count of-

(a) ONE-- Bend the upper trunk backward, raising the chest high, pulling the elbows back, and looking upward. Keep the knees straight.

(b) TWO-- Recover to the starting position.

(c) THREE-- Repeat the action of count ONE.

(d) FOUR-- Recover to the starting position.

c. Exercise 3: Squat Thrust

(1) Starting Position. Position of attention. (See fig. 3-6, C.)

(2) Cadence Moderate.

(3) Movement. A four-count exercise: at the count of—

(a) ONE-- Assume the squatting position; heels together, placing the hands flat on the ground, shoulder width apart, elbows locked and inside the knees.

(b) TWO-- Thrust the legs to the rear, assuming the front leaning rest position, body in line from head to toe, heels and toes together.

(c) THREE-- Return to the squatting position as in ONE.

(d) FOUR-- Return to position of attention.

d. Exercise 4: Side Bender

(1) Starting Position. Feet are spread more than shoulder width apart, arms are raised sideward and overhead, thumbs interlocked palms to front, fingers extended and joined, elbows locked. (See fig. 3-6, D.)

(2) Cadence. Slow.

(3) Movement. An eight-count exercise: at the count of—

(a) ONE-- Bend to left as far as possible, then recover slightly.

(b) TWO-- Again bend to the left as far possible, then recover slightly.

(c) THREE-Repeat the action of count TWO.

(d) FOUR-- Recover sharply to the starting position.

(e) FIVE-- Bend to the right as far as possible, then recover slightly.

(f) SIX-- Again bend to the right as far as possible, then recover slightly.

(g) SEVEN-- Repeat the action of count SIX.

(h) EIGHT-- Recover sharply to the starting position.

NOTE:Keep the elbows and knees locked throughout the exercise. The bend should occur to the side and not the front.

e. Exercise 5: Knee Bender

(1) Starting Position. Feet are spread less than shoulder-width apart, hands on hips, thumbs in small of back, elbows back. (See fig 3-6, E.)

(2) Cadence. Moderate.

(3) Movement. A four-count exercise: on the count of—

(a) ONE-Do a knee bend, lean trunk forward at the waist, thrust arms between legs until the extended fingers touch the ground palms to the ground, hands 6 inches apart.

(b) TWO~Recover sharply to the starting position.

(c) THREE-Repeat the action of count ONE.

(d) FOUR-Repeat the action of count TWO.

f. Exercise 6: Bottoms Up

(1) Starting Position. Front leaning rest position. A silent one-two count is used as in the pushups. (See fig. 3-6, F.)

(2) Cadence. Moderate.

(3) Movement A four-count exercise: at the count of-

(a) ONE-- With the weight on the hands, and knees locked, jump forward bringing the feet as close to the hands as possible; look to the rear.

(b) TWO--Keeping the knees locked, thrust the legs backward assuming the front leaning rest position.

(c) THREE-- Repeat the action of count ONE.

(d) FOUR-- Repeat the action of count TWO.

g. Exercise 7: Stationary Run

(1) Starting Position. Position of attention. (Sec fig. 3-6, G.)

(2) Cadence. Fast.

(3) Movement

(a) At the command of execution, start running in place at double time, lifting the left foot first. Follow the instructor as he counts two repetitions of cadence; for example: 1, 2, 3, 4--1, 2,3, 4. The instructor then gives informal commands such as FOLLOW ME, running on the toes and balls of the feet, keeping the back straight, speeding up the cadence to a sprint, raising the knees high, leaning forward at the waist, and pumping the arms vigorously.

(b) To halt the exercise, the instructor will count two repetitions of cadence as the left foot strikes the ground: 1, 2, 3, 4 - 1, 2, 3, HALT.

NOTE: When counting cadence, the instructor counts only as the left foot strikes the ground. The duration of the exercise is approximately 1 1/2 minutes.

Section III. Rifle and Log Drills

3301. RIFLE DRILL

a. Description and Objective. Rifle exercises are conditioning exercises performed with a rifle. Each drill contains six exercises (fig. 3-7) and they are numbered in a set pattern. The drill takes 15 minutes to complete. The objective of rifle drills is to exercise the arms, shoulders, and back muscles in order to develop strength and endurance, particularly in the upper body. In units without rifles, log drills may be substituted.

b. Area and Equipment. Any level area is satisfactory for conducting this drill. Each Marine completes these exercises from a standing position and no ground contact is required. Each Marine will need a rifle and, if the group exceeds a platoon in size, then the instructor will need an instructor's stand.

c. Formation. The extended rectangular formation is used in this drill. (See app. A.)
d. Starting Positions. Starting positions vary with the exercises and are explained in each exercise. As in all set conditioning drills, the command used to start the exercise is STARTING POSITION, MOVE. The following directions apply to rifle drill.
(1) In those exercises which start from the rifle downward position, on the command MOVE, execute port arms as prescribed in NAVMC 2691, Drill and Ceremonics Manual, and then assume the starting position. The command to return to the men to the position of attention at the conclusion of the exercise is POSITION OF ATTENTION, MOVE.
(2) In exercises which terminate in the rifle downward position, on the command MOVE, execute the position followed by order arms as prescribed in NAVMC 2691.
(3) In the exercises which terminate in a position other than the rifle downward position, Marines first assume the rifle downward position before executing port arms and order arms.
(4) These movements are executed without command. This procedure promotes uniformity, but precision is not expected. To be effective, rifle exercises must be strenuous enough to tire the arms, but not to the point where the arms cannot move with precision.

e. Leadership. A principal instructor demonstrates and leads the drill. He must be familiar with leadership techniques for conditioning exercises and the peculiar techniques for rifle drill.

3302. EXERCISES PERFORMED WITH RIFLES

The exercises of rifle drill are outlined in the following paragraphs,
a. Exercise 1: Foreup, Behind Back

(1) Starting Position. Rifle downward, feet together.(See fig. 3-7, A.)

Figure 3-7 is available in the illustration section at the end of the book.

(3) Movement. A four-count exercise: at the count of—
(a) ONE-- Swing the arms forward and upward to the overhead position. Inhale.

(b) **TWO**-- Lower the rifle to the back of the shoulders. Exhale.
(c) **THREE**-- Recover to position ONE and inhale.
(d) **FOUR**-- Recover to the starting position and exhale.

b. Exercise 2: Lunge Side, Turn and Bend
(1) Starting Position. Rifle downward, feet together.(See fig. 3-7, B.)
(2) Cadence. Moderate.
(3) Movement. An eight-count exercise:at the count of—

(a) ONE-- Lunge sidewards to the left, swing the rifle forward and upward to the overhead position.

(b) TWO-- Turn the trunk to the left and bend forward over the left hip. At the same time, swing the rifle to a **low** horizontal in front of the left ankle.

(c) **THREE**-- Recover to position ONE.
(d) **FOUR**-- Recover to the starting position.
(e) **FIVE, SIX, SEVEN, and EIGHT**-- Repeat on the right side.

c. Exercise 3: Foreup, Back Bend
(1) Starting Position. Rifle downward, feet together.(See fig. 3-7, C.)
(2) Cadence. Moderate.
(3) Movement. A four-count exercise:at the count of—

(a) ONE-- Swing the arms forward and upward to the overhead position.

(b) TWO-- Bend backward, emphasizing the bend in the upper back. The face is up.Keep the knees straight.

(c) **THREE**-- Recover to position ONE.
(d) **FOUR**-- Recover to the starting position.

d. Exercise 4:Up and Forward
(1) Starting PositionRifle downward, feet together.(See fig. 3-7, D.)
(2) Cadence. Fast.
(3) Movement. A four-count exercise:at the count of—

(a) ONE-Swing the arms forward and upward to the overhead position.
(b) TWO- Swing the arms forward to shoulder level.
(c) THREE-Recover to position ONE.
(d) FOUR-Recover to the starting position.

e. Exercise 5: Foreup, Full Squat
(1) Starting Position. Rifle downward, feet in narrow stance. (See **fig.** 3-7, E.)
(2) Cadence. Moderate.
(3) Movement. A four-count exercise: at the count **of**—

(a) ONE-- Swing the arms forward and upward to the overhead position.
(b) TWO-- Swing the arms down to shoulder level and assume the squatting position.
(c) **THREE**-- Recover to position ONE.
(d) **FOUR**-- Recover to the starting position.

f. Exercise 6: Arms Forward, Side Bend

(1) Starting Position. Sidestraddle, regular stance, rifle forward.(See fig. 3-7, F.)
(2) Cadence. Moderate.
(3) Movement. A four-count exercise:at the count of-

(a) **ONE--** Bend the trunk to the left. Keep the knees straight.
(b) **TWO--** Recover to the starting position.
(c) **THREE--** Bend the trunk to the right. Keep the knees straight.
(d) **FOUR--** Recover to the starting position.

NOTE:Keep the rifle on the same level as the shoulders throughout the exercise.

3303. LOG DRILL

a. Description and Objective. Log exercises are conditioning exercises performed with a log. Each drill contains six exercises and they are numbered in a set pattern. The drill takes 15 minutes to complete. The objective of log drills is to develop strength and muscular endurance and, in this instance, under maximum loads. Log exercises also develop teamwork. Log may be used in lieu of conditioning drills after the Marines have become somewhat conditioned.

b. Area and Equipment. Any level area is satisfactory for conducting this drill. Each Marine completes these exercises from a standing position and no ground contact is required. If the group exceeds a platoon in size, then the instructor will need instructor's stand. Each six-person group or eight-person group will need a log. The logs should be from 6 to 8 inches in diameter. They may vary in length from 14 feet (for 6 people) to 18 feet (for 8 people).The logs should be skinned, smoother, and dried. The 14-foot logs should weigh approximately 300 pounds and the 18-foot logs, approximately 400 pounds. Rings should be painted on the logs to indicate each person's position. When not in use, the logs should be stored on a rack to keep them off the ground.

c. Formation. All the Marines assigned to the same log team should be about the same height at the shoulders. The recommended method of dividing the platoon is to have the Marines form a single file or column with short people to the front and tall people to the rear. Have the Marine assume their positions in the column according to shoulder height, not head height. When they are in position, they are given the command COUNT OFF BY SIXES (OR EIGHTS), COUNT OFF, to divide them into six- or eight-person log teams. Each team in turn, can then proceed to the log rack, shoulder a log, and carry it to the designated exercise area. The log teams form in columns in front of the instructor. With the Marine holding the log in the chest position, have them face the instructor and ground the log at least 10 yards between columns and 10 yards between log teams within the columns.

d. Starting Positions. The Marines fall in, facing the log, with their toes about 4 inches from it. The basic starting positions and commands are as follows:(See fig. 3-8.)
(1) **RIGHT-HAND STARTING POSITION, MOVE.** It the command MOVE, move the left foot 12 inches to the left, and lower the body into a flat foot squat. Keep the back straight, head up, and arms between the legs. Encircle the far side of the log with the left hand. Place the right hand underneath the log.(See fig. 3-8, A.)

(2) LEFT-HANDED STARTING POSITION, MOVE. These commands are executed in the same manner as in paragraph a. except that the left hand is underneath the log and the right hand encircles its far side. (See fig. 3-8, B.)

Figure 3-8 is available in the illustration section at the end of the book.

(3) RIGHT SHOULDER POSITION, MOVE. At the command MOVE, pull the log upward in one continuous motion to the right shoulder. At the same time, move the left foot to the rear and stand up, facing left. Balance the log on the right shoulder with both hands.(See fig. 3-8, C.)This movement cannot be performed from the left-hand starting position because of the position of the hands.

(4) LEFT SHOULDER POSITION, MOVE. These commands should be given from the left hand starting position. At the command MOVE, pull the log upward in one continuous motion, to the left shoulder. At the same time, move the right foot to the rear and stand facing right. Balance the log on the left shoulder with both hands.(See fig. 3-8, D.)This movement cannot be performed from the right-hand starting position.

(5) WAIST POSITION, MOVE. From the right hand position pull the log waist high. Keep the arms straight and fingers laced underneath the log. The body is inclined slightly to the rear and the chest is lifted and arched. (See fig. 3-8, E.)

(6) CHEST POSITION, MOVE. This command should be given after the waist position has been assumed .On the command MOVE, shift the log to a position high on the chest, bring the left arm under the log and hold the log in the bend of the arms.(See fig. 3-8, F.)Keep the upper arms parallel to the ground.

(7) To move the log from the right shoulder to the left shoulder, the command is: LEFT SHOULDER POSITION, MOVE. On the command MOVE, push the log overhead and lower it to the opposite shoulder.

(8) To return the log to the ground from any of the above positions, the command is: STARTING POSITION, MOVE. At the command MOVE, slowly lower the log to the ground. The hands and fingers must be kept from under the log.

e. Leadership. A principal instructor demonstrates and leads the drill. He must be familiar with the leadership techniques for conditioning exercises and the peculiar techniques for log drill.

3304. EXERCISE PERFORMED WITH LOGS

The exercises of log drill are outlined in the following paragraphs. Figure 3-9 graphically explains log drill.

a. Exercise l:Two-Arm Pushup

(1) Starting Position. Right or left shoulder position. Regular stance. (See fig. 3-9, A.)

(2) Cadence. Moderate.

(3) Movement. A four-count exercise:at the count of—

- **(a) ONE--** Push the log overhead until the elbows lock.
- **(b) TWO--** Lower the log to the opposite shoulder.
- **(c) THREE--** Repeat the action of count ONE.
- **(d) FOUR--** Recover to the starting position.

b. Exercise 2: Forward Bender

(1) Starting Position. Chest position. Regular stance.(See fig. 3-9, B.)

(2) Cadence. Moderate.

(3) Movement. A four-count exercise:at the count of-

(a) ONE-- Bend forward at the waist, keeping the back and legs straight.

(b) TWO-- Recover to the starting position.

(c) THREE-- Repeat the action of count ONE.

(d) FOUR-- Recover to the starting position.

c. Exercise 3: Straddle Jump

(1) Starting Position. Right or left shoulder position, feet together, fingers interlaced on top of the log.(See fig. 3-9, **C.)**

(2) Cadence. Moderate.

(3) Movement. A four-count exercise:at the count of—

(a) ONE-- Jump to a side-straddle. Pull down on the log with both hands to keep it from bouncing on the shoulder.

(b) TWO-- Recover to the starting position.

(c) THREE-- Repeat the action of count ONE.

(d) FOUR-- Recover to the starting position.

Figure 3-9 is available in the illustration section at the end of the book.

d. Exercise 4: Side Bender

(1) Starting Position. Right shoulder position, feet regular stance.(See fig. 3-9, D.)

(2) Cadence. Moderate.

(3) Movement. A four-count exercise:at the count of—

(a) ONE-- Bend sideward to the left as far as possible, bending the left knee.

(b) TWO-- Recover to the starting position.

(c) THREE-- Repeat the action of count ONE.

(d) FOUR-- Recover to the starting position.

(4) After completing the required number of repetitions, change shoulders and execute an equal number of repetitions to the other side.

e. Exercise 5: Knee Bend

(1) Starting Position. Right or left shoulder position. Narrow stance. Fingers interlocked on top of the log.(See fig. 3-9, E.)

(2) Cadence. Slow.

(3) Movement. A four-count exercise:at the count of—

(a) ONE-- Flex the knees to a quarter-squat position.

(b) TWO-- Flex the knees to a half-squat position.

(c) THREE-- Lower the body to a three-quarter squat position.(Lean slightly forward.)

(d) FOUR-- Recover to the starting position.

NOTE:Pull forward and downward on the log throughout the exercise.

f. Exercise 6: Overhead Toss

(1) Starting Position. Right or left shoulder position, regular stance. The knees are bent to a quarter-squat.(See fig. 3-9, F.)

(2) Cadence. Moderate.

(3) Movement. A four-count exercisc:at the count of-

(a) ONE-Straighten the knees and toss the log into the air approximately 12 inches overhead. Catch the log with both hands and lower it toward the opposite shoulder. As the log is caught, lower the body into a quarter-squat.

(b) TWO-- Again toss the log into the air and when caught return it to the original shoulder.

(c) THREE-- Repeat the action of count ONE.

Section IV. Grass Drills

3401. GENERAL

a. Description and Objective. Grass drills are extremely strenuous exercises and are performed at top speed for only short periods of time. No cadence is counted but the Marines continue to execute the multiple repetitions of the command until the next command is given. The grass drill consists of two drills:Drill One and Drill Two. Each drill contains six exercises. The objective of the drills is to decrease reaction time, to develop cardiovascular endurance, and to provide a vigorous workout for all major muscles.

b. Area and Equipment. Any level area suitable for ground contact and of a size to accommodate the group is adequate. No equipment is needed.

c. Formation. All movements are executed in place. The extended, rectangular formation is recommended for a platoon- or company-size unit. The circle formation is suitable for groups of squad or section size. At the beginning of an exercise program, 2 to 3 minutes of grass drills will insure a good workout.

d. Progression. Progression is gained by gradually increasing the length of time devoted to the drills. As the physical condition of the Marines improves, the periods should be gradually lengthened to 5 minutes. As the second drill is more difficult than the first, some progression can be attained by initially executing Drill One; then as the program and the Marines progress, introduce Drill Two. To extend the duration of the drill, it may be necessary to repeat the drill.

e. Starting Position

(1) The drills are started from the GO position. Other basic positions are FRONT, BACK, and STOP. (See fig. 3-10, A.)

(a) GO. Running in place (top speed): on the toes and balls of feet, knees raised high, arms pumping, body bent forward at waist.

(b) FRONT. Prone position: elbows bent (along body), palms flat on ground directly under the shoulders, legs together and straight.

(c) BACK. Supine position (flat on back): arms extended near side on ground with palms down, legs together and straight, feet toward the stand or instructor.

(d) STOP. Football lineman stance: feet spread and staggered, left arm across left thigh, right arm straight, knuckles on ground, head up, back parallel with ground.

(2) To assume the FRONT or BACK position from the STANDING, GO, or STOP position, vigorously get into the prescribed position as quickly as possible. (See fig. 3-10, B.)

(3) To change from the FRONT to the BACK position, quickly do a pushup, move the feet several short steps to the right or left, lift the arm on the side toward which the feet move, and thrust the legs vigorously to the front. (See fig. 3-10, C.)

Figure 3-10 is available in the illustration section at the end of the book.

(4) To move from the BACK to the FRONT position, sit up quickly, place both hands on the ground to the right or the left of the legs. Move the feet several short steps to the rear on the side opposite the hands. When the feet are opposite the hands, thrust the legs vigorously to the rear and lower the body to the ground.(See fig. 3-10, D.)

f. Leadership. A warm up activity of lesser intensity should proceed grass drill. During the instructional phase and conduct of these drills, the following points should be applied.
(1) The instructor executes only GO and STOP with the Marines. This allows the instructor to supervise the drill.
(2) The commands for grass drills are given in rapid succession without the usual preparatory command.
(3) To prevent confusion, the instructor should give the commands sharply to distinguish them from comments or encouragement.
(4) As soon as the Marines know the drill, they should respond to the instructor's commands and perform all exercises vigorously and as rapidly as possible. All exercises are executed continuously until the next command is given. Insist on top speed performance; anything less is not effective.
(5) The commands peculiar to each exercise are identical to the name of the exercise.
(6) Marines are not to be required to assume the position of attention once the drills are started. To halt the drill for instructions or for rest, the command UP is used. At this command, the Marines assume a relaxed standing position. Do not demand formality. At the conclusion of a fast and vigorous 5-minute grassdrill, it is physically impossible for people to stand at attention.
(7) The sequence of commands for the execution of grass drills should occur as follows:Drill One. GO, FRONT, Bouncing Ball; GO, BACK, Bicycle; GO, Full Squatter; GO, BACK, Situps; GO, FRONT, Mountain Climber; GO, FRONT, Roll Left; GO, STOP, UP.

g. Use With Other Programs. Since grass drills can be executed in a short period of time, they may be executed where only a few minutes are available for exercise or in conjunction with another type of activity. Grass drills are an excellent substitute for running when time is a factor.

3402. GRASS DRILL ONE AND TWO

a.Grass Drill One

(1) Bouncing Ball. From the FRONT position, push up, supporting the body on the hands (shoulder-width apart) and feet. (See fig. 3-11.)Keep the back and legs in line and the knees straight. Bounce up and down by a series of short, upward springs from the hands, hips, and feet simultaneously,
(2) Bicycle. From the BACK position, raise the legs and hips. Keep the elbows on the ground and support the hips with the hands. Move the legs vigorously as if pedaling a bicycle.
(3) Full Squatter. From the STOP position, assume a full knee bend, the feet on line, hands on hips. Bounce up and down in place by short, bouncing jumps.

Figure 3-11 is available in the illustration section at the end of the book.

(4) **Situps.** From the BACK position and with arms stretched overhead, sit up, reach forward, and touch toes. Return to the supine position.

(5) **Mountain Climber.** From the STOP position, place both hands on the ground directly under the shoulders. Thrust the right leg to the rear, knee straight. The left foot should be close to the left hand, the left knee outside the left arm. Shift the weight to the hands, thrust off with the rear (right) foot and bring that foot up close to the right hand, the right knee outside the right arm. At the same time, thrust the left leg vigorously to the rear, knee straight. Continue at a fast cadence, alternating the legs.

(6) **Roll Left.** From the BACK or FRONT position, make one complete roll in the direction commanded. On completing the roll, return to the FRONT or BACK position.

b. Grass Drill Two

(1) **Legs Over.** .From the BACK position and with arms stretched overhead, palms up, raise the legs upward and then swing them backward over the head until the toes touch the ground behind the head. Return legs to the starting position. (See fig. 3-12.)

(2) **V-Up and Touch Toes.** .From the BACK position, raise the legs with the knees straight, sit up until the trunk and legs form a V, and touch the toes with the hands. Return to the BACK position.

(3) **Rocker.** .In the FRONT position, clasp the hands behind the back, arch the body, holding the head back. Start rocking, using the front part of the trunk as a rocker.

(4) **Bounce** and **Clap Hands**. The procedure is the same as for bouncing ball, but while in the air, clap the hands. This requires a more vigorous bounce or spring. (See fig. 3-12.)

(5) **Leg Spreader.** .From the BACK position, raise the legs so that the heels are 10 to 12 inches from the ground, spread them apart as far as possible, then close them together. Open and close legs as rapidly as possible.

(6) **Forward Roll.** .For forward roll from the STOP position, place both hands on the ground, tuck the head, and do one complete forward roll, keeping the legs tucked as you roll, and come back to the STOP position.

Figure 3-12 is available in the illustration section at the end of the book.

Section V.Guerrilla Exercise

3501.GENERAL

a. Description and Objective. Guerrilla exercises are individual exercises of various basic skills that are performed rapidly while moving forward in a circle formation. There are two tables of guerrilla exercises. Each table takes 15 minutes to complete. The objective of this exercise is to increase strength and endurance, aid flexibility, and develop coordination.

b. Area and Equipment. Any level area is suitable for conducting guerrilla exercises. No ground contact is required other than the hands. There is no equipment requirement.

c. Formation

(1) The circle formation (app. A) is used for guerrilla exercises. Each platoon forms its own circle and engages in guerrilla exercises under a platoon instructor. If the platoon exceeds 30 Marines, double or concentric circles may be used.

(2) When the circle is formed, the instructor steps into the center of the circle and moves clockwise in a small circle. He commands:QUICK TIME, MARCH, 1-2-3-4. (Rapid cadence of approximately 130 counts per minute. Cadence and step are maintained between exercises.)

(3) To reform the platoon after completing guerrilla exercises, the instructor halts the Marines and places the base man or platoon guide where he wishes and commands:

(a) BASE MAN (or platoon guide),POST.

(b) FALL OUT AND FALL IN ON THE BASE MAN (or platoon guide).

d. Progression. Progression may be attained by moving from table I to table II or by shortening the quick time marching periods between exercises and performing all exercises a second time.

e. Leadership

(1) To execute the exercises, the Marines continue at quick time while the instructor simultaneously explains and demonstrates the exercise to be performed, and then commands the Marines accordingly. In each instance, the preparatory command will be the name of the exercise and, in all instances, the command of execution will ***MARCH**.. To* terminate each exercise, the command *is **QUICK TIME, MARCH.** The* Marines immediately pick up the step as the instructor counts cadence.

(2) Unless specified differently, each exercise should be continued for 20 to 40 seconds depending upon the vigor of the exercise. The leader can determine the duration of each exercise by observing its effect upon the Marines.

(3) To form for double guerrillas, the commands for pairing the Marines (who are in circle formation) are-

(a) PLATOON, HALT.

(b) FROM (designate an individual), BY TWO'S, COUNT OFF(.Examplel-2; 1-2; 1-2; etc.)

(c) EVEN NUMBERS MOVE UP BEHIND ODD NUMBERS(At this time, adjust pairs according to height and weight.)

(d) YOU ARE NOW PAIRED UP FOR DOUBLE GUERRILLAS. To change the Marines position, merely command CHANGE.)
(e) FORWARD, MARCH.

f. Place in the Program. Many Marines have not had the opportunity to perform the simple skills involved in guerrilla exercises. The conduct of these exercises is a simple matter since they can be performed easily and quickly in almost any situation. The tables of exercise are applicable to all personnel. The tables can constitute a station within a 1-hour period or be completed within a separate 15-minute period.

3502.GUERRILLA TABLES

a. Table I

(1) Double Time.(See fig. 3-13, A.)Hold arms in the thrust position. Execute a double-time run, maintaining the circle formation and the prescribed distance between your fellow Marines. Duration-1 minute.

(2) All Fours.(See fig. 3-13, B.) Face downward. Support the body with the hands and feet. Walk forward hands first.

(3) Crab Walk.(See fig. 3-13, C.) Get in the sitting position, face upward, and lift the hips. Support the body with the hands and feet. Walk forward feet first.

(4) Squat Walk.(See fig. 3-13, D.) Assume a full knee bend position. Grasp the ankles (left ankle with the left hand, right ankle with the right hand). Walk forward.

(5)Broad Jump.(See fig. 3-13, E.) Jump forward on both feet in a series of broad jumps. Swing the arms vigorously to assist the jumps.

(6)Toe-Touch Walk.(See fig. 3-13. F.) Walk forward, bending at the waist and touching one hand to the toe of the opposite foot while it is on the ground. Raise the trunk to the vertical position between steps. Keep the knees straight.

(7)Bottoms-Up Walk.(See fig. 3-13, G.) Assume the front leaning rest position and move the feet toward the hands in short steps, keeping the knees locked. When the feet are as close to the hands as possible, walk forward on the hands to the front leaning rest position.

(8) Straddle Run.(See fig. 3-13, H.) Run forward, leaping to the right from the left foot and to the left from the right foot.

(9) Fireman's Carry .(See fig. 3-13,1.)

(IO)Single-Shoulder Carry(See fig. 3-13, J.) Two men execute the carries as indicated by the diagram. No. 1 man executes one type; No. 2 man executes the other.

Figure 3-13 is available in the illustration section at the end of the book.

b. Table II

(1) Double Time.(See fig. 3-14, A.)Hold arms at the thrust position. Execute a double-time run, maintaining the circle formation and the prescribed distance between your fellow Marine.Duration-1 minute.

(2) Toe-Grasp Walk.(See fig. 3-14, B.) Bend forward and grasp toes. With knees slightly bent, walk forward.

(3) Hand-Kick Walk..(See fig. 3-14, C.) Walk forward, kicking the moving foot upward on every step. At the same time, lean forward and touch the elevated toe with the hand of the opposite arm.

(4) Pike Juinping.(See fig. 3-14, D.) Jump forward and upward from both feet, keeping the knees straight, and at the same time, swing the legs forward and touch the toes with the hands at the top of each jump.

(5) Squat Jump.(See fig. 3-14, E.) Leap forward from the squatting position, with the hands on the ground and the arms between the legs. Land on the ground with hands and legs extended. Bring up the legs to the squatting position.

(6) Steam Engine..(See fig. 3-14, F.) Lace the fingers behind the neck and walk forward in the following manner: as the left leg moves forward, raise the knee high, bend the trunk forward, and touch the outside of the right elbow to the outside of the knee. Then lower the left leg and step forward on the left foot and raise the right leg. Repeat with the right leg and left elbow.

(7) Knee-Touch Walk.(Scc fig. 3-14, G.) Walk forward, bending the knees and touching the ground on each step. The knees are bent and straightened on each step.

(8)Hobble Hopping.(See fig. 3-14, H.) Hold foot behind back with opposite hand and hop forward. On the command CHARGE, grasp the opposite foot with opposite hand and hop **forward.**

(9) Cross Carry.(See fig. 3-14,1.)(10)Saddle Back Carry. (See fig. 3-14, J.) Two Marines execute the carries as indicated in the diagram. No. 1 Marine executes onetype;No. 2 Marine executes the other.

Figure 3-14 is available in the illustration section at the end of the book.

Section VI. Running and Orienteering

3601. RUNNING

a. General

(l)Description. The general form and technique for all types of running are fairly constant. (See fig. 3-15.)The head is erect, body slightly forward without bending at the waist, and the arms are at a loose thrust position alternating from front to rear in straight planes. A cross-body arm movement wastes energy .The movement of the legs and feet will be discussed in subsequent paragraphs dealing with the different types of running. Of primary importance is the fact that in all types of running, the toes should be pointed straight ahead. Toeing out is a common error in both running and walking and should be an item of individual correction.

(2) Objective. The objective of running is to develop cardiovascular endurance. Despite the fact that Marines have developed their muscle structure and the strength of their muscle tissue, unless they have developed cardiovascular endurance to a satisfactory degree, they are not entirely physically fit or combat ready. Cardiovascular endurance (wind) depends on the efficiency of the lungs and heart. The efficiency of the lungs and heart depend on the amount of oxygen the lungs can absorb with each breath inhaled and the amount of carbon dioxide the lungs can expel. The process of absorbing oxygen and expelling carbon dioxide (cardiopulmonary process) is performed by the blood that circulates through the lungs. The condition of this process will determine the amount of effort a person can exert over a period of time. Running is one of the best activities to develop this vitally important endurance.

b. Running Skills. In the development of running skills, individuals may require instruction to improve their proficiency. Some important skills to consider are-

Figure 3-15 is available in the illustration section at the end of the book.

(1) Action of the Arms. Arm action is important. Check to see that arms are held loosely and that the action is relaxed. The faster the run, the more rapid the arm action.

(2) Breathing. Allow the individual to breathe through the mouth as the body demands a large supply of oxygen. Oxygen can be inhaled in greater quantities through the mouth.

3602. TYPES OF RUNNING

a. Double Time

(1) Description and Objective. Double timing is marching at the rate of 180 steps per minute, each step being 36 inches in length. It takes practice to double time with precision in formation. The Marines should keep in step, placing their feet flat on the ground. This, however, should not be a stamping motion, but should be done with as slight a jolt as possible. Double timing is like a jog, the difference being that in a jog the feet are lifted well off the ground and the running motion is bouncy. In double timing, the feet skim the ground and there is no bounce to the ran. Double timing is a vehicle for teaching proper running form and for the development of the cardiovascular system.

(2) Area and Equipment. This type of running can be completed over a variety of surfaces. Usually a Marine uses a field or road. There is no equipment requirement.

(3)Progression. There is no set standard for alternating quick time and double time in the early conditioning of Marines. A general rule is to begin with enough quick-time marching to ensure a thorough warming up, then double time about 100 paces. Change again to quick time until the individuals have made a reasonable recovery from the running, then double time another 100 paces. The amount of double time can be increased and the quick time decreased from week to week, until the individuals are double timing about 1800 yards. This type of training should be given at least twice a week, but by no means is it adequate as the sole means of conditioning.

(4) Leadership

(a) The instructor should be to one side of the column or group and toward the rear so there is a full view of all Marines. Inexperienced instructors have a tendency to supervise from a position too far forward.

(b) Select an individual who can maintain the proper cadence to act as the guide during double-time running.

(c) There are several ways for the instructor and group to count cadence while double timing. If not contrary to local policy, learn several methods and use them for variety.

b. Wind Sprints

(1) Description and Objective. This type of running involves a series of 30- or 40-yard dashes, usually conducted in successive waves of squads. Each squad is in line and the squad leader on the right flank. Wind sprints assist indeveloping speed and cardiovascular endurance. Any flat and level area may be used which will permit the squad to form a line and run the required distance.

(2) Progression. One or two 30-yard sprints will be adequate at the beginning. As time passes, sprints can be lengthened and up to six or seven sprints may be used.

(3) Leadership

(a) At the command READY (given by squad leader), each runner assumes the sprinter's starting position. At the command GO, the squad sprints approximately 30 yards, takes 10 yards to stop, and lines up immediately with the squad leader who repeats, READY, GO, and again the squad sprints. At the conclusion of the third sprint, the squad waits until all the squads of the platoon have made three sprints. Then they all line up and the squad leaders conduct three more wind sprints in the opposite direction.

(b) Valuable time is gained by having each squad ready to go when the preceding squad has moved off its second sprint mark.

c. Cross-Country Running

(1) Description and Objective. Cross-country running is a distance run conducted on a course laid out along roads, across fields, over hills, through woods, and on any irregular ground. The cross-country ran may be utilized as a conditioner or as a competitive event; the objective is to cover the distance in the shortest possible time. The course should be 2 to 2.5 miles in length and be laid out to avoid heavy vehicular traffic. The course should be marked by directional arrows until the runners know the course. These runs build leg muscles, increase lung capacity, and develop endurance. Any local area of varied terrain is suitable.

(2) Progression. In the mass training of a large group, leaders should be stationed at the head and the rear of the column and should make every effort to keep the individuals together .After determining the abilities of the individuals in cross-country running, it is advisable to divide the unit into three groups. The poorest conditioned group starts first, and the best conditioned group, last. The starting time of the groups should be staggered so that all of them finish about the same time. In preliminary training, the running is similar to ordinary road work in that it begins with rather slow jogging, alternating with walking. The speed and distance of the run is gradually increased. As the condition of the individual improves, occasional sprints may be introduced. At first the distance run is from 1/2 to 1 mile. It is gradually increased to 2 or 2.5 miles. Well-conditioned personnel can run 2 to 2.5 miles within a 15-minute period.

(3) Leadership. Marines should not be required to take part in distance running until they have been through a progressively scheduled training program which requires a considerable amount of running. Cross-country runs should be scheduled occasionally to provide variety in the program. Cross-country running has the advantage of allowing mass participation. Interest can be stimulated by putting the runs on a competitive basis. As a single activity, short cross-country runs can be scheduled once a week, gradually increasing the distance as the physical conditioning improves.

(4)Use With Other Programs. Cross-country running can be combined with other activities such as conditioning exercises.

d. Fartlek Training

(l)Description and Function. Fartlek training is various running exercises conducted along an intense combat-like course. The running is conducted to overwork the lungs, allowing only partial recovery which is followed by another intense period of overload. This sequence is repeated for the duration of the workout. Fartlek training is a useful combat training method and general cardiovascular conditioner. The entire unit must be in uniformly good to excellent condition to qualify for this training.

(2) Progression. The following is an example of one fartlek session:

(a) Warm up by stretching 3-5 minutes.

(b) Warm up running easily 5-10 minutes.

(c) Run at a fast, steady pace for 3/4 to 1 3/4 miles (dependent on terrain).

(d) Walk/jog at a moderate pace for 5 minutes (recovery).

(e) Run easily sprinting 15 to 20 meters occasionally.

(f) Run full speed uphill for 175 to 200 meters.

(g) Warm down by running easily for 1/2 to 1 mile.

(h) Warm down by stretching 3 to 5 minutes.

(3)Leadership. To add variety, try incorporating combat movements, terrain association, a series of exercises or carrying table of equipment weapons (hit and roll). This will take some initiative in setting the course, but it allows for variety. In this manner, the fartlek course can be used as a training session or part of one. It works extremely well with highly motivated, competitive Marines.

3603.ORIENTEERING

Orienteering is land navigation over a prescribed course as a timed, competitive event. It is an excellent way to integrate land navigation training with physical conditioning and requires Marines to be skilled in both areas. Commanders adjust the training as needed to change the degree of difficulty or to emphasize either navigation or conditioning. Taken to the extreme, an orienteering exercise is conducted in full combat gear over rugged terrain against aggressors. Orienteering requires the following physical skills: endurance, speed, strength, and all-around conditioning.

Section VII. Circuit Training

3701. GENERAL

a. Description and Objective. A strength circuit consists of a series of stations where individuals in small groups exercise vigorously for a short period of time and then move (on signal) to the next station where a different form of exercise is conducted. This rotation continues until all groups move through all stations. Strength circuits contain no set or specific types of exercise stations within the circuit. The objective of circuit training is to develop strength. There are three general types of circuits.

(1) Fixed Circuit. This is a circuit in which apparatus of an immovable type (fixed into the ground) is used.

(2) Movable Circuit. This circuit consists of individual exercise apparatus which is portable and can be moved to and from the training area.

(3) Simplified Circuit. This circuit requires no equipment or apparatus.

b. Formation. The exercises are done at will, but rapid, steady, and continuous work is required of all. Each Marine's nervous and muscular system reacts differently to timed vigorous exercises. His performance should be measured on how many movements per exercise he can complete as an individual. For example, one Marine may be able to complete 5 movements, while another may be able to complete 20, and yet each is receiving the maximum benefit. All three circuits contained in this chapter are designed for platoon-sized groups. Expansion beyond this capacity requires a large amount of equipment, as each Marine in the fixed and movable types of circuits must have an item of equipment available for exercise at each station. A group larger than a platoon could be exercised through use of the simplified type of circuit; however, the group would be unwieldy and control could be a problem.

c. Place in the Program. All circuits illustrated can be completed in a 15-minute period. This feature allows the exercise of a platoonor smaller group on the circuit for a single 15-minute period, or the scheduling of the circuit as a 15-minute period within a longer period. A circuit can thus be utilized within a rotating activity system of scheduling. Choice of a circuit by the unit depends upon area, facilities, and other local factors; however, there is a circuit for every need.

3702. FIXED STRENGTH CIRCUIT

a. Description and Objective. The strength circuit is an arrangement of various types of exercise apparatus which are fixed in position. (See fig. 3-16.)Seven basic exercises are used and each exercise requires an apparatus. All apparatus of one type are positioned together to constitute a station. Each station will accommodate 10 Marines. The circuit is designed to be accomplished in 15 minutes when conducted on a time-rotation basis, as normally executed for unit training. For individual use, a Marine may complete the circuit by executing a specific number of repetitions for each exercise. In this case, the time required for completion of the circuit would vary slightly depending on the number of repetitions the Marine completed. The objective of this circuit is to provide a series of exercises which will improve and maintain the strength of the body's major muscle groups.

Figure 3-16 is available in the illustration section at the end of the book.

b. Starting Level. Marines must be thoroughly warmed up prior to participating in the circuit system. If Marines have not engaged in vigorous exercise immediately prior to starting the circuit, then an instructor should execute the following warm up exercises. These exercises should be conducted in the normal formation for set drills. Seven repetitions of each exercise will normally provide sufficient warm up. The exercises are-

(1) High jumper, Conditioning Drill 1.
(2) Bend and reach, Conditioning Drill 1.
(3) Squat bender, Conditioning Drill 1.

c. Progression. The instructor can adjust the circuit system through several methods to accommodate participants of varying physical ability. One method of adjusting an exercise is to change the method of executing the leg lift or to select a heavier weight for the barbell curls. When it becomes apparent during a unit program that the overall fitness of the group has Improved, then the exercise may be made more strenuous in two additional ways. First, the instructor can have one or more of the supplementary stations added. Second, the time spent exercising at each station can be increased in 5-second increments to a maximum of 60 seconds. Also, as Marines learn the circuit, the 45-second period for movement and instruction between stations can be eliminated, thus allowing only enough time to change stations.

d. Starting Positions. The Marine starts on any station, exercises steadily for a certain period (45 seconds initially), then moves on command to the next type of apparatus where he again exercises steadily for an equal period. The Marine continues until the required exercise is completed at each station. If it is desired to expand the number of stations in the circuit to accommodate more participants at one time, the instructor can provide four additional supplementary exercises, requiring no equipment. This will increase the amount of time required to complete the circuit.

e. Leadership. Close control of all Marines is necessary to ensure that a minimum amount of time is spent in moving them to their initial stations and in moving them between stations. One instructor can control the activity on the strength circuit. A stop watch or wrist watch with a second hand is required. When Marines arrive at the strength circuit, the instructor will form them for exercise and conduct the warm up drill. The group is then reassembled and formed into a number of files equal to the number of stations being used in the circuit. Each file is then directed to a station. As soon as all participants have reached an exercise position at a station, the instructor gives the command READY, GO. After 45 seconds of exercise, the instructor gives the command STOP, CHANGE OVER. The instructor allows Marines 45 seconds for moving to the next station and for preparing for the next exercise before the command EXERCISE is again given. In lieu of verbal commands, a whistle may be used to stop and start the exercises. For large groups, a megaphone or loudspeaker is useful.

f. Individual Conditioning Program. For the Marine working alone on the strength circuit, it would be impractical to time the exercise periods. For individual exercise, the Marine should select a number of repetitions of each exercise to complete, then rotate to the next stations after completing these repetitions. The number of repetitions selected should be at or near the maximum that the Marine is capable of doing without halting for rest.

g. Fixed Circuit Stations

(1) Primary Stations (See fig. 3-16.)

(a) Pullups/Chinups. A horizontal bar placed 8 feet above the ground is necessary for each Marine at this station. The Marine will also need a space on the bar that is 45 inches wide. On the command**EXERCISE**,the Marine grasps the bar with both palms facing either forward or to the rear, arms fully extended, and feet free of the ground and executes the exercise as described in MCO 6100.3_.The exercise is repeated as many times as possible until the command **STOP** is given. Then the Marine moves to the next exercise. If a Marine has done his maximum number of pullups/chinups prior to the command **STOP**, he will remain in the "dead" handing position until the command **STOP** and move to the next station.

(b) Twist Grip. The apparatus is a horizontal bar, free to turn, held between uprights placed 30 inches apart. The bar is 52 inches above the ground. A weight of 20 pounds is attached to the center of the bar by a light rope long enough to permit the weight to rest on the ground. The Marine stands at arm's length from the bar and grasps it with his hands on either side of the rope, palms down, thumbs under the bar. On the command **EXERCISE,** the hands are rotated so that the backs of the hands are rotated away from the body, thus winding the rope on the bar. The elbows are kept straight to ensure that the exercise is performed by the hand and forearm. When the weight is drawn up to the bar, the bar is then rotated in the opposite direction to lower the weight to the ground. This exercise is continued until the command **STOP** is given. The Marine then moves to the next station.

(c) Pulley Weights. The apparatus is a T frame with a system of pulleys that suspends a weight of about 90 pounds. The weight is attached to a light steel cable which has a drawbar attached to the other end. The Marine grasps the drawbar and sits down directly under the bar, legs extended to the front and arms extended overhead. The exercise is executed by pulling the drawbar down behind the head, then extending the arms slowly again until they are fully extended overhead. The exercise is repeated as many times as possible until the command STOP is given. The Marine then will move to the next exercise. Upon completion of the exercise, the weight is lowered slowly to the ground.

(d) Barbell Curls. A barbell is necessary for each Marine at this station. The barbell is constructed of 1 1/4-inch pipe 5 feet long, and two concrete-filled No. 10 cans. Each barbell should weigh about 40 pounds. Variance in the weight of the barbells, up to about55 pounds, will allow appropriate overload to be applied to Marines who are above average in strength or weight. The Marine grasps the bar with the palms forward and assumes a standing position with the barbell held in front of the hips, hands approximately shoulder's width apart. On the command **EXERCISE,** the Marine flexes the elbows and draws the barbell up until it touches the upper chest. The elbows remain at the sides. Breath is inhaled with the upward movement and exhaled as the barbell is lowered to the starting position. The exercise is repeated as many times as possible until the command **STOP** is given. The Marine moves to the next apparatus.

(e) Step-Up. The apparatus is a platform or ledge 18 inches high and of such size. At accommodate 10 Marines. The Marine faces the platform and on the command **EXERCISE,** steps up onto the platform, bringing his trailing foot up beside the leading foot. He then steps back down to the original position, stepping down first with the same foot he initially used in stepping up. After 10 repetitions of the exercise, he changes the order of moving the feet to use the opposite leg for stepping up. Repeat this exercise until the command **STOP** is given. The Marine then moves to the next station.

(f) Leg Lift. The apparatus is a horizontal bar constructed as described in (a). To prevent the body from swaying, a horizontal back support is added 40 inches below the horizontal bar. The arms are kept fully extended. On the command **EXERCISE,** the Marine jumps up, grasps the bar with the palms forward and the back support behind him. The exercise is executed by raising the legs to a horizontal position then slowly lowering them to the vertical position. The Marine does not flex his knees. He does not swing the legs to the rear of a vertical position to gain momentum for raising them in the next repetition of the exercise. The movement is repeated until the command **STOP** is given. The Marine then moves to the next exercise. If unable to raise his legs to a horizontal position without flexing his knees, the Marine flexes his knees and draws them up to the chest, then lowers his legs to the vertical position.

(g) Rope Climb. The rope climb is 20 to 30 feet high with five ropes suspended from a horizontal bar which forms the uppermost part of the framework. To prevent the horizontal bar from sagging and to provide safety, only five ropes are attached to it. There are two frameworks per station. The ropes are 6 feet apart. Any method may be used to climb the rope, and the Marines climb as high as possible. Marines who are proficient should climb therope several times during the time allotted. Inexperienced Marines should be cautioned to take care during descent to avoid rope burns on their hands.

(2) Supplementary Stations. The following exercises are designed to expand the basic circuit by being inserted in specific places within the system. For each supplementary station used, there should be adequate room for 10 Marines to exercise.

(a) Bent Knee Situp or Bottoms-Up. These calisthenics are designed to strengthen the abdominal muscles. These exercises will be inserted between the pullup/chinup and twist grip stations. The primary stomach exercise is the situp. In case of inclement weather or other conditions that make ground contact undesirable, the bottoms up exercise is used. In situps on the command **EXERCISE,** the Marine lies on his back (supine position) with knees flexed and both feet flat on the ground and executes the exercise (minus an assistant) as described in MCO 6100.3_.The Marine then returns to the starting position, repeating the exercise until the command **STOP** is given. The Marine then moves to the next station. In bottoms up on the commandEXERCISE,the Marine assumes the front leaning rest and executes the bottoms-up exercise as described in Conditioning Drill 3. He continues this exercise at a moderate cadence until the command **STOP** is given. The Marine then moves to the next station.

(b) Pushup. This exercise is designed to strengthen the arm- and shoulder-girdle muscles. It should be included between the twist grip and pulley weight stations. Upon the command **EXERCISE,** the Marine executes the pushup as described in Conditioning Drill 1. He continues this exercise at a moderate cadence until the command **STOP** is given. The Marine then moves to the next station.

(c) Knee Bender. This exercise is designed to build leg muscles and is included between the pulley weight and barbell curl stations. On the command **EXERCISE,** the Marine executes the knee bender as described in Conditioning Drill 3.He continues this exercise at a moderate cadence until the command **STOP** is given. The Marine then moves to the next station.

(d) Trunk Twister. This exercise strengthens the major muscles of the trunk and is included between the step-up and pullup/chinup stations. On the command EXERCISE, the Marine executes the trunk twister as described in Conditioning Drill 1. He continues this exercise at a moderate cadence until the command **STOP** is given. The Marine then moves to the next station.

3703. MOVEABLE STRENGTH CIRCUIT

a. Description and Objective. The exercises in this circuit are progressive and the course is planned to gain and hold the interest of the participating groups. The circuit consists of a series of stations, with each station designed to develop a particular group of muscles. Along with muscular development, correct posture and deep rhythmic breathing should be stressed on this circuit at all times.(See figs. 3-17 and 3-18.)

Figure 3-17 is available in the illustration section at the end of the book.

b. Equipment. The equipment is set up in files. Six files of 8 stations will accommodate a platoon

c. Formation. The platoon marches to the area where the equipment is positioned and forms a file within each lane of stations, covering down on a piece of equipment. Movements are made on the double, the important factor being that no time is wasted in getting to work.

d. Progression. Initially 40 to 45 seconds per station is adequate. As individuals become stronger the time should be increased in 5-second increments until a minute to a minute and a half is reached.

e. Leadership

(1) The leader stands in front of the barbell station and controls the rotation from this position. The leader supervises the entire group, with the assistance of several instructors who move about in the platoon correcting and encouraging the Marines.

(2) The leader starts each group but does not count cadence nor lead them through the exercises. Each individual exercises rapidly but individually.

(3) As each Marine finishes his repetitions with the barbell, he places the barbells on the ground. The leader calls, READY, followed by the command, FALL OUT ONE. All will double-time to the station directly in front of them, while the individuals on the barbell stations will do an about face to the rear station in their lane.

f. Movable Circuit Stations. The Marine can obtain best results on the movable circuit if the exercises on the various stations are given in the following manner:

(1) Station 1—The Barbell. (See fig. 3-17, 1.)The exercises at this station stress proper posture; deep, rhythmic breathing; and development of the muscles of the arms, shoulders, and upper body. An instructor teaches the proper methods of lifting before the exercises begin. The methods are to lift with the legs, to keep the back straight, and to merely grip with the hands. Two recommended exercises are given below. The instructor will use only one exercise per period .He may use either exercise. At successive periods, the other exercise is used.

(a) Exercise 1--Two-Hands Military Press. (See fig. 3-18, 1.)Grasp the barbell with both hands, knuckles up at shoulder width, and lift to the chest. Steadily press to arm's length overhead; lower to the chest resisting weight all the way. Inhale as the weight is pressed up and exhale as the weight is brought down.

(b) Exercise 2--Two-Hands Regular Curl. (See fig. 3-18, 1.) Lift the weight to the waist, with the palms of the hands out, heels together, stomach in, chest lifted and arched, shoulders back, elbows in close to the sides; inhale deeply and curl the weight to the shoulders, using the arms only, at the same time keeping the elbows close to the sides; exhale Emphasize posture and the use of the arms only. This is a very valuable exercise for the development of the biceps and the grip and should be repeated from 8 to 16 times, depending on the ability of the Marine.

(2) Station 2--The Jump Rope. (See fig. 3-18, 2.) This exercise develops strength and agility in the legs and stamina of the whole body. It makes the Marine agile on his feet and increases his footwork efficiency and timing. Each Marine should progress until able to jump rope at least 3 minutes at top speed.

(3) Station 3--The Twist Grip. (See fig. 3-18, 3.) The twist grip is an excellent exercise for the hands and forearms, and adds greatly to the Marine's ability in hand-to-hand combat. The handle is gripped and twisted, winding the rope until the weighted can is level with the height of the hands, which are held horizontal. The weight is lowered in the same manner; the Marine resists the weight all the way, occasionally stopping the twisting motion and alternately removing first one hand, then the other, from the handle. A variation of the above exercise is to wind the handle with the palms up and the arms bent and with the elbows held close in to the sides. Each Marine should maintain a good posture and keep the stomach muscles taut throughout this exercise.

(4) Station 4--The Incline Plane. (See fig. 3-18, 4,)The use of the incline plane is a very strenuous exercise and well designed for the development of the abdomen. Although 6 to 10 repetitions are sufficient for the beginners, more can be added as ability increases. Marines with hernias or recent operations will be excused from participation at this station.

(5) Station 5--The War Club. (See fig. 3-18, 5.) The war club is a simple and effective means of exercising the principal muscle groups of the body, especially those of the trunk, back, and shoulders. To gain the maximum benefit from this exercise, the Marine must keep both feet flat on the ground at all times. Throughout the exercise period, the weight is swung from arm's length as follows:

(a) As in chopping wood, first on one side, then on the other.

(b) As a batter warming up with a number of bats.

(c) In large circles, first with one hand and then with the other.

(6) Station 6--The Bicycle Ride. (See fig. 3-18, 6.) The bicycle ride is well suited to exercising many of the muscle groups of the body, particularly those of the abdomen. Vary the speed of the exercise, but keep everyone "riding" the entire period. A variation exercise may be performed by placing the legs together, raising them slowly to a height about 2 feet from the ground, and then lowering them slowly to the ground.

(7) Station 7--The Step-up. (See fig. 3-18, 7.) The step-up exercises the legs. The step-up is performed by initially stepping up with the left foot, followed by the right, then stepping down with the left foot followed by the right. Continue for 20 seconds, then change to the right foot as the lead foot for 20 seconds.

(8) Station 8--The Isometric Pull. (See fig. 3-18, 8.) Two trainees work at this station with a cable pull and perform the following exercises:

(a) Initially start with one Marine in the supine position and one Marine sitting. The sitting Marine lowers the upper body to the ground and pulls the partner up to the sitting position. The partner then performs this same action and this is continued for 20 seconds at a rapid rate.

(b) During the last 20 seconds, the same action takes place but, in this case, the Marine in the supine position resists the pull of the partner for approximately 5 seconds before allowing to be pulled up into the sitting position.

3704. CIRCUIT-INTERVAL TABLE

a. Description and Objective. The circuit-interval table is designed to develop strength and endurance within a short period of time, with no equipment requirement, through a rapid and vigorous routine of exercise. Fifteen minutes is an adequate period to execute all exercises and to secure a vigorous workout with the circuit-interval principle.

Figure 3-18 is available in the illustration section at the end of the book.

b. Formation and Starting Positions

(1) A leader forms platoons or smaller groups in an oval or circular formation with 3- to 5-yard intervals between Marines. The Marines face to the right and move forward at quick time and then double time. (See fig. 3-19, A.)After running several platoon circle laps, the leader calls out the name of an arm and shoulder exercise from the list below, orders quick time and commands, for example, PUSHUPS. On this command, all Marines immediately hit the ground and individually and rapidly begin doing pushups. No cadence is counted. (See fig. 3-19, B.) After 30 seconds of exercise, the leader commands. **QN YOUR FEET, FORWARD, MARCH** The platoon resumes the quick time cadence and the leader, when ready, gives the necessary commands for double time. The double time is continued for one or more laps and the leader calls out the name of the next exercise and the process is repeated. This continues, with running between each exercise, until every body part has been exercised.

(2) The instructor controls the running and quick time to observe the effects of the exercise upon the Marines. Cadence, step, and precision are not important to the objective and the instructor should not use them. What is important is speed and the instructor should stress this. After the exercise period is started, the Marines do not stop. This circuit method emphasizes stress and recovery, the recovery occurring during the quick time periods.

c. Activities. The leader can use the following exercises. These exercises can be repeated if necessary, during a second round. The leader can use other calisthenic exercises in the circuit-interval table.

(1) Arms and shoulders-push-ups.
(2) Stomach situps.
(3) Back-squat thrusts.
(4) Legs-bicycle (on back).

d. Progression. lhe progress is controlled by the leader, who must pace the running, quick-time movement, and exercise in such a way that Marines will receive a vigorous workout yet be able to participate throughout the 15-minute period. Marines who are in the initial stages of physical condition will not be able to double time or exercise as long as those who are better conditioned. The idea is to set a pace which can be increased during each workout, thus progressing gradually to a higher level of physical fitness.

e. Leadership. The platoon leader, platoon sergeant, or section leader can lead the group. The leader must execute the exercise with the unit to feel the effects and thereby adjust the pace.

f. Place in the Program. This activity may be scheduled whenever a short period of time is available. The only requirement is that enough space, indoors or out, be available to form the circle.

Figure 3-19 is available in the illustration section at the end of the book.

Section VIII. Basic Physical Skills and Obstacle Courses

3801. GENERAL

The purpose of this section is to list the basic military skills and the methods for their development. Many of these skills are best developed by obstacle courses but other drills are also discussed. The section explains types of obstacle courses, details of construction, and methods of negotiating the various obstacles.

3802. BASIC PHYSICAL SKILLS

a. Objective. The objective is to develop proficiency in the various military physical skills which are essential to personal safety and effective combat operations. In travel by foot over rugged terrain and in the execution of combat duties, Marines must be trained to perform basic skills, such as running, jumping, climbing, and carrying. During training, Marines will develop agility and coordination in these skills .Fast and skillful execution of these skills may mean the difference between success and failure on the battlefield.

b. Place in the Program. These skills are practiced throughout the entire program in many different activities. Many of these skills are best practiced on obstacle courses.(See par. 3803.)

c. Description. These basic skills are the minimum skills required by all Marines. The basic skills are as follows:

(1) Running. Running is used to strengthen the legs and develop cardiovascular endurance. Marines should be exposed to running in various situations: on roads, over rough ground, up and down hills, cross-country, and running over low obstacles.

(2) Jumping. In broad jumping, the take-off foot is planted firmly and the spring comes from the extension of this leg as the other leg reaches for the far side of a ditch or similar obstacle. (See fig. 3-20.) The arms are forcibly raised forward and upward to assist in propelling the body up and forward. Landing may be on one or both feet depending upon the length of the jump. In vertical jumping downward from a height, the jumper should aim his feet at the desired landing spot and he should also jump with the knees slightly bent and feet together, with the trunk inclined slightly forward. As the feet touch the ground, the shock is absorbed by bending the knees into a full squatting position. If the height is too great or the ground too hard to absorb the shock, then the jumper should forward roll or side roll thus eliminating some of the momentum.

Figure 3-20 is available in the illustration section at the end of the book.

(3) Dodging In combat situations, it is often times necessary to change directions quickly. To execute this movement while running, a lead foot is firmly planted, left foot if the direction is to the right and right foot if the direction is to the left. The opposite foot is moved toward the new direction. The knees are slightly flexed during the movement and the center of gravity is low and balanced. (See fig. 3-21.) At the time of the change of direction, the head and trunk are turned quickly in the new direction.

Figure 3-21 is available in the illustration section at the end of the book.

(4) Climbing and Surmounting. All Marines should know how to climb and surmount various types of obstacles. (See fig. 3-22.)

Figure 3-22 is available in the illustration section at the end of the book.

(a) Vertical Climbing, as in Climbing a Rope or Pole. The hands grasp the rope or pole overhead with the palms toward the face. Gripping the object, the body is pulled upward with the arms and shoulders, assisted by the feet which grip the object and assist by pushing downward. If shoulder girdle strength and body coordination are not adequate to permit alternating the hands, the arms act together in pulling upward.

(b) Climbing as in Surmounting a Wall. Ingoing over a wall, the body should be kept as close to the top as possible, since in combat operations it is important to offer as small a target as possible to the enemy. If an individual climbs a wall while carrying a rifle, both hands should be freed by slinging the rifle over the back. There are two methods commonly used for surmounting a wall of moderate height, but only one for dropping from it.The methods are as follows:

1 Run, Jump, and Vault. Approach the wall at a run, jump forward and upward at the wall and place one foot against it as high up as possible. Use the foot in contact with the wall to help push the body upward while grasping the top of the wall with the hands. Pull the body up with the arms, assisted by pressure of the foot against the wall and swing the legs over, propelling the body weight over the wall.

2 Hook and Swing. Approach the wall at a run and jump forward and upward. Hook one elbow over the wall, locking the arm in place by pulling up until the top of the wall is underneath the armpit. Grasp the top of the wall with the other hand. Draw the leg which is closer to the wall up as far toward the top as possible. Then swing the other leg over the top of the wall.The body is then carried over with a rolling motion. A variation of this leg action can be used by Marines who are unable to draw up the leg as described. While hanging with both legs fully extended, start a swinging motion with the legs together. When the legs have enough momentum, swing the outside leg over the top of the wall with a vigorous kick, then follow with the body.

3 Dropping. All drops from a height are executed in the same manner regardless of the method used to gain the top. One hand is placed against the far side of the wall while the other hand grasps the top. From this position, the body is rolled over the wall and vaulted away from it with the legs swinging clear. As the body passes over the wall and drops, it should at all times face the wall. This will keep the rifle and other equipment clear. Break the fall by retaining a grasp on the top of the wall as long as possible.

(c) Climbing Ladders and Cargo Nets. Rope ladders, stationary vertical ladders, and cargo nets employ the same general technique. The important element is to grasp the side supports firmly in the hands about shoulder height and place the feet on a rung which would cause the body to be fully extended. In movement upward, one hand is moved upward and a new grasp is secured and, at the same time, the opposite leg moves up a rung. As the knee straightens, the body is elevated. This process is repeated using the opposite arm and leg. Alternation continues in this manner until the climber reaches the objective.

(5) Traversing Horizontal Objects. The traversing of horizontal objects puts heavy stress on the arms and shoulder girdle area as the feet are usually suspended in the air with all of the body weight on the arms and shoulders.(See fig. 3-23.)

(a) Traversing Horizontal Ropes or Pipes. The hands grasp the horizontal support overhead with the palms facing. To propel the body forward, one hand is released and moved forward to secure a new grasp. At the same time, the opposite side of the body is swung forward (some people are able to "walk" in the air, keeping the body to the front and moving the legs in time with the arms as in walking on the ground). The other hand is then released and moved forward; this alternation is continued until the objective is reached.

Figure 3-23 is available in the illustration section at the end of the book.

(b) Traversing Horizontal Ladders. In this situation, the movement is the same as used in traversing a rope or pipe. The hands, however, are placed on the rung with the palms away from the face. Other than this difference the technique is the same.

(6) Crawling. Crawling in combat situations is an often used skill. Crawling may be high or low.(See fig. 3-24.)

Figure 3-24 is available in the illustration section at the end of the book.

(a) High Crawl.

In the high crawl, the Marine moves on hands and knees, moving one hand and the opposite knee and then continuing to move the hands in alternation with the opposite knee following the companion hand.

(b) Low Crawl. The Marine is in the prone position usually with the forearms and palms of the hand on the ground. He propels forward by bending the knee of one leg and pushing with the inside edge of the shoe. At the same time, the opposite arm moves forward and pulls to the rear. The body remains low and movement is continued by bending the opposite knee and pushing, and at the same time sliding the opposite arm forward and pulling. Alternation of hands and legs continues until the objective is reached.

(7) Throwing. Throwing may be executed from the kneeling or standing positions. The object to be thrown is held in the hand, and the throwing arm is bent at the elbow; the hand is then moved to the rear until the hand is behind the ear. The body is turned so that the lead foot and balance arm on the side toward the target are pointing at the target. (See fig. 3-25.) The balance arm is used to sight over and align the throwing hand and the target. When properly aligned, the elbow is moved rapidly forward until it is at a point just in front of the body where the arm is straightened and the wrist snapped. This whip motion propels the object to the target. Underhand throws secure momentum by the thrower bending his knees and swinging the throwing arm to the rear. As the knees are straightened, the arm is forcefully swung forward from the shoulder and the object released.

Figure 3-25 is available in the illustration section at the end of the book.

(8) Vaulting. Vaulting is employed to overcome low barriers or fences. (See fig. 3-26.)The object to be surmounted is approached at an angle. The hand on the side next to the obstacle is placed on the top of the obstacle and, with a straight arm, the body weight is pushed upward. At the same time, the leg on the side next to the obstacle is thrown upward and over the top followed by the other leg (side approach). In landing, the weight comes down on the leading leg first followed by regaining the balance on both legs. The free arm serves as a balance. A direct (front) approach can be used at which time both legs go over the object together.

(9) Man Carrying. There are three basic individual means of carrying personnel in combat situations and one of these methods may be used in carrying objects.

Figure 3-26 is available in the illustration section at the end of the book.

(a) Fireman's Carry."A" stands sideways in front of "B", "A" bends his knees and leans forward, placing one arm through "B's" crotch, grasps the wrist of "B's" arm, which is hanging over the shoulder, and then "A" runs forward.

(b) Saddle-Back Carry. With his back toward "B", "A" stands in front of "B". "B" mounts "A's" hips and clasps his arms in front of "A's" chest. "A" grasps "B's" thighs.

(c) Single-Shoulder Carry. "A" stands facing "B"."A" assumes a semi-squatting position. "B" leans forward until "B" lies across "A's" right shoulder. "A" clasps his arms around "B's" legs and straightens up, lifting "B" from the ground. "A" then runs forward. This method may also be used to carry heavy objects.

(10) Balancing. Balancing the body while walking or running on a narrow object when crossing obstacles is a skill which requires practice and confidence. Balance is required in negotiating a log placed across a stream, in crossing a narrow beam or rail, and in similar situations.(See fig. 3-27.) To perform this skill, place the feet on the object to be crossed, hold the arms to the sides at shoulder level, and fix the eyes on the object approximately 5 yards in front of the feet. Generally, it is not a good practice to look down at the feet. Walk the beam by placing first one foot and then the other in the center of the beam, thereby moving forward, using the arms to aid in maintaining balance.

Figure 3-27 is available in the illustration section at the end of the book.

(11) Falling. Injury can be avoided if Marines are taught to fall properly. They should know how to use the body momentum to their advantage during a fall rather than to try resisting that force. (See fig. 3-28.) If enough force is present, such as occurs during a fall while running or in umping downward from a height, individuals can extend their hands to catch the weight. At the same time, duck the head and roll forward onto the feet. The key to falling without injury from the standing position is relaxation and rolling on the outside of the leg, hip, and buttocks to take the brunt of the fall.

Figure 3-28 is available in the illustration section at the end of the book.

3803. OBSTACLE COURSES

a. General. Obstacle courses are a valuable part of physical readiness training. The challenge presented by the obstacles assists in developing and testing the basic physical skills. In many combat situations, success will depend upon the Marine's ability to perform one or more of these skills, often while carrying field equipment and when fatigued. In this section, two types of obstacle courses will be discussed-Conditioning Obstacle Course and Confidence Obstacle Course.

b. Course Safety Precautions. Commanders and course instructors should take certain precautions to prevent injury on obstacle courses. A few of the precautions are:
(1) Inspect the course for faulty construction of obstacles, protruding nails, rotten logs, condition of the landing pits, and other hazards to safety.
(2) Conduct warm up exercises before the unit runs the course.
(3) Explain and demonstrate the correct techniques for negotiating all the obstacles before allowing the Marines to try them.
(4) Give Marines at least two weeks of conditioning exercises before scheduling the obstacle and/or confidence courses.
(5) Ensure that negotiation of the higher and more difficult obstacles are under the supervision of an instructor.
(6) Do not permit individuals who have neither practiced the basic skills nor run the conditioning obstacle course to participate in the confidence obstacle course.
(7) Weather conditions may cause footing or handhold surfaces to be slippery. If such is the case, postpone training on the course.

3804.CONDITIONING OBSTACLE COURSES

a. Description and Objective. The Conditioning Obstacle Course is commonly known as the Obstacle Course. This course consists of fairly low obstacles which are designed to be negotiated quickly. The obstacles serve to test various basic skills, and running the course is a test of the Marine's physical condition. After receiving instruction and an opportunity to practice the skills, Marines run the course against time.

b. Area and Equipment
(1) Complete standardization of obstacle courses should not be attempted since topographical conditions always vary. Commanders should use ingenuity in constructing a course, making good use of streams, hills, trees, rocks, and other natural obstacles. Since the course is eventually run at high speed, it should not be dangerous.

(2) The course should be wide enough for six or eight men to run simultaneously, encouraging competition. The lanes for the first several obstacles should be wider and the obstacles themselves easier than those that follow. This avoids congestion until the contestants scatter out over the course. The last two or three obstacles should not be too difficult and should not involve high climbing. This prevents injuries and falls resulting from fatigue.

(3) The total distance of the course should range from 300 to 450 yards and include from 15 to 25 obstacles. Normally the obstacles should be 20 to 30 yards apart and arranged so that those which exercise the same groups of muscles are separated.

(4) The obstacles should be substantially built. Peeled logs, 6 to 8 inches in diameter, are ideal for many of the obstacles. Sharp points and corners should be eliminated. Landing pits for jumps or vaults should be filled with sand or sawdust to prevent injuries.

(5) The course should be constructed and marked so that it is not possible to sidestep or detour obstacles. However, it is desirable to provide alternate obstacles of varying degrees of difficulty.

(6) The course should be in the shape of a horseshoe or figure eight so that the finish is close to the start and signs should be placed to indicate the course route.

c. Leadership. Before Marines run an obstacle course, they should be instructed in the proper technique of negotiating each obstacle. In each case this technique should be explained and demonstrated in detail, with emphasis on avoiding injury. Every individual should be given an opportunity to practice on each obstacle until he becomes reasonably proficient at negotiating it. Before the course is run against time, it is advisable for individuals to make several runs at a slower pace. During such practice or trial runs, the instructor should observe the performances and make appropriate corrections. Marines should never be permitted to run the course for time until they have practiced on all obstacles. The best method of timing the runners is to have the timer stand at the finish and call out the minutes and seconds as each individual finishes. If several watches are available, each wave may be timed separately. If only one watch is available, the different waves should be started at regular intervals, such as every 30 seconds. If an individual fails to negotiate an obstacle, a previously determined penalty should be exacted.

d.Types of Obstacles

(1) Jumping-Type Obstacles. These obstacles may be ditches which are cleared with one leap, trenches which the individuals can jump into, heights which require jumping downward, or hurdles which an individual can jump over. (See fig. 3-29.)

Figure 3-29 is available in the illustration section at the end of the book.

(2) Dodging-Type Obstacles. Obstacles of this type are usually mazes consisting of posts set in the ground at irregular intervals. (See fig. 3-30.)The intervals between posts should be rather narrow so that the Marines must pick their way carefully through and around them. Lane guides may be constructed to guide the Marines to dodge and change direction. Obstacles may be put into a maze pattern to cause the Marines to change direction.

Figure 3-30 is available in the illustration section at the end of the book.

(3) Vertical Climbing and Surmounting Type Obstacles. These obstacles may be climbing ropes, either plain or knotted and 1-5 inches in diameter; cargo nets or walls 7 or 8 feet high; or vertical poles 6 to 8 inches in diameter and 38 feet high.(Seefig. 3-31.)

Figure 3-31 is available in the illustration section at the end of the book.

(4) Horizontal Traversing-Type Obstacles. Horizontal obstacles may be pipes, beams, ladders, or ropes. (See fig. 3-32.)

(5) Crawling Type Obstacles. Obstacles which require crawling may be constructed of large pipe sections, low rails, and wire. (See fig. 3-33.)

Figure 3-32 is available in the illustration section at the end of the book.

(6) Vaulting-Type Obstacles. Obstacles of 3 to 3.5 feet in height such as low walls or fences may be used as a vaulting obstacle.(See fig. 3-34.)

(7)Balancing-Type Obstacles. Beams, logs, and planks may be used as balancing-type obstacles.(See fig. 3-35.)These items may be used to span water obstacles and dry ditches, or raised off the ground somewhat to simulate natural depressions.

Figure 3-33 is available in the illustration section at the end of the book.

3805.CONFIDENCE OBSTACLE COURSES

a. Description and Objective. The Confidence Obstacle Course is commonly known as the Confidence Course. This course is composed of higher and more difficult obstacles than those used in the conditioning course. The confidence obstacle course is designed to give Marines confidence in their mental and physical capacities and to cultivate their spirit of daring. They are encouraged but not compelled to negotiate this course and the course is not run against time. The negotiation of a confidence course, however, is strenuous enough to be an excellent physical conditioner. Marines should NEVER attempt to take the obstacles at high speed and should not compete for speed. The obstacles vary from fairly easy to extremely difficult. Some are of considerable height to accustom Marines to climbing such heights without fear. Considerable emphasis is placed on obstacles that train and test an individual's balance.

Figure 3-34 is available in the illustration section at the end of the book.

b. Area and Equipment

(1) The confidence course accommodates four platoons, one platoon at each group of six obstacles. The course should be made up of about 24 obstacles, numbered and marked as follows:! to 6, white numbers on red background; 7 to 12, black numbers on a white background;

13 to 18, white numbers on a blue background; and 19 to 24, white numbers on a black background.

(2)A few simple pieces of equipment should be provided for individuals who do not have the strength, courage, or ability to negotiate the obstacles.

Figure 3-35 is available in the illustration section at the end of the book.

c. Formation. The obstacles should be divided into groups of six, and each group is designated by a different color. Each platoon starts at a different color. Individuals are separated into groups of 8 to 12 at each obstacle. At the starting signal from the company commander, they proceed numerically through. Anyone may skip an obstacle who is afraid to try. Individuals proceed from obstacle to obstacle until time is called, then assemble as ordered.

d. Leadership. If the Marines are new to the confidence course, an instructor will demonstrate or will give a brief orientation at each obstacle, including an explanation and demonstration of a method of negotiating it. Marines are encouraged to try the various obstacles, but they are not compelled to do so. No compulsion is to be used. The manner of negotiating any obstacle is left to the discretion of the Marine. However, the instructor assists anyone who experiences difficulty. Instructors must supervise closely at all times to prevent injuries, as some of the obstacles are quite high. Also, some of the obstacles should not be used when slippery or wet. The example of instructors and especially selected demonstrators will serve to inspire the individuals to greater effort.

e. Negotiating the Obstacles. Although personnel need not conform to any one method of negotiating the obstacles, there should be some uniformity in the approach to them. A general method of negotiating the obstacles is indicated below.

(1) Red Group This group contains the first six obstacles. (See fig. 3-36.)

(a) The Belly Buster Individuals may vault, jump, or climb over. Warn them that the log is not stationary.

(b) Reverse Climb. Climb the reverse incline and go down the other side to the ground.

(c) The Weaver. Move from one end of the obstacle to the other by weaving the body under one bar and over the next.

(d) Hip-Hip. Step over each bar, either alternating legs or using same lead leg each time.

(e) Balancing Logs. Step up on log, retaining the balance, walk or run along log.

(f) Island Hopper. Jump from one log to another until the obstacle is negotiated.

(2)White Group. This group is composed of the second six obstacles. (See fig. 3-37.)

(a) Tough Nut. Step over each X in the lane.

(b) Slide for Life. Climb the tower, grasp the rope firmly, and swing the legs upward. Hold the rope with the legs to distribute the weight between them and the arms. Braking the slide with the feet and legs, proceed down the rope. Warn Marines that there is danger of getting rope burns on their hands. When the rope is slippery or wet, this can be a dangerous obstacle.

(c) Low Belly Over. .Mount the low log and jump onto the high log, both arms grasping over the top of the log, the stomach area in contact with it. Swing the legs over the log and lower the body to the ground.

(d) Belly Crawl Move forward under the wire, belly down, to the end of the obstacle.

(e) The Dirty Name Mount the low log and jump to or reach the higher logs in succession, then jump or drop to the ground. Warn the Marines about the height of the final log.

(f) The Tarzan Mount the lower log and walk the length of it and each successive, higher log until reaching the horizontal ladder. Grasp two rungs of the ladder and swing the body into the air. Negotiate the length of the ladder by releasing one hand at a time and swing forward, grasping a more distant rung.

(3) Blue Group. This group is formed by the third group of six obstacles. (See fig. 3-38.)

(a) High Step over. Step over each log, alternating the lead foot or using the same lead foot.

(b) Swinger. Climb onto the swinging log and over to the ground on the opposite side.

(c) Low Wire. Move under the wire on the back, using the hands to raise the wire to clear the body.

(d) Swing, Stop and Jump. Gain momentum with a short run, grasp the rope, and swing the body forward to the top of the wall. Release the rope while standing on the wall and jump to the ground.

(e) Six Vaults. Vault over the logs, using one or both hands.

(f) Easy Balancer. Walk up one inclined log and down the one on the other side to the ground.

(4) Black Group. The last group is formed by the final six obstacles. (See fig. 3- 39.)

(a) Incline Wall. Approach the underside of the wall, jump up and grasp the top and pull the body up and over. Slide or jump down the incline to the ground.

(b) Skyscraper. Jump or climb to the first floor, climb up the corner posts or assist each other to any desired floor. Descend to the ground in any desired manner.

(c) Jump and Land. Climb up the ladder to the platform and jump to the ground.

(d) Confidence Climb. Climb the inclined ladder to the vertical ladder. Go to the top of the vertical ladder, then down the other side to the ground.

(e) Belly Robber. Step on the lower log and assume the prone position on the horizontal logs. Crawl over the logs to the opposite end of the obstacle.

(f) The Tough One Climb the rope or pole on the higher end of the obstacle, then go down the ladder and across the log platform. Climb over or between the logs at the end and go down the rope or pole to the ground. Vault over the final log.

Figure 3-36 is available in the illustration section at the end of the book.

Figure 3-37 is available in the illustration section at the end of the book.

Figure 3-38 is available in the illustration section at the end of the book.

Figure 3-39 is available in the illustration section at the end of the book.

Section IX Individual Exercise Programs

3901.GENERAL

Marines at times will **be** stationed on independent duty and consequently will be responsible for their own physical fitness program. This section will assist them in understanding the need for exercise and will aid in the planning and execution of an individual exercise program. Exercise activities included are the Bench Conditioner, 6-12 Plan, Weight Training, and Isometric Contraction. Keeping physically fit is a problem that faces every Marine. Even though we are frequently engaged in training that requires some physical effort, in many cases, it is not enough to prepare us to meet the intense physical demands of combat. Attaining a satisfactory level of physical readiness is not an insurmountable objective. Available time appears to be the most difficult obstacle to the development of physical readiness. In most cases, regular physical training programs are centralized, requiring the individual to temporarily leave the work area. The problems involved in setting an hour aside two or three times each are numerous. However, most of us can devote a few minutes each day to physical fitness with little, if any, impact on our daily work schedule, especially if it does not require us to leave our work area.

a. Type of Program. There are many good physical fitness programs available to the individual or group. Regardless of the type or duration, to be effective, the program must contain exercises that are strenuous and are challenging to the individual. Space will not permit the inclusion of all available means of individual exercise. The programs selected for this chapter have met the requirement of minimum space and minimum time.

b. Need to Augment Program. These programs are quite strenuous and will develop a satisfactory level of physical readiness. However, if the Marine desires additional development of endurance, it is recommended that he supplement these programs with a 15- to 30-minute period of wind sprints and double timing on an alternating daily basis.

c. Progressive Training. If Marines are performing duties which require little or no physical activity, they must plan a physical conditioning program that assures a moderate beginning, moderate but steady progression, and sufficient warmup before starting the vigorous exercise. To avoid organic of bodily harm, a Marine activity without adequate warmup. He should conduct conditioning programs on a daily basis over an extended period of time, never on an unduly accelerated or crash basis.

3902.THE BENCH CONDITIONER

a. Description and Objective. The bench conditioning program uses a modified bench to employ both isotonic (moving) and isometric (stationary) exercises are as the nucleus of the program. The exercises are designed to develop strength and endurance in all the major muscle groups of the body. The principles of progression, overload, and balance are employed when the exercises are performed properly.

b. Area and Equipment. The conditioning apparatus can be constructed in any unit motor pool with welding equipment found in most salvage yards.(See figs. 3-40, 3-41.)Additionally, there are available any number of commercially produced apparatus available. The important thing is that in utilizing this equipment the following exercise routine be adhered to.

c. Starting Level and Progression.. The program consists of two tables, each with 10 exercises. The Marine controls progression by required repetitions or, in some cases, by application of maximum effort. Each table can be completed within 15 minutes.

d. Starting Position. To start the program, the Marine begins with the Table 1 and executes each exercise for the required number of repetitions as indicated. The Marine controls the starting level and progression. When the Marine executes the maximum repetitions for Table I within a 15-minute exercise period, he progresses to Table II. To maintain this level of development, the Marine should also execute the maximum repetitions for Table II within a 15-minute exercise period. The Marine should keep substitution of exercises to a minimum. However, it he completes a full 15 minutes of strenuous exercise and exercises all muscles, then there should be no appreciable difference in the overall development.

e. Bench Conditioning, Table I

(1) Exercise 1: Side-Straddle Hop. This is a two-count warmup exercise done at moderate cadence. The starting position is the position of attention. On count ONE jump slightly into the air, swinging the arms out to the sides and up to a vertical position, hands touching.(See fig. 3-40, A.)At the same time, spread the feet wider than shoulder-width apart. On count TWO, using a slight flexing of the knees and ankles, jump slightly into the air and return to the starting position by swinging the arms back down to the sides. Twenty repetitions of this exercise is the standard dosage throughout the program.

(2) Exercise 2: Hand Walk. Remove the lower horizontal bar. Adjust the upper horizontal bar so that it is high enough to permit a "dead hanging" position, with the feet off the ground.(See fig. 3-40, B.j From the "dead hanging" position, release one hand and drop the arm to the side of the body. Then raise that arm and regrasp the horizontal bar. Release the bar with the other hand and drop that arm to the side. Repeat this as many times as possible.

(3) Exercise 3: Situps. Lie down with the fingers interlocked and placed behind the head. Hook the toes under the foot braces. Raise the trunk and upper body to an upright sitting position, twisting it to the left and then forward and downward until the right elbow touches the left knee.(See fig. 3-40, C.)Lower the body to the starting position .Sit up again but twist the body to the opposite direction as before, touching the left elbow to the right knee. Again lower the body to the starting position. The starting dosage is 20 situps. The Marine should continue the progression until he has attained 40 situps.

Figure 3-40 is available in the illustration section at the end of the book.

(4) Exercise 4: Double Step-upStarting at one end of the bench, step up onto the bench, and walk across it. Step down from the other end; turn around and repeat the process to return to the starting point.(See fig. 3-40, D.)Each return to the starting point constitutes a repetition. The starting dosage is 20 repetitions. Maximum dosage is 35 repetitions. This exercise should be done at a rapid cadence.

(5) Exercise 5: Isometric Bar Lift. Adjust the lower bar so that it is slightly higher than the beltline. Placing the feet on the footplates at the base of the frame, grasp the lower bar so that the hands are spread shoulder-width apart. Assume a crouched position and lift with maximum effort using the arms, back and legs.(See fig. 3-40, E.)Starting dosage is 4 repetitions of a stress time of 5 seconds followed by a 5-second rest prior to the next repetition. The Marine obtains progression by lengthening stress periods to 6 and later 7 seconds. Do not increase the number of repetitions.

(6) Exercise 6: Knee Lift. Adjust the upper bar to the same height used in Exercise 2.Adjust the lower bar so that is stops rearward movement of the hips when the "dead hanging" position is assumed.(See fig. 3-40, F.) Keeping the arms extended, flex the legs and raise the knees as high as possible. Hold this position for 5 seconds, then return to the starting position. After 2 seconds in the starting position, raise the knees again. Each return to the starting position constitutes one repetition. The dosage is five repetitions. Progression is the same as in Exercise 5.

(7) Exercise 7: Isometric Pull. Adjust the lower horizontal bar to a position where it is slightly higher than the beltline. Grasp the handles and pull outward.(See fig. 3-40, G.)Apply maximum effort and hold for approximately 5 seconds. Relax for 5 seconds between repetitions; perform four repetitions. Moving the body closer to or farther away from the bar will change the stress from the upper arms to the forearms. Progression is the same as in Exercise 5.

(8) Exercise 8: Isometric Compression. Maintain the position as in exercise 7.(See fig. 3-40, H.)Grasping the handles in the same manner, press in **with** maximum effort and hold for approximately 5 seconds. Relax for 5 seconds between repetitions. Progression is the same as in Exercise 5.

(9) Exercise 9: Isometric Press. Remove the lower horizontal bar. Adjust the upper horizontal bar until it is about 6 inches lower than the extended arms can reach. Stepping on the footplates at the bottom of the frame, grasp the bar with both hands and push up.(See fig. 3-40,1.)Keep both the legs and arms slightly flexed and the back straight. Apply maximum effort for 5 seconds then relax for 5 seconds. Complete four repetitions. Progression is the same as in Exercise 5.

(10) Exercise 10: Pushups. Grasping the foot braces with both hands, assume the front leaning rest position.(See fig. 3-40, J.)Keeping the back and legs straight, lower the body until the chest is lower than the **hands,** then return to the starting position. The Marine should complete the maximum possible number of repetitions.

f. Bench Conditioning, Table II. There is no limit on the maximum number of repetitions attainable in Exercises 3, 4, and 6 of Table II. The only limit imposed is that the entire program of 10 exercises outlinted in either table should not exceed 15 minutes.

(1) Exercise 1: Side-Straddle Hop. This is a two-count warm up exercise done at a moderate cadence. The starting position is the position of attention. On count ONE, jump slightly into the air, swinging the arms out to the sides and up to a vertical position, hands touching.(See fig. 3-41, A.)At the same time, spread the feet wider than shoulder-width apart. On count TWO, using a slight flexing of the knees and ankles, jump slightly into the air and return to the starting position by swinging the arms back down to the sides. Twenty repetitions of this exercise is the standard dosage throughout the program.

(2) **Exercise 2: Pull-up.** Adjust the horizontal bar so that is is high enough to permit a "dead hanging" position with the feet off the ground. Grasp the bar with both hands, palms facing forward. By flexing the arms, raise the body to a position where the chin is higher than the bar.(See fig. 3-41, B.)Then lower the body to the "dead hanging" position. Repeat as many times as possible.

(3) **Exercise 3: Bench situps.** Sit on the bench and hook the feet under the foot braces. With the fingers interlocked behind the head, lean back until the head touches the floor.(See fig. 3-41, C.)Return to the starting position. The starting dosage is 15 situps.

(4) **Exercise 4: Step-Up.** Face the bench and step up on it with one foot, bringing the trailing foot up next to the leading foot. Step back down again, leading with the same foot used first in stepping up.(See fig. 3-41, D.)Perlorm half of the total repetitions, then change the sequence of moving the feet to use the other leg in stepping up, and repeat the same amount of exercise. The starting level is a total of 40 step-up. **This exercise should be done at a rapid cadence.**

(5) **Exercise 5: Isometric Bar Lift**\djust the lower bar so that it is slightly higher than the beltline. Placing the feet on the footplates at the base of the frame, grasp the lower bar so that the hands are spread shoulder-width apart. Assume a crouched position and lift with maximum effort using the arms, back, and legs.(See fig. 3-41, E.)Starting dosage is four repetitions of a stress time of 8 seconds followed by a 5-second rest prior to the next repetition. Progression is obtained by lengthening stress periods to 10 seconds. Do not increase the number of repetitions.

Figure 3-41 is available in the illustration section at the end of the book.

(6)**Exercise 6: Leg Lift.** Adjust the bars and assume the starting position as shown in figure 3-41, F. Keeping arms and legs extended, raise the legs to a horizontal position and hold in that position for 2 seconds. Then lower the legs slowly to the starting position. Five repetitions is the starting level.

(7) **Exercise 7: Isometric Pull.** Adjust the lower horizontal bar so that it is slightly higher than the beltline. Grasp the handles and pull outward.(See fig. 3-41, G.)Apply maximum effort and hold for approximately 8 seconds. Relax for 5 seconds between repetitions; perform four repetitions. Moving the body closer to or farther away from the bar will change the stress from the upper arms to the forearms. The Marine obtains progression by lengthening the stress period to 10 seconds.

(8) **Exercise 8: Isometric Compression.** (See fig. 3-41, H.)Maintain the position as in exercise 7.Grasping the handles in the same manner, press in with maximum effort and hold for approximately 8 seconds. Relax for 5 seconds between repetitions; perform four repetitions. The Marine obtains progression by lengthening the stress period to 10 seconds.

(9) **Exercise 9: Isometric Pres.** Remove the lower horizontal bar. Adjust the upper horizontal bar until it is about 6 inches lower than the extended arms can reach. Stepping on the footplates at the bottom of the frame, grasp the bar with both hands and push up.(See fig. 3-41,1.)Keep both the legs and arms slightly flexed and the back straight. Apply maximum effort for 8 seconds, then relax for 5 seconds. Complete four repetitions. The Marine obtains progression by lengthening the stress period to 10 seconds.

(10) **Exercise 10: Inclined Pushup.** Assume the front leaning rest position with the feet on the bench.(See fig. 3-41, J.)Keeping the back and legs straight, lower the body until the nose touches

the ground. By extending the arms, raise the body to the starting position. Repeat as many times as possible.

3903. THE 6-12 PLAN

a. Description and objective. The 6-12 plan of physical fitness has been developed to assist in regulating quantity and progression and to provide a convenient set of exercises. This is a basic program and will take 18 weeks to complete if you follow the moderate progression as to the time prescribed for each level of achievement. The time can be shortened as explained below. This plan consists of six basic exercises a day which can be completed in 12 minutes. There are six tables of six exercises each, thus allowing you to progress from table to table. The plan is progressive, fits any age group, contains balance and variety, and applies the principle of overload in a safe and gradual manner. Begin at Table I, Progression Guide, with the number of repetitions as indicated by age.

b. Formation. If just starting an exercise program, do not rush through the first table. Remember, individuals should remain at each level for about a week before moving upward. The time allotment stated for each exercise at the bottom of the tables is a guide; some people may take more and some less time on the individual exercises. At the end of a one-week period, when the individual can comfortable perform the six exercises in 12 minutes, move on to the next level. To a certain degree, the individual must be the judge of his ability to progress form level to level and table to table. If starting with a certain degree of fitness, some of the beginning tables may present little challenge.

c. Starting Level **and Progression.** There are three levels of achievement for each age group, indicated as A, B, and C. Start at the C level for the appropriate age group. At the end of a one-week period, or when the individual can do all exercises at that level within 12 minutes, progress to the B level. At the end of the second week, or when the individual can accomplish that level within 12 minutes, progress to the A level. At the conclusion of the third week or when the individual can achieve the A level within the time limitation, move on to table II.(See Tables I through VI, Progression Guide.)

d. Maintenance Level. Attempt to work through all six tables. If this proves to be too difficult, then maintain exercise at the—

(1) A-level on Table IV, Progression Guide it in the 45 to 49, 50 to 59, or over 60 age group.

(2) A-level on Table V, Progression Guide if in the 17 to 29, 30 to 39, or 40 to 44 age group.

e. Precautions. To achieve the maximum benefit, perform each exercise exactly as specified. Read the descriptions and study the illustrations. Do not slight the movements. Use a sensible approach and follow these four points as they apply before starting or during your exercise program.

(1) If you have the slightest doubt about your ability to participate in this exercise program, consult a physician.

(2) Stop immediately if you notice unusual breathlessness or chest pain while taking part in these exercises. If these conditions persist, consult a physician.

The following Tables of Progression Guides and Exercises are the progression of the 6-12 Plan program.

Table I. Progression Guide is available in the illustration section at the end of the book.

Table I. Exercises.is available in the illustration section at the end of the book.

Table II. Progression Guide.is available in the illustration section at the end of the book.

Table II. Exercises.is available in the illustration section at the end of the book.

Table III. Progression Guide.is available in the illustration section at the end of the book.

Table III. Exercises.is available in the illustration section at the end of the book.

Table IV. Progression Guide.is available in the illustration section at the end of the book.

Table IV. Exercises.is available in the illustration section at the end of the book.

Table V. Progression Guide.is available in the illustration section at the end of the book.

Table V. Exercises.is available in the illustration section at the end of the book.

Table VI. Progression Guide.is available in the illustration section at the end of the book.

Table VI. Exercises.is available in the illustration section at the end of the book.

(3) Unless you have exercised regularly and know yourself to be in good physical condition, start at table I with the C level appropriate to your age.

(4) If you are out of shape, admit that fact to yourself. Hide your pride; after all, you are in the privacy of your own quarters. Set your goal for the longer, steadier pull toward fitness. Resist the urge to pass over the lower numbered tables to find a table that will test your fitness. You are not trying to test, but rather to develop.

3904.WEIGHT TRAINING

Weight or barbell training should not be confused with the more common types of weight lifting used as a competitive sport. Weight lifting is designed to develop specific muscle groups so that the individual is capable of lifting a large amount during a single lift. In contrast, weight training is the systematic development of all the major muscle groups by the use of calisthenics reinforced with weight to provide resistance.

3905.WEIGHT LIFTING

a. Description and Objective. The weight lifting program is progressive and applies the principle of overload in a safe, gradual manner. The exercises of the table can be completed in 15 minutes. The objective of the exercises is to develop strength and muscular endurance, and muscle tone of the five major muscle groups:legs, arms, back, trunk, and shoulder girdle.(See fig. 3-42.)

b. Warmup. A warm up exercise is important to prepare the body for the more vigorous exercises that are to follow. Ten repetitions of the high jumper exercise are excellent for a warm up period. Also, an exercise which will require fast body movement is needed.To provide such exercise, 3 to 5 minutes of rope skipping is recommended to increase the individual's development potential.

c. Formation. Marines should take care in the completion of these weight lifting exercises. They should ensure that their back is straight during the lifting phase of all exercises. When exercises require assuming the standing position with the weight, they should always grasp the weight while in a squatting position and then rise to a standing position.

d. Progression. Each exercise has a starting number of repetitions and specified pounds of weight. After each 4th or 5th day of exercise, the Marine should increase the repetitions by one until he has reached the maximum of 10 repetitions. At this time, the Marine should increase the weight by 5 pounds. Then he should repeat the process again with the initial number or repetitions.

e. Weight Training

(1) Exercise 1:Squat(See fig. 3-42, A.)Starting level—six repetitions, 50 pounds (commonly called the flatfoot deep knee bend). Place the bar upon the shoulders. Stand with feet about 18 inches apart. Keeping the feet flat, lower the body into the low squat position. Stand erect and repeat. Exhale while lowering into the squat position and inhale while standing. This constitutes one repetition.

(2) Exercise 2:Waist Bender(.See fig. 3-42, B.)Starting level-6 repetitions, 40 pounds. Assume the standing position with the bar across the shoulders and the feet a shoulder-width apart .Bend forward at the waist until the upper body is parallel to the ground; return to the starting position. Each return to the upright position constitutes one repetition.

(3) Exercise 3:Curl (See fig. 3-42, C.)Starting level—6 repetitions, 40 pounds. Grasp the barbell with the palms facing to the rear and assume the standing position with the feet a should-width apart. With the barbell held in front of the hips, flex the elbows and lift the weight until the bar touches the upper chest. Lower the barbell back to the hip level position. Inhale deeply with the upward movement and exhale on the downward movement. Each time that the bar touches the chest will constitute one repetition.

(4) Exercise 4:Side Bender(.See fig. 3-42, D.)Starting level—6 repetitions per side, 40 pounds. Assume the standing position, with the bar across the shoulders, with feet a shoulder-width apart. Bend to the left as far as possible and return to the starting position. Repeat six times and then execute the same procedure to the right for six repetitions.

(5) Exercise 5:Standing Press(See fig. 3-42, E.)Starting level—6 repetitions, 45 pounds. Grasp the bar with the palms facing forward and assume the starting position. Curl the weight to the upper chest position. Inhale deeply and press the bar upward to an overhead position. Exhale while lowering the bar to the chest position. Each time that the bar is pressed upward constitutes one repetition.

(6) Exercise 6:Upward Row. .see fig. 3-42, F.)Starting level—6 repetitions, 40 pounds. Grasp the bar, hands close together, palms to the rear, and assume the standing position. Starting with the bar held in front of the hips, flexing the elbows and the shoulder girdle muscles, lift the bar straight up to an overhead position. Inhale deeply while lifting the bar. Exhale while lowering the bar to the hip position. Each time that the bar returns to the hips constitutes one repetition.

(7) Exercise 7:Shoulder Curl. See fig. 3-42, G.)Starting level—6 repetitions, 25 pounds. Grasp the bar, palms down, and assume the standing position. Keeping the elbows locked, curl the bar, pivoting the arms at the shoulders until the bar is in an overhead position and as far to the rear as possible. Return the bar in the same manner to the hip position. Each time that the bar returns to the hip position constitutes one repetition.

Figure 3-42 is available in the illustration section at the end of the book.

3906.ISOMETRIC EXERCISING

a. Description and Objective. Isometric exercising is the application of maximum effort during an exercise period. The isometric principle is to apply force gradually over a 5- to 10- second period until the maximum effort is applied. Relaxation follows for approximately 5 seconds and then force is applied again. The Marine continues this process at the prescribed level for each exercise for a period of 15 minutes or less. The objective of isometric exercises is to create muscle growth. Isometric exercises are the fastest means of creating muscle growth. They are founded on the fact that a muscle will grow only so fast regardless of the type or duration of the activity. The principle of overload—that the muscles develop commensurate with demand—reinforces that fact. However, isometrics will not develop cardiovascular or muscular endurance. Consequently, Marines who choose to use isometrics in their exercise routine must also include running or some other type of aerobic activity.

b. **Area and Equipment** isometric exercises may be designed to be performed with or without equipment.

c. Use With Other Programs. The isometric exercises presented in this section will provide a variety from which to choose. In addition to the exercises contained in this section, the application of isometric force is used in some of the exercises in section II.

d.Door Frame Exercises. The following exercises are designed for use with a standard door frame found in all offices or barracks.(See fig. 3-43.)

(1) Exercise l:Arm Press. Stand in the doorway with the legs straight, knees locked. Using the arm muscles, press hard upward against the top of the door frame. Repeat for three repetitions applying gradual effort to maximum contraction.

(2) Exercise 2:Leg Press. Stand in the doorway with the hands on the top of the door frame, elbows locked .With knees bent, press hard with the leg muscles .Repeat for three repetitions beginning with a gradual effort and increasing to maximum contraction. A low platform may be necessary to reach the top of the door frame and still maintain a bent knee position.

(3) Exercise 3:Side Press. Extend both arms to the side of the doorway. Palms are shoulder high, facing outward. With both arms, press hard against the sides of the door frame. Repeat for three repetitions. Begin gradually and increase to maximum contraction.

(4) Exercise 4:Lateral Raise. Extend both arms to the sides of the doorway, arms down, palms facing inward. With the back of the hands, press hard against the sides of the door frame .Repeat for three repetitions. Begin with a gradual effort and increase to maximum contraction.

Figure 3-43 is available in the illustration section at the end of the book.

(5) Exercise 5:Neck Press?lace the forehead against the door frame, hands clasped behind the back Using the neck muscles, press hard against the door frame. Repeat for three repetitions, then reverse position so that the back of the head is resting on the door frame. Again do three repetitions .Begin gradually with both exercises and increase to maximum contraction.

(6) Exercise 6:Door Pull Stand facing the edge of the open door and grasp the doorknobs. Pull outward with both arms (if doorknobs are not available grasp the edge of the door).Applying outward pressure, move the body toward and away from the door .Repeat the three repetitions. Begin with gradual effort and increase to maximum contraction.

Chapter 4
COMBAT WATER SURVIVAL
Section I. Marine Corps Water Survival Program

4101.MILITARY SWIMMING

Military swimming emphasizes strokes that result in staying power rather than speed. Marines need to swim easily, quietly, and with adequate vision. Strokes should be used that allow the Marine to carry basic combat equipment and to tow or push a wounded buddy while keeping the face out of the water to allow breathing. The best strokes to use are the side stroke and the breast stroke. Marines must be proficient in floating, drown-proofing, and in using their basic combat equipment to make flotation devices. They must be drilled on using the Kapok and Mae West buoyancy compensators and on emergency egress from helicopters, landing craft, and amphibious assault vehicles.

4102.PROGRAM DEVELOPMENT

Commanders of the Landing Force Training Commands (Atlantic/Pacific) are tasked with developing and maintaining the individual water survival and swimming training program. They are responsible for training all Marine Corps water safety and survival instructors and integrating this training in to their combat readiness training program. All Marines are deployable and as soldiers of the sea must constantly practice the skills necessary to achieve mission accomplishment. Commanders must be imaginative in program development, integrating basic skills development which enhances the total mission accomplishment. An example is teaching waterproofing of the Marine's backpack so that it can be used as a flotation device in streams or rivers. Once learned, this skill would be practiced first in a pool and then in a stream or a river. Routine inspections during deployments would ensure that backpacks could be used as flotation devices.

4103.QUAL STDS AND TRAINING GUIDELINES

The inherent nature of Marine Corps operations and training requires that Marines achieve an ability to survive in water. Water survival and swimming training is designed to reduce fear of water, instill self-confidence, and develop a Marine's ability to survive in water. As a minimum, every Marine should be qualified as swimmer, third-class (S3).This qualification should be met during recruit training or Officer Candidate School. Once qualified, a Marine need not be requalified unless requalification at a specific level (S3 or higher) is required by unit mission, military occupational specialty or duty assignment, or for other reasons determined by the commander. The following classifications and standards apply to Marine Corps water survival and swimming training. Abbreviations will be used for service, medical, and training record entries citing MCO 1510.29A, Individual Water Survival and Swimming Training, as authority.

a. Qualification standards and test procedures are described in paragraph 4104.

b. Swimmer classification and abbreviations are:

(1) Unqualified - UQ.

(2) Swimmer, third-class (S3).Minimum aquatic skill level for all Marines.

(3) Swimmer, second-class (S2)-

(a) Minimum standard for all naval aviators, naval flight officers, and aircrewmen with written waiver from the commanding officer.

(b) Requirements for military occupational specialty 1803 (assault amphibian vehicle officer), 1833 (assault amphibian vehicle crewman), 1302 (engineer officer), 1371 (combat engineer), 1381 (shore party specialist), and 1379 (engineer operations chief).

(4) Swimmer, first-class (S1)-

(a) Naval aviator, naval flight officer, and aircrewman qualification requirement.

(b) Requirements for military occupational specialty 0321 (reconnaissance man), 8652 (reconnaissance man, parachute jump qualified), 8653 (reconnaissance man, self-contained underwater breathing apparatus qualified), 8654 (reconnaissance man, parachute/self-contained underwater breathing apparatus qualified), 9952 (self-contained underwater breathing apparatus Marine (officer/enlisted)), and 9953 (parachutist/self-contained underwater breathing apparatus Marine (officer/enlisted)).

(5) Water Survival Qualified (WSQ).Ultimate water survival/swimming goal for Marines.

c.Instructor and assistant instructor classifications are-

(1) **American Red Cross Advanced Lifesaver.** Individuals who are currently certified as American Red Cross advanced lifesavers may assist qualified instructors during Marine Corps water survival/swimming training and testing. Certification is valid for 3 years.

(2) American Red Cross Water Safety Instruction. Individuals who are currently certified as American Red Cross water safety instructors may conduct swimming training and qualify individuals as 3d, 2d, and lst class swimmers. Certification is valid for 2 years.

(3) United States Marine Corps Water Safety and Survival Instruction. Those personnel certified as water safety and survival instructors by one of the landing force training commands may conduct Marine Corps water survival and swimming training and American Red Cross swimming and water safety training. These instructors may also participate as assistant instructors during the conduct of landing force training command-sponsored water safety and survival instructor courses. Certification is valid for2 years.

(4) American Red Cross Water Safety Instructor Traine. Individuals qualified who have been certified as instructor trainers by the American Red Cross may train American Red Cross water safety and basic swimming instructors. Certification is valid for 2 years.

(5) United States Marine Corps Water Safety and Survival Instructor Traine. Individuals qualified who are certified as instructor trainers through special training courses conducted jointly by the American Red Cross and the Marine Corps, under the direction of the Commandant of the Marine Corps, may train water safety and survival instructors and American Red Cross water safety and basic swimming instructors. Certification is valid for 2 years.

d. Training Guidelines Instructors engaged in training Marines as swimmers will use the current edition of the American Red Cross swimming and water safety courses instructors' manuals as the authoritative textbooks for swimming procedures, strokes, breaks, and holds.

4104.WATER SURV/SWIM QUAL STD AND TEST PROCS

a. The following qualification standards and test procedures are applicable to Marine Corps water survival and swimming training. The utility uniform without cover, boots/oxfords or socks will be worn for all swimming tests through swimmer, first class. The utility uniform with boots/oxfords will be worn as prescribed for the water survival qualification.

(1) Swimmer, Third-Class (S3) To qualify as a swimmer, third-class, a Marine must enter the water feet first from a minimum height of 5 feet and remain afloat for 5 minutes. During this time the Marine must swim 50 yards using any stroke or combination of strokes. This test should be taken by all Marines as early as possible in their initial training. Those unable to pass this test will be classified as UQ and should be given instruction in fundamental swimming skills. Those who barely meet requirements are swimmers who need help and should receive additional instruction in fundamental swimming skills, whenever possible.

(2) Swimmer, Second-Class (S2.) To qualify, a Marine must enter the water feet first from a minimum height of 10 feet and remain afloat for 10 minutes. The Marine must swim 100 yards and use the three basic survival swimming strokes (side stroke, backstroke, and breast stroke) for a minimum distance of 25 yards each.

(3) Swimmer, First-Class (S1)A prerequisite to qualification as swimmer, first-class, is successful completion of the test for swimmer, second-class. To qualify as a swimmer, first-class, a Marine must be able to do each of the following:

(a) Approach a person of approximately the same size while in the water, demonstrate a release, get that person in a carry position, and tow the "distressed swimmer" 25 yards.

(b) Enter water feet first and immediately swim underwater for 25 yards. Swimmer is to break the surface twice for breathing during this distance at intervals of approximately 25 feet.

(c) Remove trousers in water, inflate for support, and remain motionless for a minimum of 1 minute.

(d) Swim 220 yards using any survival swimming stroke or any combination of survival strokes.

(4) Water Survival Qualification (WSQ)A prerequisite to water survival qualification is successful completion of the test for swimmer, first-class. To be water survival-qualified, a Mannemust:

(a) Enter water by jumping from a height of 10 feet.

(b) Stay afloat in full utility uniform (less cover) for 1 hour. Boots/oxfords will be removed after 5 minutes in the water but will be retained by the swimmer.

(c) Traverse 75 yards in deep water in full utility uniform (less cover) with boots/oxfords.

b. All water survival and swimming training will include instruction in the following:

(1) Employment of safety lines or other comparable expedients to assist in swift stream/river crossings.

(2) Adverse physiological effects caused by cold water (hypothermia) and the precautionary measures to be taken prior to exposure to such an environment.

(3) Employment of a standard and expedient flotation devices in a water survival situation.

(4) Removal of individual field equipment after unscheduled water entry. Entry will be made from a minimum height of 10 feet with a complete set of field equipment, properly packed as an expedient flotation device and waterproofed. The ALICE pack will then be removed, the rifle laid across the top of the pack, the helmet remaining or put back on the individual's head, and the individual will traverse 25 yards to the side of the pool or water's edge.

c. The additional water survival qualification (flight physiology training) and requalification required of personnel assigned to flight status will be conducted in accordance with OPNAVINST 3710.7_,Promulgation of NATOPS General Flight and Operating Instructions. Swimming qualification (first-class, SI) is a prerequisite to water survival (flight physiology) training and is a one-time requirement.

Section II. Water Sports and Competitive Activities

4201.GENERAL

The commander, when developing the combat readiness program, needs to look carefully at the mission. If the mission requires Marines who must perform mission-essential water-related activities of long duration, he may want to look to water sports/competitive activities as a method of developing endurance. Endurance and confidence can only be developed by extended time in the water. While defined training can accomplish this goal, the goal can be enhanced through training which is integrated with the physical conditioning program. This approach will provide variety and enthusiasm. Two activities which will accomplish this goal are water relays and water polo. The important consideration is that the unit achieve total participation by all members of the unit.

4202.WATER RELAYS

Development of water relays will generally conform to the guidelines in chapter 5, section IV. When contemplating relays, the commander should review chapter 5 and adapt the principles to the size of the unit and the facility to be used.

4203.WATER POLO

The official rules provided in this paragraph will need to be adapted to the unit's unique situation, its size, and the size of the facility. Commanders should make every effort to have all members of the unit participate.

a. NCAA Rules This style of play is based on two popular sports: swimming and basketball. Playing is done on the surface of the water by teams of seven players each. The size of the playing area is 25 yards by 14 yards, which is the size of the typical high school or college indoor pool. The deeper the water, the better. If the entire playing area is deep-6 feet or more—this is ideal, although most indoor pools have a shallow end. At each end of the pool is a goal. In deep water, the goal is 10 feet across and 3 feet high. In shallow water, where the goalie can stand on and jump from the bottom, the goal is 10 feet across and 5 feet high. One of the seven players on each team is the goalie. He is the only player on the team who can stand on or jump from the bottom (if it is shallow enough to permit this), catch and pass the ball with both hands at the same time, or hit the ball with a clenched fist. However, the goalie may not swim across or pass the ball across the middle of the pool.

b.Equipment. The goalie on the visiting team must wear a white cap with the number 1 clearly marked on it; his teammates, who are designated as guards and forwards, must wear white caps numbered 2 through 7, with the substitutes wearing white caps numbered 8 through 21.The goalie on the home team must wear a dark-colored cap with the number 22 clearly marked on it; his teammates designated as guards and forwards, must wear similar dark-colored caps numbered 29 through 42.The guards and forwards on each team may swim freely up and down the pool as they see fit, interchanging positions as often as they wish. They may not stand on or jump from the bottom, touch the ball with more than one hand at a time when catching or passing or shooting it, or enter inside the opposing team's 2-yard line unless preceded by the ball. The ball is similar to a soccer ball, except that it is yellow and covered with a waterproofing substance that makes it easier to handle with one hand.

c. **Playing Time.** In intercollegiate competition the game consists of four 7-minule quarters, the teams changing ends after every period of play. In high school competition the game consists of four-,five-, or six-minute quarters, depending on the ages of the participants. Here again the teams change ends after every period. There is a very brief interval between each quarter. At the start of each period the teams line up at their respective ends, and with ablast of the whistle, the head referee throws the ball into the middle of the pool. The fastest swimmers on each team then race to gain possession.

d. Officials. The head referee walks along one side of the pool watching for infractions. He is aided by the assistant referee, who patrols the opposite side of the pool. The referees are both equipped with whistles and two official's flags. These flags should be 12 inches square and should correspond in color to the caps worn by the respective teams. Whenever a referee sees an infraction, he blows his whistle and signals with his flag indicating that the player nearest to the ball with the appropriate colored flag will be given possession of the ball. Whenever the whistle is blown, all players should react by looking immediately to see which color flag is being waived which will indicate which team should be moving into the defense. Other officials are a timer and a scorer. As in all sports, it is important that the officials be skilled and competent.

e. **Technical Fouls.** The following are some of the common technical fouls: starting before the referee blows his whistle to open play; holding onto or pushing off from the side of the pool during play; taking hold of the ball underwater when tackled by an opposing player; swimming inside an opponent's 2-yard line unless preceded by the ball; touching the ball with both hands at the same time (goalie excepted); standing, walking on or jumping from the bottom when taking an active part in the game. When a technical foul occurs, thereferee blows the whistle and with his flag awards possession. The player on the team awarded possession nearest the point of infraction then has 5 seconds to put the ball into play by passing to a teammate or by dropping the ball into the water and swimming after it.

f. Personal Fouls. The following are some of the more common personal fouls: committing any of the technical fouls mentioned above for the purpose of scoring or preventing a goal; holding, ducking, pulling, pushing off from, swimming over, or impeding the arm or leg movement of any opponent who is not touching the ball; and splashing water in the face of an opponent. When a personal foul occurs, the referee will follow the same procedure as with a technical foul. While doing this, the referee will also call out clearly and loudly the number of the player who was guilty of the infraction and a personal foul will be marked against that player by the official scorer. When a player accumulates five personal fouls, he fouls out of the game and must be replaced by a substitute.

g. Penalty Shot. A penalty shot can be awarded by either of the officials or the scorer when:

o An offensive player inside the opponents' 4-yard line not touching the ball is held, ducked, pulled-back, kicked, or struck.
o A team has accumulated a total of ten personal fouls. When the former occurs, the head or assistant referee should immediately blow the whistle and by holding a two-flagged stick in a vertical position above his head, signal that a penalty shot has been awarded. When the latter occurs, the scorer should use a buzzer located at the scorer's tabic to signal that one team has accumulated ten personal fouls, thereby entitling the other team to a penalty shot. A penalty shot is taken from the 4-yard line in front of the goal. All players except the defending goalie must leave the 4-yard line until the shot is taken. No player can be within 1 yard of the shooter. After ascertaining that the shooting player is on the 4-yard line and the goalie is onThegoal line, the referee will ask the shooter to lift up the ball. When he docs, the referee will give a sharp blast from the whistle. At the whistle, the shooter must shoot without delay and without any faking at the goal. The goalie may try to block the shot. If he does, or the shot is otherwise missed, it is immediately in play and action continues.

h. When a Goal is Scored. When a goal is scored, either from a shot taken by a player out in the field or by penalty shot, the defending goalie must then pass the ball to the nearest referee. The referee will then pass the ball back to the goalie, who puts it into play by a pass to one of his teammates as soon as possible.

Section III. The Battle Swimming Test

4301.GENERAL

The battle fitness test is the evaluation contained in the Guidance for Basic Warrior Training (BWT) as part of the Basic Warrior Training Concept Plan.

4302.THE BATTLE SWIMMING TEST

The battle swim test consist of the following:

a. Be able to climb a three meter high board, while wearing a field uniform, carrying a M-16, web gear, and four canteens. Jump off blindfolded, recover and swim 100 meters with the M-16.

b. Repeat the same process, without the blindfold, and upon entering the water a second time, recover and retrieve all submerged gear.

c. Remain afloat and tread water for five minutes.

d. Demonstrate a knowledge of how to cross a stream using field expedient techniques and a single strand rope bridge.

e. Successfully complete helicopter egress training.

Chapter 5
COMPETITIVE CONDITIONING ACTIVITIES
Section I. Organization of Competitive Activities

5101.LEADERSHIP OF COMPETITIVE ACTIVITIES

a. Place in the Program. Competitive conditioning activities consist of dual combatives, relays, team contests, and team sports in which individuals or teams compete against an opponent to win. Competition is one of the best ways of maintaining interest in the physical training program during the sustaining stage. Organized competition provides enjoyable, vigorous physical activity that has proved to be one of the best supplements to conditioning drill activities. The benefits of competition are the development of aggressiveness, unit pride and identity, teamwork, and the will to win. These activities help develop cardiovascular and muscular endurance, strength, and coordination. Scheduling competitive activities in an orderly and progressive manner is desirable. The progression should be from relays to dual combatives, to team contests, and finally to team athletics. These activities should be conducted as part of the program after a basic level of conditioning has been developed, usually during the slow-improvement stage of conditioning. Muscles and joints should be strengthened by preconditioning to withstand the strain placed upon them by sudden stops and turns, body contact, bearing of weight, and falls. Competitive activities, however, should not be allowed to dominate the physical training program.

b. Area and Equipment. Some of the competitive activities included in this chapter require specific types of areas and equipment; others do not. The area requirement can usually be satisfied on available training fields. When items of equipment are required or specific courts or field layouts are to be marked off, such information is included.

c. Leadership. The principal factor for success as a leader of competitive activity is an energetic, dynamic, enthusiastic approach. The leader's attitude is reflected by the group, so the commander must carry on the activity in a snappy and vigorous manner. Confidence on the part of the commander will create an impression of decisiveness and certainty. Confidence grows out of experience and a thorough knowledge of the activity. Mastery of subject is the first step in developing confidence, assurance, and poise.

(1) The following suggestions are offered for leaders of competitive combatives, contests, and sports:

(a) Get the activity underway quickly by selecting and teaching only the essentials.

(b) Use rules to add to the enjoyment of the activity and do not allow them to interfere with the spirit of competition.

(c) Stop the activity before interest begins to lag.

(d) In team contests, clearly distinguish sides by attempting to maintain unit identity.

(e) Always insist on fair play, enforcing the rules impartially.

(2) The following procedure is recommended for presenting a competitive activity.

(a) Name the activity.

(b) Briefly explain the objective of the activity and give only the pertinent rules.
(c) Have a demonstration at slow speed and answer questions.
(d) Organize groups into teams and appoint captains.
(e) Arrange teams in the proper starting positions.
(f) Conduct the activity.

d. Competitive Units. Units for competition should be the same organizations in which Marines train: battalions, companies, platoons. In most situations, the unit is the squad.

e. Provisions for Instruction. One of the most effective methods of maintaining interest and participation in competitive activities is to provide instruction in activities with which most Marines are unacquainted. Such instruction can be conducted during regular physical training periods. Careful planning is required to keep all individuals continuously engaged in vigorous activity. The materials in the following sections should be used as a guide for instructional purposes.

f. Officiating and Control. Every effort should be made to provide good officials for all competitive activities. Poor officials quickly cause dissatisfaction among participants in team activities and create a situation which can turn an organized activity into a brawl. Each company should have or develop several competent and qualified officials available for games on company and platoon levels.

5102.MILITARY FIELD MEETS

When units reach the latter part of the slow-improvement and sustaining stage of training, interest in the program may lag. A change in course content can arouse the desire to participate. An event which does not require a high degree of skill yet demands strenuous activity is ideal for the military field meet. This is a series of team contests conducted on a station-to-station basis during a given period of time. Team contests carried on simultaneously provide essential training, vigorous exercise, and stimulating competition in an atmosphere that is enjoyable for all. Because of the healthy rivalry that a field meet arouses, it is an excellent form of inter-unit competition.

a. Objective and Advantages. The objective of the military field meet is to provide activity for everyone in the participating units. Activities are chosen that will develop aggressiveness, teamwork, a will to win, and competitive spirit, and that will stimulate interest and build esprit de corps. The military field meet can **be** included as part of the physical training program or as part of the off-duty recreational program. It is a form of contest that can be conducted in nearly all circumstances because it can be easily modified, requires a minimum amount of equipment, and can be readily organized.

b. Level of Competition. The military field meet is flexible. It can be adjusted to large or small groups. A company-size unit is the most desirable, but it may also be administered within a larger unit. If it is conducted within a company, the participating units will be squads. If it is held within a larger unit, platoons compose the teams. A larger-size unit requires more extensive organization and administration.

c. Selection of Events. In organizing a military field meet, select events that are simple and easy to administer. All rules and regulations should be clearly understood by everyone. No event should require previous practice. In selecting the events, consider the interest and capabilities of the Marines and available equipment and facilities. Select events which will require that all members of the units participate. Events should not be dominated by the athletes; rather, MOS-related events which have significance in producing skills and combat readiness should be used.

d.Equipment and Facilities. The site must be large enough to permit events to be grouped about a central control point. Use available facilities such as parade decks, Softball fields, or tracks. A public address system is desirable at the control point for the initial orientation of teams and for subsequent announcements of time lapses, cumulative scores, and final standings. A tally board is necessary so that all units can constantly monitor how the events are progressing.

e.Administrators. Efficient administration of the military field meet depends on the referees, judges, and scorers. For these positions, choose individuals within the units who have had athletic or officiating experience. Prior to the day of the event, all administrators should be briefed and assigned a specific task in order that they may become familiar with the rules and organization of the contests they will conduct. The following officials should be available:

(1) A primary instructor or supervisor who is in charge of the control center and who is responsible for the successful operation of the athletic carnival. The supervisor must have an assistant to act as a timer and scorer.

(2) One assistant instructor in charge of each event. This person should be-

(a) Familiar with the rules of the event.

(b) Effective as a leader to ensure proper supervision and control over the participating teams.

(c) Enthusiastic to provide proper motivation.

(d) Self-confident of the ability to judge infractions of the rules. The assistant instructor must be fair in judgment and penalize without hesitation when infractions occur.

(3) Runners between event stations and the control point. These individuals collect and deliver scores.

f. Team Organization for Competition. The size of the teams is determined by the level on which the field meet is organized. Maintaining the integrity of the unit promotes esprit de corps, but this does not preclude grouping two squads into one team.

g.Conduct of Events

(1) The assistant instructor at each station takes charge of the group and gives a brief explanation of the major rules of the event. The assistant instructor speaks clearly and distinctly from a position to be seen and heard by all. A short demonstration is desirable if it will help clarify the event.

(2) The assistant instructor should make certain that teams can be clearly distinguished by the use of, for example, T-shirts and fatigue jackets, or caps and no cap, or colored jerseys. The assistant instructor should teach a whistle response (teams stop play immediately upon hearing whistle), get the event started as quickly as possible, and make any necessary corrections as the contest progresses. The rale of good officiating is to use a minimum of calls, yet maintain control of the contest. Penalize when necessary, but refrain from disqualifying contestants or teams.

(3) Keep the activity moving as rapidly as possible. When the central control point sounds the whistle to stop the play, all competition ends immediately. The assistant instructor then assembles the group, forwards the team scores to the central point, and, upon a signal from the central control point, rotates the teams to their next station. It is essential that rotation and orientation of teams be quick and orderly to reduce time spent between contests.

(4) Upon completion of the final event, assistant instructors move their teams to the central control point for the announcement of winning teams, presentation of awards (if any), and final critique.

h. Scoring System. The system for determining the winner of the military field meet should be simple and efficient. At the completion of competition, the scorer totals the points that each team has scored in all contests. The scorer then subtracts the number of points scored against a team (penalty points for nonparticipation,etc.) from the total. The resulting scores are placed in a column with the highest score at the top and the lowest at the bottom. The team with the highest total is the winner. This type of scoring system encourages a team to ensure total unit participation.

Section II.Combatives

5201.DESCRIPTTON AND OBJECTIVE

Combatives are strenuous, short competitive contests in which two individuals attempt to overcome each other in a bout of skill and strength. These contests help to develop Marines' resourcefulness, confidence, strength, agility, coordination, and will to win. Any level ground area can be used. However, extremely hard ground should be avoided as some of the combatives require ground contact. A whistle is needed to control the bouts during the competition since voice commands may go unheeded. The extended rectangular formation is used for dual combatives. Combatives are conducted on an informal basis. Marines are allowed to remain at ease between activities and are allowed to brush themselves off after being on the ground. To get the most out of combatives, individuals must be urged to overcome their opponents as quickly as possible as would be required in combat.

a. Benefits for Marine Training. The possibility of close contact with an enemy in combat faces Marines at all times. They must be trained to react aggressively and violently. Combatives may be used as an introduction to such hand-to-hand contact and should be followed by hand-to-hand combat training. Marines enjoy competition, and this type of activity is a welcome change from formal conditioning activities.

b. Instructor Responsibility

(1) The instructor tells the individuals that all combatives begin and end on the whistle signal. The instructor demonstrates each activity before having it performed, explaining it in simple terms.

(2) After stopping one activity, the instructor gets everyone in place for the next.

(3) The instructor must closely supervise combatives to insure that contestants do not use unfair or unsportsmanlike tactics. To avoid unnecessary injury, instructors must see that the bouts are closely controlled and opponents equally paired. Adjustments should be made in apparent cases of mismatched abilities.

5202.COMBATIVES TABLES

There are three tables of combatives, each of which can be completed in 15 minutes. These tables become progressively more difficult from lower to higher numbers.

a.Combatives Table I

(1) Open-Hand Slap Boxing(See fig. 5-1, A. individuals assume a boxer's stance, palms open, fingers extended and joined. Each contestant tries to slap the opponent about the head and upper body with the open hand. This is a good warm- up activity.

(2) Wrist Tug-of-War(See fig. 5-1, B.)Two individuals sit on the ground with the soles of their feet in contact. Each grasps the opponent's wrists so that the hands are directly over the feet. At the whistle, each individual tries to pull the opponent from sitting to standing position.

(3) Arm-Lock Wrestling.(See fig. 5-1, C.)Two individuals sit back-to-back with legs spread and arms locked at the elbows. Each right arm is inside the opponent's left. Each tries to force the opponent's left arm or shoulder to the ground. The individual who first wins three bouts is the winner.

(4) Bulling(See fig. 5-1, D.)Two individuals assume the Westmoreland wrestling hold, each grasping the opponent's neck with the right hand and the opponent's right elbow with the left hand. Each tries to force the other to move one foot by pushing, pulling, or otherwise manipulating. The individual who first wins two boutsis the winner.

(5) Indian Wrestling(See fig. 5-1, E.)Two individuals lie on the ground, side by side, with their heads in opposite directions. They link right elbows. On the instructor's signal or by mutual agreement, each individual raises the right leg approximately straight and far enough to engage the opponent's heel. To start the contest, each person usually raises the leg three times rhythmically and on the third time engages the opponent's heel. Each tries to roll the other over backward. The right leg is used for three bouts, then the left leg for three bouts.

Figure 5-1 is available in the illustration section at the end of the book.

b. Combatives Table II

(1) Wrist Bending.(See fig. 5-2, A.) Opponents pair off, face each other, raise their arms forward, and, with palms forward, interlock their fingers. At the starting signal, each individual attempts to bend the opponent's wrist. The arms are kept up and forward and are not allowed to swing around and down to the sides. The individual who first wins two bouts is the winner.

(2) Back-to-Back Push.(See fig. 5-2, B.)Two individuals stand back-to-back with elbows locked. Each right arm is inside the opponent's left arm. At the starting signal, each pushes backward, trying to move the opponent. Opponents are not allowed to lift and carry each other; only pushing is permitted. The one who pushes the opponent the farthest wins the bout. The individual who first wins two bouts is the winner.

(3) Hop and Pull Hand.(See fig. 5-2, C.) The individuals are matched in pairs. Each grasps the opponent's right hand and, hopping on the right foot, attempts to pull the opponent off balance. Contestants automatically lose if they touch their free hand or their lifted foot to the ground. For successive bouts, they alternate hands and feet.

(4) Westmoreland Wrestling. (See fig. 5-2, D.)Each contestant grasps the back of the opponent's neck with the right hand and the opponent's right elbow with the left hand. In this position, each attempts to pull, push, or force the opponent to touch the ground with any part of the body other than the feet. The individual who first wins two bouts is the winner.

(5) Crab Fight.(See fig. 5-2, E.)Two individuals sit on the ground facing in opposite directions with their hands on the ground behind them. At the whistle, they raise their hips and push with their shoulders and bodies, each trying to make the other's hips touch the ground. The individual who first wins two bouts is the winner.

Figure 5-2 is available in the illustration section at the end of the book.

c. Combatives Table III

(1) Hand Wrestling.(See fig. 5-3, A.) Two individuals stand facing each other. Their right feet are forward and braced side by side. They grasp right hands for the first bout, left hands for the second bout. Each pulls, pushes, makes sideward movements, and otherwise maneuvers to force the opponent to move one or both feet from the original position. The contestant who first wins two bouts is the winner.

(2) Back-to-Back Tug.(See fig, 5-3, B.) Two individuals stand back-to-back with both arms linked at the elbows. Each has the right arm inside the opponent's left arm. At the starting signal, each attempts to pull the opponent. Lifting and carrying are permitted. The contestants must maintain their original direction and keep their arms linked. After a predetermined time, the player pulled or carried the farthest is the loser.

(3) Wrestling to Lift off Feet.(See fig. 5-3, C.) Contestants face each other. Each places the right arm under the left arm of the opponent and around the body. The left arm is over the opponent's right shoulder. Each tries to lift the other off the ground. The individual who first wins two bouts is the winner.

(4) Arm Pull Between Legs.(See fig. 5-3, D.) Two individuals are paired off, back-to-back. Each bends forward and, extending the right arm between the legs, grasps the opponent's right wrist. At the starting signal, each person attempts to pull the opponent After a predetermined time, the player who has pulled the opponent the farthest is the winner of the bout. The one who first wins two bouts is the winner. Repeat with the left hand and then both hands.

(5) Rooster Fight.(See fig.5-3, E.) Each contestant grasps the left foot with the right hand from behind, and the right arm with left hand. Each hops on the right foot, and, by shoulder-butting the opponent or by feinting and sudden evasions, forces the opponent to let go of the foot or arm. The contestant who first wins two bouts is winner.

Figure 5-3 is available in the illustration section at the end of the book.

Section III .Relays

5301. DESCRIPTION AND OBJECTIVE

Relays are races in which each member of a team runs one leg of the race. The team effort decides the winner. Relays provide stimulating competition and contribute to the conditioning of personnel. They also develop aggressiveness, team spirit, and the will to win. Relays should be dispersed throughout the program for short periods of time to provide a change of activity.

a. Team Organization. Relays are conducted most efficiently in platoon-size groups. Teams of equal size must be organized. Competitive spirit is encouraged and teams are organized better by basing teams on units such as squads, crews, or sections. Team captains should be designated. Extra Marines may be used as officials. The number of individuals on a relay team should be limited to squad size. If larger teams are used, the runners will spend too much time awaiting their turns and too little time actually participating. Two to six teams are ideal for relay competition. It is difficult to keep track of winners when more teams compete.

b. Administration of Relays

(1) The time spent on any one relay should be relatively short. If one team achieves a substantial lead in a long relay, the competitive spirit and enthusiastic participation of the other teams may decrease. Several short relays are generally better than one long relay.

(2) To maintain competitive spirit throughout a number of relays, determine the teams that win, place, and show in each relay and their total points for all relays. This can be done by awarding points to all teams on the basis of position at the finish of each relay. The team with the greatest number of points is the winner of the entire set of relays.

(3) Difficulties commonly encountered in conducting relays may be avoided by the following procedures:

(a) The last player in a relay race should be conspicuously identified. For example, the last runner can tie a handkerchief around the head or arm, take off the shirt, put on a hat or take it off, or use some other means.

(b) Another way to keep track of the progress of the race is to have each player sit or squat as soon as each is finished.

(c) Judges at the starting line can keep the runners from starting too soon.

(d) To prevent contestants from turning before they run the full distance, they should be required to run around a peg, pole, or assistant instructor.

(e) Batons, handkerchiefs, tent pegs, or other objects should be passed from one runner to the next when relays are run on a circular track.

(4) Before a unit's first participation in a relay, inform participants of the rules and scoring system. Violation of the rules should not result in disqualification. Instead, impose point penalties. A point penalty is imposed by subtracting a point from the team total at the conclusion of the relay.

(5) Careful administration will prevent most violations. For efficient conduct of relays, follow this procedure:

(a) Announce the name of the relay.

(b) Form the individuals in relay position.

(c) Briefly explain the relay and the rules for running it.
(d) Demonstrate.
(e) Have a definite finish line, and ensure that all know where it is.
(f) Answer questions.
(g) Conduct the relay.
(h) Determine winner and award points.

5302. RELAY TABLES

a. Events. The following relays are grouped into a table of activities. Each relay table can be completed in 15 minutes. Thus, relays can be used as a sole activity or as a part of a longer period. Each table is planned for a platoon-size group (30-60Marines). Adequate warmup for participants is provided by conducting several repetitions of exercises 1 and 2 of a conditioning drill. The recommended relay tables require an area 40 by 60 meters in size. Each table provides a variety of activity. The tables are progressive in the overload applied and should be scheduled in numerical order although not necessarily on successive days.

b. Relay Table I
(1) **60-Meter Lane Relay** (See fig. 5-4, A.) Each team is assembled in single file behind the starting line. On signal, the first individual of each team runs to the turn-around line 30 meters away, then runs back and touches the next teammate waiting at the starting line. The winning team is the first team to get its last member across the finish line. If an individual starts before being touched by the preceding runner, the team may be penalized.
(2) Wheelbarrow Race.(See fig. 5-4, B.) The players of each team pair off and line up in single file behind the starting line. The first individual walks on the hands while the partner grasps the ankles. They advance to the distance line (25 meters) behind which they exchange positions and return to the starting line. The rear individual must always hold the partner's ankles. After the first pair returns across the starting line, the next pair starts.

Figure 5-4 is available in the illustration section at the end of the book.

(3) Squad Front Relay .(See fig. 5-4, C.) The teams form in a line along the starting line with a 10-foot interval between teams. The members of each team lock elbows so that they are linked together. At the starting signal, the teams run to the distance line (20meters) where the left flank individual acts as a pivot. The team swings around on the pivot and returns to the base line. If a team breaks its links, it must reform before continuing. The first team to completely cross the base line intact is the winner.
(4) Crab-Walk Race.(See fig. 5-4, D.) The players of each team line up in single file. The first person of each team assumes the crab-walk position with feet forward on the starting line, hands on the ground behind it. At the starting signal, contestants move forward to the distance line (10 meters). They touch the line with their feet and then return to the starting line in the reverse position with the head and hands leading. The second person may not start until the first individual touches the finish line.

c.Relay Table II

(1) 100-Meter Lane Relay.(See fig. 5-5, A.)This relay is conducted exactly as the first relay of table 1 except that the start and turn-around line are 50 meters apart. This relay provides progression in sprinting.

(2) Frog-Jump Relay. (See fig. 5-5, B.) Each team lines up in single file. The first individual assumes a squatting position on the starting line. At the starting signal, the contestant progresses to the distance line (15 meters) and back by leaping forward, catching the weight on the hands, and bringing up the legs to the squat position for the next leap forward.

(3) Simple Relay. (See fig. 5-5, C.)Each team lines up in single file. Place a marker on the distance line (20 meters) in front of each team. Each team member, behind the first one, grasps the belt of the individual ahead. At the starting signal, each team runs as a unit to the marker, circles it, and returns to the starting line. The first team to completely cross the starting line intact is thewinner.

(4) Fireman's Carry Relay.(See fig. 5-5, D.)The players of each team line up in pairs, one behind the other. One individual in each pair carries the partner to the distance line (30 meters), using the fireman's carry. At the distance line, individuals exchange places and return to the starting line. As a variation, the person to be carried lies on the ground. The one carrying lifts the partner to the proper position. This relay may be performed with the other carries described in guerrilla exercises in chapter 3.

d.Relay Table III

(1) 200-Meter Circle Relay .(See fig. 5-6, A.)A course is laid out in a circular, rectangular,or oval pattern that is 200 meters around. Each team provides one runner on the starting line. On signal, the runner races around the 200-meter track and touches the next teammate, waiting at the starting line, who runs the same course. Each team member runs one lap of the course.

Figure 5-5 is available in the illustration section at the end of the book.

(2) Bear-and-Crab Race.(See fig. 5-6, B.) Each team lines up in single file. At the signal to start, the first contestant in each column assumes the bear-walk position (hands on ground in front), walks to the distance line (15 meters), and then runs back to the starting line. The returning contestant touches off the second individual and goes to the rear of the line. The second contestant assumes the crab-walk position (hands on ground behind) and crab-walks with feet leading to the distance line. This contestant also runs back to the starting line and touches off the next individual, who walks bear fashion. The rest of the members of each team alternate in this manner. The relay ends when the first person is back at the head of the line.

(3) Pilot Relay.(See fig. 5-6, C.) The players are grouped in threes, arms interlocked at the elbows, and outside players facing backwards. The middle person runs forward; the two outside individuals run backward. They run to the turning point (15meters), where they start back, this time with the middle person running backward and the two outside individuals running forward. The next set of three players starts when the first set crosses the starting line.

(4) Saddle-Back Relay.(See fig. 5-6, D.) Mark two parallel lines 15 meters apart. Each team selects a rider. The remaining members of each team count off. The even-numbered players from each team form in single files behind one line and the odd-numbered players from each team form in single files behind the other line directly across from their teammates. At the starting signal, the rider mounts the back of the first player of the team who carries the rider across the other line. There the rider changes mounts to the first player in the second line without touching the ground. This person carries the rider to the next player waiting in the first line. The relay continues until all of the mounts have carried the rider. A rider who falls off must mount again at the point of the fall. A rider who falls in changing mounts must get back on the original mount before making the change.

e. Relay Table IV

(1) 100-Meter Circle Relay (See fig. 5-7, A.)A course is laid out in either a circular, rectangular, or oval pattern that is 200 meters around. Each team is divided in half with each half positioned at starting lines on opposite sides of the track. Each runner races halfway around the track and touches a teammate who completes the lap. Each runner then waits in file at the halfway point until touched by the runner and then completes the second half of the lap. The first team to return all runners to their original starting line is the winner.

Figure 5-6 is available in the illustration section at the end of the book.

(2) In-and-Out Relay.(See fig. 5-7, B.) Each team lines up in a file with players 2 meters apart. At the starting signal, the first player runs back through the column in a zigzag fashion. He alternates, going to the right of one teammate and to the left of the next. Upon completing the run, the player lines up 2 meters behind the last one in the column. As soon as the first runner has passed the second person, the latter starts to run. This continues until all the players have realigned their original order. The team that finishes first is the winner. It may be desirable to have this relay continue until all individuals have run through their entire team two or three times in succession.

(3) Circle Race.(See fig. 5-7, **C.)** Each team forms a circle and holds hands with all individuals facing out except one who faces in and is the "driver."At the starting signal, the teams race to the distance line (20 meters) and back, keeping the circle intact. All the individuals in the circle must completely cross the distance line. The "driver" gives directions and orders. When the circle breaks, it must be reformed before it can continue. The first team completely over the starting line is the winner.

(4) Horse-and-Rider Relay.(See fig. 5-7, D.)Each team lines up in a single file. At the signal to start, the second player in each column leaps upon the back of the first who carries the rider across the distance line (30 meters). At the distance line, the rider dismounts and runs back to the starting line. There the rider picks up the third individual in the column, and carries this player to the distance line where the first player has remained. This continues until the last person is carried across the distance line.

Figure 5-7 is available in the illustration section at the end of the book.

Section IV. Team Contests and Athletics

5401. DESCRIPTION AND OBJECTIVE

Team contests are competitive activities in which Marines as a team compete with another team to win. They are guided by simple rules and organization. The function of team contests is to provide competition and an opportunity for body contact and to contribute to the development of physical readiness. In competing and working together as a team, individuals develop aggressiveness, the will to win, and teamwork. Team athletics deserve a prominent place in the physical training program because they contribute to increased combat efficiency. Because of the competitive nature of athletics and their natural appeal, individuals take part with enthusiasm. Athletic teams formed at intramural and higher levels are a strong unifying influence and provide one of the best means of developing esprit de corps.

a. Area and Equipment A level training field is sometimes the only area required. Many contests need no equipment. In contests requiring equipment, the need is for standard items such as logs, balls, nets, goals, and similar types of equipment. Specific requirements for area and equipment are listed with each contest description.

b. Progression. Team contests of a strenuous nature should be introduced after a basic period of conditioning has been completed and individuals are in the slow improvement stage of conditioning. Progression can take place from the less active to the more vigorous contests, and then from the non-contact to the contact or combative-type contests.

c. Necessity for Preconditioning Individuals must undergo conditioning prior to participation in athletics. Muscles, organs, joints, and ligaments not accustomed to stress and strain from sudden stops and starts, falls, body contact, rapid turns, prolonged running, and other rigors of athletic competition are subject to injury. Although athletics should not be introduced until players are physically prepared, there is still opportunity to engage in competition through lead-up contests. Individuals learn many of the skills required for athletics while participating in team contests.

d. Benefits. Athletics are beneficial primarily to sustain interest in the program and to maintain an achieved level of physical fitness. Athletics are a supplement, not a substitute for other types of conditioning activities which should continue. All of the desirable traits of physical fitness cannot be developed through athletics, yet their contribution is significant. For athletics to make a proper contribution to physical conditioning, the selected sports must bevigorous. The team contests recommended here are carefully selected for their simplicity, aggressiveness and applicability to squad- teaching teamwork. They are generally strenuous and involve the whole team throughout the entire duration of the contest.

5402. TEAM CONTESTS

a. Pushball

(1) **Players.** 10 to 50 players on a side. (See fig. 5-8.)

(2) Equipment A large pushball 5 to 6 feet in diameter.
(3) Area. A field 240 to 300 feet long, 120 to 150 feet wide. Mark a center line in the middle of the field parallel to the end lines. Mark a line 45 feet on either side of this center line and parallel to it, extending it across the width of the field. Mark another parallel line 15 feet from each end line, extending it across the width of the field.
(4) Game. Four 10-minute quarters are played. Give 2-minute rests between quarters and 5-minute rests between halves. The object of the game is to propel the ball over the opponent's goal line by pushing, rolling, passing, carrying, or any other way except kicking.

(a) The ball is placed on the center line with the opposing captains 3 feet from the ball. The rest of the players are 45 feet from the ball, on their half of the field. On the referee's starting whistle, the captains immediately play the ball with their respective teams coming to their assistance.

(b) At quarter time, the ball remains dead for 2 minutes at the spot where it was when the quarter ended. At half time, the teams exchange goals. The play is then started as it was in the beginning.

Figure 5-8 is available in the illustration section at the end of the book.

(c) Players may use any means of interfering with an opponent's progress except striking and clipping. Clipping is throwing the body across the back of an opponent's leg, while the opponent is running or standing. Legal use of force may be applied to all opponents whether or not they are playing the ball. For striking an opponent, the offender is removed from the game. The team penalty is half the distance to their goal. The penalty for clipping is the same.

(d) When any part of the ball goes out of bounds, it is a dead ball. The teams line up at right angles to the side lines and 3 feet apart at the point where the ball went out. The referee then tosses the ball between the teams.

(e) When, for any reason, the ball remains in one spot for more than 10 seconds, the referee declares the ball dead. The ball is then put into play as it is for an out-of-bounds situation.

(5) Scoring. A goal is scored when the ball, or any part of it, is propelled across the opponent's end line. A goal counts 5 points. The team scoring a goal has the privilege of trying for a point after the goal. To try for this extra point, the ball is placed on the opponent's 5-yard (or 15-foot) line. The teams line up on either side of this line separated by the width of the ball. Before the whistle blows, one player on each team may place hands on the ball. On the referee's signal, the ball is put into play for 1 minute. If any part of the ball is driven across the goal line in this 1-minute period, the offensive team scores 1 point. The defending team may not score during the opponent's try for the extra point.

b. Line Rush
(1) Players. Any number up to 50 on each side. (See fig. 5-9.)
(2) Equipment. None.
(3) Area. A field, 75 by 100 feet.
(4) Game. One team lines up behind one goal line and the other in midfield. On the starting signal, the team standing behind the goal line seeks to cross the goal line at the opposite end of

the field within 30 seconds. The team in the center seeks to prevent it by catching and holding the runners. At the end of 30 seconds, the teams change.

(5) Scoring. Count the number of individuals who have crossed the far goal at the end of 30 seconds. After each team has had from three to five tries, the scores are added and the winner declared. A player scores 1 point when any part of the body is across the goal line.

Figure 5-9 is available in the illustration section at the end of the book.

c. Human Tug-of-War

(1) Players. 10 to 20 on a team.(See fig. 5-10.)

(2) Equipment. None.

(3) Area.40 to 60 feet.

(4) Game. Draw a line in the center of the area. Divide the players into two equal teams. Place them in single file on opposite sides of the center line facing each other. Each individual places the arms around the waist of the teammate in front. The two leaders of the opposing teams grasp each other around the waist. On signal, the teams try to pull each other over the center line within 30 seconds.

(5) Scoring. The team pulled across the center line loses. If neither team is pulled over the center line, but one team breaks its file, that team loses the match.

(6) Variation. Use a 3/4- or 1-inch rope. Space the leading players on each team 10 feet apart, each holding one end of the rope. The team pulled across the center line loses.

Figure 5-10 is available in the illustration section at the end of the book.

d. Master of the Ring

(1) Players. Any number. (See fig. 5-11.)

(2) Equipment. None.

(3) Area. A clearly marked circle large enough to contain all the players.

(4) Game. All the players stand inside the circle. At the signal, all players attempt to throw each other out of the circle. All tactics are fair except unnecessary roughness. When any part of the body touches across the line, the player is out and leaves the circle at once. Several officials are needed to spot the players who cross the line.

(5) Scoring The player who remains in the circle when all the others are out is the master of the ring.

(6) Variation 1. The players are divided into two equal teams. Each team is clearly marked. On signal, each team tries to throw the opponents out of the circle. The winning team is the team that eliminates all opponents from the circle.

(7) Variation 2. The players are divided into two equal teams. Each team sends only one individual into the circle. When one person has been forced out of the circle, the losing side sends in only one player. The team which eliminates all opponents is the winner.

(8) Variation 3.A pit, approximately 4-feet deep, is used rather than a circle on level ground. The contest may then be an individual or team activity.

Figure 5-11 is available in the illustration section at the end of the book.

e. Log Pivot Circle. Space teams far enough apart so each team can pivot in a circle without colliding with any other team. Each team holds a log in the bend of the arms in front of the chest. (See fig.5-12.) At the command **CIRCLE RIGHT, MOVE**,the left-flank individual holds the pivot while the log is carried around 360 degrees in a counter- clockwise motion, back to the original position. This movement may also be performed to the left by facing in the other direction and pivoting clockwise(**CIRCLE RIGHT, MOVE**).Other commands may be used such as: **CIRCLE HALF-RIGHT, CIRCLE HALF- LEFT** and so on. The first team to complete the prescribed movement is the winner.

Figure 5-12 is available in the illustration section at the end of the book.

f. Log-Rolling Race. Each team tries to roll its log a measured distance by pushing it with the hands and driving the body forward with the legs. (See fig. 5-13.)The first team to get the entire length of the log across the finish line wins.

Figure 5-13 is available in the illustration section at the end of the book.

g. Prone Pushing Contest.. Two teams lie prone, facing each other with a log between them. (See fig. 5-14.)Both teams place their hands against the log, keeping their arms straight. Then, by driving with the legs, each team attempts to push the other a measured distance to the rear.

Figure 5-14 is available in the illustration section at the end of the book.

h. Shuttle Relay Race. Each team in this race divides into two parts, A and B. Members of part A run 50 yards with a log held under their right arms. (See fig 5-15.)At the distance line, they give the log to their teammates in part B who bring it back to the starting line. The team pair finishing first is the winner.

Figure 5-15 is available in the illustration section at the end of the book.

5403.CROSS-COUNTRY AND DISTANCE RUNNING

a. Value of Running. Long-distance running gives some benefits that cannot be obtained in the same degree from any other sport. It builds powerful leg muscles, increases lung capacity, and develops endurance. For these reasons, cross-country and distance running should be included in the physical training program. These sports require only a few miles of open space which is generally available. Short cross-country runs and middle-distance runs can be used to supplement other activities, particularly team sports or the sports that develop precision or agility rather than endurance. Short cross-country runs can be scheduled once a week, gradually increasing the distance as the physical condition of the runners improves.

b. Cross-Country Runs. A cross-country run is a distance run held on a course laid out along roads, across fields, over hills, through woods, and on any irregular ground. A flat cinder or dirt track is not a suitable surface for cross-country running. Some runs are as long as 6 miles while others are as short as 3 or 4 miles. If cross-country running is being used to supplement other activities, the 3-mile course is long enough for most. Cross-country runs should be used only after individuals reach the sustaining stage of conditioning. This type of running should then be scheduled occasionally to provide variety. Cross-country running has the advantage of allowing mass participation. Interest can be stimulated by putting the runs on a competitive basis.

c. Practice Methods. Conditioning is essential to distance and cross-country running. Championship distance running depends on stamina, which can be developed only through constant training. An individual of only average ability can become an outstanding distance runner by steady and careful training. Hiking is an excellent method for getting into condition before the season opens. Long walks build up leg muscles. During the first month of the season, training should be gradual, starting with short distances, and increasing day by day. At first the legs will become stiff, but the stiffness gradually disappears if running is practiced for a while every day. To prevent strain, it is essential to limber up thoroughly each day before running.

(1) In training a large group, leaders should be stationed at the head and the rear of the column. They should make every effort to keep the runners together. After determining the abilities of the unit in cross-country running, it is advisable to divide them into three groups. The poorest-conditioned group is started first, the best-conditioned group, last. The starting time of the groups should be staggered so that all of them come in at about the same time.

(2) In preliminary training, running is similar to ordinary road work in that it begins with rather slow jogging, alternating with walking. The speed and distance of the run is gradually increased. As conditioning improves, occasional sprints may be introduced. Atfirst, the distance run is from one-half to 1 mile. It gradually increases to 2 or 3 miles. On completing runs, require runners to walk for 3 or 4 minutes before stopping to permit a gradual cooling off.

d. Facilities and Equipment. There should be at least one timer with a stopwatch (preferably three) for timing the runners. A course 3 or 5 miles long should be measured and marked by one of the three following methods:

(1) Fasten directional arrows to the top of tall posts and place them at every point where the course turns. Such signs should also be placed at every other point where there may be doubt as to the direction of travel.

(2) Place a lime line on the ground over the entire course.

(3) Place flags, clearly visible to the runners.

- **(a)** A red flag indicates a left turn.
- **(b)** A white flag indicates a right turn.
- **(c)** A blue flag indicates the course is straight ahead.

e. Rules/Scoring. A cross-country team consists of seven individuals, unless otherwise agreed. In dual meets, a maximum of 12 on one team may be entered, but 7 or fewer are scored. First place scores 1 point, second place 2, third place 3, and so on. All who finish the course are ranked and tallied in this manner. The team score is then determined by totaling the points scored by the first five of each team to finish. The team scoring the least number of points is the winner. If fewer than five (or the number determined prior to the race) finish, the places of all members of that team are disregarded. If two or more teams score the same number of points, the event is called a tie.

5404.TEAM ATHLETICS

Team athletics add variety to the combat readiness program. At no time should they be allowed to replace intramural activities or to provide practice time for intramural or base teams. If team athletics are integrated into the program, the goal should be total participation and competition. Therefore, the games selected can be played by the skilled or the unskilled. They require limited equipment, yet will teach competitiveness, aggressiveness, and teamwork while they improve conditioning.

a.Soccer. Soccer is one of the best athletic activities for developing endurance, agility, leg strength, and great skill fulness in using the legs. The most popular sport in the world, soccer is the national game of many European, Asian, and Central and South American countries. In recent years, it has become popular in the United States. A soccer ball is the only equipment needed for the game, and Marines can learn to play it easily. Players do not need much skill to participate, but the amount of skill they can develop is almost unlimited.
(1) Place in the Program. Soccer should be introduced during the latter part of the slow improvement stage of physical training. It can be used as a competitive activity in the sustaining stage. It is primarily a spring or fall sport. Any level field is suitable for competition. The boundaries for the soccer field are similar to the dimensions for a football field. (See fig. 5-16.) Goal posts are essential, but they are easily constructed and usually temporary so that they may be removed when not in use.

Figure 5-16 is available in the illustration section at the end of the book.

(2) Basic Skills

(a) Passing. Passing with the feet is the primary means of moving the ball. Short passes are easier to control. Emphasis should be continually placed on the skill of passing.

(b) Dribbling. The ball is dribbled by a series of kicks with the inside or outside of the foot. Do not kick with the toe. Keep the head over the ball when kicking and propel it only a short distance at a time. Keep it close to the feet. If a player lets a ball move far from the feet, an opposing player can easily take it away.

(c) Instep Kicking. The instep kick is the basic soccer kick. The toe does not come in contact with the ball. The toe is pointed downward, and the instep (the area under the shoe laces) is applied to the ball with a vigorous snap from the knee.

(d) Inside-of-the-Foot Kicking. The ball is kicked with the inside of the foot, and the leg is swung from the hip. The toe is turned outward, and the sole of the foot is parallel with the ground as the foot strikes the ball. This kick is used for short passes and for dribbling.

(e) Foot Trapping. The foot trap is the method of stopping the ball by trapping it between the ground and the foot. Place the sole of the foot on top of the ball at the instant it touches the ground, keeping the foot relaxed. This is an effective way to stop a fast-moving ball.

(f) Shin Trapping. .The shin trap is a method of stopping the ball with the shins. Stand just forward of the spot where the ball should strike the ground and allow it to strike the shins in flight or on the bounce. Use either one or both legs from the knee down so as to absorb the bounce.

(g) Body Trapping. .Intercept the ball with any part of the upper body except the arms and hands. Keep the body relaxed and inclined toward the ball. To keep the ball from bouncing, move backwards from it as it strikes the body. This drops the ball at the feet in position for dribbling or passing.

(h) Heading. Heading is a technique for changing the direction of the flight of a ball by butting it with the head. Tense the neck muscles and jump up to meet the ball. Butt the ball with the forehead at about the hairline to reverse its direction. Use the side of the head to deflect it to the side.

(3) Offensive and Defensive Positions. Forwards usually play on the offensive half of the field. Fullbacks usually play on the defensive half of the field. Halfbacks are the backbone of the team. They move forward on the offense and back on defense. The goalkeeper almost always remains within a few feet of the goal.

(4) Abridged Rules

(a) A soccer team is composed of 11 players. (See fig. 5-16.)

(b) The player propels the ball by kicking it with the feet or any part of the legs, by butting it with the head, and by hitting it with any portion of the body except the arms or hands.

(c) The goalkeeper is the only individual allowed to place hands on the ball, but only in the goalkeeper's area. The term "hands" includes the whole arm from the point of the shoulder down.

(d) A goal is made by causing the ball to cross completely the section of the goal line lying between the uprights and under the cross bar.

(e) Each goal scores 1 point for the team scoring the goal.

(f) The penalty for a foul committed anywhere on the playing field (except by the defensive team in its penalty area) is a free kick awarded to the opposing team.

(g) All opponents must be at least 10 yards from the ball when a free kick is taken.

(h) The penalty for a foul committed by the defensive team in its penalty area is a penalty kick.

(i) A penalty kick is a free kick at the goal from the spot 12 yards directly in front of the goal. The only players allowed within the penalty area at the time of the kick are the kicker and the defending goalkeeper.

(j) An official game consists of two 30-minute halves,

(k) Teams change goals at the end of every quarter.

(l) In the event of a tie, an extra quarter may played. This may be followed by a sudden death period. If the tie still exists, the game can be decided by penalty kick.

(m) After a team has propelled the ball across a side line and out of play, the ball is put back into play by a throw-in from the side line by a member of the opposing team. The ball is thrown in from the point where it crossed the side line as it went out of bounds. The ball must be thrown in with both hands using an over-the-head motion, keeping both feet on the ground.

(n) When the offensive team propels the ball across the defensive team's goal line but not in the goal, the defensive team is awarded a goal kick-a free kick taken from within its own goal area. The ball must come out of the penalty area to be in play.

(o) When the defensive team causes the ball to go behind its own goal line, excluding the portion between the goal posts, the offensive team is awarded a corner kick—a free kick taken by a member of the offensive team at the quarter circle, at the cornerflag post nearest to where the ball went behind the goal line. The flag post must not be removed.

(p) The game is started and, after a goal has been scored, is resumed by placing the ball in the center of the midfield line. Players must be on their side of the line until the ball is kicked. The ball must be kicked forward and must move at least 2 feet to be legal. The first kicker may not touch the ball twice in succession at the kick off. The opposing team must be 10 yards from the ball until it moves.

b.Speedball. Speedball offers vigorous and varied action with plenty of scoring opportunities. It is easy to learn and provides spontaneous fun. Little equipment is needed-a ball is all that is absolutely necessary. Speedball combines the kicking, trapping, and intercepting elements of soccer; the passing game of basketball; and the punting, drop kicking, and scoring pass of football.

(1) Place in the Program.Speedball, like soccer, should be introduced into the physical training program during the latter part of the toughening stage and used as a competitive activity in the sustaining stage. Speedball may be played any time weather permits but is primarily a spring or fall activity.

(2) Method of Play.The game is played as follows:

(a) Two teams of 11 each play the game under official rules, but any number of players may successfully constitute a team.An inflated leather ball, usually a soccer ball, is used. The playing field is a football field with a football goal post at each end. (See fig. 5-17.)

(b) The game starts with a soccer-type kickoff. The kicking team tries to retain possession of the ball and advance it toward the opposite goal by passing or kicking it. Running with the ball is not allowed, with the result that there is no tackling or interference. After the ball touches the ground, it cannot be picked up with the hands or caught on the bounce but must be played as in soccer until it is raised into the air directly from a kick. Then the hands may be used.

(c) When the ball goes out of bounds over the sidelines, it is given to a player of the opposing team and is put into play with a basketball throw- in. When it goes over the end line without a score, it is given to a player of the opposing team who may either pass or kick it onto the field.

(d) When two opposing players are contesting possession of a held ball, the official tosses the ball up between them as in basketball.

(e) Points are scored by kicking the ball under the crossbar of the goal post, drop kicking the ball over the crossbar, or completing a forward pass into the end zone for a touchdown.

Figure 5-17 is available in the illustration section at the end of the book.

(3) Basic Skills

(a) Shared Skills. Skills include kicking, passing, heading, and trapping (from soccer); punting, dropkicking and forward passing (from football); and passing, receiving, and pivoting (from basketball).

(b) Kickups. The kickup is a play in which a player lifts the ball into the air with the feet so as to legally play the ball with the hands. The kickup generally makes the transition from ground play to aerial play. The technique of making the play depends upon whether the ball is rolling or stationary. If the ball is rolling or bouncing toward the player, the foot is held on the ground with the toe down until the ball rolls onto the foot. Then the foot is raised, projecting the ball upward. If the ball is stationary, the player rolls it backward with one foot. Then the player places a foot where the ball will roll onto it. The player then raises the foot, projecting the ball as before. If a ball is rolling away, the player should stop it with a foot and play it as a stationary ball. The player can also raise the ball by standing over it with a foot on either side. The player presses the feet against the ball and jumps into the air, propelling the ball into the hands.

(4) Offensive and Defensive Play. The positions of players in speedball are much the same as in soccer. However, some of the positions have different names. Each team has 11 players. The forward line is composed of five players, the right end, right forward, center, left forward, and left end. The second line consists of right halfback, fullback, and left halfback. In the next line is the right guard and left guard. The player who defends the goal is the goal guard. The strategy employed in speedball during offensive play is very similar to that of soccer. There are two types of defensive formations in speedball: man-for-man and position defense. Man-for-man defense is recommended for beginning players.

(5) Abridged Rules

(a) Field. A speedball field is 100 yards long with additional 10-yard end zones on each end and 55 yards wide, the same as a regulation football field.

(b) Players. Eleven players are on a team. The goal guard has no special privileges.

(c) Time. Ten-minute quarters. Two minutes between quarters. Ten minutes between halves. Five minutes for extra overtime periods.(Begin the first overtime by a jump ball at center, same goals; change goals in the event of a second overtime period.)

(d) Winner of Toss. The winner of the toss has the choice of kicking, receiving, or defending a specific goal.

(e) Kickoff. A kick off starts the first and second half of a game and starts play after a score. The kick off is made from any place on the 50-yard line. The kicking team must be behind the ball when it is kicked. The receiving team must stay back of its restraining line (10 yards distance) until the ball is kicked. The ball must go toward the opposing team's goal before it can be played. Kick off out of bounds goes to the opponent at that spot. A kick off touched by the receiving team before going out of bounds, with no impetus added, still belongs to the receiving team. A kick off, in possession and control of the receiving team and then fumbled out of bounds, belongs to the kicking team at that spot.

(f) Starting Second and Fourth Quarters. The ball is given to the team that had possession at the end of the previous quarter, from out of bounds, as in basketball.

(g) Second Half. The first-half receiving team kicks off at the beginning of the second half.

(h) Fly and Ground Ball. The most characteristic playing rule of speedball is the difference between a fly ball (or aerial ball) and a ground ball. A player is not permitted to touch a ground ball with the hands and must play it as in soccer. A fly ball is one that has risen into the air directly from the foot of a player (for example, a punt, drop kick, place kick, or kickup).Such a ball may be caught with the hands provided the catch is made before the ball strikes the ground again.

(i) Kickup and Overhead Dribble. A kickup is a ball that is so kicked by a player that he can catch it himself. A bounce from the ground may not be touched with the hand because it has touched the ground since being kicked. This rule prohibits the ordinary basketball dribble, but one overhead dribble (throwing the ball into the air and advancing to catch it before it hits the ground) is permitted.

(j) Out of Bounds. If a team causes the ball to go out of bounds over the sidelines, a free throw-in (any style) is given to the opposing team. When the ball goes over the end line without scoring, it is given to the opponents who may pass or kick from out of bounds at that point.

(k) Tie Ball. In case two players are contesting the possession of a held ball, even in the end zone, a tie ball is declared. The ball is tossed up between them.

(6) Scoring

(a) Field Goal(3 points).A soccer-type kick, in which a ground ball is kicked under the crossbar and between the goal post from the field of play or end zone, is a field goal. (A punt going straight through is not a field goal for it is not a ground ball. The ball must hit the ground first.) A drop kick from the field of play that goes under a crossbar does not count as a field goal. A drop kick from the end zone that goes under the crossbar counts as a field goal;if it goes over the crossbar, it is ruled as a touch back, and the other team takes possession of the ball.

(b) Dropkick(2 points). A scoring drop kick must be made from the field of play and go over the crossbar and between the uprights. The ball must hit the ground before it is kicked (usually with the instep).

(c) End goal(1 point). This is a ground ball which receives its impetus (kicked or legally propelled by the body) from any player, offensive or defensive, in the end zone and which passes over the end line but not between the goal posts.

(d) Penalty Kick(1 point).A ball kicked from the penalty mark that goes between the goal posts and under the crossbar is a penalty kick. The penalty mark is placed directly in front of the goal at the center of the goal line.

(e) Touchdown(1 point). A touchdown is a forward pass from the field of play completed in the end zone. The player must be entirely in the end zone. If the player is on the goal line or has one foot in the field of play and the other in the end zone, the ball is declared out of bounds. If a forward pass is missed, the ball continues in play, but it must be returned to the field of play before another forward pass or drop kick may be made.

(7) Substitutions. Substitutions may be made any time when the ball is not in play. A player who is withdrawn may not return during that same period.

(8) Timeout. Three legal timeouts of 2 minutes each are permitted each team during the game.

(9) Fouls

(a) Personal Foul. Kicking, tripping, charging, pushing, holding, blocking, and unnecessary roughness of any kind, such as running into an opponent from behind, are personal fouls. Kicking at a fly ball and thereby kicking an opponent is a personal foul. Four personal fouls disqualify a player from the game.

(b) Technical Foul. Illegal substitution, more than three timeouts in a game, unsportsmanlike conduct, and unnecessarily delaying the game are technical fouls.

(c) Violation. Traveling with the ball, touching a ground ball with the hands or arms, double overhead dribble, violating tie ball, and kicking or kneeing a fly ball before catching it are violations.

(d) Penalties. The offended player shall attempt the kick.

(e) Officiation of Fouls

1 Fouls in the field of play allow no follow-up while fouls in the end zone always allow follow-up.

2 On penalty kicks, with no follow-up, only the kicker and goalie are involved.

3 On penalty kicks, with a follow-up, the kicking side is behind the ball and the defending side behind the end line or in the field of play. No one is allowed in the end zone or between the goal posts except the goal guard. The kicker must make an actual attempt at goal and cannot play the ball again until after another player plays it.

c. Volleyball. Volleyball is a popular sport. The game entails much physical activity, yet it is not strenuous. It is, therefore, a game for young and old alike, for beginners and for skilled players. It may be played indoors or outdoors on any type of terrain. As an informal activity, volleyball can be played by any number; as an organized activity, it provides, as few other sports do, a game for 12 to play in a limited area. While volleyball requires no great skill to play, it permits a high degree of proficiency. Players naturally get more enjoyment when they know the game and play it well. For this reason, instruction in the basic skills should be provided.

(1) Place in the Program Volleyball may be used occasionally as a competitive activity during the sustaining stage. Though it is a year- round sport, it should be included in the physical training program only when a more strenuous activity is impractical. Volleyball is an excellent off-duty activity.

(2) Instruction. Usually during the first three or four classes, a 10- to 15- minute period of instruction, followed by scrimmage, is enough to teach the basic skills, rules, and techniques of volleyball. More time, if available, can be given to teaching basic skills, but emphasize competitive play rather than formal instruction. Divide the unit into 6-person teams, organized on the basis of ability .Teams should be as nearly equal as possible.

(3) Basic Skills

(a) Passing the High Ball. The chest pass is the most effective method of playing the ball. To receive the ball, the knees are fixed, and the body is tilted forward. The elbows are raised sideward to a point in line with the shoulders. The wrists are extended in line with the forearm. The arms, wrists, and hands are rotated inward. To pass the ball, the hands are chest-high, thumbs pointing inward. The fingers are flexed, forming a cup, allowing them to contact the ball. On contact with the ball, the wrists are snapped while the fingers and elbows are pushed upward, sending the ball upward. A high ball is much easier to handle than a low one.

(b) Passing the Low Ball. A ball that is lower than the waist is one of the easiest to hit, but it is also a frequent cause of fouls (holding or carrying the ball). The best position for handling a low ball is to have the knees flexed and arms flexed at the elbows and rotated so the thumbs are pointing outward, the palms up. When the fingers contact the ball, the entire body reacts in a lifting motion. The arms and hands swing upward in a scooping action. It is important that the fingers, not the palms, contact the ball, and that the ball is batted, not thrown.

(c) Underhand Serve. Take a position behind the back line facing the net, left foot forward, holding the ball in the palm of the left hand. The left knee is flexed; the right knee is straight. Swing the right arm back and at the same time move the left hand (holding the ball) across the body in line with the right hip. Then swing the right arm forward hitting the ball off of the left hand with the palm of the right hand, raising the hips and arching the back in the same motion. Be certain to swing the right arm in a straight line, or the ball will be difficult to control. When the opposition is in formation, the server should try to place the ball in the right or left back area, and not near the net.

(d) Setup. A setup is a ball hit into the air near the net by one player, so that a teammate may hit or "spike" it sharply downward into the opponent's court. The chest pass is the best pass to use. The ball is sent approximately 10 feet into the air toward the spiker so it will descend from 4 to 20 inches from the net.

(e) Spiking. The spike is a leap into the air and a sharp downward hitting of the ball into the opponent's court. A spiker must be able to spring easily from the floor, judge the movement of the ball, and strike it with a downward movement of the arm. To jump from the floor, step off with one foot and jump with the other. Stand with the right or left side to the net, facing the setup player. Much depends upon the setup player to place the ball in the proper position. The spiker jumps into the air and strikes the ball above its center so as to drive it downward. A snapping movement of the arm and wrist will drive the ball forward and downward with power and control. Aim for a weak spot in the opponent's defense.

(f) Blocking. The block is a defense technique used to prevent a spiker from driving the ball across the net. It is an attempt by one or more defensive players at the net to block a hard-hit shot by using the force of the ball to send it immediately back into the opponent's court. In an effective block, forwards on the defensive team spring into the air at the time of the spike, placing both hands and arms in the expected path of the ball. An effective block tends to upset the offense and presents another element for the spiker to worry about. To be effective, the blocker must anticipate the path of the ball and time the block with the spike.

(4) Abridged Rules

(a) The volleyball court is 30 feet wide by 60 feet long. (See fig. 5-18.)

(b) The top of the net is 8 feet high.

(c) A volleyball team consists of 6 players.

(d) A match consists of the best two out of three games.

(e) The first team scoring 15 points wins the game, provided that they have 2 points more than their opponents.

(f) A deuce game is a game in which both teams score 14 points. The game is continued until one team obtains a 2-point advantage over the other.

(g) Only the serving team can score. If the serving team commits a fault, it loses the serve to the opposing team.

(h) The team members who receive the served ball rotate one position in a clockwise direction.

(i) The ball is put into play by serving from behind the back line.

(j) A served ball touching the net results in the loss of the serve. At any other time during play, a ball touching the net is still in play.

Figure 5-18 is available in the illustration section at the end of the book.

(k) The ball is out of play when it touches the ground or goes outside one of the boundary lines.

(l) All line balls are good.

(m) The players must hit or bat the ball; they may not throw, lift, or grasp it.

(n) A player may not touch the ball with any part of the body below the knees.

(o) A player may not play the ball twice in succession. In receiving a hard-driven spike, a defensive player may make several contacts with the ball even if they are not simultaneous. All such contacts, however, must constitute one continuous play, and all must be above the knees.

(p) The ball may be touched no more than three times on one side of the net before being returned across the net to the opposing team.

(q) A player must not touch or reach across the net.

(r) A player may touch the line under the net, but may not cross it.

Chapter 6
EVALUATION OF PERFORMANCE DURING TRAINING

6001.OBJECTIVES OF COMBAT READINESS

The objectives of the Marine Corps physical fitness program are-

a. To contribute to the health and well-being of every Marine through regular exercise and health education.

b. To develop Marines who are physically capable of performing their duties in garrison and in combat.

c. To develop Marines with a reserve level of physical fitness that will help them win in combat.

d. To provide a medium for developing the self-confidence of the individual Marine and thereby enhancing overall discipline, morale, esprit de corps, unit efficiency, and the desire to excel within the Marine Corps.

6002. THE PHYSICAL FITNESS TEST

The physical fitness test (PFT) has distracted attention from the proper goal of the physical fitness program. Simply stated, the goal of the physical fitness program is the success of Marines in combat. Unit commanders must not allow the PFT to become an end in itself. The PFT should be used within the commander's program of combat readiness training as a standard, an index of the physical fitness of individual Marines at a given time and place. The PFT is not difficult, and it must be kept in perspective. All athletes should be first class, and first class is attainable for all Marines. The conduct of the PFT is described in Marine Corps Order 6100.3_.

6003. OTHER METHODS OF EVALUATION

The commander should use other ways of evaluation to provide variety and imagination. Intangible objectives, such as confidence and aggressiveness, are hard to measure. The use of either inspection or observation is necessary to evaluate objectives of this nature. The commander has several other methods available when evaluating the physical condition of the command. The additional methods of inspection, observation, medical examination, and testing are summarized as follows:

a. Formal Inspections. Formal inspections, using inspection officers and standardized rating criteria, may help evaluate unit physical fitness.

b. Daily Observation. Routine observation of physical performance and appearance can serve as an indicator of an individual's or unit's physical readiness. However, mere observation is not a totally reliable or accurate means of evaluation.

c. **Physical Examination.** Medical examination may detect any individual disability or detrimental physical condition. It may guide in application of remedial, therapeutic, or limited exercise programs.

d. Other Physical Fitness Tests. Some other tests are described in detail in this manual for information purposes. Commanders are cautioned, however, not to place too great an emphasis on testing.

6004. RESPONSIBILITY

The commander is responsible for the physical fitness of the command and for the measurement and evaluation of its physical readiness. The goal of combat readiness training remains to prepare Marines and units for the physical demands of combat. The bulk of the commander's combat readiness training program should be such activities as obstacle and confidence courses, endurance courses, and progressive load-bearing marches which will prepare Marines for combat. Group activities which are competitive and combative in nature are advantageous and add imagination and enthusiasm to the program. Activities such as orienteering, which is not only physically demanding but also competitive and which teaches essential subject skills, are excellent and motivate all concerned. A commander's goal should be to provide an integrated training program that is so aggressive and demanding of Marines and leaders alike that a PFT administered at any time will reflect that each and every Marine is both physically and mentally prepared for the demands of combat.

6005. COMMANDER'S PHYSICAL FITNESS PROGRAM

The Commandant of the Marine Corps has directed that the physical fitness test shall be the universal measure of adequate individual physical fitness. The focus of a commander's physical fitness training program, however, should not be to prepare Marines to pass the physical fitness test, but to prepare them for the physical rigors of combat.

a. Administration. The physical fitness program requires each Marine to spend a minimum of 3 hours per week in physical fitness training, to be tested semiannually, and to obtain a minimum standard of third class. The attainment of a higher score is a laudable individual goal which should be encouraged but should not become a unit objective. Overemphasis could be detrimental to training required to develop the complete Marine. All commanders should exercise sound judgment as well as positive, aggressive leadership in striving toward the training objective.

b. Variety of Activities. The program must include the following elements:
(1) Physical Readiness Training. physical readiness training is a complete physical training program which develops and maintains the strength, endurance, and physical skills needed to sustain the individual during combat.
(2) Remedial Physical Conditioning. Remedial physical conditioning is a process by which physically substandard Marines are conditioned to meet prescribed standards.
(3) Competitive Conditioning Activities. Competitive conditioning activities consist of teams or individuals competing against an opponent to win. This involves a combination of sports and military skills designed to foster competitive attitudes and develop unit pride and esprit de corps.
(4) Occupational Conditioning. Marching long distances, dry net training, etc., should not be ignored as a means to fulfill the time requirement.

c.Time. In order to develop the desired level of physical fitness, unit commanders are advised that a minimum program of 3 hours per week is mandatory. The program must encompass at least three exercise periods weekly. Each period should include calisthenics, running, or other forms of vigorous activity .The minimum desirable length for each exercise period is 60 minutes. Shorter periods may be authorized by the unit commander if dictated by the local training situation. Unit commanders are encouraged to use the normal working day to satisfy this training requirement. However, commanders are authorized to conduct required physical fitness training during off-duty hours when the mission, workload, personnel status or other significant factors preclude fulfilling it during duty hours.

d.Action. Marine Corps Order 6100.3_ requires that commanders-

o Establish and maintain an effective physical fitness program which ensures that all Marines maintain an acceptable level of physical fitness.

o Ensure that all individuals are medically qualified prior to participating in the physical fitness program.

o Establish a minimum physical fitness program of at least 3 hours per week. Physical training may be authorized on an individual basis at the discretion of the unit commander.

o Ensure that all Marines participate in physical fitness conditioning activities commensurate with their medical qualifications and limitations.

o Conduct physical fitness testing for all Marines.

o Place all Marines who fail to pass the PFT on a daily command-supervised remedial physical conditioning program until they pass the test.

o Ensure that results of physical fitness testing are entered on section A of the fitness report. Comments are placed in section C in connection with attainment of superior physical performance or a medical excuse from the PFT.

6006. OTHER FORMAL TESTS

Commanders must realize that the best index of combat readiness is personal observation of performance on conditioning marches of 15 miles while under load. This manual presents three additional indicators of unit fitness. These indices are the physical readiness test, which was formerly the Marine Corps standard and is now routinely administered at Officer Candidate School; the airborne trainee physical fitness test which many Marines who attend U.S. Army formal schools are required to take; and the battle fitness test, the ultimate test of the combat readiness of Marines. Commanders may wish to include these tests or portions thereof within their combat readiness programs.

6007. PHYSICAL READINESS TEST

a. Purpose. The purpose of this test is to measure the Marine's ability to meet the minimum standard of physical readiness for duty in the field. This test is currently administered to all male officer candidates at Officer Candidate School, MCCDC, Quantico, VA. Female candidates do not take this test.

b. Test Area. Administration of this test requires an area suitable for crawling, jumping, and running 160 yards to include a trench that is 8 feet wide.

c. Equipment. The equipment required for this test includes—
o A platform, rail on a ledge, or a box 18 inches high,
o A 20-foot climbing rope,
o A stopwatch.

d. Conduct of Test. Adequate timers and other supervisory personnel must be provided by the company staff. All events should be conducted in a single session of one morning or afternoon. The events may be run in any sequence.

e. Uniform. The uniform for the physical readiness test is as follows: full utilities, cartridge belt, belt suspender straps, two canteens (full of water), rifle, and helmet. Depending on the heat condition, the PT shirt may be worn in place of the utility jacket. All equipment will remain secure throughout the test. Except for emergencies, canteens are not to be emptied.

f. Events
(1) Event 1: Climbing Uphill (Step-ups)This event simulates marching uphill at a rapid and steady pace. The Marine stands in front of the step. On the command GO, the Marine places either foot on the step and steps up. Hand pressure on the knee may be used if desired. The Marine then stands erect and steps backward and down, on one foot at a time. The same leg may be used for each repetition or the legs may be alternated. For 100 points, Marines will complete 60 repetitions in the maximum allotted time of 1 minute and 20 seconds (80 seconds).
*(2) Event 2: Rope Climb. T*his event resembles entering and leaving a hovering helicopter, using ropes in house-to-house fighting, and ascending and descending landing nets. The Marine assumes a sitting position on the ground at the bottom of the rope with hands at the highest point that can be reached. On the command GO, the Marine jumps up and climbs to the top. The Marine must touch the 20-foot mark with one hand, then descend without sliding or dropping. Marines must reach the 20-foot mark in the maximum allotted time of 30 seconds or less.
(3) Event 3: Evacuation. This event simulates reaching and evacuating a wounded Marine under fire. The Marine assumes the prone position at a distance of 50 yards from a casualty who is of approximately equal weight. On the command GO, the Marine springs up and covers the distance in a sprint. The Marine lifts the casualty to a fireman's carry and returns to the starting point. Assistance by the casualty is permitted. The tested individual must carry all equipment; the casualty has none. Marines must negotiate the course in a maximum time of 47 seconds or less.
(4) Event 4.'Advance by Fire and Maneuven. This event simulates advancing and assaulting during an attack. The Marine assumes the prone position at the starting point. On the command GO, the Marine then alternately crawls, runs in a zigzag manner, rolls, jumps, and assumes different firing positions for the next 130yards. Before reaching the objective, the Marine must "hit the deck," roll over, and assume the firing position three times. While assuming the firing position, the Marine must pull back the charging handle, take aim, and squeeze the trigger before rising from the ground. After rising from the ground for the third time,the Marine continues running and jumps to clear the 8-foot trench in one leap. The Marine then finishes the course in

the maximum allotted time of 2 minutes. The Marine is allowed multiple attempts only within the specified time limit.

(5) Event 5: Forced March. This event is a combat run of 3 miles without halts and against time. The Marine must double-time. Marines must complete the 3-mile course in a maximum allotted time of 30 minutes or less.

g. Event Standards

(1) To successfully pass the physical readiness test, the Marine must pass each event with a minimum of 80 points for each. Points are then assessed for each event in accordance with the scoring tables. (See fig. 6-1.)

(2) If a Marine fails one event, the highest possible score that the Marine may receive for the entire test is 79 percent. (See example 1 in fig. 6-2.) If two events are failed, the highest possible score for the entire test is 69 percent, regardless of the total points accumulated. (See example 2 in fig. 6-2.) If three events are failed, the highest possible score is 59 percent. Failure in four or more events will result in a score of zero for the entire test.

(3) To convert the total points for all five satisfactorily passed events to a final percentage score, total the accumulated points for all five events and divide by five. (See example 3 in fig. 6-2.)

Figure 6-1 is available in the illustration section at the end of the book.

Figure 6-2 is available in the illustration section at the end of the book.

6008. AIRBORNE TRAINEE PHYSICAL FITNESS TEST

a. Use and Composition of the Test.

(1) Use. The airborne trainee physical fitness test is a means of determining the physical ability of the applicant for acceptance to and retention in the airborne course of instruction.

(2) Test Events The test battery consists of five events as follows: chinups, knee bender, pushups, situps, and an endurance run.

b. Method of Scoring and Standards

(1) Scoring. The Marine will be scored by a trained scorer who is thoroughly familiar with the minimum standards for the test events. The Marine will be scored on a pass or fail basis. The performance on each event may be recorded on the reverse side of DA Form 705 (Physical Fitness Testing Record).The examinee records the personal information on the face of the card as specified on the first, second, and third lines. The face of the card is identified by writing diagonally across the lower half of the card "Airborne Trainee PPT—See reverse side."Using line 7 on the reverse side of the scorecard, the scorer enters the test title in the block entitled "(Other (Specify))" and completes the required information on the remainder of that line. The "Remarks" section can be used to record the test event titles and the applicant's performance on each test event.

(2) Standards. To successfully pass the test, the examinee must reach the standard in each test event. The standards follow:

(a) Chinups

(b) Knee bender 80(2-minute period)

(c) Pushups 45(2-minute period)

(d) Situps45(2-minute period)

(e) Endurance runs:2 miles in 15:54 minutes or less in athletic gear and 4 miles in 32 minutes in utilities and running shoes.

c. Uniform for Testing

(1) Examinees. The prescribed uniform for test participation is boots and the work uniform of the season. No headdress is worn. When climatic conditions permit, jackets or outer shirts may be removed.

(2) Officials. Scorers and other test officials should be uniformly and distinctively dressed for contrast with Marines being tested.

d. Test Administration Procedure

(1) Preparation for Test. The administration of the test to a large group makes it mandatory that the test effort be organized and efficiently operated. All testing is not completed with large test groups; on certain occasions, individuals and small groups are tested. Care must be exercised to administer the test uniformly and to standardize the conduct of all elements of the test. Regardless of the size of the test group, the following elements of sound test administration should be included:

(a) An orientation to include the purpose, method of administration, scoring of the test, preparation of the score card, and required standards.

(b) A correct demonstration of each event to ensure that there is no misunderstanding of the proper form and required standards.

(c) Completion of all five test events in one test period with all Marines taking the events in the same sequence.

(d) Adequate rest periods between the test events to allow for recovery before the next event.

(2) Method of Administration with a Large Group. With a minimum of 12 lanes per test event, 14 officials can administer the test battery to 150 or 200 Marines in 2 hours. The officials are designated as follows: one officer in charge, one demonstrator,and 12 scorers. If more or fewer Marines are to be tested, a greater or lesser number of officials will be required. The following procedure is recommended:

(a) Conduct an orientation and ensure the examinees have properly completed their score cards.

(b) Assign Marines to lanes and caution them to remain in the same lane order throughout the test.

(c) Explain and demonstrate the chinup event, administer it, and score it. Then proceed to the knee bender and pushup events, and administer them in the same manner.

(d) Grant a 5- to 10-minute rest period after the pushup event. Advise against excessive consumption of water during the break period.

(e) Explain and demonstrate the situp event, administer it, and score it as prescribed. Then move to the run area, explain the running event, and complete it.

(f) Retain the scorecards at the completion of the running event.

(3) Method of Administration with a Small Group. A similar procedure is followed for the testing of individuals and small groups. The informality usually associated with small groups must not conflict with sound test administration. With fewer examinees, fewer officials are required.

e. Description and Explanation of Test Events

(1) Test Event 1:Chinups

(a) Purpose. This event is devised to test arm and shoulder flexor strength.

(b)Equipment There is one horizontal bar per lane, made of plumber's pipe or a gymnasium horizontal bar 1 1/2 inches in outside diameter. The bar must be rigidly supported at a height of 8 feet above the ground. The upright supports must be 5 feet apart. There must be a movable stand at each bar for short Marines to stand on to reach the bar.

(c) Officials There is one scorer per lane.

(d) Organization. The examinees, holding their scorecards, stand in order behind the restraining line in their respective lanes. The scorers take each scorecard when the Marine is called forward for the test.

(e) Starting Position The bar is grasped with the palms turned toward the face, the thumbs underneath the bar. The body is fully extended in a "dead" hanging position with the arms straight and the feet above the ground.

(f) Movement. Pull the body directly upward until the chin is placed over the bar. Lower the body until the elbows are completely straight and the body is again in the "dead" hanging position. Repeat as many times as required.

(g) Instructions. Explain and demonstrate the fully extended "dead" hanging position with the proper grasp. Show that the chin is placed over the bar at the top of the movement and that the arms are fully extended, the elbows completely straight, at the bottom of the movement (the hanging position). Explain that the body must be kept from swinging and that it is permissible to raise the legs and flex the hips when pulling up, but that any kicking, bicycling, or jerking motion with the trunk or legs is not acceptable. Inform the examinees that no penalty is exacted for hanging on the bar to rest in the bottom position but that this is not to their advantage. Tell them that half-completed chinups are not counted, and that the scorer will repeat the number of the last correct chinup when incorrect execution is detected.

(h) Administration and Scoring. Caution the examinees to assume the "dead" hanging position and wait for the scorer's command to begin. The scorer is at the examinee's left with a clear view of the bar. If the examinee begins to swing widely, the scorer should stop the swinging by extending the left arm across the front of the examinee's body, being sure not to hinder the execution of the chinup. The scorer counts aloud the number of chinups correctly executed. When a chinup is not correctly executed, the scorer repeats the number of the last correct one. The scorer records the number of correct chinups on the scorecard and returns the card to the examinee.

(2) Test Event 2:Knee Bender

(a) Purpose. This event measures the strength and endurance of the leg muscles.

(b) Equipment. None.

(c) Officials There is one scorer per lane.

(d) Organization. Marines stand in numerical order behind the restraining line in their respective lanes. The scorer takes each scorecard when the Marine is called forward for the test.

(e) Starting Position. The feet are spread less than shoulder width apart, hands on hips, thumbs in the small of the back, elbows back.

(f) Movement. Do a knee bend and at the same time bend slightly forward at the waist and thrust the arms between the legs until the extended fingers touch the ground. The hands are about 6 inches apart. The bend is approximately a three-quarters bend. From this knee bend position, recover to the starting position by moving the body upward, straightening the knees, and returning the hands to the hips. Repeat as many times as required.

(g) Instructions. Explain and demonstrate the correct starting position. Be certain examinees understand the correct knee bend and that only the tips of the fingers touch the ground. Tell them the scorer will repeat the number of the last correct knee bender when incorrect execution is detected. Some of the common errors are failure to correctly bend the knees, failure to touch the ground, and failure to assume the completely erect position after the bend has been executed.

(h) Administration and Scoring. The scorer stands to one side so as to see that the knees are properly bent and the fingers touch the ground as prescribed. From this position, the scorer can view the examinee to see that a properly erect position is assumed after each knee bend. The scorer counts aloud the number of correctly executed knee bends. When a knee bend is done incorrectly, the scorer repeats the number of the last correct one. The scorer records the number of correct knee bends on the scorecard and returns it to the examinee.

(3) Test Event 3:Pushups

(a) Purpose. Pushups measure arm and shoulder extensor strength.

(b) Equipment. None.

(c) Official. There is one scorer per lane.

(d) Organization. Marines stand behind the restraining line in their respective lanes until the scorer calls on them to perform. The scorer takes each scorecard when the Marine comes forward.

(e) Starting Position. The front-leaning rest position is the starting position. The body is straight from head to heels, palms are flat on the ground directly underneath the shoulders, and elbows are straight and locked. The body weight is supported on the hands and toes throughout the event.

(f) Movement. Bending only the elbows, lower the body in one straight plane until the chest touches the scorer's hand. Straightening and locking the elbows, raise the body in one straight plane, returning to the original front-leaning rest position. Repeat as many times as required, keeping the body in a straight line from head to heels.

(g) Instructions. Explain and demonstrate that the arms are straight at the beginning and completion of the movement and that the chest must touch the scorer's hand, but the stomach and thighs must not touch the ground. Also explain that the whole body must be maintained in a straight line as it is lowered and raised; that is, there is to be no breaking at the hips or shoulders so that any body part is lowered or raised in advance of the other or as a separate segment .Dipping or rolling through the shoulders is illegal, as is lowering or raising the body with one arm or shoulder at any time. Resting is not permitted during repetitions. Instruct

Marines that the scorer will repeat the number of the last correct pushup when incorrect execution is detected.

(h) Administration and Scoring. It is recommended that Marines assume a prone position while placing their feet and hands in the proper positions. This permits them to rest while the scorer gets into position and, at the same time, provides a feel of the body in a straight plane from head to heels. The scorer lies on the right hip and side to the right of the examinee. This gives the scorer a clear view of the examinee's body to see any errors. The palm of the scorer's right hand rests flat on the ground underneath the lowest part of the examinee's chest. By keeping the right forearm flat on the ground at an angle in front of the examinee's right arm, the scorer's position will not prevent the examinee from lowering the body completely. The scorer's left hand is free to test the straightening of the elbow at the completion of the movements and to point out body segments being lowered or raised separately. When in position and ready, the scorer has the examinee assume the starting position and begin doing pushups. The scorer counts aloud the repetitions done correctly and repeats the number of the last correct pushup if one is done incorrectly. There is no penalty if the contour of the examinee's body causes the hips to protrude slightly out of line, provided that the whole body is raised and lowered simultaneously. The scorer enters the number of repetitions on the scorecard and returns it to the examinee.

(4) Test Event 4: Situps

(a) Purpose. This event primarily measures abdominal strength.

(b) Equipment. None.

(c) Official. There is one scorer per lane.

(d) Organization Marines stand behind the restraining line in their respective lanes until the scorer calls on them to perform. The scorer takes each scorecard when the Marine comes forward.

(e) Starting Position. The examinee lies flat on the back with knees flexed, both feet flat on the ground. The correct angle of the thighs to the ground is 45 degrees. If the heels are too near the buttocks, the applicant will not be able to sit up. Legs are spread shoulder width apart. The examinee interlaces fingers and places them behind the head in contact with the ground. The feet are not held by another person.

(f) Movement. Bend forward at the waist and raise the upper body until the head is directly over the knees. Heels are not to leave the ground. Elbows remain in the same plane to the head and body throughout the event. The upper body is slowly lowered to the starting position until the head touches the ground. Repetitions are done at a slow cadence with no rest periods.

(g) Instructions. Explain and demonstrate the correct starting position and the proper execution of the situps to be sure that Marines understand the movement. Warn them that their knees must remain flexed during each situp, the heels cannot leave the ground at any time, and they may not roll up on one side and push up with one elbow. Tell them they must do the repetitions at a slow cadence with no rest periods. Instruct Marines that the scorer will repeat the number of the last correct situp when incorrect execution is detected.

(h) Administration and Scoring. When the performer is in position and ready, the scorer has the performer assume the starting position and begin doing situps. The scorer counts aloud the correct executions. When a situp is improperly done, the scorer repeats the number of the last correct one. No situp is credited if the hands are unclasped from behind the head, if the back is used to bounce up from the ground (which means the shoulders would not touch the ground), or if one shoulder or elbow is used to push up. The examinee is not penalized if the heels slide forward slightly so long as the knees remain flexed and the heels maintain contact with the ground. The scorer enters the number of repetitions on the scorecard and returns it to the examinee.

(5) Test Event 5:Endurance Run

(a) Purpose. This event measures cardiovascular endurance.

(b) Equipment One stopwatch or watch with a sweep second hand.

(c) Area. A large training field on which a quarter-mile track has been staked out or a level road over flat terrain may be used as a running surface. A 1-mile route is designated with wooden stakes marking the start point, finish point, and one-quarter mile intervals.

(d) Officials For large groups, a scorer times the event and controls the conduct of the run, and a guide runs with the group and sets the pace.

(e) Organization The run is conducted with groups of Marines in a column formation. Company-sized units may run at the same time with platoons serving as running groups. The scorer issues the command to assume double-time.

(f) Starting Position. Marines are assembled in proper column formation (column of twos, threes, or fours, as appropriate to the size of the group) with short men to the front. When all is ready, the column is moved forward a short distance before the running period is started.

(g) Movement. At the command DOUBLE-TIME, MARCH, the examinees retain their places in the column formation and execute the command. Length of steps is about 40 inches. The scorer has the group execute the run. The formation is maintained during therun.

(h) Instructions. Marines are instructed to maintain formation while running and are informed that the guide will set the proper pace. They will be instructed in the command used to control the column in the execution of the test. The scorer should announce the 4-minute, 2-minute, 1-minute, and 30-second remaining time intervals.

(i) Administration and Scoring. The event may be administered as previously prescribed to a large group, to several Marines, or to an individual. An individual examinee usually does not require a guide or pacer. If the event is administered on a training field, the scorer may stand in the center as the runner(s) circles about the field. This method of administration relieves the scorer of running with each group to be tested. Scoring is based on successful completion of the run as prescribed. The scorer should announce the remaining times as prescribed in (h).

6009. BATTLE FITNESS TEST

The battle fitness test is a physical fitness evaluation which is currently under development as part of the basic warrior training concept plan.

Chapter 7
THE HUMAN BODY

7001.GENERAL

To implement an effective physical conditioning program, a leader must understand how the human body functions and how exercise affects the body .This chapter provides only an introductory discussion of anatomy and functioning.

a. Body Functioning During the Stages of Conditioning. For more on the stages of conditioning, see chapter 1.

(1) During the toughening stage, the waste products of muscle activity (lactic acids) collect more rapidly than the blood can remove them. This acid waste builds up in the muscle tissue and irritates the nerves in the muscle fiber, causing pain and stiffness. As the exercise program continues, more blood is carried through the muscle, removing the waste materials more rapidly and eventually causing the soreness to disappear.

(2) During the slow improvement stage, the blood circulation in the muscles increases, and the body as a whole becomes a more efficient machine. The improvement is rapid in the first few weeks, but as a higher level of skill and conditioning is reached, the improvement becomes less noticeable. The body reaches its maximum level of performance after 6 to 10 weeks and should then be maintained at this peak.

(3) During the sustaining stage, it is possible to maintain this state of conditioning through 15 to 20 minutes of exercise a day, but the exercise must be quite strenuous.

b. Diet. Regular exercise increases the appetite. If the desire for greater amounts of food is satisfied by a balanced diet, the body benefits.

(1) There are two main types of foods: body-building and energy- producing.

(a) Body-building foods consist of proteins, which build up tissue and repair wear and tear.

(b) Energy-producing foods include carbohydrates and fats. Carbohydrates provide a quick source of energy, while fats act as a reserve store of energy.

(2) In addition, food contains vitamins, mineral salts, and water. During hot weather and strenuous training periods, the body requires greater fluid intake.

(3) Diet should be supplemented with proper rest to allow the digestive system to digest the lood.

(4) Occasionally, especially during early stages of conditioning, strenuous exercise may cause vomiting. Although vomiting is not a frequent occurrence, it is not usually a cause for concern.

7002. SYSTEMS OF THE BODY

The systems of the body include the skeletal, muscular, circulatory, respiratory, endocrine, digestive, genitourinary, and nervous systems. Each has a different function, but all must work in cooperation with one another to insure a sound body. Of these systems, the first four are the most affected by exercise and are discussed in succeeding paragraphs.

7003. THE SKELETON

a. Bones. The skeleton (see fig. 7-1) is composed of about 206 bones, which-

o Provide structure for the body and support for the attachment of muscles,

o Protect vital organs, such as the brain, lungs, and heart,

o Manufacture red blood cells, which carry oxygen through the body.

Figure 7-1 is available in the illustration section at the end of the book.

b. Joints. A joint is a place of union between two or more bones. Joints can be-

o Immovable, such as in the face and head.

o Slightly movable, such as in the vertebrae or sternum.

o Freely movable, such as in the shoulder, hip, knee, ankle, and elbow. The bones in a freely movable joint are held in place by ligaments. Freely movable joints are of the greatest importance in physical training because exercise improves their mobility and stability.

c. Cartilage and Ligaments. The joints in the body, where bones connect, are supported by cartilage and ligaments.

o Cartilage is a tough, elastic, translucent tissue that acts as a shock absorber or buffer between bones. Examples are the discs between the vertebrae, the tissue attaching the ribs to the sternum, and buffers in the knee joints. Damaged cartilage does not heal.

o Ligaments are connective tissue that bind bones together. They contract and extend but are not elastic. Consequently, ligaments can be torn or strained. Damaged ligaments will heal to some extent.

7004. THE MUSCLES

a. General. Muscles are tissue, or an organ composed of tissue, which contract or extend to produce movement. Muscles are of three classifications: involuntary, voluntary, and cardiac. The voluntary and cardiac muscles are the most affected by physical training; physical training has little effect on involuntary muscles.(Seefigs. 7-2, 7-3, 7-4.)

o People have no control over the involuntary muscles, an example of which is the diaphragm,

o Voluntary muscles are the larger skeletal muscles which cause movement and which the individual can control.

o Cardiac muscle is found only in the heart and for all practical purposes is an involuntary muscle. Physical training has a significant effect on the fitness of cardiac muscles.

b. Muscle Structure. Muscle is composed of cells which are grouped into bundles called fibers. Groups of fibers form muscles. These bundles of fibers are held in place by sarcolemma, a thin, sheath-like material that surrounds the muscle bundles and secretes a fluid that lubricates the muscle tissue. The fused ends of the sarcolemma form the tendons which attach the muscles to the bones.

c. Attachment of Muscles. The arrangement of muscles on the skeleton provides the proper angle of pull to make movement possible. Voluntary muscles are usually attached to the skeleton in two places.

d.Action of Muscles. To produce motion and to do work a muscle usually shortens its fibers. The movement may be flexion such as bending the arm at the elbow, or extension, such as the straightening the arm. Some muscles raise the arms or legs, others depress the raised limb. Some muscles have the primary function of rotating the trunk from side to side. In addition, muscles act as stabilizers as well as prime movers.

Figure 7-2 is available in the illustration section at the end of the book.

Figure 7-3 is available in the illustration section at the end of the book.

Figure 7-4 is available in the illustration section at the end of the book.

7005. FUNCTIONING OF THE SKELETON AND MUSCLES

a. General. The skeleton and muscles function in close coordination. Muscles move the body, and bones support the muscles.

b. Effect of Exercise on Bones. Continuous exercise, particularly among younger people, usually strengthens the bones, allowing them to withstand greater stress and strain. Bones which are not used lose much of their strength, a factor which should be considered when individuals return to physical training after a prolonged period of inactivity. People in this category should refrain from activities which might result in bone breakage before the bone is exercised back to normal condition. The condition known as "march fracture" is usually a result of this condition.

c. Muscular Strength. When exercised vigorously to improve strength, muscle grows in size. In general, the larger the muscle, the stronger the muscle. Furthermore, trained muscle functions more efficiently than untrained muscle. Trained muscles contract more vigorously and with less effort. Regular and strenuous exercise also toughens muscle, making it firmer and able to stand more strain.

d. **Muscular Endurance.** Muscles naturally become fatigued during continuous, repeated work. Through specific exercises, such as chinups or situps, local muscle groups become exhausted long before an individual fatigues. Training to develop muscular endurance enables people to continue a relatively heavy load of exercise over a long period of time. Lengthened exercise periods improve muscular endurance.

e. Muscular Coordination. Improved speed and strength result in part from improved muscular coordination. An unskilled individual may use irrelevant muscles to perform a particular activity, increasing the amount of work without increasing the mechanical output. This increase in skill is highly desirable, but it should be offset by greater effort in duration or intensity to compensate for the loss in overload due to increased skill.

f. Muscular Fatigue. When the rate of work is increased, the energy required is proportionately much greater than the increase in rate. For example, if an individual doubles running speed, the amount of power demanded to do this is increased eight times.

g. Circulation in Muscles Regular, strenuous exercise causes the creation of new capillaries and the opening of inactive, latent capillaries, increasing blood circulation within the muscle by as much as 400 percent. This increases the supply of food materials and oxygen to the muscle, improving its endurance. This process takes about 8 to 12 weeks of regular conditioning in young adults and longer as age increases.

7006. THE CIRCULATORY AND RESPIRATORY SYSTEMS

a. The Circulatory System. The circulatory system transports blood to all parts of the body, removes waste products for disposal, and delivers protecting and repairing substances where needed. The circulatory system includes the heart and blood vessels.

(1) The Heart. The heart is a pump which forces blood through the blood vessels to the parts of the body. The heart is a little larger than a fist and is located in the left center of the chest between the lungs.

(2) Blood Vessels. The vessels carrying blood away from the heart are the arteries. They eventually divide into capillaries, the very small vessels through which diffusion and osmosis take place. The capillaries gradually increase in size, forming the veins which carry blood back to the heart.

b. The Respiratory System. The respiratory system performs the function of breathing, whereby oxygen is inhaled and carbon dioxide is exhaled. The respiratory system consists of the mouth, nose, trachea, lungs, and diaphragm.

o Trachea. The trachea, or windpipe, is a hollow, tube-like structure that carries air from the mouth to the lungs.

o Lungs. The lungs are elastic bags in the chest where the exchange of oxygen and carbon dioxide takes place.

o Diaphragm. The diaphragm is a thin, sheet-like muscle just below the lungs. During inhalation, the diaphragm flattens out and lowers, allowing the lungs to expand and fill with air. During exhalation, the diaphragm raises into a dome shape, helping to expel air from the lungs.

7007. CARDIOVASCULAR FUNCTIONING

a. General. Cardiovascular functioning is the combined functioning of the circulatory and respiratory systems. The chief organs of these systems, the heart and lungs, function together to provide oxygen to the body.(See fig. 7-5.)

Figure 7-5 is available in the illustration section at the end of the book.

b. Heart Action. The heart, the organ which propels blood through the blood vessels, is the chief organ of cardiovascular endurance. A person tires quickly if the heart lacks the capacity to circulate the blood through the body. The heart is a muscular organ and is developed by exercise just like any other muscle. The heart cannot be exercised alone; any exercise that puts a sufficient load on the heart, such as running or cycling, will also exercise other body parts. A conditioned heart tends to beat more slowly but pump more blood when at rest. This is known as an increase in "stroke volume,"a desirable condition because it enables the heart to pump more blood with a slower contraction rate.

c. Functions of the Lungs. Strenuous and regular exercise can improve the efficiency with which the lungs transmit oxygen to the blood by as much as 25 percent In a poorly conditioned person, some of the alveoli (air sacs) within the lungs are closed or collapsed. The forced breathing created by exercise over a period of weeks causes the air sacs to slowly expand, increasing the ability to absorb oxygen.

d. Relationship of Heart and Lungs. The heart pumps carbon dioxide-laden blood through the pulmonary artery to the lungs. In the lungs the carbon dioxide is exchanged for oxygen and the purified blood is returned to the heart by way of the pulmonary vein. The heart then pumps the blood through the aorta for circulation throughout the body. Blood moving into the muscles exchanges oxygen for carbon dioxide, after which time it is ready for the return trip to the heart.

e. Cardiovascular Functioning in High Altitudes. Marines to be employed in high altitudes should be acclimated in a similar area for 10 to 14 days prior to employment. Persons not accustomed to the rarefied air of higher altitudes tire more quickly and may collapse after rapid physical exertion. Air is much less dense at high altitudes than at sea level. Persons accustomed to sea level inhale only about 80 percent of the oxygen they are used to no matter how hard they breathe. Further, people accustomed to sea level or moderate altitude simply do not have enough red corpuscles in the blood to fulfill their needs at high altitudes. The red blood cells will increase over time.

f. Symptoms of Cardiovascular Malfunctioning. The brain is the first organ to react to a lack of oxygen. Unconsciousness results when the brain is denied sufficient oxygen. "Blacking out" is actually a defense mechanism to keep the body alive.

APPENDIX A
INSTRUCTOR HINTS AND INSTRUCTOR TRAINING

1. General. This appendix informs instructors on administration of exercise periods, commands, the extended rectangular and circle formations, methods of instruction and conduct of exercise activities, explanation of basic positions associated with the various activities, and instructor training.

2. Aim of the Physical Fitness Program. The aim of the physical fitness program is to prepare Marines physically for war. This can be conveniently split into two phases.

a. Preparation Phase

(1) Development of all-around physical fitness.

(2) Development of purposeful physical skills.

(3) Development of mental alertness.

(4) Development of character and leadership.

b. Maintenance Phase. Once the preparation is completed, the program must progress to a maintenance program. This phase revolves around the need to maintain the trained Marine's physical fitness for a particular role. This will include—

(1) Maintenance of a high standard of all-around physical fitness.

(2) Application of purposeful physical skills to combat skills.

(3) Development of leadership and self-confidence.

(4) Development of mental toughness and endurance.

3. Physical and Mental Development

a. Marines need to be developed physically and mentally. The physical training (PT) instructor must ensure that the program accomplishes this.

b. The system of the body functions best in response to progressive exercise. This means that Marines should master the easier exercises before they attempt the more difficult ones.

c. The body and mind are linked inseparably and must be considered as one in physical fitness training. Physical exercises assist in developing the mind.

d. The confidence in oneself and in one's comrades attained by physical achievement helps to promote the indomitable will to win. The Marine must be taught to realize this. It will help during arduous training, which at all times must be imaginative, realistic, and filled with enthusiasm. All physical fitness training must be associated with developing the qualities required of a Marine.

4. Planning and Preparation of ProgramsThe following factors must be considered when compiling a program (see also ch. 2)~

a. Aim or desired skill level.

b. Duration of the complete training program and number of physical training lessons.

c. Purpose of included exercises,drills, or practical applications.

d. Beginning/inventory fitness levels.

e. Medical facilities, medical personnel, and Marine Corps or local command regulations governing their proximity to the training.
f. Number of Marines to be trained.
g. Availability of instructors, assistants, and facilities.
h. Apparatus and equipment available.
i. Alternative training areas for varying weather conditions,
j. Stages of progression and tests of achievement.

5. Five Basic Principles of Program Design
a. Regularity. Regularity of exercise is far more important than the amount of exercise performed on a given day. A Marine should exercise a minimum of three times per week. Human muscle begins to atrophy after 3 days without exercise.
b. Balance. The program must be balanced in terms of total physical fitness. Imbalance is created when, for example, an individual merely runs. While the cardiovascular system is improved, general upper body strength may be lacking. Conversely, a person who just does weight training may develop strength, but not cardiovascular endurance.
c. Overload. If total body fitness is to be developed, then the muscular and cardiovascular systems must be made to perform more work than previously. This is done by increasing the frequency, intensity, and duration of the exercise. The body adapts to the amount of stress placed upon it and becomes stronger and more efficient.
d. Progression. To be successful, any type of training program must be progressive. This is an area where untrained persons conducting PT can cause injury to Marines resulting in failure of the PT program. Close supervision must be made of untrained instructors. The result of progressing too fast is that the class begins to reject PT because it is no longer enjoyable or motivating. The program should start at a slow pace, and the pace should increase as proficiency improves. Average persons will encounter three stages of progressiom the toughening stage, the slow improvement stage, and the sustaining stage. (See ch. 1.)
e. Variety. Some programs fail because the training becomes boring. Variety is essential and perhaps presents the greatest challenge to the commander and staff.

6. Grouping. Providing for different levels of physical fitness is particularly recommended in the early stages of conditioning. Older Marines and those in poorer physical condition should be expected to attain a group level of fitness, but they should be given more time to do it.
a. Homogeneous Grouping One simple method of providing for the difference in levels is to group them according to their condition. A two-group classification would divide Marines into highly conditioned and average groups. A finer classification could be obtained by dividing them into three groups—a highly conditioned, a moderately conditioned, and an unconditioned group.
b. Performance or Age Grouping. The segregation of Marines into different exercise groups should be based on physical fitness test scores or on the level of fitness they demonstrate. They may also be grouped at first according to age. A common classification by ages is under 30, 30 to 34, and 35 and above.

7. Preparatory Commands and Commands of Execution. The prepreparatory command describes and specifies what is required. The command of execution calls into action what has been prescribed. All preparatory commands are given with a rising inflection. The interval between commands is long enough to permit the average Marine to understand the first one before the second one is given.

8. Extended Rectangular Formation. The traditional formation for carrying on many physical training activities is the extended rectangular formation (fig. A-1). This formation is more compact than any other. It is the best type to use for large numbers of Marines because it is simple and easy to assume.

a. For the formation of oneplatoon, the base man paces off five paces from the stand, faces left and moves five paces, halts, and again faces left. With the base man positioned facing the stand, the platoon leader then commands. **FALL OUT AND FALL IN ON THE BASE MAN**. At this command, all Marines run to the designated area and reform. This procedure is preferred to marching the unit into position. If more control is desired, the unit may march at double time to the vicinity of the base man and then be directed to fall out and fall in on him. Much valuable time is wasted in the field by needless maneuvering of Marines at quick time in an effort to position the platoon or unit on the exact spot for the exercises.

b. A company-size unit assumes the extended rectangular formation from a column of threes or fours at normal intervals between squads. This extension can also be executed from a company mass without interval between platoons. In extending either a platoon or company-size unit, take your place at the head of the column and command:
(1) TAKE INTERVAL TO THE LEFT, MARCH. At this command, the Marines in the right flank file stand fast with arms extended sideward. All others turn to the left and run forward at double time. After taking a sufficient number of steps, all Marines face the front with both arms extended sideward. The distance between finger-tips is about 12 inches, and dress is right.

Figure A-1 is available in the illustration section at the end of the book.

(2) ARMS DOWNWARD, MOVE. At this command, the arms are lowered smartly to the sides.
(3) LEFT, FACE
(4) EXTEND TO THE LEFT, MARCH. At this command, those in the right flank file stand fast with arms extended sideward. All others turn to the left and run forward at double time. Spacing is the same as in (1) above, and dress is right.
(5) ARMS DOWNWARD, MOVE. Same as in (2) above.
(6) RIGHT, FACE.
(7) FROM FRONT TO REAR, COUNT OFF. At this command, the leading Marine in each column turns the head to the right rear, calls of **ONE** and faces the front. Successive Marines in each column call off in turn **TWO, THREE, FOUR, FIVE** in the same manner.
(8) EVEN NUMBERS TO THE LEFT, UNCOVER. At this command, each even-numbered Marine stride jumps sideward to the left, squarely in the center of the interval. In doing this, each one swings the left leg sideward, jumps from the right foot to the left foot, and smartly brings the right into position against the left.

c. To assemble the unit, you commandASSEMBLE TO THE RIGHT, MARCHAt this command, all return to their original position in the column at double time and reform on the base man.

d.It is recommended that the area for grounding equipment and arms be at the edge of, or well away from, the area to be used for exercising. To conserve time and ensure proper position of the unit, the base man or, if the unit is composed of several platoon-sizegroups, the various base men may precede the unit and establish their positions in relation to the instructor's stand.

9. Circle Formation.The circle formation is effective for conducting various exercise activities (fig. A-2). This formation has an advantage in that supervision of all Marines is facilitated and a moving formation is available which provides control. Guerrilla exercises, grass drills, and some forms of running are examples of activities which are more easily conducted in the circle formation than in the extended rectangular formation.

a. It is not advisable to have more than 60 Marines in a circle. When more must be accommodated, separate circles should be used.Concentric circles may be employed to reduce the size of the circle or to accommodate more persons. If concentric circles are formed, the different circles are made by designating squads for each circle.Each additional circle requires more Marines than the one inside it.For example, one squad of a platoon may form the inner circle and theremaimng three squads the outer circle.

b. When a platoon is to form a circle, the commands ai€IRCLEFORMATION, MARCH, FOLLOW MEUpon this command, the right flank squad of the column moves forward at double time with the leader of the platoon group gradually forming a circle in a counterclockwise direction. Each succeeding file falls in behind that on the right. After the rough outline of the circle is formed, the leader commandsPICK UP A 5-YARD INTERVAL.

Figure A-2 is available in the illustration section at the end of the book.

c.The group may be halted and faced toward the center, or, if instruction is not necessary, the exercise activity may be executed without stopping the platoon.

lO.Leadership Techniques

a. Unless you experience all the exercises, you cannot appreciate how arduous they are, what movements are most strenuous and difficult, where the errors in performance are likely to occur, and what the proper cadence should be.

b. You must give everyone careful supervision and participate in the exercises to show that you can do them. When you participate, your assistant instructors should supervise because it is difficult for you to supervise and exercise at the same time.

c. Marines should never be kept too long in one position, especially a constrained one. They should never have to perform an exercise more times than they can do it without losing the proper form. You should insist upon proper form in the execution of all exercise activities.Even slight deviations from the proper form reduce the value of the exercise.

d. Avoid long explanations.As a rule, it should be necessary to give a full explanation of new exercises only. Explain the most essential features of an exercise first; add details later. Too many details at one time are more likely to confuse than to assist them.Minor corrections should be made to the entire class while the exercise is in progress (for exampHEADS **UP,** KNEES STRAIGHT). If necessary, follow this direction by the name of the person who is particularly at fault. If a Marine requires special attention, givethat person separate instruction by an assistant instructor to avoidwasting the time of the group.

e. The heavy demand on your voice can be lightened by training assistant instructors to assume some of the instruction. Using mass cadence is also an effective method of lessening the demand on your voice.

“Illustrations”

DAY	TIME	ACTIVITY
MONDAY	60 MIN	RUNNING ACTIVITY
TUESDAY	60 MIN	COMBATIVES
WEDNESDAY	60 MIN	CONDITIONING & LOG DRILLS
THURSDAY	60 MIN	CONDITIONING & RELAYS
FRIDAY	60 MIN	RUNNING & CONDITIONING
SATURDAY	60 MIN	RUNNING ACTIVITY

Figure 2-1. Sample A- -Single Activity.

DAY	TIME	ACTIVITY	REMARKS
MONDAY	20 MIN 20 MIN 20 MIN	RIFLE DRILL COMBATIVES RUNNING	PROGRESSIVE ACTIVITY SYSTEM
TUESDAY	30 MIN	GUERRILLA EXERCISES TABLE I	SINGLE ACTIVITY
WEDNESDAY	30 MIN	RIFLE DRILL	SINGLE ACTIVITY
THURSDAY	30 MIN	CIRCUIT INTERVAL TRAINING	SINGLE ACTIVITY
FRIDAY	20 MIN 20 MIN 20 MIN	STRENGTH CIRCUIT CONDITIONING DRILL CROSS-COUNTRY RUN	ROTATION ACTIVITY
SATURDAY	60 MIN	WARM UP RUN AND GRASS DRILL	SINGLE ACTIVITY

Figure 2-2. Sample B- -Various Time Periods, Activities, and Systems.

DAY	TIME	ACTIVITY
MONDAY	45 MIN	RUNNING – CROSS-COUNTRY
TUESDAY	75 MIN	TEAM ATHLETICS – TOUCH FOOTBALL
WEDNESDAY	45 MIN	RUNNING – CROSS-COUNTRY
THURSDAY	20 MIN 40 MIN	LOG DRILL DUAL COMBATIVES
FRIDAY	90 MIN	SPEED MARCH
SATURDAY	90 MIN	TEAM CONTESTS

Figure 2-3. Sample C--Sustaining Activities.

Route	Kilometers from SP	Recommended rate of march (kmph)	Remarks
SP: RJ 8th Div Rd—Superhighway	......	4	Bridge; hard surface; two guides.
RJ Jamestown—Hersey Rd	2.25	4	Traffic heavy; two guides; Jamestown Rd bears to right.
RJ Jamestown—Yankee Rd	7.10	4	Two guides
RJ Jamestown—Lightning Rd	10.50	4	Two guides
RJ Jamestown—Sunshine Rd	12.40	4	Light traffic; one guide Sunshine Rd; gravel, poor traction when wet.
RP; RJ Sunshine—Sedan Rd	14.80	4	Two guides

Figure 3-1.Example of Route Reconnaissance Report.

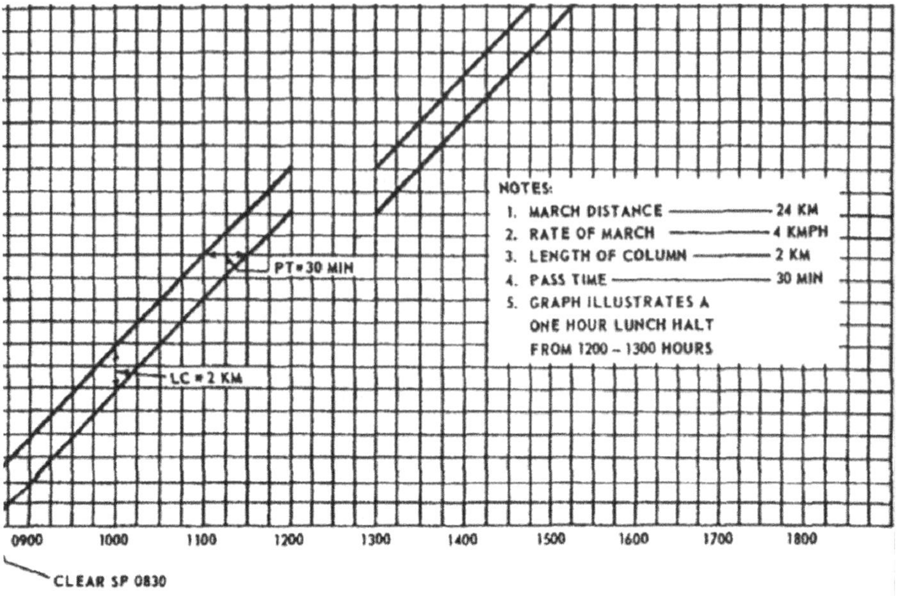

Figure 3-2.Road Movement Graph.

Figure 3-3.The Pace-Setter.

	SINGLE FILE	COLUMN OF TWO
2 m/MAN APART	2 4	1 2
5 m/MAN APART	5 4	2 7

GAP DISTANCES			
SERIAL:	(3-1) × 100 m	=	200 m
MARCH UNIT:	(12-1-2-) × 50 m	=	450 m
	TOTAL COLUMN GAP		650 m

Figure 3-4.Conditioning Drill One.

Figure 3-5.Conditioning Drill Two.

Figure 3-6.Conditioning Drill Three.

Figure 3-7. Rifle Drill.

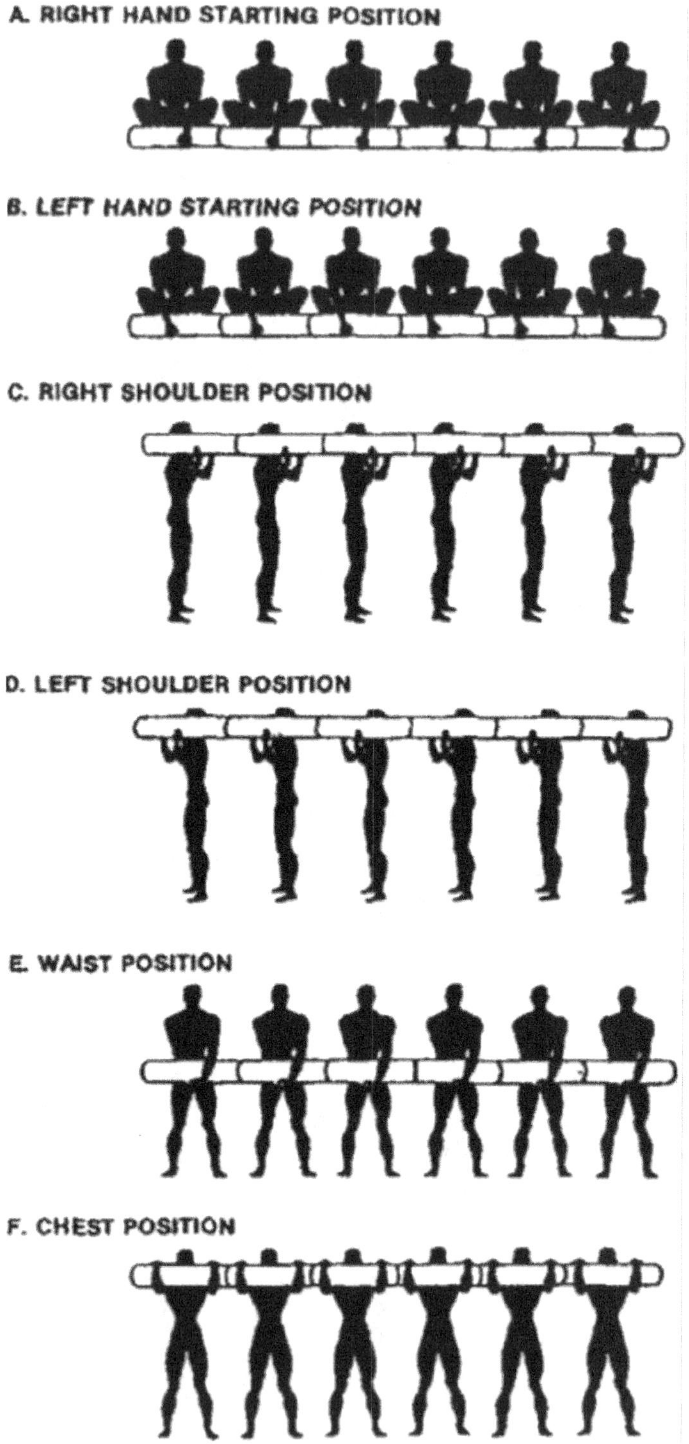

Figure 3-8. Starting Positions.

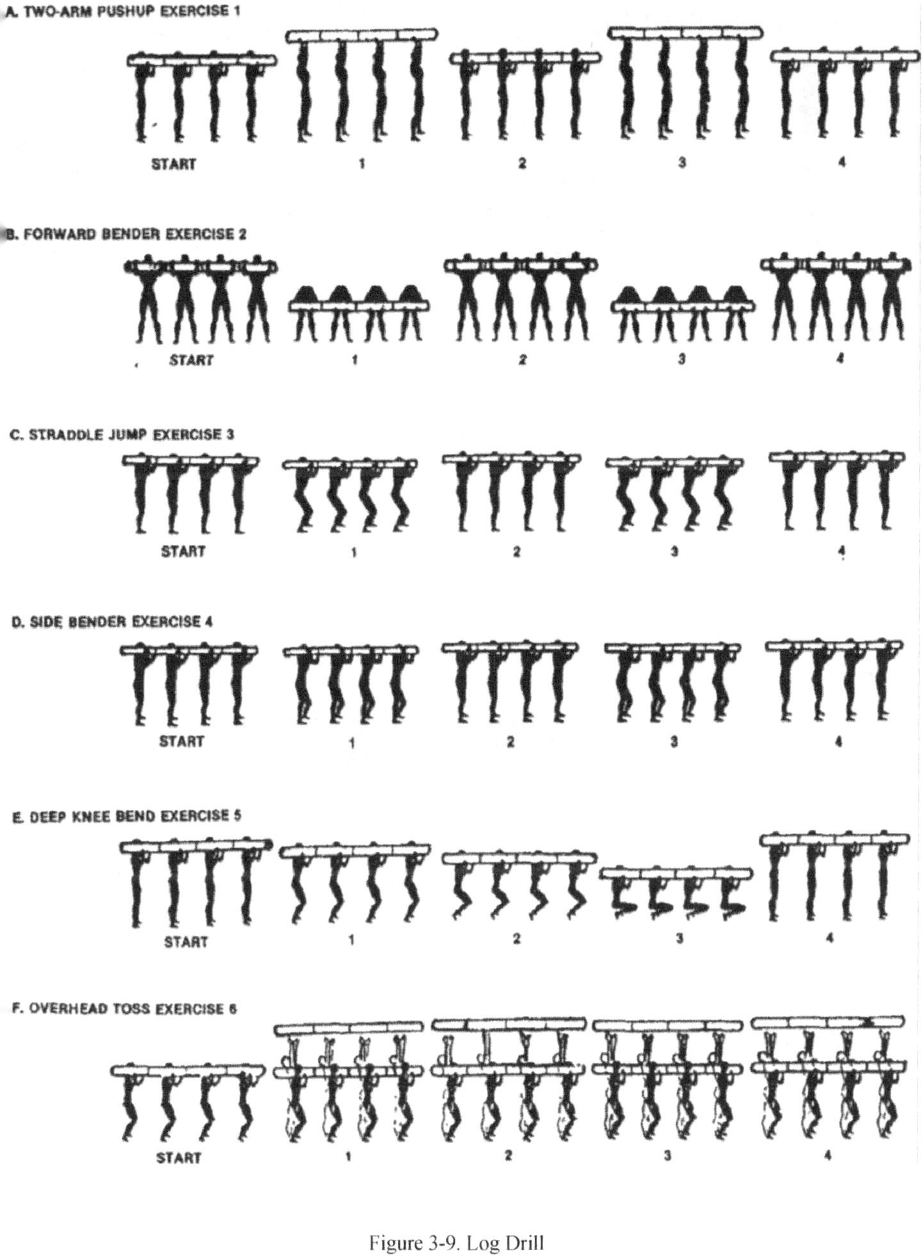

Figure 3-9. Log Drill

Figure 3-10.Basic Positions for Grass Drill.

Figure 3-11.Grass Drill One.

Figure 3-12.Grass Drill Two.

Figure 3-13.Guerrilla Table I.

Figure 3-14.Guerrilla Table II.

Figure 3-15.Proper Running Form.

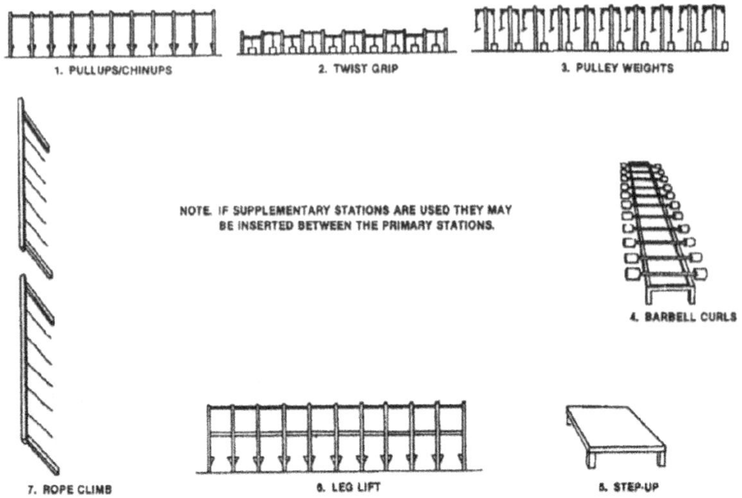

Figure 3-16. Fixed Strength Circuit.

STATION	ITEM	NO.	SPECIFICATION
1	BARBELL		1¼-INCH PIPE 5 FEET LONG WITH CONCRETE FILLED NO. 10 CANS
2	JUMP ROPE		¼- OR ⅜-INCH ROPE, 10 FEET LONG
3	TWIST GRIP		HANDLE 12 INCHES LONG, ROPE 4 FEET LONG, NO. 10 CAN CONCRETE FILLED.
4	INCLINE PLANE		¾-INCH PLYWOOD PLATFORM 2 FEET WIDE AND 6 FEET, 6 INCHES LONG, ELEVATED 10 INCHES AT ONE END. STRAP TO HOLD FEET DOWN.
5	WAR CLUB		HEAD IS 6 BY 12 INCHES, HANDLE IS 14 INCHES LONG BY 1¼ INCHES IN DIAMETER, ABOUT 20 POUNDS.
6	BICYCLE RIDE		PLYWOOD BOARD OR PLATFORM 2 BY 3 FEET WITH 2 BY 2 RUNNERS.
7	STEP – UP		A BOX OR STURDY PLATFORM 18 INCHES HIGH, 18 INCHES WIDE, 24 INCHES LONG.
8	ISOMETRIC PULL		TWO HANDLES 12 INCHES LONG WITH 4 FEET (BETWEEN HANDLES) OR LIGHT WIRE CABLE OR ¼-INCH ROPE

Figure 3-17.Movable Strength Circuit.

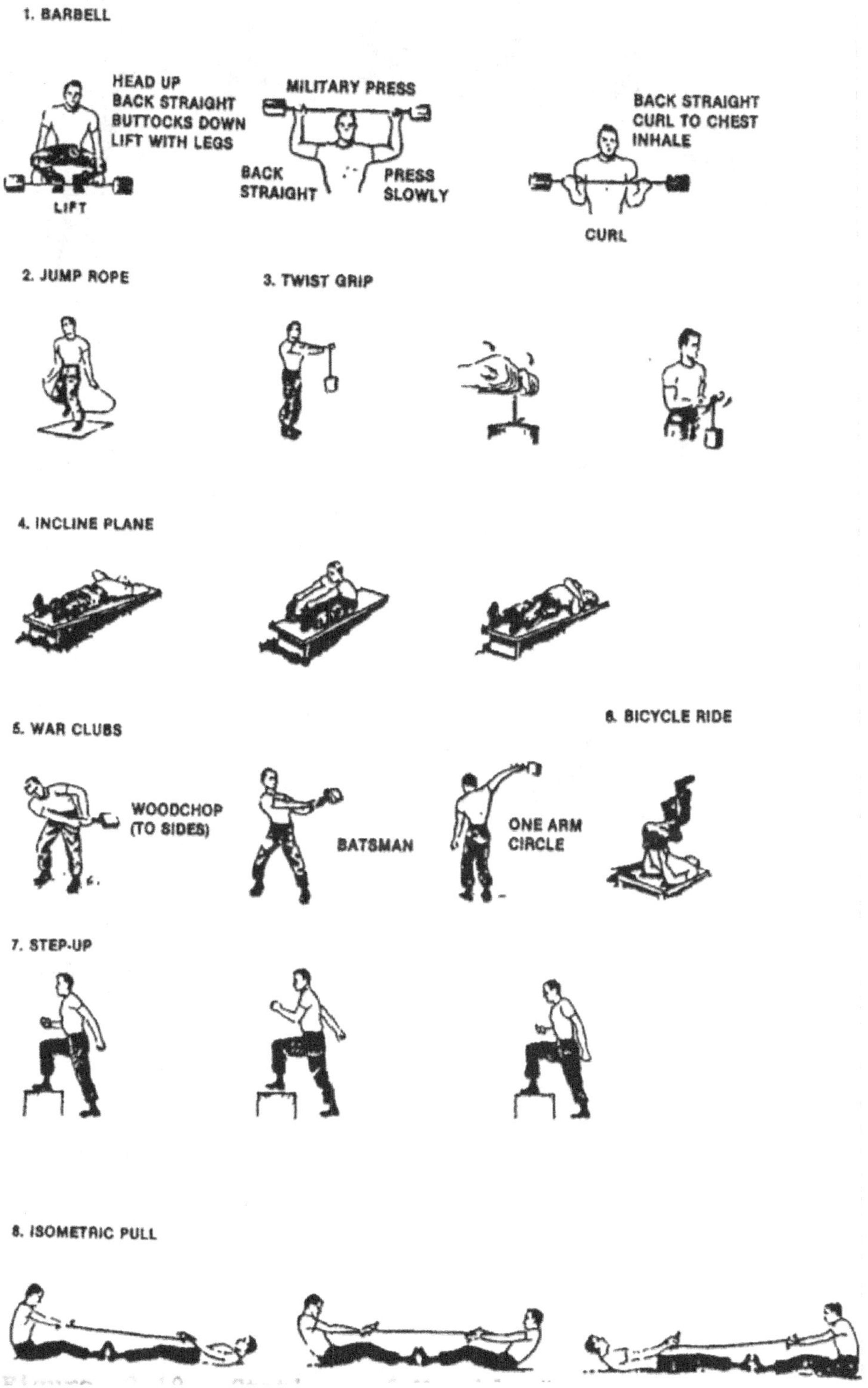

Figure. 3-18.Stations of Movable Strength Circuit.

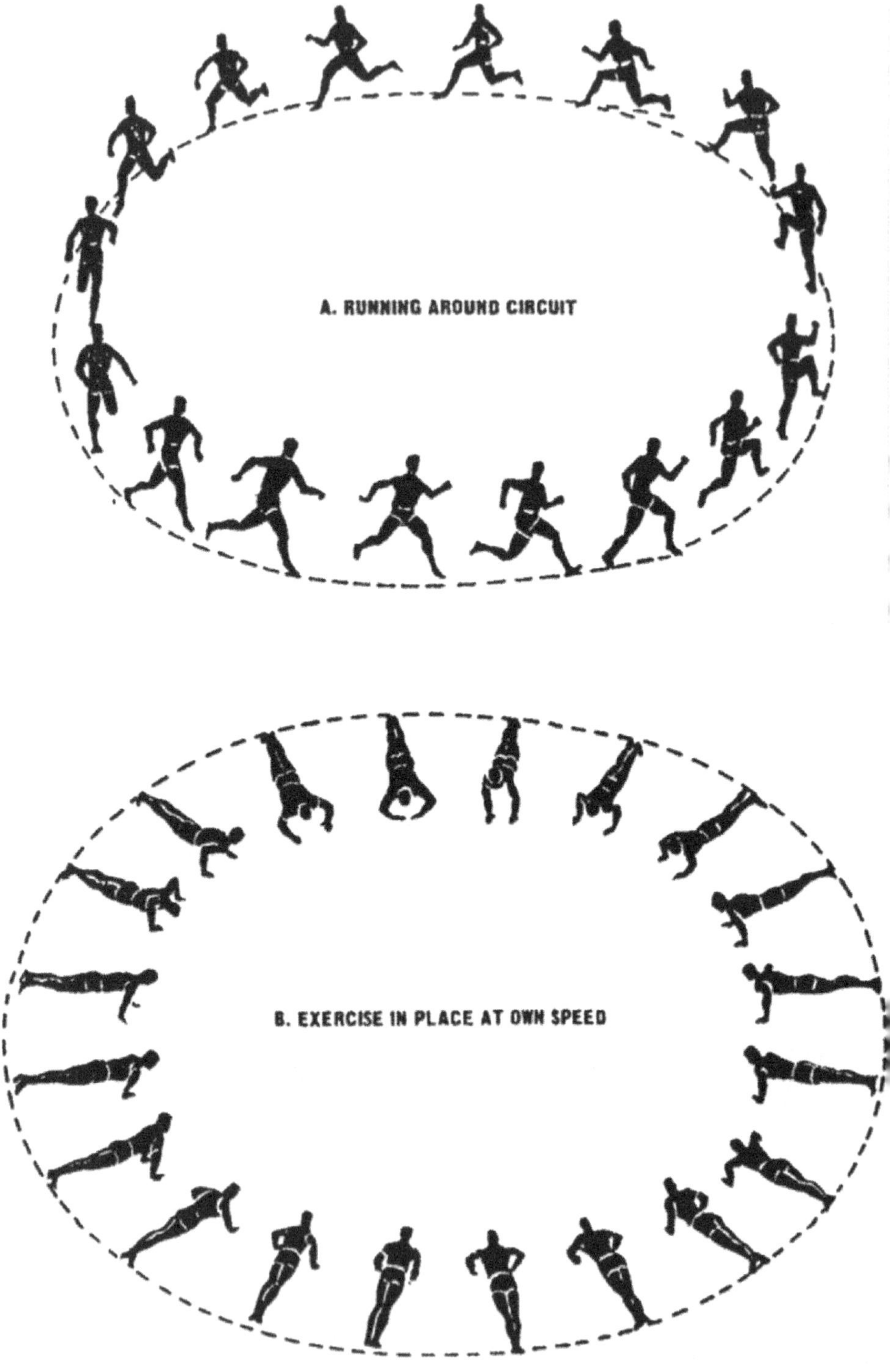

Figure 3-19. Circuit-IntervalTable.

Figure 3-20. Jumping.

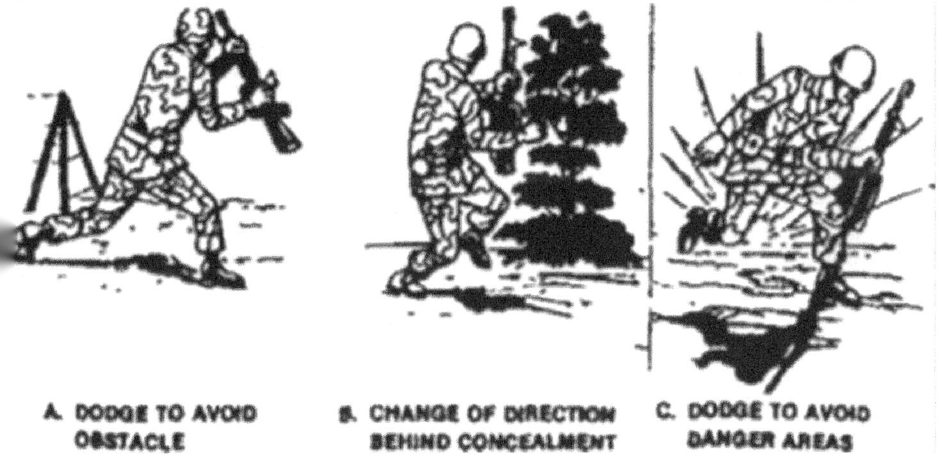

Figure 3-21.Dodging.

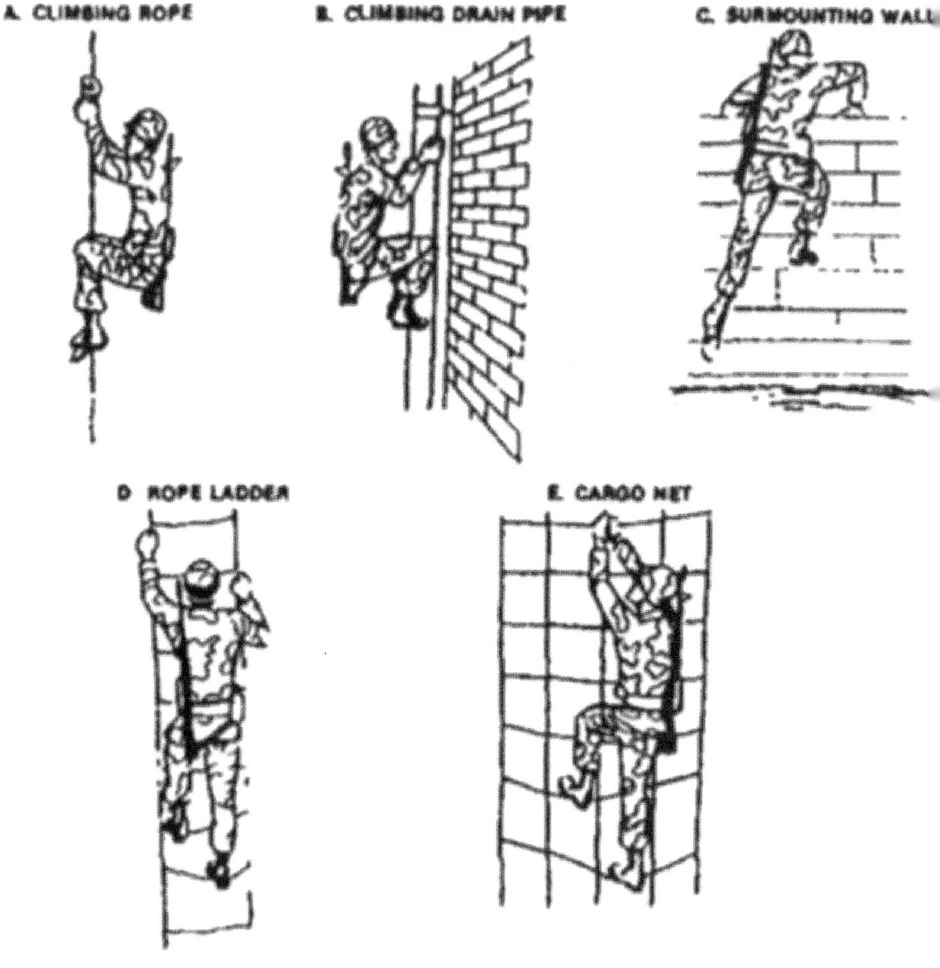

Figure 3-22.Climbing and Surmounting Vertical Objects.

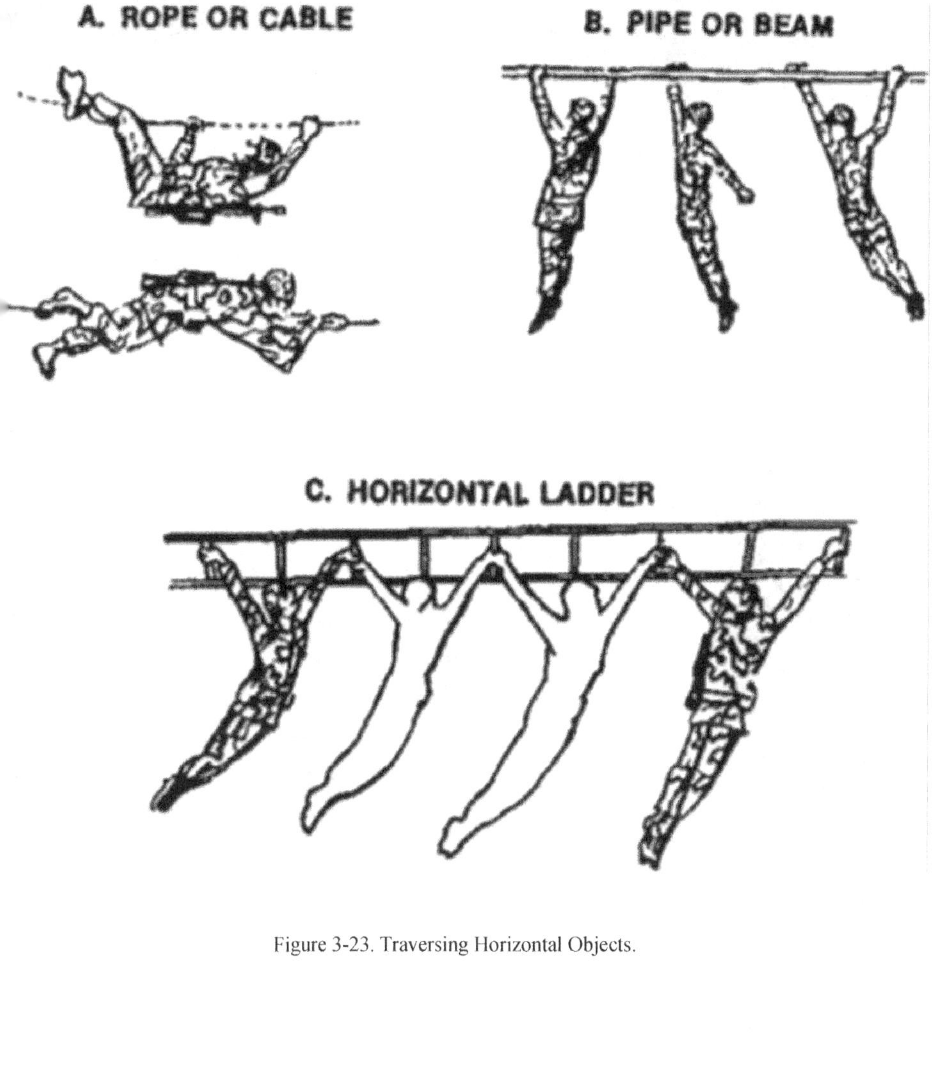

Figure 3-23. Traversing Horizontal Objects.

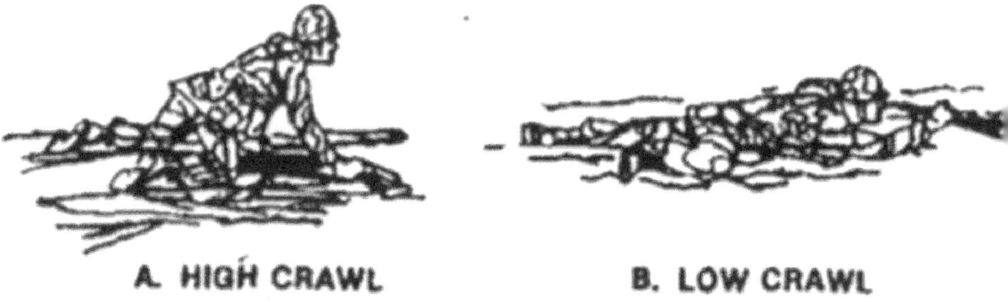

Figure 3-24.Crawling.

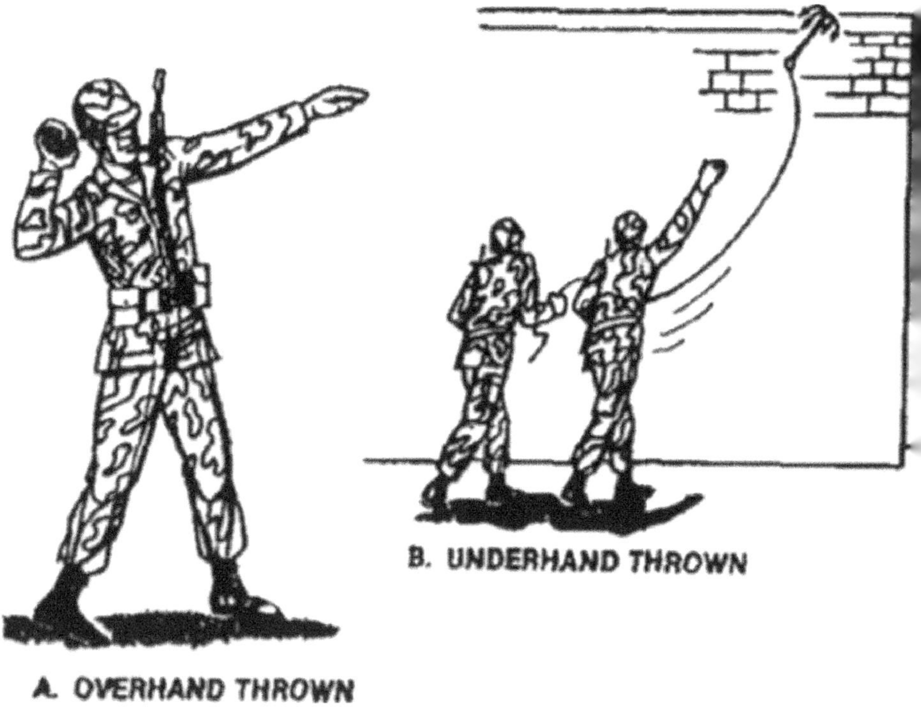

Figure 3-25.Throwing.

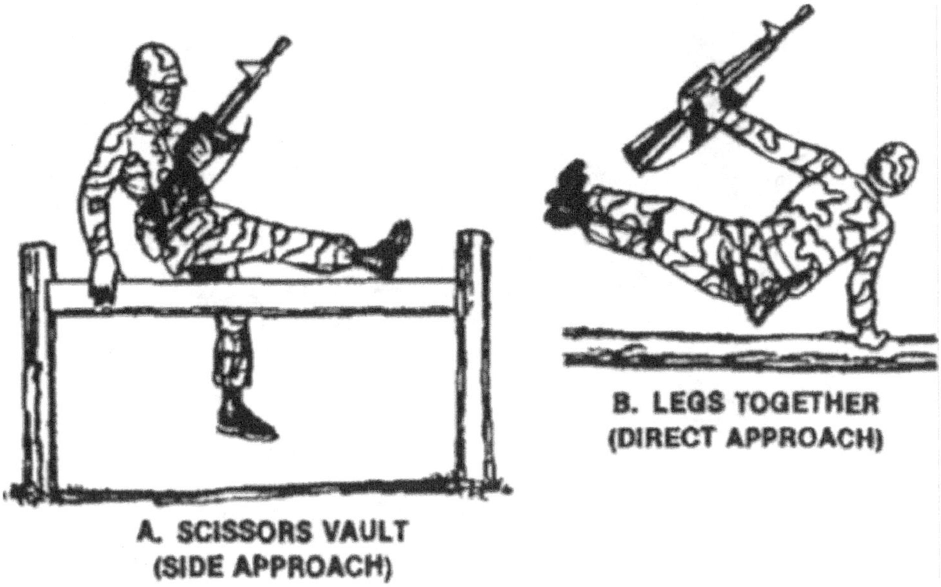

Figure 3-26. Vaulting

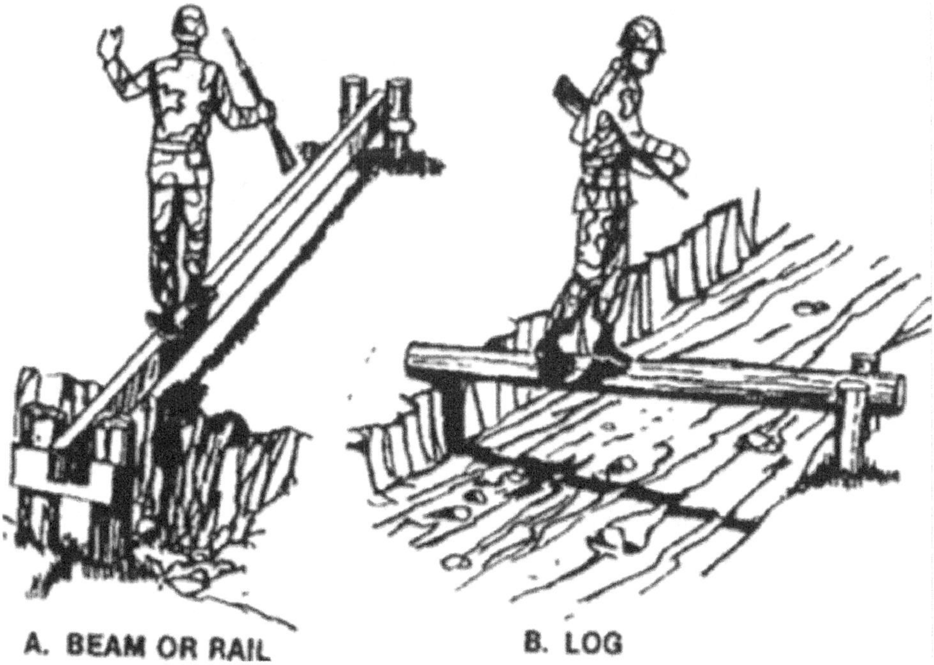

Figure 3-27.Balancing.

Figure 3-28.Falling.

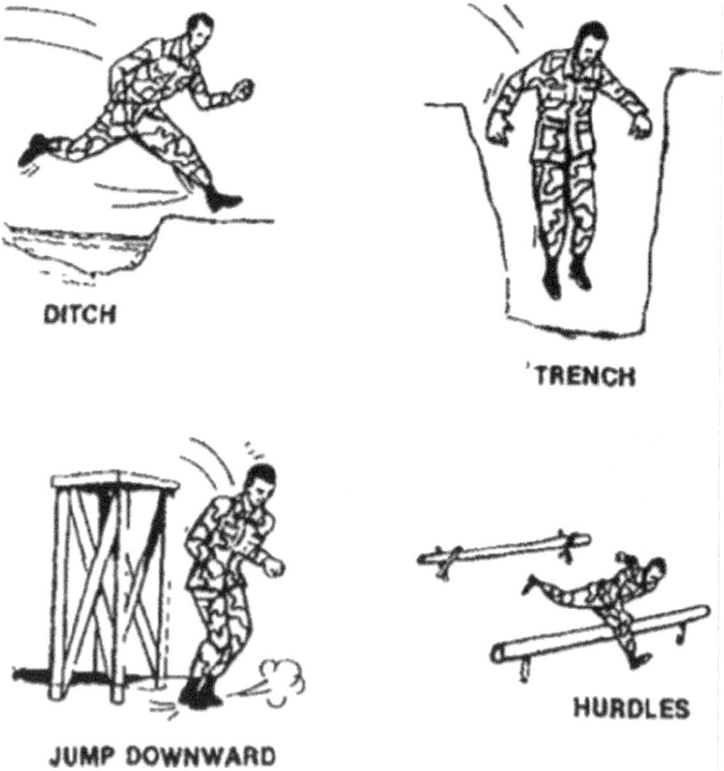

Figure 3-29.Jumping-Type Obstacles.

LANES TO GUIDE CHANGE OF DIRECTION

MAZES TO CAUSE CHANGE OF DIRECTION

Figure 3-30.Dodging-Type Obstacles.

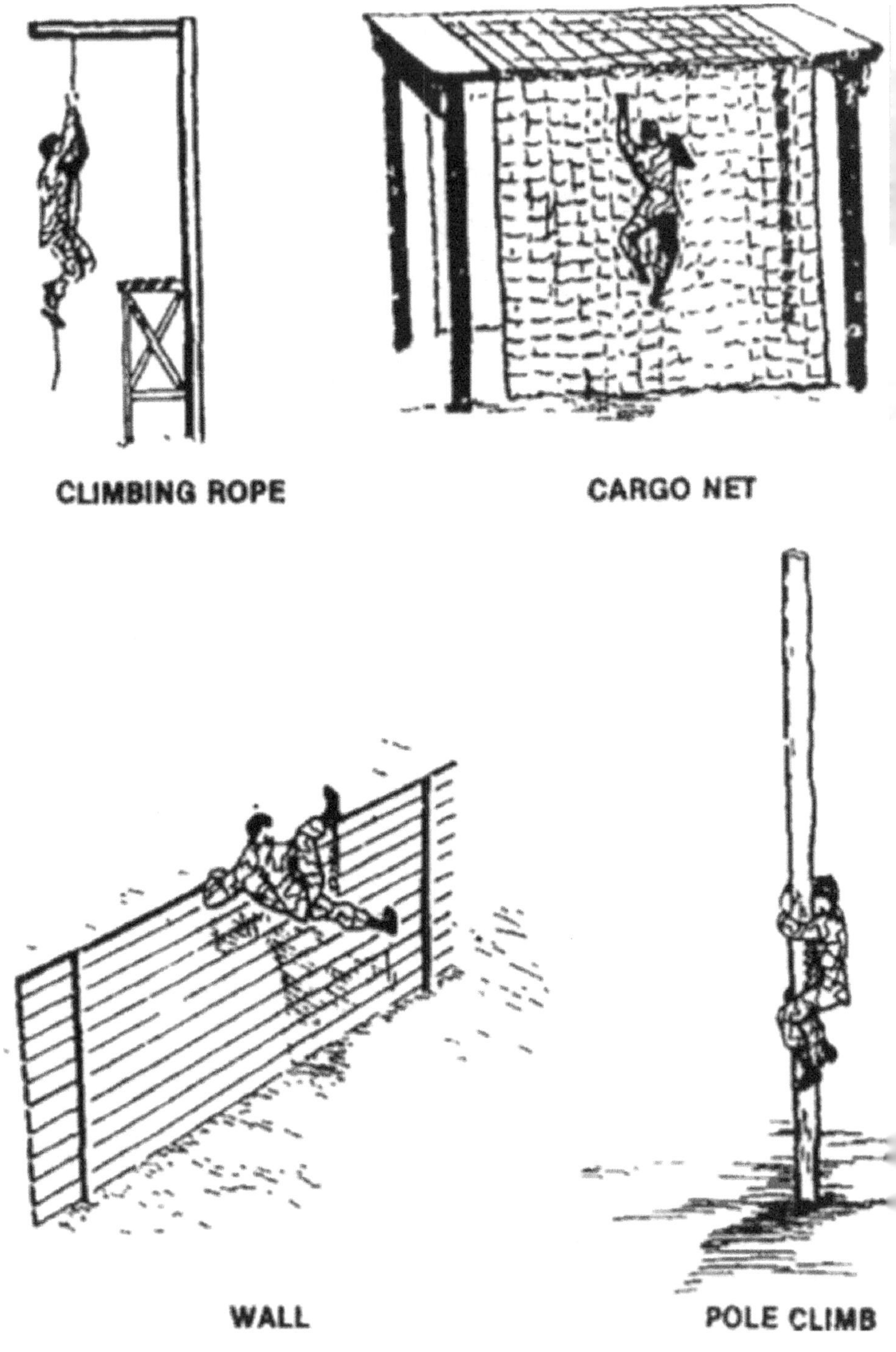

Figure 3-31. Vertical Climbing Obstacles.

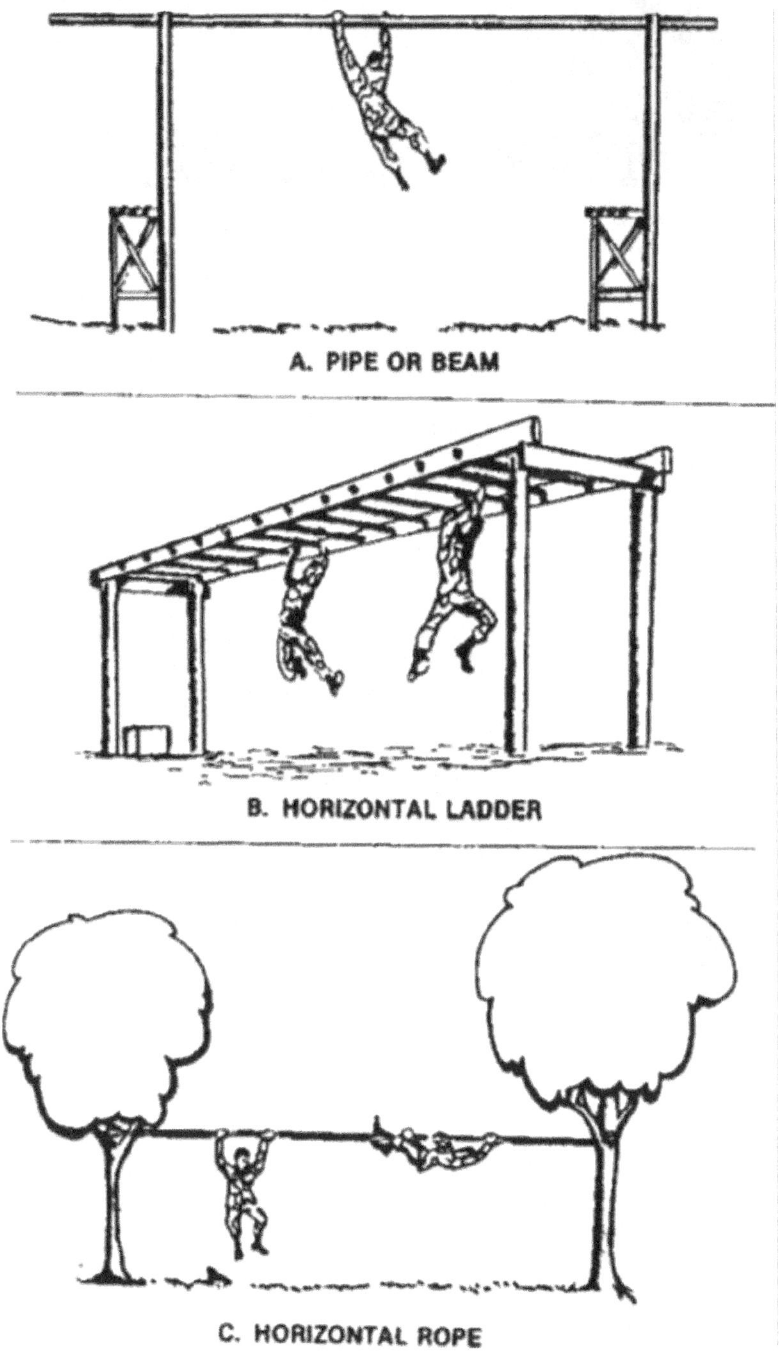

Figure 3-32.Horizontal Type Obstacles.

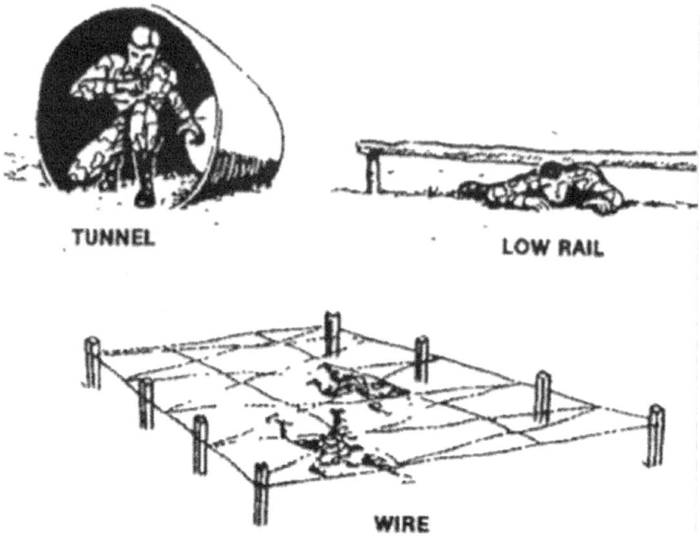

Figure 3-33.Crawling-Type Obstacles.

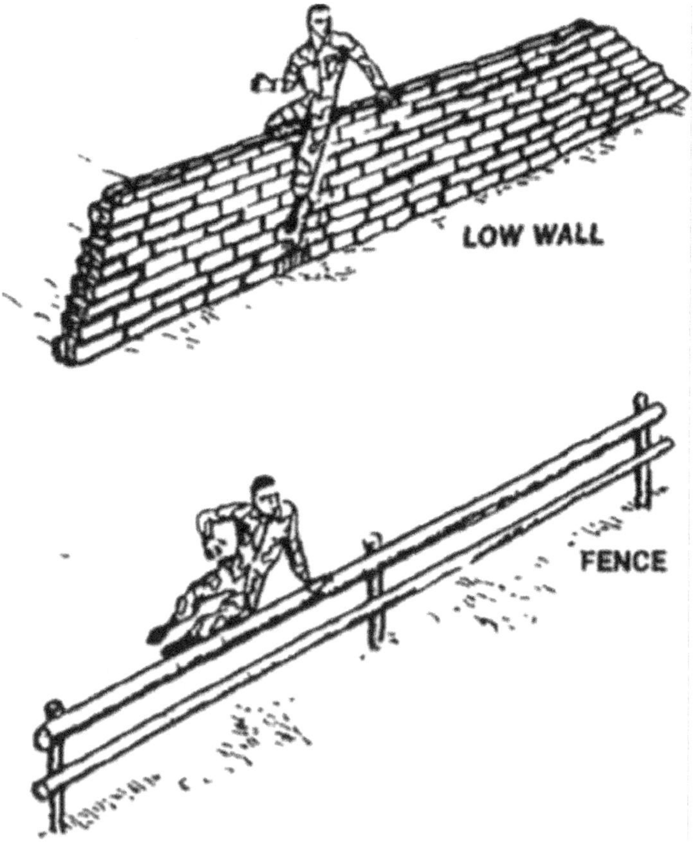

Figure 3-34.Vaulting Obstacles.

Figure 3-35.Balance-Type Obstacles.

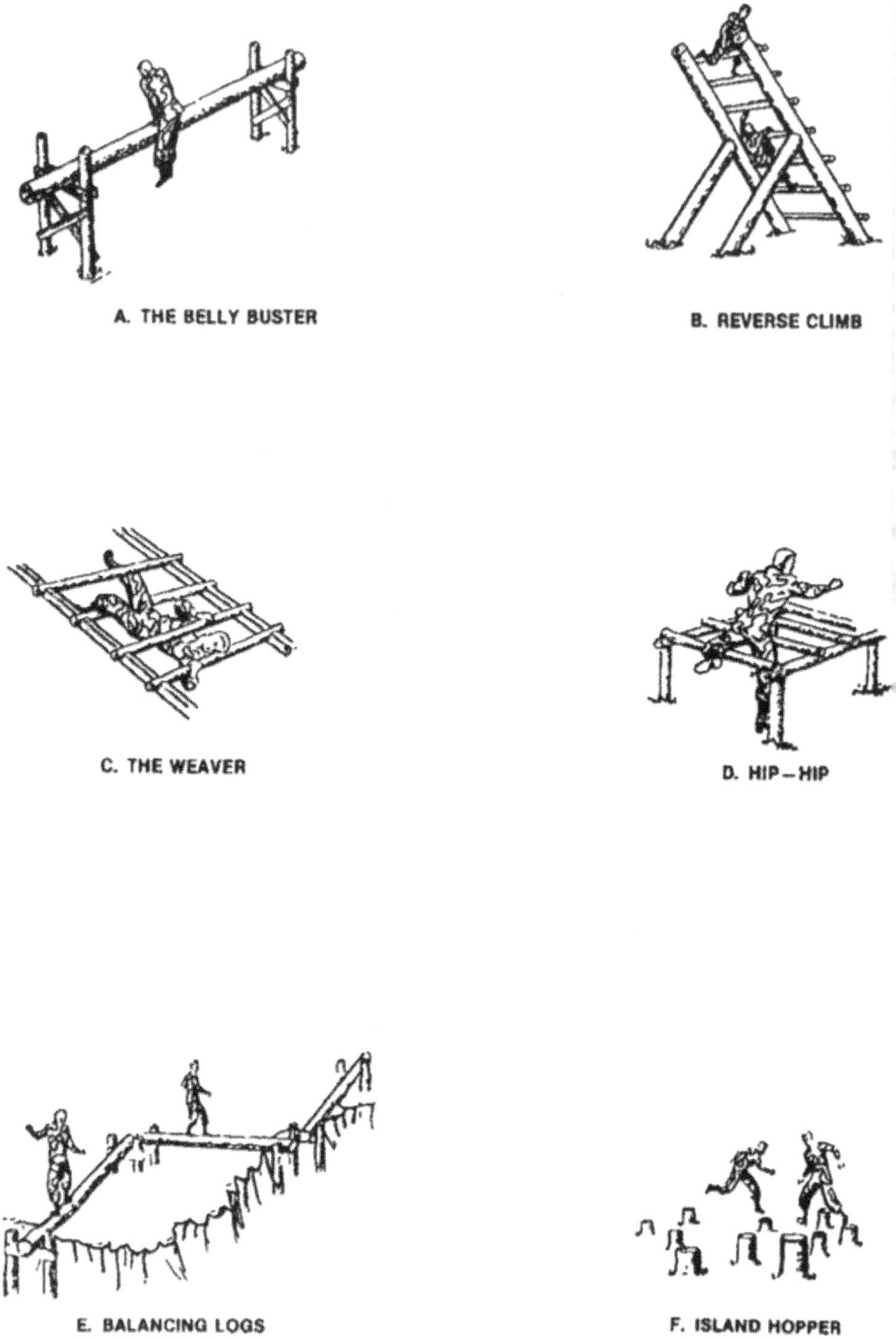

Figure 3-36.Red Group.

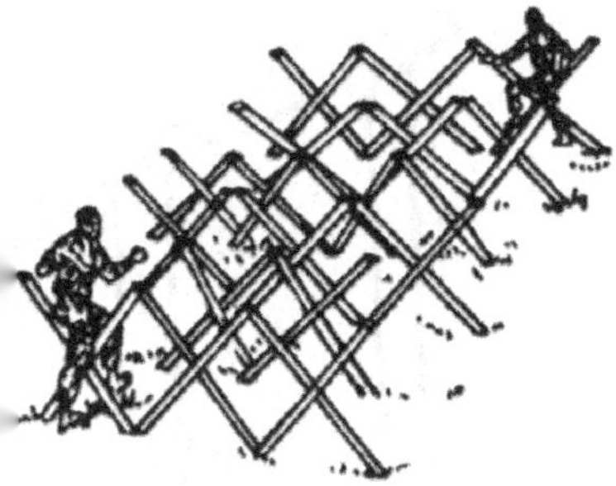

A. THE TOUGH NUT

B. SLIDE FOR LIFE

C. LOW BELLY OVER

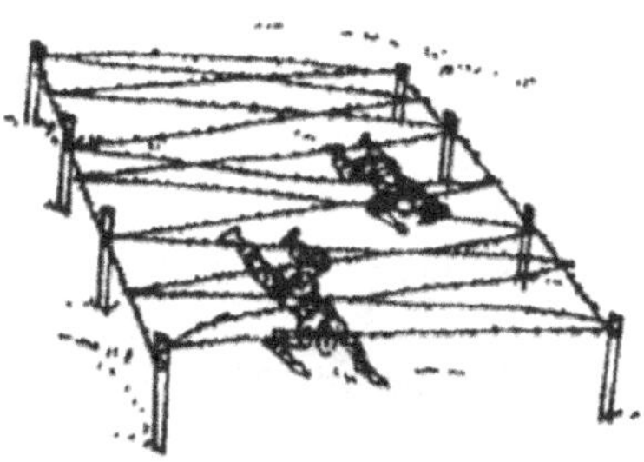

D. BELLY CRAWL

E. THE DIRTY NAME

F. THE TARZAN

Figure 3-37. White Group.

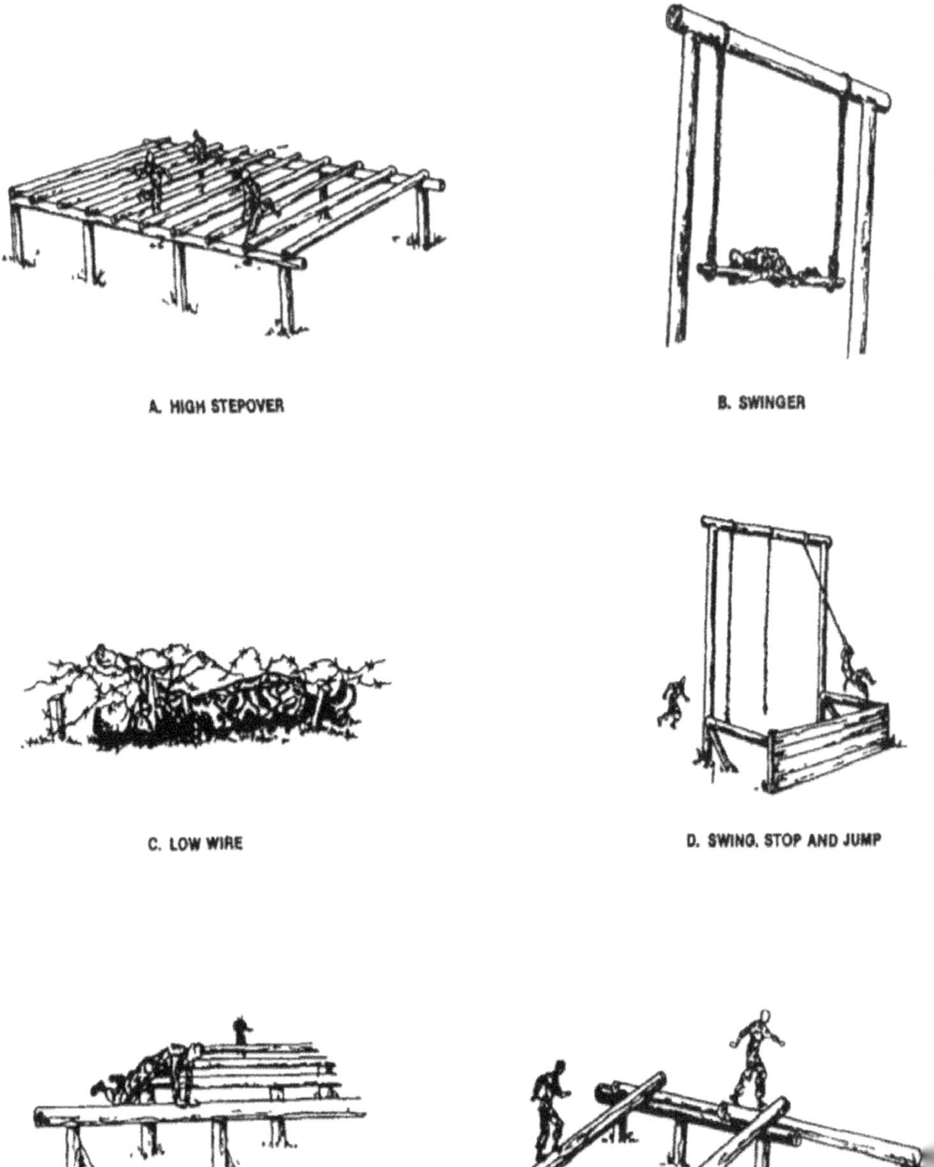

Figure 3-38.Blue Group.

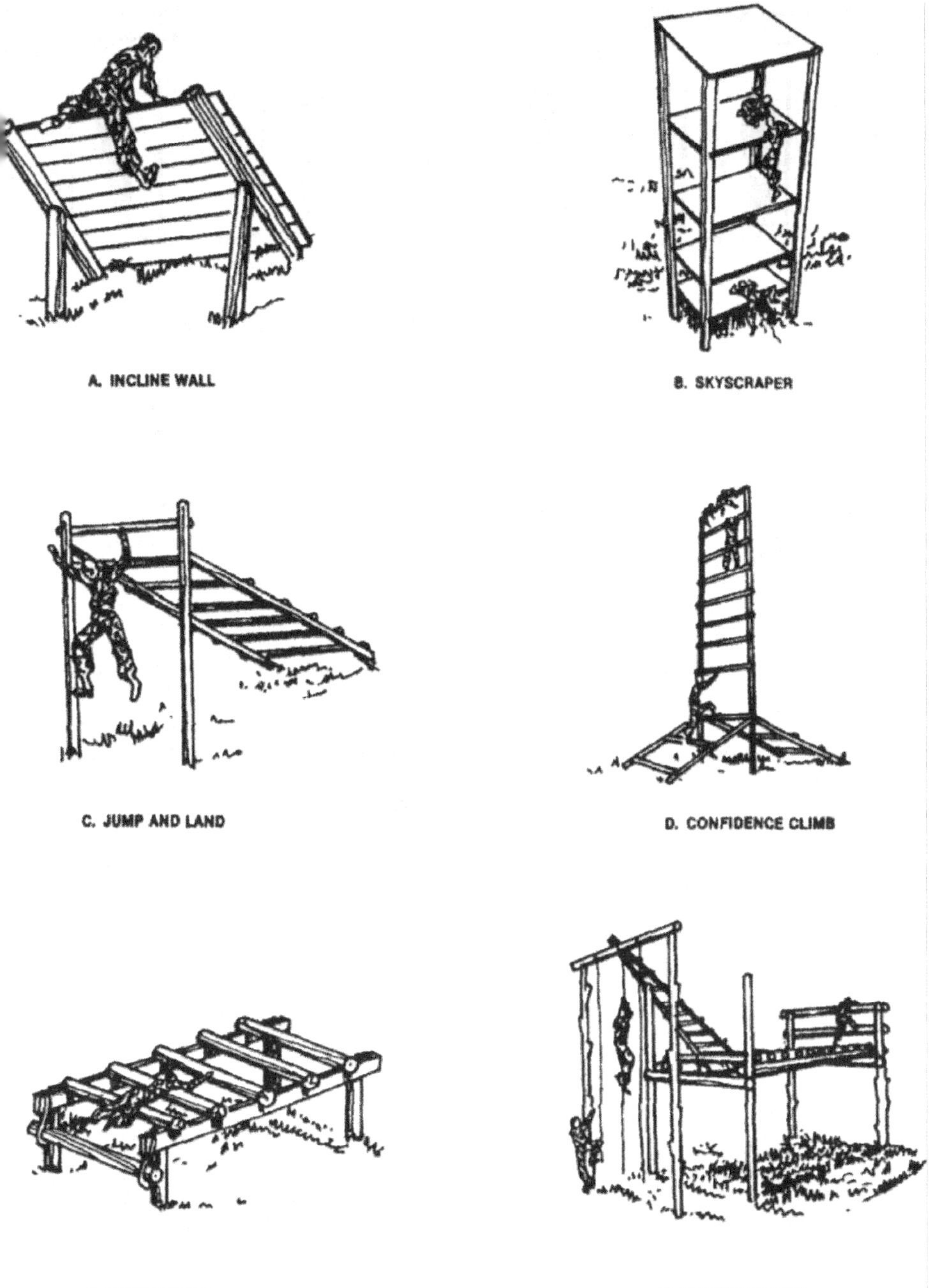

Figure 3-39.Black Group.

Figure 3-40. Bench Conditioning, Table I.

Figure 3-41. Bench Conditioning, Table II.

PROGRESSION GUIDE

AGE GROUP	LEVEL	EXERCISES 1	2	3	4	5	6
17	A	15	18	14	15	15	250
to	B	13	16	13	13	13	235
29	C	11	14	12	11	11	[illegible]
30	A	13	14	12	13	13	200
to	B	11	13	11	11	11	185
39	C	9	12	10	9	9	165
40	A	11	11	10	11	11	150
to	B	9	10	9	9	9	135
44	C	7	9	8	7	7	120
45	A	9	8	8	9	9	100
to	B	7	7	7	7	7	90
49	C	5	6	6	5	5	80
50	A	7	6	6	7	7	75
to	B	5	5	5	5	5	70
59	C	3	4	4	3	3	60
60	A	4	5	4	4	4	50
and	B	3	4	3	3	3	40
over	C	2	3	2	2	2	30
Minutes for each exercise		2	1	1	1	2	5

Table I. Progression Guide.

Table I. Exercises.

PROGRESSION GUIDE

AGE GROUP	LEVEL	EXERCISES 1	2	3	4	5	6
17	A	17	17	17	9	19	300
to	B	15	15	15	8	17	270
29	C	13	13	13	7	15	245
30	A	15	15	15	8	17	235
to	B	13	13	13	7	15	210
39	C	11	11	11	6	13	190
40	A	13	13	13	7	15	175
to	B	11	11	11	6	13	155
44	C	9	10	9	5	11	135
45	A	11	11	11	6	13	125
to	B	9	9	9	5	11	110
49	C	7	7	7	4	9	100
50	A	9	9	9	5	11	95
to	B	7	7	7	4	9	85
59	C	5	5	5	3	7	75
60	A	6	7	7	4	9	70
and	B	5	5	5	3	7	60
over	C	4	4	4	2	5	50
Minutes for each exercise		1	1	1	1 1/2	1 1/2	6

Table II. Progression Guide.

Table II. Exercises.

PROGRESSION GUIDE

AGE GROUP	LEVEL	EXERCISES 1	2	3	4	5	6
17	A	10	19	19	16	10	350
to	B	9	17	17	15	9	315
29	C	8	15	15	14	8	280
30	A	9	17	17	14	9	270
to	B	8	15	15	13	8	240
39	C	7	13	13	12	7	210
40	A	8	15	15	12	8	200
to	B	7	13	13	11	7	180
44	C	6	11	11	10	6	160
45	A	7	13	13	10	7	150
to	B	6	11	11	9	6	135
49	C	5	9	9	8	5	120
50	A	6	11	11	8	6	115
to	B	5	9	9	7	5	105
59	C	4	7	7	6	4	95
60	A	5	9	9	7	5	90
and	B	4	7	7	6	4	80
over	C	3	5	5	4	3	70
Minutes for each exercise		1 1/2	1	1	1 1/2	1	6

Table III. Progression Guide.

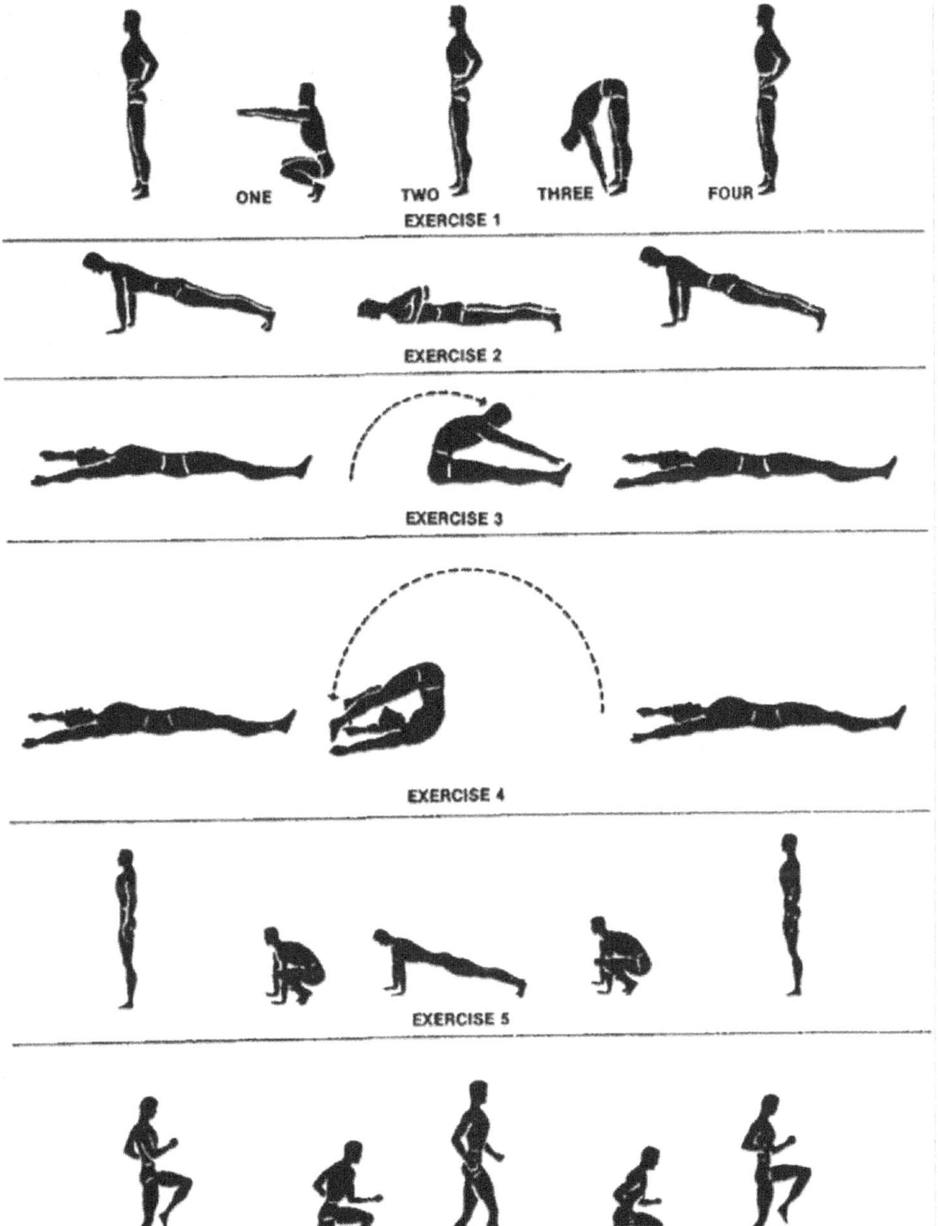

Table III. Exercises.

PROGRESSION GUIDE

AGE GROUP	LEVEL	EXERCISES 1	2	3	4	5	6
17	A	12	9	12	24	25	400
to	B	11	8	11	22	23	380
29	C	10	7	10	21	21	360
30	A	11	8	11	23	23	305
to	B	10	7	10	21	21	290
39	C	9	6	9	20	20	275
40	A	10	7	10	20	21	225
to	B	9	6	9	18	18	215
44	C	8	5	8	16	14	205
45	A	8	6	8	16	14	175
to	B	7	5	7	14	14	165
49	C	6	4	6	12	12	155
50	A	6	5	6	13	13	135
to	B	5	4	5	11	11	130
59	C	4	3	4	10	10	120
60	A	5	4	5	10	10	100
and	B	4	3	4	9	9	95
over	C	3	2	3	8	8	90
Minutes for each exercise		1	2	1	1	1	6

Table IV. Progression Guide.

Table IV. Exercises.

PROGRESSION GUIDE

AGE GROUP	LEVEL	EXERCISES 1	2	3	4	5	6
17	A	14	13	28	14	30	450
to	B	13	12	27	13	28	430
29	C	12	11	26	12	26	410
30	A	12	12	25	12	26	350
to	B	11	11	24	11	24	330
39	C	10	10	23	10	22	310
40	A	11	11	23	11	23	250
to	B	10	10	21	10	21	240
44	C	9	9	19	9	19	230
45	A	9	9	20	9	20	200
to	B	8	8	18	8	18	190
49	C	7	7	16	7	16	180
50	A	7	7	16	7	16	170
to	B	6	6	14	6	14	155
59	C	5	5	12	5	12	140
60	A	6	6	12	6	12	115
and	B	5	5	11	5	10	110
over	C	4	4	9	4	9	105
Minutes for each exercise		2	1	1	2	1	5

Table V. Progression Guide.

EXERCISE 1

EXERCISE 2

EXERCISE 3

EXERCISE 4

EXERCISE 5

EXERCISE 6

Table V. Exercises.

PROGRESSION GUIDE

AGE GROUP	LEVEL	EXERCISES 1	2	3	4	5	6
17	A	17	15	32	32	35	500
to	B	16	14	30	30	33	480
29	C	15	13	28	28	31	460
30	A	15	13	30	30	31	400
to	B	14	12	28	28	29	380
39	C	13	11	26	26	27	360
40	A	13	10	27	27	27	310
to	B	12	9	25	25	25	285
44	C	11	8	23	23	23	265
45	A	11	9	23	23	23	250
to	B	10	8	21	21	21	230
49	C	9	7	19	19	19	210
50	A	9	8	19	19	19	200
to	B	8	7	17	17	17	190
59	C	7	6	15	15	15	175
60	A	8	7	15	15	17	140
and	B	7	6	13	13	15	130
over	C	5	5	10	10	12	120
Minutes for each exercise		2	1	1	1	1	6

Table VI. Progression Guide.

Table VI. Exercises.

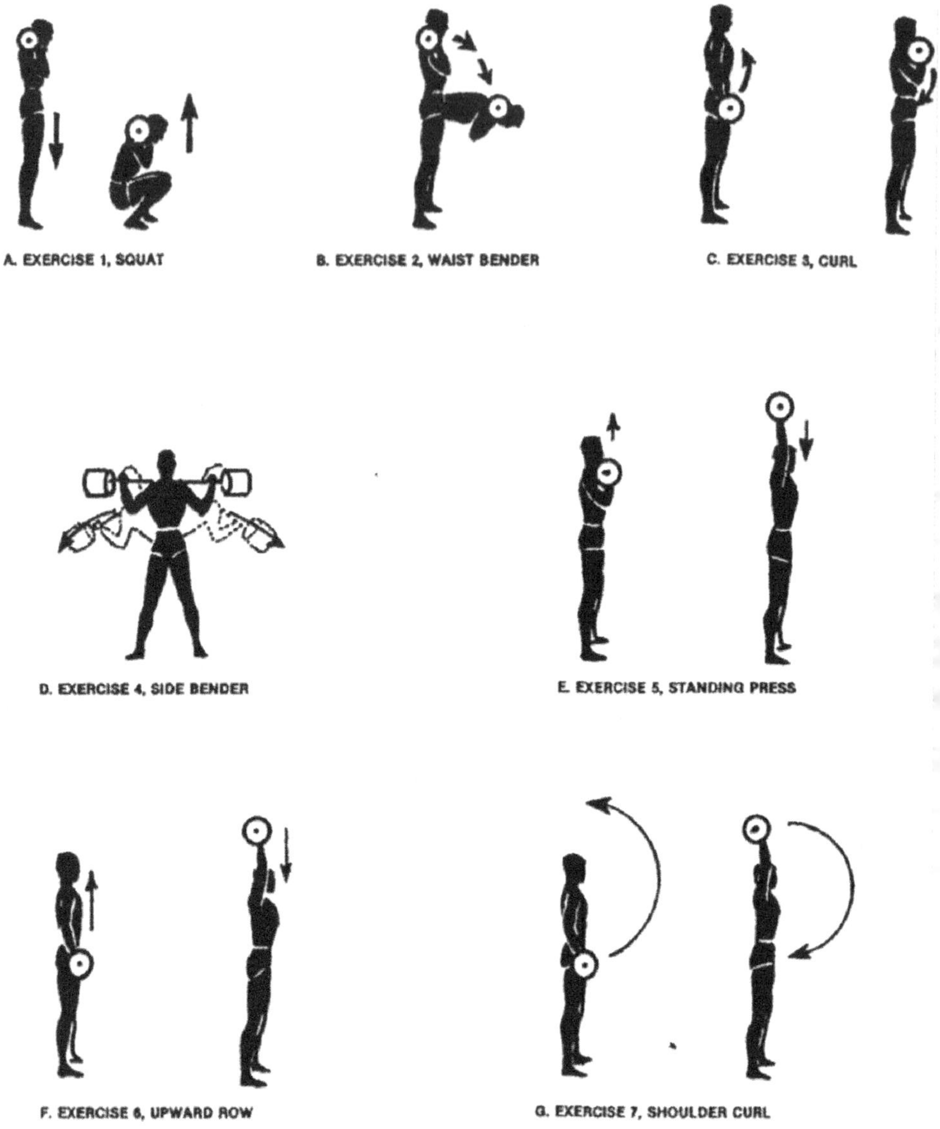

Figure 3-42. Weight Training.

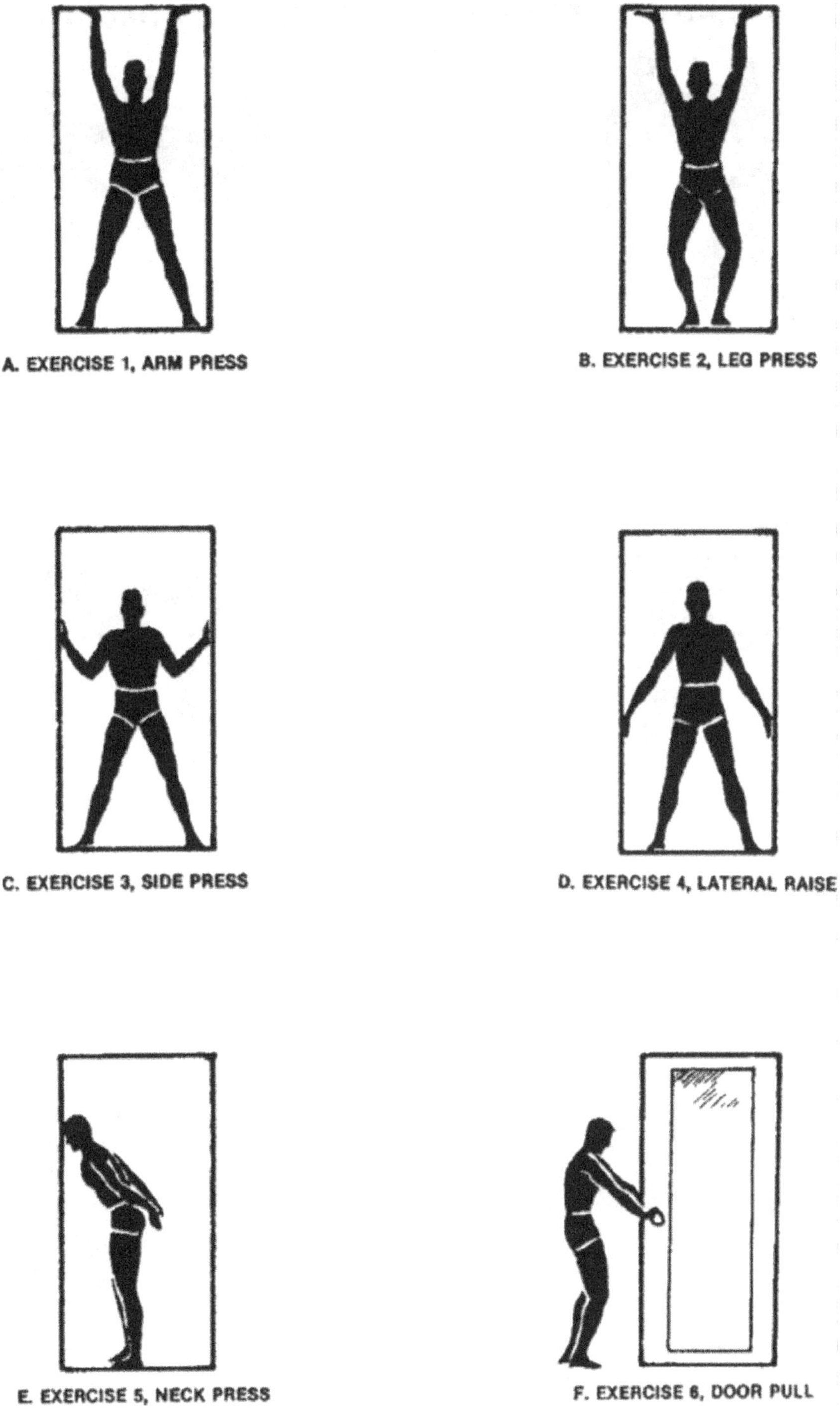
A. EXERCISE 1, ARM PRESS

B. EXERCISE 2, LEG PRESS

C. EXERCISE 3, SIDE PRESS

D. EXERCISE 4, LATERAL RAISE

E. EXERCISE 5, NECK PRESS

F. EXERCISE 6, DOOR PULL

Figure 3-43. Isometric Exercises.

Figure 5-1.Combatives Table I.

Figure 5-2.Combatives Table II.

Figure. 5-3.Combatives Table III.

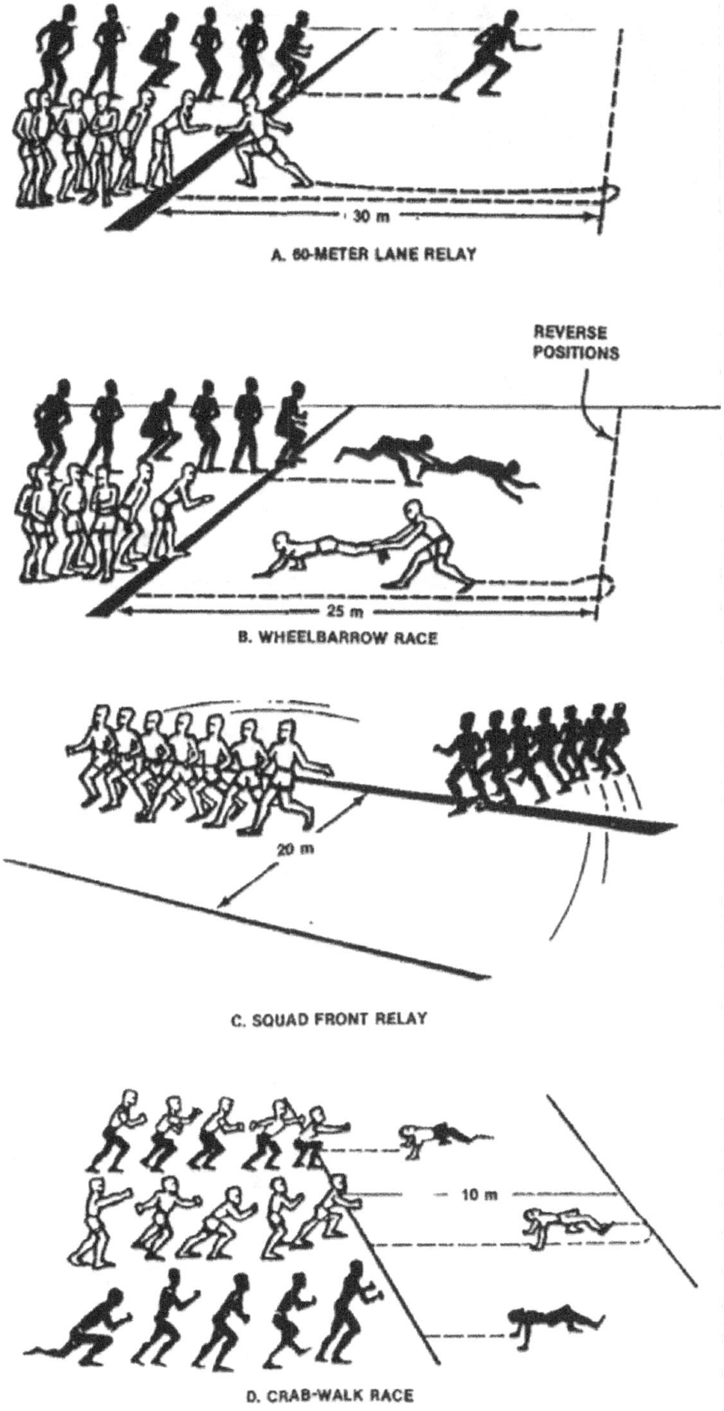

Figure 5-4.Relay Table I.

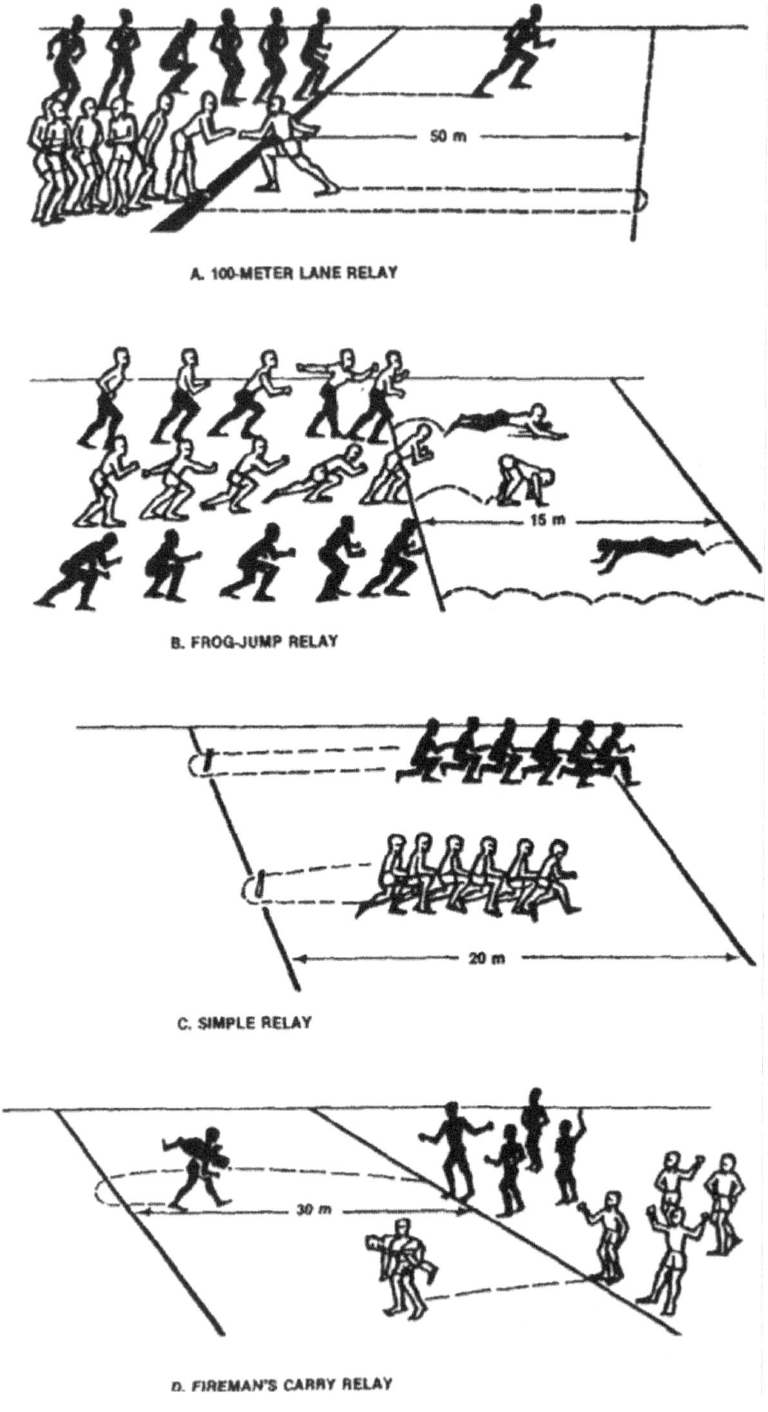

Figure 5-5.Relay Table II.

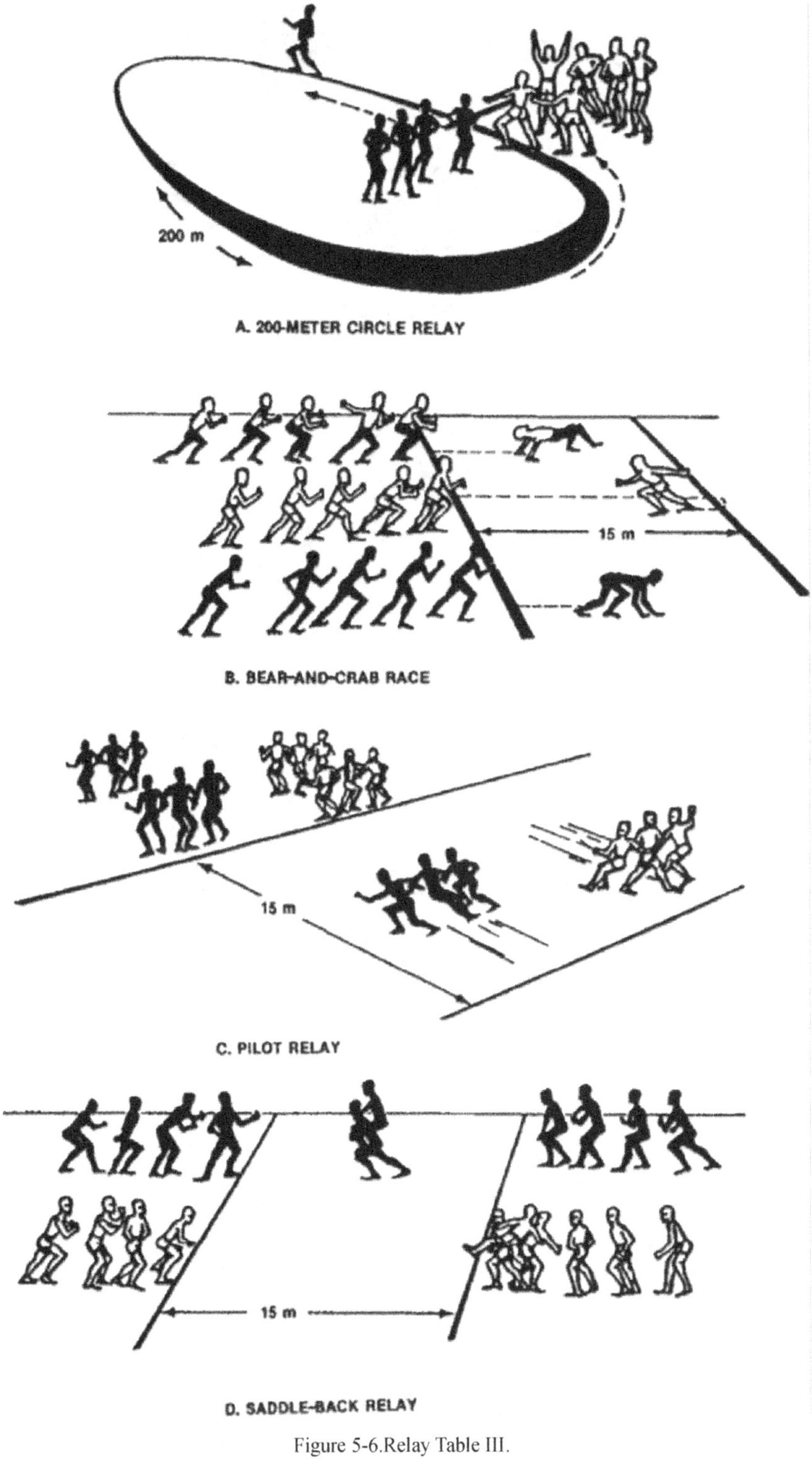

Figure 5-6.Relay Table III.

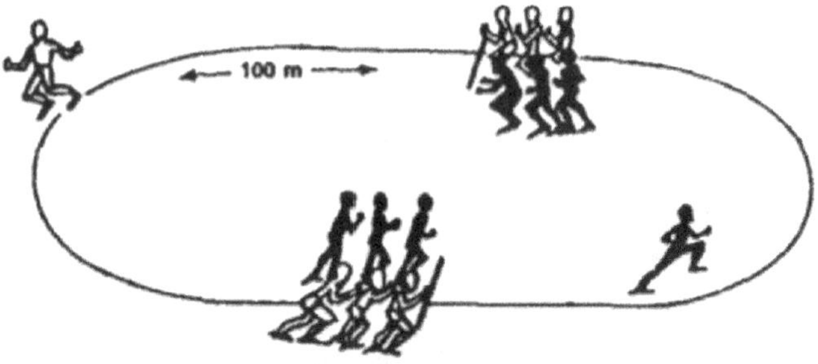

A. 100-METER CIRCLE RELAY

B. IN-AND-OUT RELAY

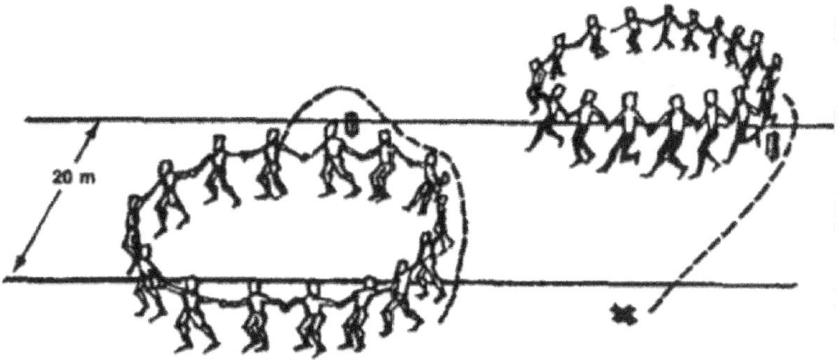

C. CIRCLE RACE

D. HORSE-AND-RIDER RELAY

Figure 5-7.Relay Table IV.

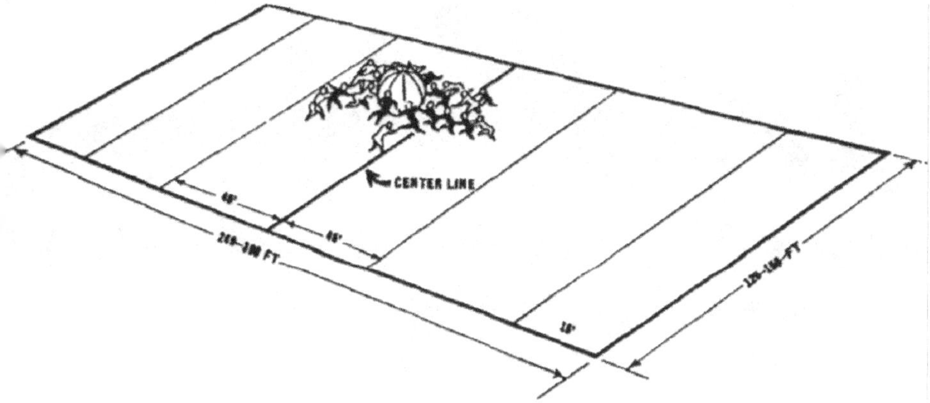

Figure 5-8.Pushball.

Figure 5-9.Line Rush.

Figure. 5-10. Human Tug-of-War.

Figure 5-11. Master of the Ring.

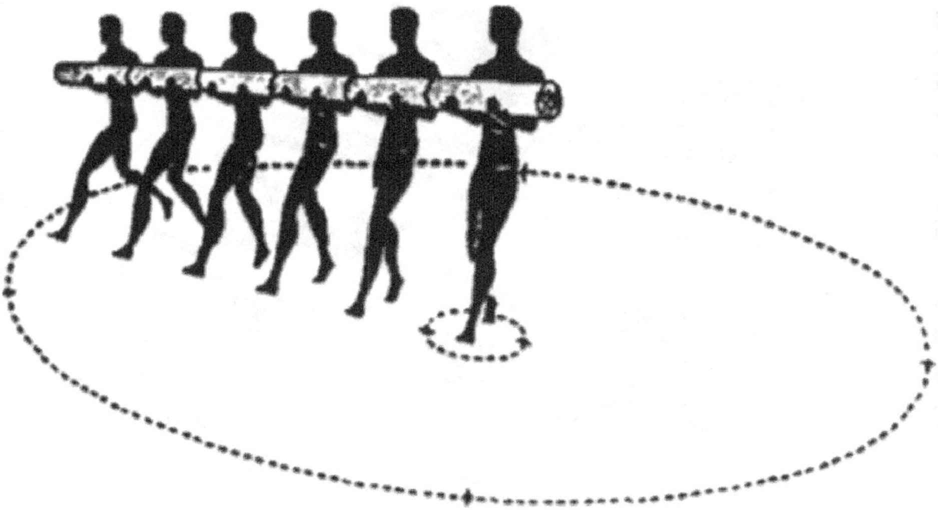

Figure 5-12.Log Pivot Circle.

Figure 5-13.Log-Rolling Race.

Figure 5-14. Prone Pushing Contest.

Figure 5-15. Shuttle Relay Race.

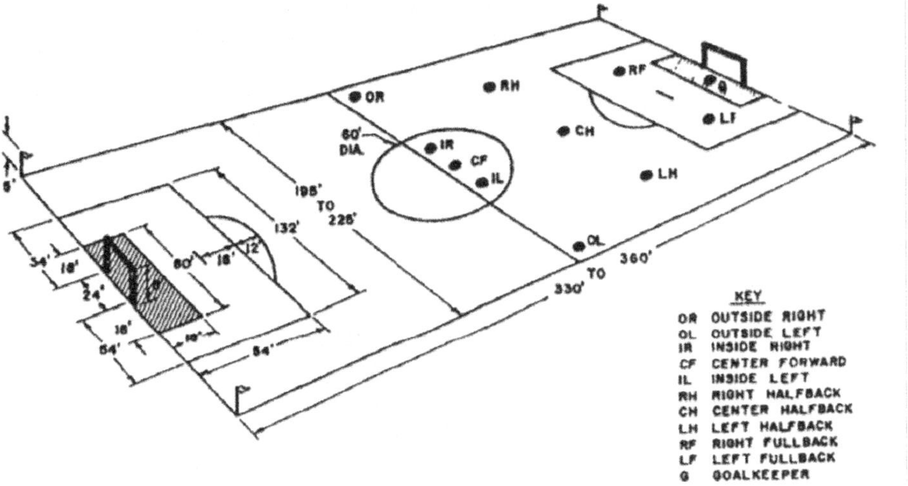

Figure 5-16. Soccer.

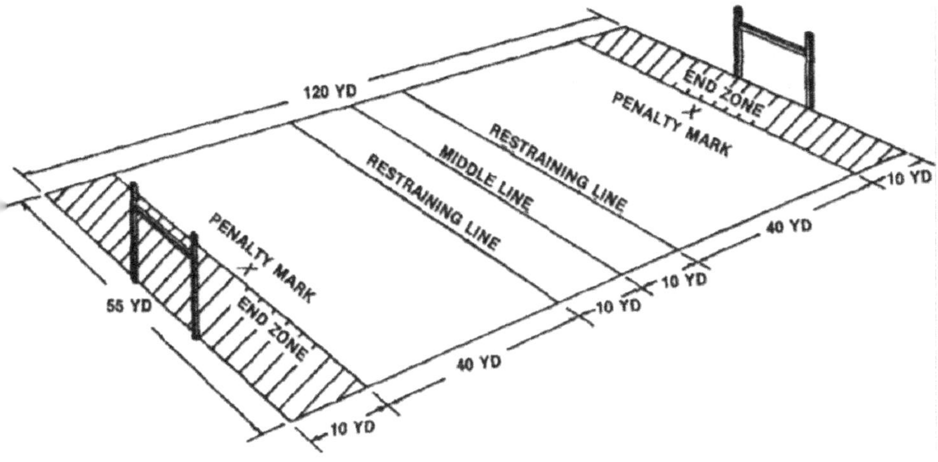

Figure 5-17.Speedball Field.

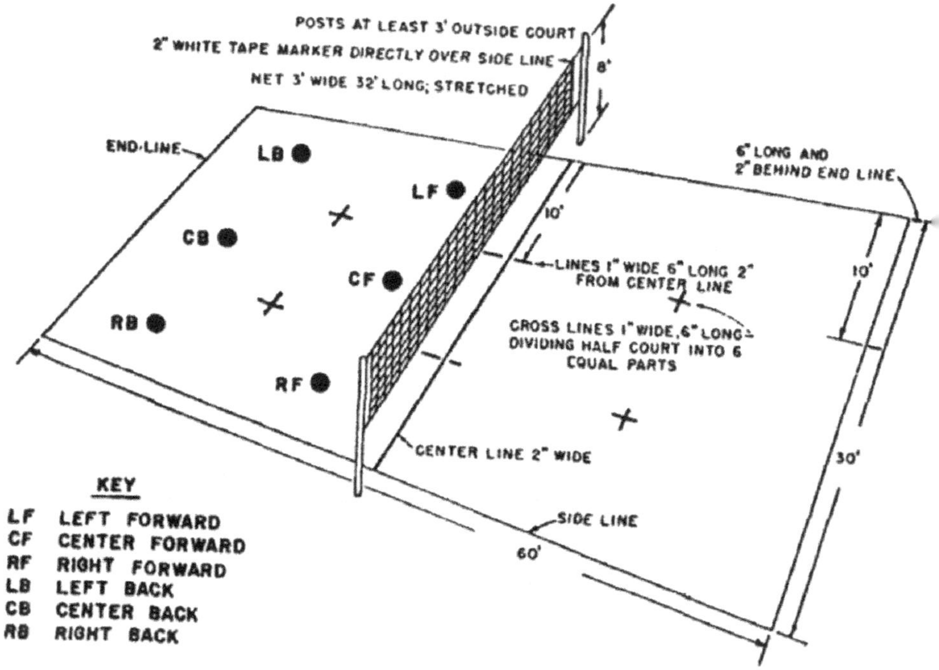

Figure 5-18. Volleyball.

PTS	UPHILL CLIMB	ROPE CLIMB	EVACUATION	FIRE AND MANEUVER	FORCED MARCH
100	1:20-BELOW	0:10-BELOW	0:25-BELOW	1:20-BELOW	26:00-BELOW
99	1:21-1:22	0:11		1:21-1:22	26:01-26:12
98	1:23-1:24	0:12	0:26	1:23-1:24	26:13-26:24
97	1:25-1:26	0:13		1:25-1:26	26:25-26:36
96	1:27-1:28	0:14	0:27	1:27-1:28	26:37-26:48
95	1:29-1:30	0:15		1:29-1:30	26:49-27:00
94	1:31-1:32	0:16	0:28	1:31-1:32	27:01-27:12
93	1:33-1:34	0:17		1:33-1:34	27:13-27:24
92	1:35-1:36	0:18	0:29	1:35-1:36	27:25-27:36
91	1:37-1:38	0:19		1:37-1:38	27:37-27:48
90	1:39-1:40	0:20	0:30	1:39-1:40	27:49-28:00
89	1:41-1:42	0:21	0:31	1:41-1:42	28:01-28:12
88	1:43-1:44	0:22	0:32	1:43-1:44	28:13-28:24
87	1:45-1:46	0:23	0:33	1:45-1:46	28:25-28:36
86	1:47-1:48	0:24	0:34	1:47-1:48	28:37-28:48
85	1:49-1:50	0:25	0:35	1:49-1:50	28:49-29:00
84	1:51-1:52	0:26	0:36	1:51-1:52	29:01-29:12
83	1:53-1:54	0:27	0:37	1:53-1:54	29:13-29:24
82	1:55-1:56	0:28	0:38	1:55-1:56	29:25-29:36
81	1:57-1:58	0:29	0:39	1:58-1:58	29:37-29:48
80	1:59-2:00	0:30	0:40	1:59-2:00	29:49-30:00
79	2:01-2:02	0:31	0:41	2:01-2:02	30:01-30:12
78	2:03-2:04	0:32	0:42	2:03-2:04	30:13-30:24
77	2:05-2:06	0:33	0:43	2:05-2:06	30:25-30:36
76	2:07-2:08	0:34	0:44	2:07-2:08	30:37-30:48
75	2:09-2:10	0:35	0:45	2:09-2:10	30:49-31:00
74	2:11-2:12	0:36	0:46	2:11-2:12	31:01-31:12
73	2:13-2:14	0:37	0:47	2:13-2:14	31:13-31:24
72	2:15-2:16	0:38	0:48	2:15-2:16	31:25-31:36
71	2:17-2:18	0:39	0:49	2:17-2:18	31:37-31:48
70	2:19-2:20	0:40	0:50	2:19-2:20	31:49-32:00
69	2:21-2:22	0:41	0:51	2:21-2:22	32:01-32:12
68	2:23-2:24	0:42	0:52	2:23-2:24	32:13-32:24
67	2:25-2:26	0:43	0:53	2:25-2:26	32:25-32:36
66	2:27-2:28	0:44	0:54	2:27-2:28	32:37-32:48
65	2:29-2:30	0:45	0:55	2:29-2:30	32:49-33:00

Figure 6-1.Physical Readiness Test Scoring Table for Male.

	EXAMPLE 1	EXAMPLE 2	EXAMPLE 3
Event #1	100 pts	100 pts	99 pts
Event #2	100 pts	99 pts	86 pts
Event #3	85 pts	75 pts (F)	80 pts
Event #4	90 pts	79 pts (F)	100 pts
Event #5	79 pts	85 pts	90 pts
Pts	454 pts	438 pts	455 pts
Final Score	79.0% (F)	69.0% (F)	91.0%
(F) = Failure			

Figure 6-2.Scoring Examples.

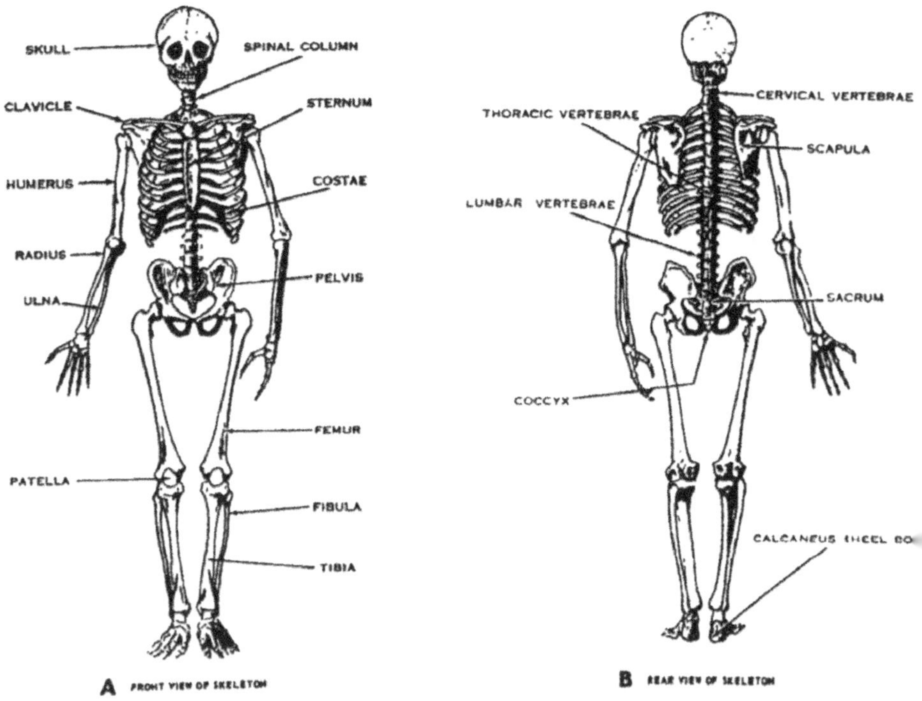

Figure 7-1.Front and Rear Views of Skeleton.

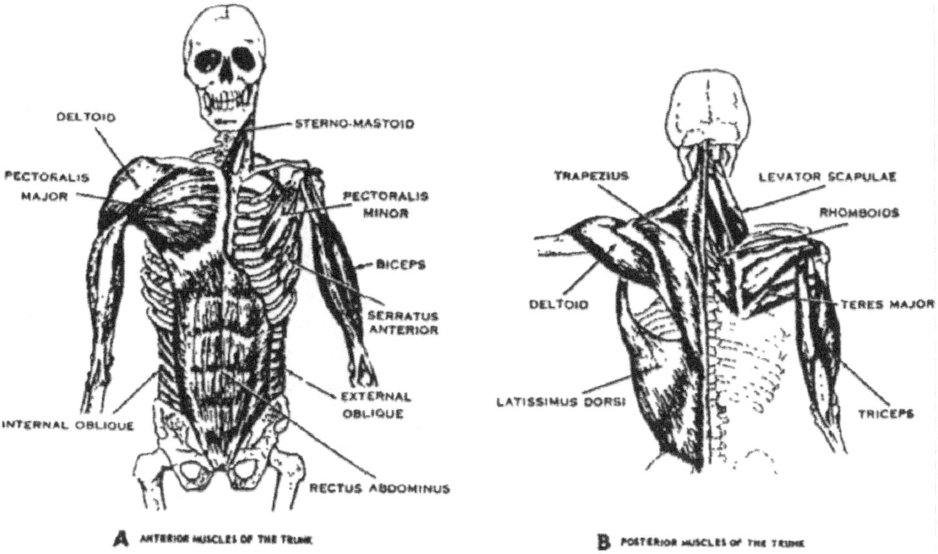

Figure 7-2.Muscles of the Trunk.

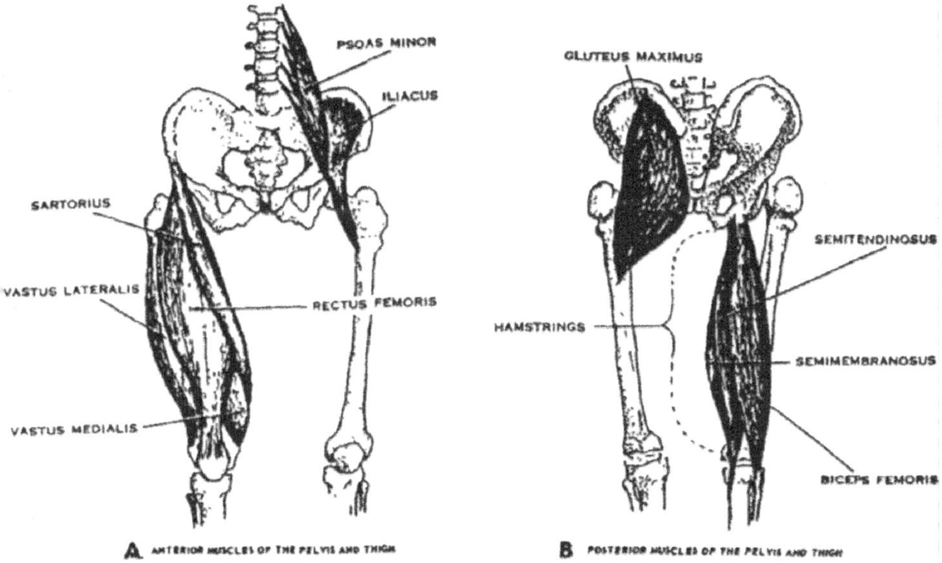

Figure 7-3.Anterior and Posterior Muscles of the Pelvisand Thigh.

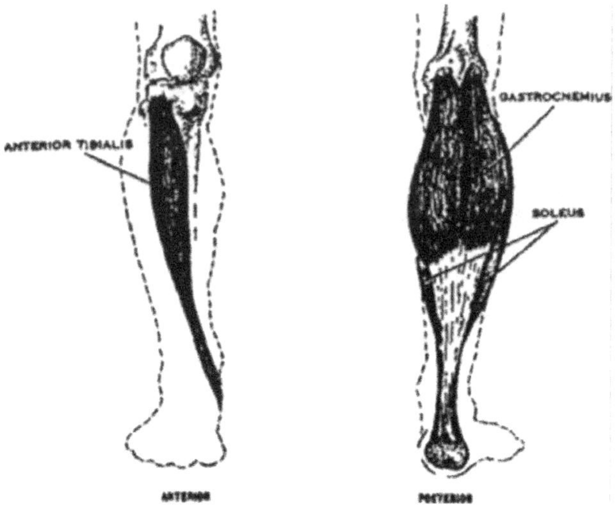

Figure 7-4. Muscles of Lower Leg.

Figure 7-5.Circulation of Blood from Heart to the Body.

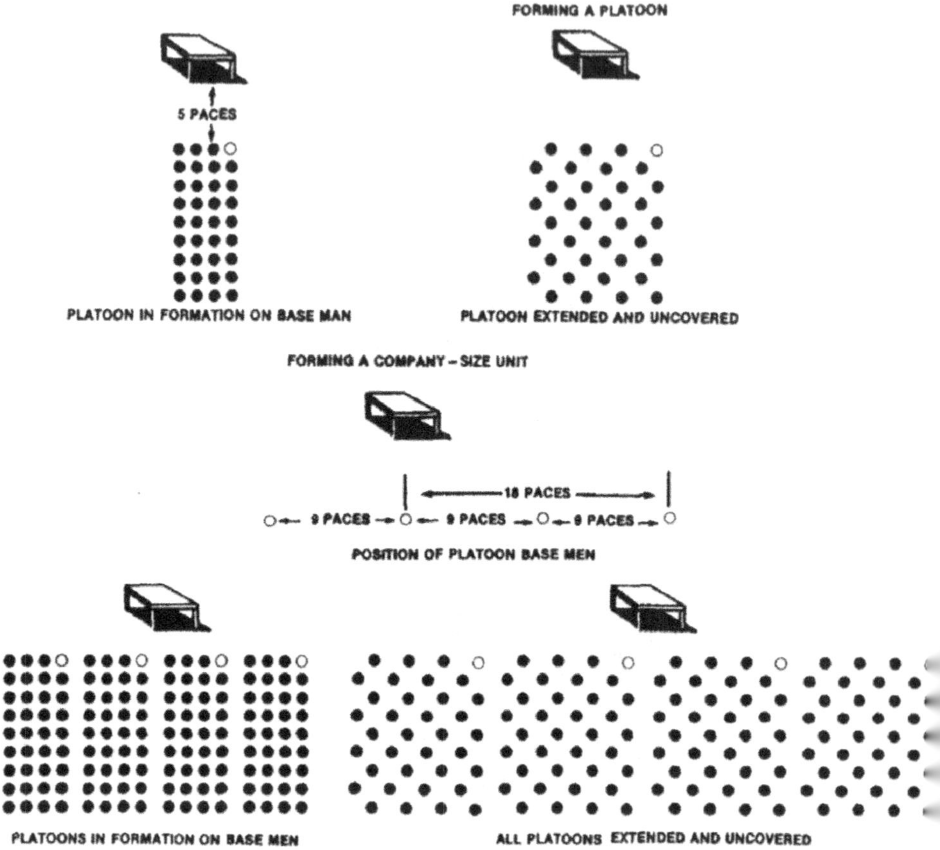

Figure A-l. Forming the Extended Rectangular Formation.

Figure A-2.Circle Formation.

Military Operations on Urbanized Terrain

Table of Contents

Chapter 1

Modern Urban Warfare

"... the likelihood is high that in the future, the National Command Authorities will again commit Marines to missions in urban areas."
A Concept for Future Military Operations on Urbanized Terrain

1001. Importance of Urban Areas. Throughout history, military planners have viewed cities as centers of gravity. As such, in war, cities are something to be either protected or taken away, depending upon one's perspective (MCDP 1, *Warfighting*). Cities house the population centers, transportation hubs, seats of government, sources of wealth, centers for industry, information networks, and key nodes of communication within a nation. Recent forecasts based on population statistics and the worldwide migration trend from agrarian to industrialized societies predict that 85 percent of the world's population will reside in urbanized areas by the year 2025. As the world trend toward urbanization increases, the military significance of cities is likely to increase proportionally.

Urbanized areas, themselves, may be significant sources of future conflict. Cities historically are where radical ideas ferment, dissenters find allies, mixtures of people cause ethnic friction, and discontented groups receive media attention. Adversaries may focus on the capture of radio and television stations in an attempt to influence public opinion and attain their political goals. Our political leaders may take advantage to neutralize or stabilize some extremely volatile political situations, or to provide assistance to allies in need of support, by deploying U.S. forces into urban environments,

1002. The Marine Corps Role in Urban Warfare. As the Nation's force in readiness, forward deployed with expeditionary forces, Marines must be prepared to fight on urbanized terrain. In the past two decades, MAGTFs ranging in size from MEFs (Saudi Arabia, Desert Shield/Desert Storm; Somalia, Restore Hope) through Marine expeditionary units (MEUs) (Beirut, Lebanon; Grenada, Urgent Fury; Somalia, Eastern Exit and Restore Hope) have participated in MOUT. The task-organization and combined-arms aspect of the MAGTF makes it well suited for combat on urbanized terrain.

The results of geographical studies show that 60 percent of politically significant urban areas outside allied or former Warsaw Pact territory are located along or within 25 miles of a coastline; 75 percent are within 150 miles; 87 percent are within 300 miles; 95 percent are within 600 miles; and all are within 800 miles. U.S. embassies and diplomatic facilities are primarily located in cities where the host country's political and economic leadership is concentrated. The Marine Corps will continue to play a prominent role in future evacuations of U.S. citizens, as well as the conduct of peace, counterinsurgency, and contingency operations centered on urbanized areas.

Today's Marine air-ground task forces (MAGTFs) are deployed as part of naval expeditionary forces (NEFs) that maintain a *global forward presence for rapid crisis response*. These integrated combined-arms forces are part of the Nation's proven contingency and naval power projection force. Therefore, Marines may find themselves rapidly deployed and employed in actions across the spectrum of military operations. Many of these trouble spots will likely be located in or around large urban centers.

In the years since World War II, the United States has employed military force more than 200 times. Of these, four out of five involved naval forces, and the majority of the naval efforts included Marines embarked in amphibious ships. The reasons are straightforward: availability and adaptability. Availability derives from the loiter time of forward deployed forces embarked on amphibious shipping. Adaptability comes from the Marine Corps' MAGTF organization, doctrine, training, and equipment, which prepare us for expeditionary missions from the sea in support of a variety of missions, including forcible entry. Enhancing our adaptability are the maritime prepositioning forces (MPFs). MPFs provide a rapid buildup of combat and logistics equipment that is joined with Marines on a distant shore, creating a substantial combat force. Despite our availability and adaptability, the prospect of urban warfare combined with an amphibious assault is a complex task which requires special preparation.

At the outset of a developing situation, forward-deployed expeditionary forces can move quickly within range of a crisis that threatens the political stability of a country. Urban intervention operations must often be planned and executed in a matter of hours or days (rather than weeks or months) to take advantage of the internal turmoil surrounding a developing crisis. Navy and Marine forces should anticipate deployment to urbanized areas on a day-to-day basis.

1003. Distinguishing Features of Urbanized Terrain. Urbanized terrain is a complex and challenging environment. It possesses all of the characteristics of the natural landscape, coupled with manmade construction, resulting in an incredibly complicated and fluid environment that influences the conduct of military operations in unique ways.

Military operations on urbanized terrain (MOUT) is defined as *all military actions planned and conducted on a topographical complex and its adjacent terrain where manmade construction is the dominant feature. It includes combat in cities, which is that portion of MOUT involving house-to-house and street-by-street fighting in towns and cities* (Marine Corps Reference Publication (MCRP) 5-12A, *Operational Terms and Graphics*). MOUT effects the tactical options available to a commander. A built-up area is *a concentration of structures, facilities, and populations, such as villages, cities, and towns,* that form the economic and cultural focus for the surrounding area. (MCRP 5-12A)

a. Cities. Cities are centers of finance, politics, transportation, communication, industry, and culture. They generally have large population concentrations ranging from tens of thousands to millions of people. Because of their psychological, political, or logistical value, control of cities have often been the scenes of pitched battles.

(1) Operations in built-up areas are normally conducted to capitalize on the operational or tactical significance of a particular city. In developing nations, control of only a few cities is often the key to the control of national resources. The side that controls a major city usually has a psychological advantage, which can be enough to significantly affect the outcome of a countrywide conflict.

(2) The abundance of guerrilla and terrorist operations conducted in built-up areas (e.g., Santo Domingo, Caracas, Belfast, Managua, and Beirut) demonstrates the importance many insurgent groups place on urban warfare.

(3) In the past 40 years, many cities have expanded dramatically, losing their well-defined boundaries as they extended into the countryside. New transportation systems (highways, canals, and railroads) have been built to connect population centers. Industries have grown along those connectors, creating "strip areas." Rural areas, although retaining much of their farmlike character, are connected to the towns by a network of secondary roads.

b. Multiple Avenues of Approach. Urbanized terrain is a unique battlespace that provides both attacker and defender with numerous and varied avenues of approach and fields of fire. The urban battlespace is divided into four basic levels: *building*, *street*, *subterranean,* and *air*. Operations can be conducted from above ground, on ground level, inside buildings, or below the ground. Most operations will include fighting on all levels simultaneously.

(1) Building Level. Buildings provide cover and concealment; limit or increase fields of observation and fire; and canalize, restrict, or block movement of forces, especially mechanized forces. They provide optimum perches for snipers and antiair weapons. Buildings also provide antitank weapons optimum positioning to allow engagement from above, exploiting an inherent weakness found in most armored vehicles.

(2) Street Level. While streets provide the means for rapid advance or withdawal, forces moving along streets are often canalized by buildings and have little space for off-road maneuver. Because they are more difficult to bypass, obstacles on streets in urbanized areas are usually more effective than those on roads in open terrain.

(3) Subterranean Level. Subterranean systems are easily overlooked but can be important to the outcome of operations. These areas may be substantial and include subways, sewers, cellars, and utility systems (Figure 1-1 on page 1-4). The city of Los Angeles alone has more than 200 miles of storm sewers located under the city streets. Both attacker and defender can use subterranean avenues to maneuver to the rear or the flanks of an enemy. These avenues also facilitate the conduct of ambushes, counterattacks, and infiltrations. (See Appendix E)

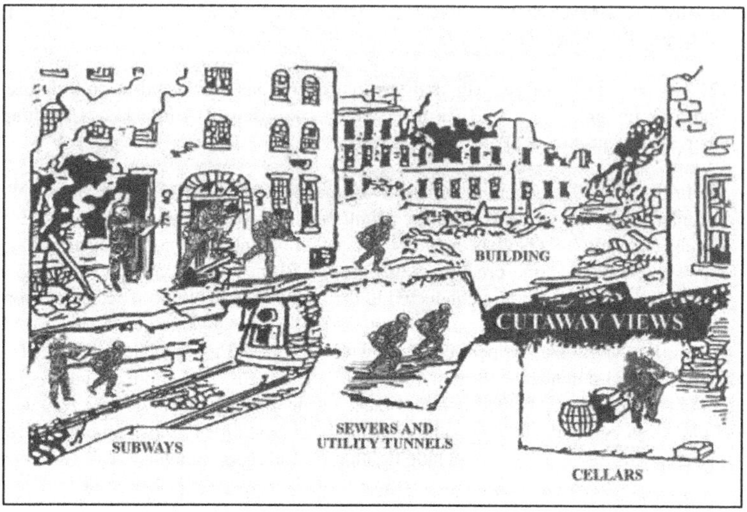

Figure 1-1. Subterranean Systems

(4) Air Level. The air provides another avenue of approach in urbanized areas. Aviation assets can be used for high speed insertion or extraction of troops, supplies, and equipment. While aviation assets are not affected by obstacles on the streets, they are affected by light towers, signs, power lines, and other aerial obstructions. They are also vulnerable to the man-portable surface-to-air missile threat, crew served weapons, and small arms fire.

c. Categories of Built-Up Areas. Built-up areas are generally classified as:

- Villages (populations of 3,000 or less)
- Strip areas (industrialized zones built along roads connecting towns or cities)
- Towns or small cities (populations of up to 100,000 and not part of a major urban complex)
- Large cities with associated urban sprawl (populations in the millions, covering hundreds of square kilometers).

d. Characteristics of Urbanized Areas. A typical urban area consists of combinations of the *city core*, *commercial ribbon*, *core periphery*, *residential sprawl*, *outlying industrial areas*, and *outlying high-rise areas*. Each of the urban area's regions has distinctive characteristics that may weigh heavily in planning for MOUT. Most urbanized areas resemble the generalized model shown in Figure 1-2.

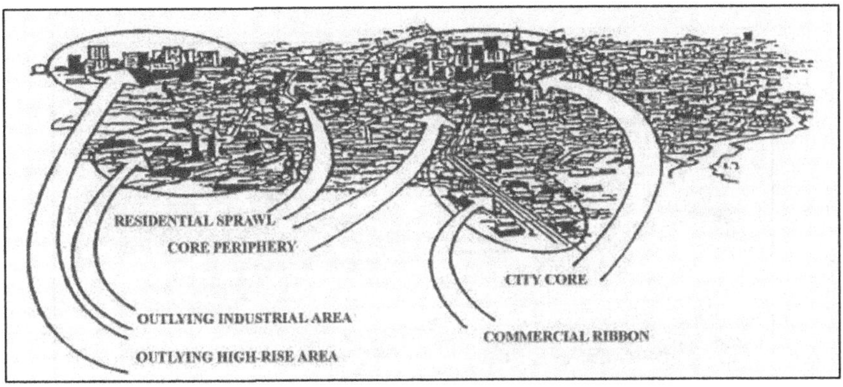

Figure 1-2. Typical Urban Area

(1) City Core. In most cities, the city core has undergone more recent development than the core periphery. As a result, the two regions are often quite different. Typical city cores are made up of high-rise buildings which vary in height. Modern urban planning for built-up areas allows for more open spaces between buildings than in old city cores or in core peripheries (Figure 1-3).

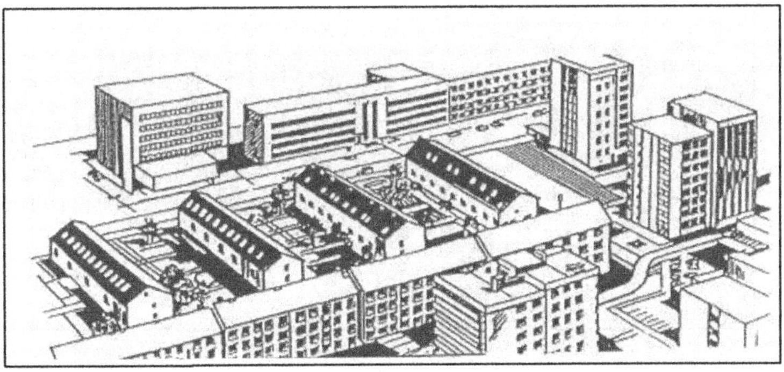

Figure 1-3. City Core

(2) Commercial Ribbon. Commercial ribbons are composed of rows of stores, shops, and restaurants that are built along both sides of major streets through built-up areas.

Typically, such streets are 25 meters wide or more. The buildings in the outer areas are uniformly two to three stories tall—about one story taller than the dwellings on the streets behind them (Figure 1-4).

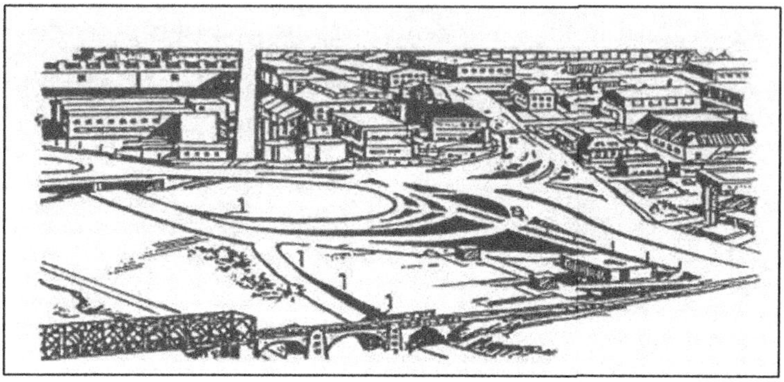

Figure 1-4. Commercial Ribbons

(3) Core Periphery. The core periphery generally consists of streets 12 - 20 meters wide with continuous fronts of brick or concrete buildings. The building heights are fairly uniform—2 or 3 stories in small towns, 5 to 10 or more stories in large cities (Figure 1-5).

Figure 1-5. Core Periphery

(4) Residential Sprawl. Residential sprawl areas consist mainly of low houses or apartments that are one to three stories tall. The area is primarily composed of detached dwellings that are usually arranged in irregular patterns along streets, with many smaller open areas between structures (Figure 1-6).

Figure 1-6. Residential Sprawl

(5) Outlying Industrial Areas. These areas generally consist of clusters of industrial buildings varying from one to five stories in height. Buildings generally vary dramatically in size and composition to match the needs of the particular businesses they house. Industrial parks are good examples of this category (Figure 1-7).

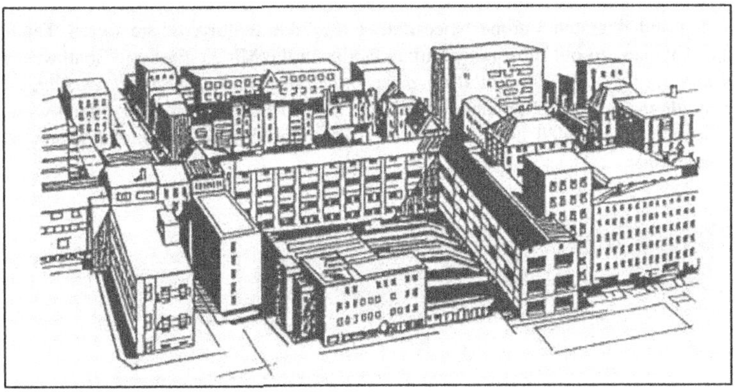

Figure 1-7. Outlying Industrial Areas

(6) Outlying High-Rise Areas. These areas are similar in composition to city core areas, but may be composed of clusters of more modern multistory high-rise buildings in outlying parts of the city. Building height and size may vary dramatically (Figure 1-8). Generally, there is more open space between buildings located in the outlying high-rise areas than is found within the city core area.

Figure 1-8. Outlying High-Rise Area

1004. Modern Battles for Urbanized Terrain. Urban warfare is as old as war itself. Since man began building villages, he has fought battles in and around them. Geography, politics, and economics dictate that cities will continue to be an objective of armies in warfare. From the armies that invaded and liberated Europe twice during the 20th century, to the forces that fought in Korea and Vietnam, to our most recent urban battles in the Middle East and Southwest Asia, *the basic principles of combat in built-up areas have essentially remained unchanged in this century*. While the principles remain the same, the introduction of helicopters, fixed-wing aircraft, armor, and precision-guided munitions (PGMs) has altered some of the techniques associated with urban combat.

Military Operations on Urbanized Terrain

Twenty two modern urban battles are discussed to illustrate the trends, dominant factors, and principles of combat in urbanized areas. (The majority of this material is extracted from the 1987 study, *Modern Experience in City Combat,* produced by Abbott Associates.)

• Aachen	• Manila
• Arnhem	• Ortona
• Ashrafiyeh	• Quang Tri City I
• Ban Me Thuot	• Quang Tri City II
• Beirut Port/Hotel (I)	• Seoul
• Beirut 1982 (II)	• Sidon
• Berlin	• Stalingrad
• Cherbourg	• Suez City
• Hue	• Tel Zaatar
• Jerusalem	• Tyre
• Khorramshahr	• Zahle

A brief description of each battle is provided to illustrate that battle's significance.

a. Stalingrad (1942 - 1943). The tenacious Soviet defense of Stalingrad cost the attacking Germans dearly in every way and set up conditions for a decisive counteroffensive. This classic urban battle involved large forces and resulted in innovative urban combat techniques and the creation of the highly successful storm groups (task-organized assault units). (Length of battle: greater than 30 days) (Casualties: 1,630,000+)

b. Ortona (1943). In this Italian town, determined resistance by a battalion of the elite German 3rd Parachute Regiment against Canadian Army attackers demonstrated the difficulty of overcoming a well-prepared defense. The Canadians were unfamiliar with urban combat and had to develop urban fighting techniques during the battle. After the town was largely destroyed and the defender had extracted a high cost in time and casualties from Canadian forces, the German parachute battalion withdrew. (Length of battle: 6 - 13 days) (Casualties: estimated in the hundreds)

c. Aachen (1944). The battle for Aachen, Germany, in the fall of 1944 developed during the U.S. First Army's offensive to breach the Westwall fortifications and the vaunted Siegfried Line. Aachen, the ancient capital of Charlemagne, had symbolic political and psychological significance to both the Germans and Americans. Furthermore, it was the first city on German soil to face an assault by the Allies. This first major battle on German soil foreshadowed bitter resistance against the American attackers in subsequent battles. The German defenders surrendered only after the city was destroyed. Although the U.S. Army had achieved a clear tactical victory, the German defense of Aachen cost the U.S. First Army valuable time and delayed the planned attack to the Rhine river. (Length of battle: 14 - 30 days) (Casualties: 8,000+)

d. Arnhem (1944). On September 17, 1944, Operation Market-Garden, the largest airborne operation in history, was launched in the Netherlands. The plan was to land three airborne divisions to seize key bridges along a 100-kilometer-long corridor through which allied mechanized forces would pass as the first step in the final offensive into Germany. The British 1st Airborne Division made a surprise landing near the Dutch city of Arnhem in order to seize a bridge over the Rhine river for advancing British forces. An unexpected German armor force counterattacked and eliminated all footholds, virtually destroying the British division before a linkup could occur. (Length of battle: 6 - 13 days) (Casualties: estimated in the thousands)

e. Cherbourg (1944). By June 17, 1944, U.S. forces advancing toward Cherbourg from the Normandy beachhead succeeded in cutting off defending German forces in the Cotentin Peninsula. Four German divisions withdrew to a perimeter surrounding Cherbourg. After much fighting, particularly in strongpoints outside the city, the German garrison surrendered to the Americans. Unfortunately for the Americans, the port facilities were destroyed which denied their early use by Allied forces. (Length of battle: 6 - 13 days) (Casualties: estimated in the thousands)

f. Berlin (1945). The long, bloody Soviet offensive to seize the German capital city effectively concluded the last battle of World War II in Europe. Bitter fighting occurred, but the defense was never well coordinated due in part to poor preparation by the Germans. (Length of battle: 14 - 30 days) (Casualties: estimated in the thousands)

g. Manila (1945). Japanese Army troops evacuated Manila under pressure from advancing American forces, but the local Japanese naval commander independently decided to hold the city at all costs. Despite defending Manila with poorly trained and equipped personnel, the determined resistance resulted in a high number of casualties to attacking U.S. forces as well as the destruction of the city and much of its population. (Length of battle: 14 - 30 days) (Casualties: 22,000+)

h. Seoul (1950). Following the Inchon landing, U.S. and Republic of Korea (ROK) forces recaptured the South Korean capital from the North Koreans. The fighting was unusual in that combat was largely centered on seizure of street barricades rather than buildings. (Length of battle: 6 - 13 days) (Casualties: Marines, 2,383; others, estimated in the thousands)

i. Jerusalem (1967). Israeli forces seized Jerusalem in a well prepared and well executed operation. Despite an uncoordinated Jordanian defense, Israeli casualties in this battle were the highest of those encountered during the Six Day War. Regular Jordanian forces withdrew during the latter stages of the battle, effectively ending organized resistance. (Length of battle: 48 hours to 5 days) (Casualties: Israeli forces, 400+; Jordanian forces, estimated in the hundreds)

j. Hue (1968). On January 31, 1968, the North Vietnamese Army (NVA) and Viet Cong (VC) forces launched their Tet Offensive at targets throughout South Vietnam. As part of this operation, two NVA/VC regiments and two sapper battalions conducted a surprise attack and seized part of the walled city (Citadel) of Hue. The NVA/VC held this part of the walled city for about three weeks against determined U.S. and South Vietnamese forces before finally succumbing. The battle for Hue is considered one of the most intense and savage battles of the Vietnam War. (Length of battle: 14 - 30 days) (Casualties: Marines, 433; others, 5000+)

k. Quang Tri City I and II (1972). An objective of the North Vietnamese 1972 winter-spring offensive was the capture of Quang Tri, the northernmost major city in South Vietnam. The NVA overwhelmed the Army, Republic of Vietnam (ARVN) defenders (I). Later, the city was recaptured (II) by a smaller ARVN force using extensive artillery and air support. The large conventional forces involved on both sides made Quang Tri I and II the major urban battles of the Vietnam War. (Length of battle: Quang Tri I, 6 - 13 days; Quang Tri II, 30 days or greater) (Casualties: battles combined, 30,000+)

l. Suez City (1973). Israeli Defense Forces (IDF) attempted to seize this Egyptian city before the anticipated United Nations (U.N.) cease-fire to end the Yom Kippur War. IDF armored shock tactics led to disaster against a well-prepared Egyptian defense. High casualties forced the IDF to withdraw. (Length of battle: less than 24 hours) (Casualties: Israeli forces, estimated 100 - 500; others, unknown)

m. Ban Me Thuot (1975). This South Vietnamese highlands town was the first strategic city to fall in the final North Vietnamese general offensive in 1975 South Vietnamese forces were surprised and overwhelmed. The fall of Ban Me Thuot resulted in a rout that the North Vietnamese exploited to achieve total victory in Vietnam. (Length of battle: 24 - 48 hours) (Casualties: estimated in the hundreds)

n. Beirut I (1976). When the Lebanese civil war broke out in the spring of 1975, combat in the capital city of Beirut assumed a central role. The battle for Beirut I was a series of small, local operations between largely irregular Christian and Muslim forces fighting over control of the hotel and port districts. Combat was not decisive, but led to changes in the boundary, called the "Green Line." This separated the antagonists and lead to the stagnation of the Lebanese conflict. (Length of battle: greater than 30 days) (Casualties: estimated in the hundreds)

o. Tel Zaatar (1976). Lebanese Christian attackers encircled and methodically besieged this Palestinian camp before overcoming its defenders with a final assault. (Length of battle: greater than 30 days) (Casualties: estimated in the hundreds)

p. Ashrafiyeh (1978). The Syrian forces occupying portions of Lebanon faced a complex political situation in which the power of the Christian militia was seen as a clear threat to stability. In an attempt to weaken the Christian militia by an attack on their center of power, the Syrians laid siege to the Christian militia stronghold of East Beirut (Ashrafiyeh). This urban battle was essentially an artillery bombardment without air attacks. Syria failed to break the will of the defenders and final positions remained unchanged. (Length of battle: greater than 30 days) (Casualties: estimated in the hundreds)

q. Khorramshahr (1980). Iranian regular forces initially evacuated this port city in the face of an Iraqi offensive. Irregular Iranian forces, however, continued to fight. They offered prolonged resistance and inflicted heavy casualties. Iraq eventually won this battle, but at a high cost in time and resources that ultimately served to halt the entire offensive against Iran. The intensity of fighting during the battle for Khorramshahr earned the city the nickname, "City of Blood." (Length of battle: 14 - 30 days) (Casualties: Iraqi, 3,000 - 9,000; Iranian, estimated in the thousands)

r. Zahle (1981). Syria laid siege to the Lebanese regular forces and militia in this key crossroads town. Fighting was inconclusive and ended in a negotiated settlement whereby the Lebanese defenders evacuated the town. (Length of battle: greater than 30 days) (Casualties: estimated in the hundreds)

s. Beirut II (1982). The siege of Beirut culminated the Israeli campaign to evict the Palestine Liberation Organization (PLO) from Lebanon. Fighting under domestic and world political pressures, the IDF besieged the PLO, selectively applying heavy ground and air firepower in conjunction with psychological warfare and limited-objective ground operations. The fighting resulted in a negotiated PLO evacuation from the city. (Length of battle: greater than 30 days) (Casualties: 2,300+)

t. Sidon (1982). Israeli forces easily seized this PLO southern headquarters during the invasion of Lebanon. The IDF was fully prepared for major urban combat using lessons learned from earlier battles, but resistance was unexpectedly light as PLO forces had largely withdrawn from the city. (Length of battle: 48 hours to 5 days) (Casualties: unknown)

u. Tyre (1982). The Israeli attack on the PLO in this Lebanese coastal city was well planned, with excellent intelligence on the target. All branches of the IDF participated in an operation that included naval fire support and amphibious landings. PLO resistance was uncoordinated and easily overcome. (Length of battle: 48 hours to 5 days) (Casualties: Israeli forces, 120+; others unknown)

1005. Modern Urban Battle Analysis and Observations. The 20th-century urban warfare experience has caused us to reevaluate old factors and consider new developments that affect the way we fight in this environment. The following discussion, based on the historical analysis of these select urban battles, illustrates the importance of maneuver warfare and combined arms philosophies in the urban environment. Factors that have had an impact on the manner in which urban warfare has been conducted are:

Intelligence is imperative to success in urban warfare. "Maneuver warfare requires a firm focus on the enemy" (MCDP 2, *Intelligence*). Few subsequent tactical changes can overcome the far-reaching impact of a major intelligence error.

a. Intelligence. The historical review of modern urban battles discloses that the attacker will usually win. Failures to win generally reflect classic military errors not characteristically unique to cities. *However, of the battles studied, battles lost were attributed to errors in* *initial intelligence*. The battles of **Arnhem** and **Suez City** probably would never have occurred had the attacker known the strength and locations of the defender's forces. At **Stalingrad**, the attacking Germans were aware of the defending forces facing them in the Sixth Army's zone, but the buildup of Soviet forces in other areas of the front was not anticipated and resulted in tactical surprise at those points, diluting the German offensive push to seize the city.

Surprise is a combat multiplier and can substantially reduce the cost of urban warfare.

b. Surprise. Surprise is a combat multiplier and can substantially reduce the cost of urban warfare. It can be achieved through deception, stealth, and ambiguity (MCDP 1-3, *Tactics*). Surprise was achieved by the attacker at **Aachen** and **Ban Me Thuot** and by the defender at **Suez City**. Surprise can be an important asset to increase leverage, but, as the failure at **Arnhem** shows, not necessarily a decisive one. When surprise is employed as a means to overcome other disadvantages, it is important to maintain accurate intelligence. In urban areas, tactical surprise by the attacker can be used to preempt effective defensive preparation of a city.

c. Combined Arms. The MAGTF must capitalize on one of the key means for gaining advantage in maneuver warfare — the use of combined arms (MCDP 1-3, *Tactics*). The use of combined arms places the enemy in a dilemma. Any action the enemy takes to avoid one combat arm makes him more vulnerable to another. An analysis of categories of weapons systems found in a MAGTF helps to illustrate it is a warfighting organization well suited for MOUT:

Combat in urban areas is primarily a small unit, infantry intensive operation

(1) Infantry. Combat in urban areas is primarily a small-unit, infantry intensive operation. Restrictions on maneuver, particularly for mechanized units, increases oportunities for infiltration. Urban combat requires small-unit leadership, initiative, and skill. Decentralized actions and difficulties in command , control, and communications are typically encountered. Built-up areas, like close terrain found in other operational environments, are generally considered to be most suited for operations conducted by infantry. Infantry units can be organized, trained and equipped to negotiate urbanized terrain that restricts observation, fields of fire, and mechanized movement.

(2) Armor. The role of armor in urban warfare can be significant. Of the 22 battles studied, armor participated in 21. In three-fourths of these battles, organic tank support was a central element when special assault teams were employed. *Overall, special assault units supported by tanks were more successful than any other task organization.*

The use of tanks to the attacker inside a city has been effective only when they were protected by infantry. Tanks in support of infantry act as an "assault gun" that delivers concentrated, sustained fires to reduce held strongpoints.

The U.S. experience in **Hue** demonstrated the key role that armor can play on a combined-arms team fighting inside the city. The Marines' most effective weapons during the battle were the M48A1 Patton tank and the M-50 Ontos. Both were protected by infantry. The M48A1, with its 90-mm main gun, was used extensively to reduce fortified positions. The Ontos, an armor-protected tracked vehicle mounting six 106-mm recoilless rifles, was highly effective against concrete and steel structures. The munitions of these armored systems provided breaches that the infantry could exploit.

Armor providing direct-fire support inside a built-up area requires protection by infantry.

In contrast, during the **Suez City** battle, Israeli armor forces attacked on "armor thrust avenues" into the city, outpacing their armored personnel carrier (APC) mounted paratroop/infantry support. The Egyptian defenders lacked organic artillery (except limited antiaircraft artillery (AAA) and mortars) and had no air support and virtually no armor support. The Egyptians prepared "kill zones" on the principal avenues down which the IDF armored forces attacked. As the lead IDF armor battalion entered the second of the three road intersection objectives, the Egyptians engaged with Sagger missiles, RPGs, ZSU-23 antiaircraft guns, antitank grenades thrown from balconies, and small arms. All of the tank commanders in the lead battalion were killed or wounded. Disabled vehicles blocked the road. Vehicles veering into the narrow side streets became trapped and were destroyed. *The lack of infantry to protect the armor proved disastrous to the attacking armor battalion.*

(3) Artillery. Artillery has played an important role in most major urban conflicts. At **Aachen,** U.S. forces combined infantry with antitank teams and artillery (in a direct-fire role) down to the squad and fire team level. Also, artillery firing shells with "delay" fuzes in an indirect-fire role were used to penetrate one or more floors before exploding, thus driving the enemy to the ground where infantry and armor could attack. Artillery was also positioned to fire *perpendicular* to the direction of movement of assaulting forces. Thus, fratricide from artillery range errors was alleviated.

Artillery has two distinct roles: outside the built-up area to isolate or prevent isolation with indirect-fire; and within the built-up area to provide direct-fire support.

Artillery proved most useful for interdicting enemy supplies, enemy evacuation, and the movement of reinforcements in the enemy's rear; for physically and psychologically harassing the enemy; and in direct-fire roles within a city.

Artillery employed in the indirect-fire role has been effective in disrupting defenders in half of the studied battles. However, some indirect-fire roles have proved more effective than others in urban combat. Artillery was most effective in the interdiction of supplies, enemy evacuation, movement of reinforcements in the enemy's rear (outside the city), and for indiscriminate physical and psychological harassment of the enemy. Artillery used as an indirect-fire siege weapon, as was done at **Ashrafiyeh** and **Zahle**, proved ineffective. Artillery can also cause problems for the attacker. The rubble resulting from indirect artillery fires can create considerable obstacles for the attacker while providing the defender with obstacles, materials, cover, and concealment.

Artillery employed in the direct-fire role proved useful in the reduction of strongpoints. Self-propelled artillery was used effectively inside **Aachen** and **Stalingrad** and more recently by the IDF at **Beirut II**.

(4) Mortars. The mortar is the most used indirect-fire weapon in urban combat. The mortar's high angle of fire allows the round to reach the street level accurately without being masked by surrounding structures. During the battle for **Hue** city, the most effective fire support provided to Marines was the indirect fire from 60-mm, 81-mm, and 4.2-inch mortars. Approximately 20,000 rounds of high explosive (HE) 60-mm and 81-mm mortar ammunition were expended during the battle. The 4.2-inch mortar was used primarily to deliver riot control agent munitions. It was discovered that 4.2-inch riot control agent munitions could be fired with great effectiveness through the tops of buildings to drive the enemy out. When the enemy emerged from their concealed positions, HE rounds were fired to complete the attack.

(5) Antiaircraft Artillery. Antiaircraft artillery (AAA) was extremely useful in a ground fire role in some urban battles. AAA was used only rarely in World War II, and then generally against assaulting personnel rather than against structures. AAA has been used more frequently in more recent events, but against buildings rather than people. *The high rates of fire of modern AAA make it excellent in terms of shock and destructive potential.* However, ammunition supply can be a problem because the volume of fire. During **Hue**, the Duster, an AAA vehicle mounting twin 40-mm guns, was provided to the Marines by the U.S. Army. These guns were extremely effective in the suppression of enemy positions due to the lethality of the 40-mm round and the quantities in which it could be delivered. During the **Suez City** battle, the Egyptians used the ZSU-23 antiaircraft gun in the direct-fire role against armored columns. This gun proved to be frightening and effective. Its high volume of explosive power created shock among IDF armor personnel akin to the shock that the IDF armor thrust tactics were designed to create. During **Sidon,** the IDF used 20-mm antiaircraft Vulcans in the direct-fire role against ground targets with great success.

(6) Aviation. Historically, aviation assets have played an important role in helping to isolate the objective and to interdict the flow of the defender's supplies and reinforcements. However, aviation has been relatively ineffective when not used in conjunction with ground forces. In past wars, bombing operations have been used in attempts to reduce the defender's will to resist and destroy their physical capabilities. In the majority of urban battles, aerial bombing, by itself, did not erode the defender's will to resist, nor did it significantly degrade the defender's military capabilities.

Aviation plays an important role in interdicting movement of the defender's supplies and reinforcement into the built-up area.

The Marine air-ground task force is employed as a combined arms team. As such, the aviation combat element will not usually be called upon to conduct independent operations in MOUT. It will function as part of a MAGTF and, regardless of operating on or over urbanized terrain, will still execute the six functions of Marine aviation (i.e., offensive air support, assault support, antiair warfare, electronic warfare, air reconnaissance, and control of aircraft and missiles) in support of the MAGTF.

In future urban warfare, aviation will be even more effective due to advances in fixed- and rotary-wing aircraft, unmanned aerial vehicles, precision guided munitions, improved munitions, communications, sensors, and targeting systems. Our battle study already indicates a trend toward more extensive aviation participation in MOUT. As an example, the IDF in **Beirut II** employed bombing by fixed-wing aircraft using cluster bomb units, "smart" bombs, phosphorous, and other munitions. Attack helicopters operated on the outskirts of the built up area with impunity, and medical evacuation (MEDEVAC) proceeded swiftly and efficiently using helicopter support.

d. Combat Forces. Whether attacking or defending, the size of the force relative to the enemy can be critical to success. When provided with adequate forces, the attacker can isolate and encircle the defender and prevent a breakout or linkup. The defender can use them to create a mobile defense or to create strong reserves for counterattacks.

Other factors that impact on the size of forces required are the degree of surprise achieved and the firepower utilized (aviation, armor, artillery, mortars, etc.). They should be weighed against the sophistication of the prepared defense. Ample consideration should be given to the local population, degree of external support, and utilization of existing services (communications, water, etc.).

Regardless of the size or quality of defensive forces, the defender usually extracts large costs from the attacker in time, resources, and casualties.

In the historical examples reviewed, the defender was usually outnumbered by the attacker, the quality of the defender's available forces was inferior, and defeat of the forces defending the city was usually certain. However, regardless of the size or quality of defensive forces, the defender can extract enormous costs to the attacker in time, resources, and casualties. As was seen at **Khorramshahr**, the Iranian defenders, outnumbered 4 to 1, still held the city for 26 days. Although the Iranian defenders eventually lost the city, its defense allowed the remaining Iranian forces time to organize and redeploy. Furthermore, the winter rains that followed the battle turned much of the region into a sea of mud and largely halted further Iraqi efforts. The Iraqi army's offensive thrusts into Iran lost momentum as a result of the defense of Khorramshahr.

e. Special Assault Teams. In these battle studies, "shock units" or "special assault teams" have been used by attackers (and often by defenders) with great success. These organizations were characterized by the integration of combined arms at the battalion level and below. Control was decentralized to lower echelons (down to the squad level in some cases). Assault teams typically contained infantry with various combinations of armor, artillery, and engineers. Hence, all combat arms should plan, train, and develop common tactics, techniques, and procedures for use on urbanized terrain. Before the battle of **Aachen**, U.S. ground forces conducted intensive urban training with antitank teams and artillery pieces down to the squad and fire team levels. In contrast, the defender, in most cases having to fight with whatever forces were locally available, had no opportunity to conduct combined-arms training.

The use of combined arms warfare on urbanized terrain is imperative.

f. Time. In most cases, successful conclusion of an urban battle took two to three times longer than the initial estimates. Consequently, the additional tome resulted in the expenditure of more logistics and the loss of more personnel then initially anticipated. This often had adverse affects on the overall campaign. Well-planned urban defense, even if the defender is isolated or lacking in aviation, armor, or artillery weapons, can be time consuming to the attacker. Time can allow the defender to reorganize, redeploy, or marshal resources in other areas.

Generally, urban warfare is time consuming.

Three battles where time played an unanticipated critical role in the attacker's strategic timetable were **Aachen**, **Khorramshahr**, and **Stalingrad**. In these battles, the defenders delayed the attacker longer than was estimated, resulting in the modification of operational or strategic plans.

g. Isolation. *The attacker won all urban battles where the defender was totally isolated.* Even the partial isolation of the defenders resulted in attackers enjoying a success rate of 80 percent. Conversely, attackers won only 50 percent of the battles in which defenders were not significantly isolated, and those victories came at great cost.

No single factor is more important to the attacker's success than isolation of the urban area.

In most urban battles, some form of isolation occurred as a result of the attacker's actions. While it is unrealistic to envision complete isolation of a city until enemy forces to the rear of the city are pushed far beyond its outermost boundaries, *total isolation does not appear necessary.* The key to the attacker's success is in stemming the *unimpeded* flow of manpower, supplies, and weapons to replace the defender's losses.

h. Cost. The cost of conducting urban warfare is relative to the percentage of total expended resources, the time elapsed, and the results achieved. The cost to the attacker was considered high in the majority of urban battles. A high cost does not necessarily imply that the results were not worth the price. *The attacker and defender must thoroughly evaluate the overall cost prior to committing to an urban battle.*

From the offensive point of view, several factors can be associated with cost. First, and most importantly, isolating the urban area is critical to the attacker's success. Second, overwhelming superiority is needed if all costs are to be minimized. Third, the operation should be carefully planned. Fourth, intelligence is invaluable. Knowing where and how the city has been prepared for defense is also important. Fifth, attacking forces should understand the unique nature of urban combat. Clear tactics, techniques, and procedures for urban combat is required. Every aspect, from taking a building to using destructive technology and coordinating combined arms, must be thoroughly understood. Careful consideration of these five factors can minimize the cost of urban warfare to the attacker.

Attacker cost was generally high in casualties, time, and resources in the majority of urban battles studied.

From the standpoint of the defender, the critical variable is the defensive preparation of the city. Defensive preparations should include measures to prevent isolation of the city by the attackers. The capture of a prepared city can be made to be extremely costly. Preparations can include creating kill zones, clearing fields of fire, constructing canalizing obstacles, establishing reinforcing and fall-back positions, decentralizing command and control, and organizing multiple movement routes above ground, between rooftops, and below ground in subways and sewers. Artillery and aviation support can also be valuable force multipliers in defense.

i. Rules of Engagement (ROE). The nature of the military operation may restrict our use of weapons. The majority of urban battles since 1967 (such as **Beirut II**, **Hue**, **Jerusalem**) have had one or more of the following restrictions imposed on the attacking force:

(1) Minimizing civilian casualties and/or collateral destruction in order to:

- Avoiding alienation of the local population
- Reducing the risk of adverse world or domestic opinion
- Preserving facilities for future use
- Preserving cultural facilities and grounds.

(2) Limiting the use of specific ground or air weapons.

j. Logistics. Timely combat service support, particularly in the areas of ammunition resupply and casualty treatment and evacuation, is a critical element in MOUT.

(1) Logistics Support. Historically, combat in urban environments has seen a dramatic increase in the amounts of Class IV (shoring, sandbags, concertina wire), Class V (ammunition), and Class VIII (medical material). Unique items, such as rope, grappling hooks, and ladders are required for operations on urbanized terrain. Intense close-quarter combat requires a continuous flow of ammunition, particularly small-arms, tank, antitank, mortar, and artillery ammunition, as well as mines, grenades, and demolition explosives. Medical supplies must be readily available to treat the anticipated increase in casualties. Once battle the has been initiated, combat forces will require continuous supply.

Urban warfare requires a flexible, balanced logistics system capable of sustaining the close-quarter fighting evolutions within the built-up area.

One method of providing continuous logistics support is to establish a "push system." Essentially, this system pushes supplies to fighting units without their having to request them. A "push system" should negate the inherent delays of a "pull system," which requires units to request supplies then await their arrival. Optimally however, a balanced approach should be taken to sustain engaged forces (MCDP 4.). Combat service support, in terms of its timeliness and anticipation of the true needs of the combat units, should sustain the tempo of operations. It should avoid the delivery of unwanted or unnecessary supplies. To achieve this balanced approach, logisticians must develop flexible support plans that readily adapt to the ebb and flow of urban combat.

Mobile aid stations and surgical teams should be positioned as far forward as the tactical situation allows and be provided with dedicated evacuation vehicles.

(2) Health Service Support (HSS). Responsive treatment and evacuation plans should be established to handle the expected increase in casualties. Both immediate treatment and evacuation are critical in maintaining the morale and confidence of forces engaged in urban warfare. To meet casualty treatment and evacuation needs, mobile aid stations with surgical teams should be placed as far forward as the tactical situation allows. Dedicated vehicles and aircraft for the evacuation of casualties should be provided.

In addition to casualties resulting from physical injury on the battlefield, there are often considerable psychological casualties in modern battles, particularly urban battles. This can be attributed to the intense, sustained, and close-quarter combat associated with urban warfare. In many battles, attacking troops were not properly trained for urban warfare and felt inadequate in their ability to fight on equal terms with the defender. Many felt moral or ethical dilemmas associated with collateral damage and civilian casualties. Finally, the constant threat of being killed from any quarter coupled with the extreme fatigue of fighting in an urban environment can create psychological casualties. Leaders at all levels should be attuned to the symptoms associated with psychological casualties in order to get effected individuals prompt treatment so that they can be returned to their units.

1006. Implications of Urban Warfare. The commander charged with making decisions needs to understand the operational and strategic implications of a tactical struggle in an urban area (MCDP 1-1, *Strategy*, and MCDP 1-2, *Campaigning*.). Three urban battles (**Stalingrad**, **Hue**, and **Beirut II**) illustrate the importance of seeing beyond the tactical nature of the battle.

Commanders must see beyond the immediate tactical implications of fighting an urban battle.

The battle for **Stalingrad** had major operational and strategic implications. The entanglement of German forces at Stalingrad bought time for the Soviets to mobilize and prevent a major combined effort of two German Army Groups in the south. While Stalingrad resulted operationally in the destruction of the German Sixth Army, the strategic results were even greater. Stalingrad caused a complete change of German strategy in the east. Hitler made major changes in his General Staff, and from this point on in the war, he was a man estranged from his military leadership. The loss of enough men and equipment to field one-fourth of the German Army shook the foundations of the Third Reich.

The Battle for **Hue**, although only one of over one hundred different attacks of the Tet Offensive of 1968, had a negative impact on the will of both the American people and their political leadership. Hue marked a revolution in the coverage of war by modern mass media. It was the first time Americans could and watch an ongoing battle from their living room on the evening news. Hue was a television bonanza for almost a month. When North Vietnamese leadership directed that Hue be held for at least seven days, it was clearly *not* their intent to win a tactical battle, but to strike at the *strategic center of gravity*—in this case, the will of the American people. Although the battle for Hue was a tactical victory for the U.S., the North Vietnamese had achieved their strategic goal of making the American public question the costs associated with the war.

During the battle of **Beirut II**, the IDF's objective was to drive the PLO from Lebanon. The ability of the PLO to leverage the media to gain an advantage was one of the most significant weapons in their arsenal. Despite an Israeli tactical victory, the costs in image, prestige, allies, and, most importantly, its own national will were enormous. The results of this battle eventually caused a change of political leadership at the highest levels of government.

1007. Key Insights. Analysis of modern urban battles provides insight into the major factors affecting the course and outcome of combat. Consideration of these key insights are required in order to effectively and efficiently plan and execute MOUT:

a. MOUT is infantry intensive.

b. A tactical battle may have far-reaching implications. No longer are the strategic, operational, and tactical levels overlap, creating the situation where tactical actions can have operational and possibly strategic repercussions (MCDP 1-1, *Strategy*, and MCDP 1-2, *Campaigning*).

c. Commanders at all levels must understand the impact that media representation will have on the accomplishment of operational and strategic objectives.

d. Maneuver warfare doctrine must be applied to the environment.

e. Intelligence is imperative to success in urban warfare.

f. Surprise is a combat multiplier.

g. Armor, artillery, and aviation are effective at the outer perimeter of built-up areas for causing isolation or preventing reinforcement.

h. Armor operating inside a built-up area must be protected by infantry.

i. Artillery providing direct fire inside a built-up area can be effective in the reduction of strongpoints.

j. As force ratio increases in favor of the attacker, combat duration decreases.

k. Urban warfare is time consuming.

l. Isolation of an urban defender ultimately ensures his defeat.

m. Attack of an urban area is costly to the attacker in terms of resources and casualties.

1008. Necessity for Preparation. Since MAGTFs are usually forward-deployed forces, it is inevitable that MAGTFs will be tasked to deploy to urban areas and conduct military operations. Fighting in this environment will be violent, close, and personal. Fighting in villages, towns, and cities are likely to become more frequent for military forces as the United States responds to global crises. Forces will be expected to exercise adaptability in responding to a wide variety of missions in the world's built-up areas. Marines may soon find themselves operating in the conceptual "three block battlefield" within a city grid: feeding refugees in one block, separating combatants in the adjoining block, and returning hostile fire in a third block.

MOUT is infantry intensive. The remaining chapters and appendices of this publication focus on how the GCE fights and integrates supporting elements in an urban environment.

Chapter 2

Offensive Operations

Urban combat isolates and separates units. Operations are reduced to a series of small-unit actions, placing a premium on small-unit leadership, initiative, and skill.

2001. Introduction. The Marine Corps' manuever warfare philosophy guides the conduct of offensive operations in MOUT. This warfighting philosophy serves to guide commanders through the development, planning, and execution of missions. A detailed discussion of maneuver warfare can be found in MCDP 1, *Warfighting*, MCDP 1-3, *Tactics*, MCWP 3-1, *Ground Combat Operations* (under development), and other Marine Corps doctrinal publications.

This chapter focuses only on offensive operations in an urban environment where collateral damage is of minimal concern. It describes tactics, techniques and procedures to be employed for seizing and clearing buildings and other urbanized areas. Chapter 6 addresses some considerations associated with fighting under constraints and restraints on urbanized terrain.

Section I

Planning

2101. Considerations. Identification of the adversary's centers of gravity and critical vulnerabilities will allow the commander to focus his efforts on those portions of the built-up area essential for mission accomplishment.

a. Reasons for Attacking a Built-Up Area. A commander considers the following before deciding to attack a built-up area:

(1) Tactical Advantage. Cities control key routes of commerce and provide a tactical advantage to the commander who controls them. Control of features such as bridges, railways, and road networks can have a significant impact on future operations. Urbanized areas may be used by the enemy as a base of operations from which they launch their own offensive operations. It may be advantageous to attack those bases and separate the enemy from their support infrastructure.

(2) Political Advantage. The political importance of a built-up area may justify the use of time and resources to liberate it. Capturing a city could destroy the seat of local and national government. At the very least, it could deal the enemy a decisive psychological blow.

(3) Economical Advantage. The destruction or capture of key industrial and commercial cities with the resulting denial of production and distribution of equipment and supplies strikes at the enemy's future ability to wage war. The requirement for a logistics base, especially a port or airfield, may play a pivotal role in the enemy's ability to continue the conflict. Capture of such cities may prove extremely beneficial to the attackers, who can use these resources to their advantage.

(4) Potential Threats to Operations

(a) Enemy Threat Too Great To Bypass. Though the terrain around a built-up area may facilitate its bypass, the enemy within that urbanized area may remain a threat capable of interdicting lines of communications. This may require the enemy force to be contained or destroyed.

(b) Terrain Does Not Allow Bypass. The urbanized area may sit between two natural slopes on the avenue of approach and thus require capture in order to secure the main supply route. Additionally, the urbanized area, itself, may sit on dominating terrain that threatens combat support and CSS elements.

b. Reasons for Not Attacking a Built-Up Area. The commander considers the following reasons for not attacking a built-up area:

(1) Built-Up Area Not Required To Support Future Operations. The attacker may have adequate supply support and resources established at another site from which combat forces can be supported. Therefore, during the estimate process, commanders may assess that the urbanized area is not necessary to support future operations.

(2) Enemy Not a Threat. The commander may decide to bypass upon determination that no substantial threat exists in the built-up area that could affect the unit's ability to accomplish its mission.

(3) Time or Risk Unacceptable. The commander's intent may dictate that speed of movement is essential to the mission. Because MOUT can be time-consuming, the commander may choose to bypass the urbanized area to maintain tempo. Furthermore, the potential for numerous casualties, the expenditure of critical resources, or the restrictions placed on attacking forces may result in unacceptable risks to the commander's primary mission.

(4) Declared an Open City. The area may have been declared an "open city" because it is undefended or of religious or historical significance. By international agreements, open cities are demilitarized and must be neither defended nor attacked. (See HR, 25, Annex to Hague Convention No. IV, *Embodying Regulations Respecting the Laws and Customs of War on Land*, 18 October 1907.) The attacking force must assume civil administrative control and treat the civilians as noncombatants in an occupied country. The defender must immediately evacuate and cannot arm the civilian population. A city can be declared open only before it is attacked. Other reasons for not defending could be the presence of large numbers of noncombatants, hospitals, or wounded personnel or the city's cultural, religious, or historical significance.

2102. Commander's Estimate. Once assigned the mission to attack a built-up area, a thorough mission analysis is conducted. Marines follow the same planning process in MOUT as for other operational environments by evaluating the mission, enemy, terrain and weather, troops and support available, and time available (METT-T). Based on METT-T and application of maneuver warfare concepts, the commander can decide on a plan that defeats the enemy by exploiting identified critical vulnerabilities. The commander may not have sufficient forces to fight everywhere and so must focus on the enemy and concentrate resources to decisive ends. This may involve making conscious decisions to accept risk at some place or time and to seize opportunities at others. The challenges of an urbanized environment dictate an even greater need to fully understand maneuver warfare.

This section addresses METT-T considerations as they pertain to the offense in urban warfare. METT-T is covered in detail in MCRP 3-11.5A, *Marine Troop Leader's Guide.*

a. Mission. Mission analysis is the first step of the estimate process. One of the important mission analysis considerations in urban conflict is the scope of clearance required to accomplish the mission. Commanders and planners should ask themselves the following questions:

- Do I need to clear every building?
- Should I clear only certain blocks?
- Should I only control certain areas?
- What level of protection is required for my lines of communication?

b. Enemy. Maneuver warfare focuses on the enemy. The intelligence preparation of the battlefield (IPB) process should be used to analyze the enemy and terrain in detail. (See Field Manual (FM) 34-130, *Intelligence Preparation of the Battlefield.*) The commander decides whether threat forces are conventional or unconventional.

(1) Conventional Forces. Most potential adversaries have adopted techniques of urban combat from either the United States or the former Soviet Union. Therefore, potential adversaries will build their urban defense to counter the attack of a combined arms force. Potential enemy forces will normally organize their defense in depth. Prepared strong points will usually form the perimeter of a larger defense while reserves locate in a separate position within the perimeter. You can expect ambushes to fill gaps in the perimeter while dummy strongpoints are established to deceive the attacker. Positions for securing the entrances to and exits from underground structures and routes are normally established. Security positions will normally be located forward of first-echelon defensive positions.

(2) Unconventional Forces. Urban areas have become a haven for unconventional forces. The large noncombatant population provides cover and concealment for unconventional force operations. Conventional forces operating in MOUT will normally be placed under restrictive ROE to minimize collateral damage. Unconventional forces will often use our restrictive ROE and the noncombatant population to their advantage when devising an urban defense. (See FM 34-130, FM 7-98, *Operations in a Low-Intensity Conflict*, and Chapter 6 of this manual.)

c. Terrain and Weather

(1) Terrain. Offensive operations should be tailored to the urban environment based on a detailed analysis of urbanized terrain (see Appendix I). Commanders and subordinate leaders use **KOCOA** (**k**ey terrain, **o**bservation and fields of fire, **c**over and concealment, **o**bstacles, and **a**venues of approach) to identify important terrain factors:

(a) K: Key Terrain. Key terrain surrounding an urban area can facilitate entry or deny escape. Within the city, airports or airfields, stadiums, parks, sports fields, school playgrounds, public buildings, road junctions, bridges, or industrial facilities may be key terrain.

Critical public buildings are identified during the terrain-analysis phase of an IPB. Hospitals, clinics, and surgical facilities are important because the laws of war prohibit their attack when not being used for military purposes other than medical support. The locations of civil defense, air raid shelters, and food supplies are critical in dealing with civilian affairs. Additionally, population size, locations, and density; density of the built-up area; firefighting capabilities; the location of hazardous materials; police and security capabilities; civil evacuation plans; and key public buildings should be identified.

(b) O: Observation and Fields of Fire. Urbanized terrain is characterized by restrictive observation and fields of fire. Weapon ranges can be greatly reduced because of buildings and other manmade structures. On the other hand, high ground or tall buildings can provide perches which enhance line of sight (LOS) for observation and communications as well as for individual and crew-served weapons. This includes man-portable surface-to-air missiles.

(c) C: Cover and Concealment. Buildings, sewers, and subways can provide excellent cover and concealment for enemy and friendly forces. They also provide covered and/or concealed maneuver routes within the built-up area. The civilian population can also offer cover and concealment to enemy forces.

(d) O: Obstacles. Natural or manmade obstacles restrict or deny maneuver within the urban area. Bridges, walls/fences, canals, streams, rivers, as well as rubble created by the effects of weapons should be thoroughly analyzed. Construction sites and commercial operations such as lumberyards, brickyards, steelyards, and railroad maintenance yards are primary sources of obstacle and barrier construction materials.

These sites can also supply engineers with materials to strengthen existing obstacles or to set up antitank hedgehogs or crib-type roadblocks.

(e) A: Avenues of Approach. Avenues to the city should support maneuver and be concealed either by terrain, darkness, smoke, or a combination of the three. Avenues that canalize or choke maneuver, due to the density of built-up area or natural terrain, should be identified and avoided. Roads, rivers, streams, and bridges provide high-speed avenues for movement. **Generally, military maps do not provide enough detail for urbanized terrain analysis.** They usually do not show the underground sewer system, subways, underground water system, mass transit routes, and power plants. Local maps of intracity road networks and subway systems and city hall/department of public works blueprints of all city buildings (if available), coupled with aerial photos, should be used.

(2) Weather. As in any military operation, weather affects equipment, terrain, and visibility, but its greatest impact is on the individual Marine. Snow, ice, dust, wind, rain, humidity, and temperature extremes reduce human efficiency. Weather extremes coupled with stress and the physical strain of urban combat can be minimized with effective small-unit leadership. Weather factors include:

(a) Precipitation. Rain or melting snow may flood basements and subway systems. This is especially true when automatic pumping facilities that normally handle rising water are deprived of power. Flooding makes storm and other sewer systems hazardous or impassable. In an NBC environment, chemical agents can be washed into underground systems by precipitation. As a result, these systems may contain chemical agent concentrations that are much higher than surface areas and thus become contaminated "hot spots." Hot spot effects become more pronounced as agents are absorbed by brick or unsealed concrete sewer walls.

(b) Fog. Many major cities experience fog, especially those located in low-lying areas and along canals or rivers. Industrial and transportation areas are the most likely to be affected by fog because of their proximity to waterways. Fog adversely affects vision and optical aids. It may also be used to help conceal friendly movement.

(c) Inversion Layers. Air inversion layers are common over cities, especially cities located in low-lying "bowls" or in river valleys. Inversion layers trap dust, smoke, chemical agents, and other pollutants, reducing visibility and often creating a greenhouse effect, which raises ground and air temperatures.

(d) Temperature. Built-up areas are often warmer than surrounding open areas during both summer and winter. This difference can be as great as 10 to 20 degrees hotter throughout the year.

(e) Wind Effects. Generally, wind chill is not as pronounced in built-up areas. However, the configuration of streets, especially in the city core and outlying high-rise areas, can cause wind tunneling. This increases the effects of the wind along streets that parallel the wind direction, while cross-streets remain relatively well protected. Wind tunneling can have a negative impact on aviation support within the urbanized area by creating dangerous wind effects down streets and between buildings.

(f) Day/Night Differences. Night and periods of reduced visibility favor surprise, which in turn, may facilitate infiltration, detailed reconnaissance, attacks across open areas, seizure of defended strongpoints, and reduction of defended obstacles. Because of the difficulties of night navigation in restrictive terrain (usually without reference points and in close proximity to the enemy) forces may consider using simple maneuver plans with easily recognizable objectives.

Rotary-wing aircraft are best utilized at night when operating on urbanized terrain. The most effective method of protecting helicopters from a successful engagement by hostile ground forces is to remain unseen. During night flights, the use of night vision goggles (NVGs) should be weighed carefully. Artificial lights can render NVGs partially ineffective and can increase the possibility of mid-air collisions by degrading the goggles to the point where nearby aircraft and aerial obstructions are not seen.

(g) Aviation Weather Considerations. Weather conditions can be a critical factor in determining the amount of aviation support the aviation combat element can provide in an urbanized area. Aviation forces should consider the following:

- Presence or absence of fog, industrial haze, low clouds, heavy rain, and other factors that limit visibility for pilots
- Illumination and moon phase/angle during night vision goggle operations
- Ice, sleet, and freezing rain that degrade aerodynamic efficiency
- Updrafts and downdrafts in the urban canyon that may present an unpredictable hazard
- High temperatures and/or high-density altitudes that degrade aircraft engine performance and lift capability
- High winds (large gust spreads) and crosswinds, to include the tunneling effect, that may create localized and unpredictable hazards to aviation
- Weather conditions that create hazards on pick-up zones and LZs, such as blowing dust, sand, or snow.

d. Troops and Support Available. The commander must be aware of all of his combat power (e.g., types of weapons systems, their numbers, and their capabilities) when performing analysis for operations in a built-up area. In the initial planning phase, force size must be evaluated in relation to urban size, enemy forces, and mission assignment. In the attack of a built-up area (population 100,000+), the GCE of a MEF would be a Marine division. However, in the initial stages of an amphibious or MPF operation, a reinforced infantry battalion or regiment may provide the initial forces. No matter what the size GCE, operations will be conducted by task-organized, reinforced battalions and their companies/platoons. Therefore, by using the infantry battalion as the basis for projecting the size of an overall force, the number of regiments or divisions required to secure a built-up area can be determined. Whatever size MAGTF is required for operations in an urbanized area, much of the fighting may be generally conducted by small teams of Marines. Consequently, our success in urban fighting may largely depend upon small unit and individual tactical skills. Specifically, leaders at all levels must analyze the following factors:

- Number and type of available units
- Task organization for urban combat (See Appendix A.)
- Availability of critical weapons systems
- State of training and discipline (training for urban warfare is imperative)
- Strength in terms of men and materiel in relation to enemy and size of built-up area
- Aviation support available
- State of maintenance and supplies
- Available combat support (based on availability and mission requirements)
- Available CSS (critical logistical and maintenance items)
- Host nation support available. (If the city's occupants are friendly to the attacker, then support in intelligence, deception, and diversion may be possible.)

e. Time Available. The following issues should be considered when analyzing the time available for an attack on urbanized terrain:

- A significant amount of time is required for clearing buildings, blocks, or axes of advance.
- Marines tire more quickly when clearing buildings because of stress and additional physical exertion.

- Adequate time should be allocated for thorough reconnaissance, planning, and rehearsals. Ultimately, this can result in higher tempo in the execution of the plan.

Once an estimate of the situation (using METT-T) has been conducted, planning for offensive operations can proceed. It is important to remember that the initial estimate of the situation must be continually updated throughout the planning process and the conduct of operations to account for changes in METT-T. (See MCWP 5-1 *Marine Corps Planning* [under development])

2103. Phases of the Attack. Attacks are categorized as either hasty or deliberate. Both hasty and deliberate attacks should take advantage of as much planning, reconnaissance, and coordination as time and the situation permit. Regardless of the size of the attacking force or of the objective to be secured, the phases of the attack (i.e. reconnoiter, isolate, secure foothold, and control the objective) remain constant.

a. Deliberate Attack. A deliberate attack *is a type of offensive action characterized by preplanned coordinated employment of firepower and maneuver to close with and destroy or capture the enemy* (Jt Pub 1-02). It is a fully coordinated operation that employs the MAGTF's combined arms team against the enemy's defense. It is used when enemy positions are well prepared, when the built-up area is large or severely congested, or when the element of surprise is lost. Given the nature of urbanized terrain, the deliberate attack of a built-up area is similar to the assault of a strongpoint (see MCWP 3-1, *Ground Combat Operations* [under development]). The deliberate attack of a built-up area is conducted in the following phases:

(a) Phase I: Reconnoiter the Objective. Intelligence gathering and reconnaissance/ surveillance are critical to the planning process and success of the operation. All available intelligence resources should be used to gain vital information on the enemy. Whenever possible, a leader makes a personal reconnaissance of the objective area to collect first-hand information regarding the area to be attacked. A thorough reconnaissance yields a wealth of information to use in the continued development of the plan. Avenues of approach, observation posts (OPs), supply routes, and the emplacement positions of direct- and indirect-fire weapons systems are all examples of information that may be amassed during the reconnaissance of the objective area. Composition and structure of buildings and roadbeds, cover and concealment opportunities, and other information not apparent in a map study may have a significant impact on the plan.

(b) Phase II: Isolate the Objective. The objective can be isolated by seizing natural and man-made features that dominate the area. Isolation may also be accomplished by coordinated use of supporting arms to seal off enemy lines of communication. This phase may be conducted simultaneously with Phase III (securing a foothold). Figure 2-1 shows a task-organized, reinforced infantry battalion isolating the objective.

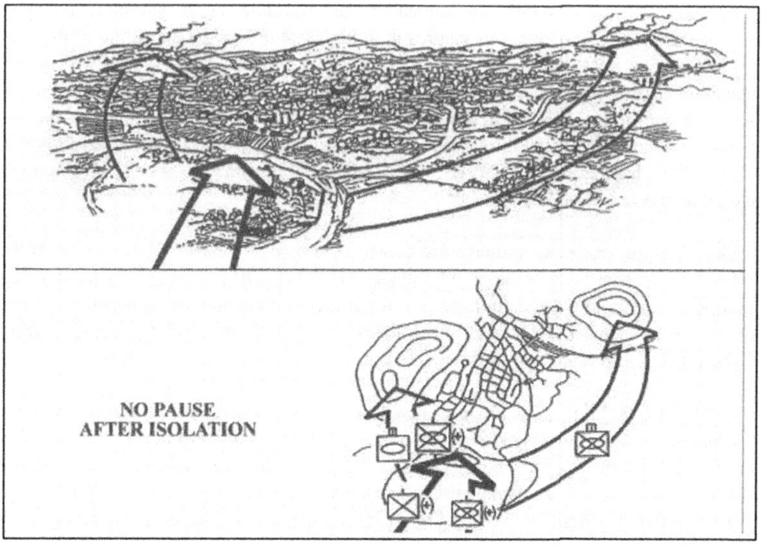

(c) Phase III: Secure a Foothold. Once the objective is isolated, a foothold should be secured as soon as possible in order to maintain tempo. The seizure of a foothold provides the attacking force with a position from which they can continue the assault through the objective area. The assault is supported by direct and indirect suppressive fires and smoke. Figure 2-2 (pg. 2-10) shows the battalion objective being isolated. One company is providing suppressive fires while another company seizes an initial foothold. Fire support assets are providing suppressive fires on the objective.

Figure 2-1. Isolation of the Objective Area

(d) Phase IV: Seizing the Objective. Once a foothold is seized and consolidated, supporting forces move to the built-up area to support the seizing of the objective area. To maintain tempo, the transition between the phases should be seamless. Once the foothold has been established, forward units continue the attack through the objective area. Supporting units assist as required. The momentum of the assault is continued until the objective area is cleared or controlled.

The assault force should establish limited objectives to ensure that the attacking forces do not get strung out along the axis of advance. Gaps may give the enemy the opportunity to infiltrate along the line of advance or make isolated friendly forces vulnerable to attack.

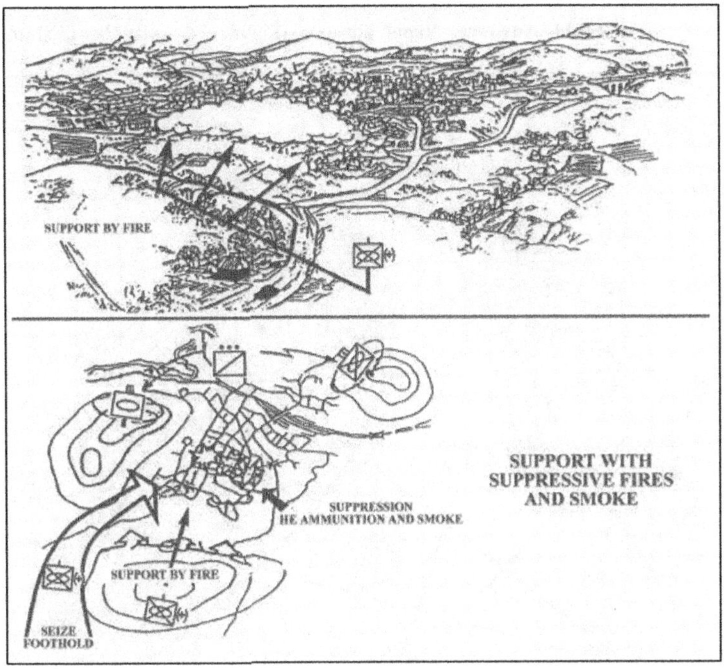

This phase also includes consolidation of the objective area and reorganization for future taskings. By setting limited objectives, the attacking forces have an opportunity to reorganize and defend against counterattacks while maintaining momentum. Once the limited objectives have been secured, the urbanized area may be sectored for detailed clearing. Clearing by sectors may help in the allocation of forces and in the prevention of fratricide. In addition to defensive preparation against counterattack, preparation should be made for follow-on missions (e.g., restoration of civilian facilities, marking and clearing minefields, etc.).

Figure 2-2. Seizing a Foothold

Seizing may consist of a systematic house-by-house, block-by-block advance through the entire zone of action, or it may be a rapid advance through a lightly defended area to seize a key objective. The difference between the two techniques is made by the commander on the basis of mission requirements and METT-T analysis. Seizing an objective in urbanized terrain requires detailed planning, coordination, decentralized command and contol, and small-unit execution. It is a continuation of the assault until all mission requirements are met.

(1) Rapid Advance. Rapid advance (Figure 2-3) may be used to maintain

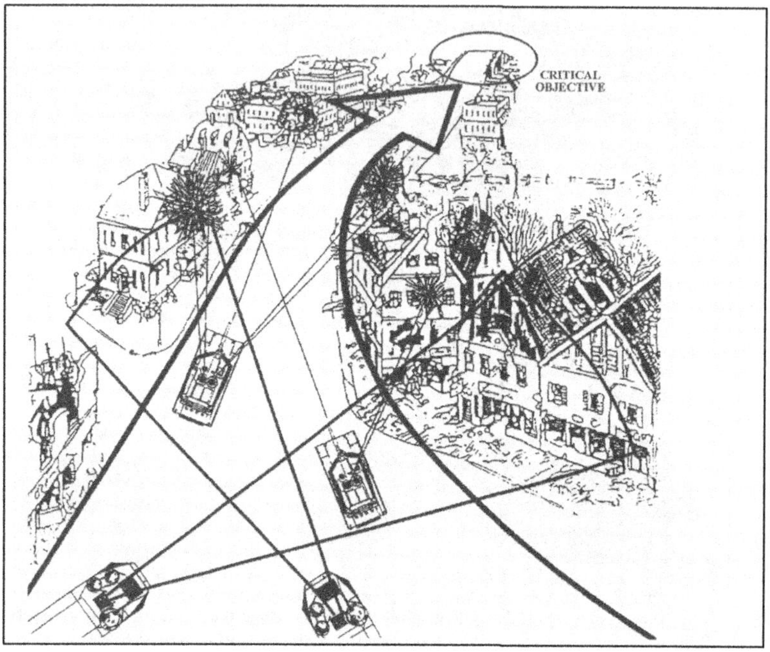

momentum through the zone of action and to seize key objectives. Some enemy positions may be isolated or bypassed in order to maintain tempo. This procedure may be used when:

- A key objective is identified (an enemy critical vulnerability or an objective that supports the friendly scheme of maneuver)
- Time is critical in reaching the objective
- Only selected buildings on the route of advance require clearing.

Figure 2-3. Rapid Advance En Route to a Key Objective

(2) Systematic Clearance. Systematic clearance (figure 2-4 on pg. 2-12) involves the deliberate reduction of all significant enemy positions throughout the objective area. It is usually used when time is not a critical factor.

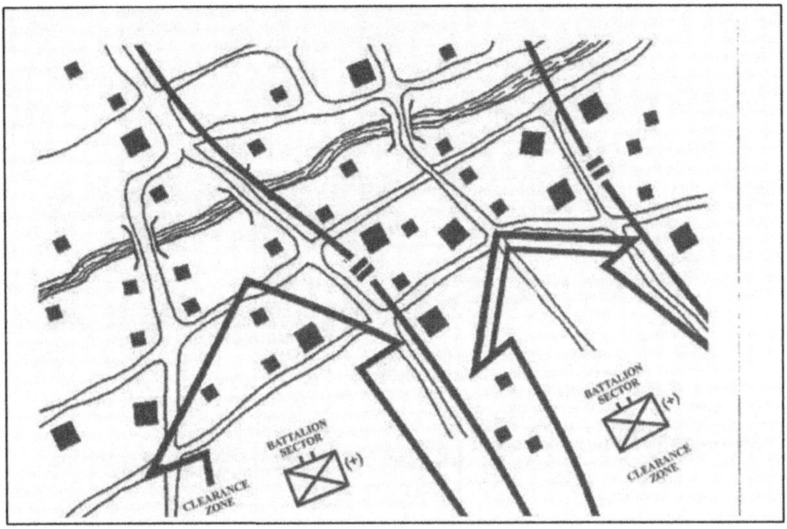

b. Hasty Attack. A hasty attack is, *in land operations, an attack in which preparation time is traded for speed in order to exploit an opportunity* (Jt Pub 1-02). Normally, hasty attacks are conducted as a result of unexpected contact during movement through the built-up area or when an opportunity for exploitation presents itself. Upon contact, the commander immediately deploys, returns fire, reports the contact and situation, develops the situation, and chooses a course of action. The hasty attack follows the same four phases as the deliberate attack; however, the time between preparation and response is compacted.

2104. Warfighting Functions. The warfighting functions (WFs) encompass all activities in the battlespace. Planners consider and integrate the warfighting functions when analyzing how best

Figure 2-4. Systematic Clearance Within Assigned Sectors

to accomplish the mission. The WFs include maneuver, intelligence, fires, logistics, command and control, and force protection.

a. Command and Control. The commander positions himself where he can best observe and influence the battle. Positioning is relative to the level of command. For example, a squad leader locates where he can observe and direct his fire teams, while the battalion commander locates where he can observe and direct his companies.

Close-quarter fighting in and around buildings makes command and control exceedingly difficult. Commanders and small-unit leaders rely on guidance and intent provided by higher

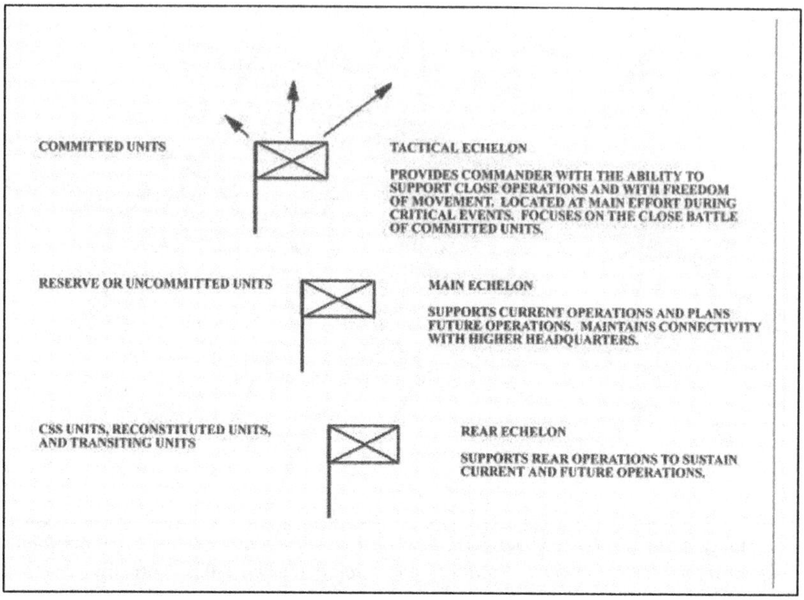

headquarters to facilitate decentralized execution while maintaining control and direction. In close-quarter fighting, commanders may have to rely upon messengers to communicate with their units and higher headquarters.

At battalion and higher levels, command and control is accomplished through three command echelons, which can be collocated or echeloned across the battlefield. The three echelons are tactical, main, and rear (see fig. 2-5). Each echelon may have a combat operations center (COC) capability. The COCs are the facilities (buildings, vehicles, or tents) used by the commander or staff at each echelon to plan, direct, control, and coordinate operations.

Figure 2-5. Tactical, Main, and Rear Echelons

(1) Tactical Echelon. The tactical echelon is normally collocated with the main echelon until the situation dictates separation. When the tactical echelon separates, designated command group members displace forward. Normally, the commander will move with the tactical echelon. The tactical echelon must have the communications and mobility necessary to allow the commander to exercise command. This echelon is usually small in size so the commander and command group can quickly displace forward to observe and influence actions. Organization of the tactical echelon should be addressed in the unit's standing operating procedures (SOPs).

(2) Main Echelon. The main echelon contains a COC, which is the nerve center for command and control of battalion through division-sized organizations. A COC consists of the command group (primary staff members with support personnel) and possesses the necessary mobility and communications for command and control of tactical operations. The main echelon COC monitors the battle, coordinates fire support, issues orders, maintains communications with subordinates and higher and adjacent commands, conducts current and future tactical plans, and assists the commander and subordinate commanders as appropriate. On urbanized terrain, the main echelon COC is normally located in buildings that provide good communications with subordinates and higher and adjacent commands. The command post (CP) is where the commander is physically located. Therefore, the CP can be located at any of the three echelons based on the commander's presence.

(3) Rear Echelon. The rear echelon's COC consists of the logistical and administrative personnel required to plan, coordinate, and execute logistics support. During urban warfare, the rear echelon COC is usually closer to combat forces than normal as it uses the city's captured buildings and facilities to provide cover and concealment for stocking supplies, conducting maintenance repair, and treating the injured.

Attack of a built-up area requires detailed planning due to restrictive terrain, close proximity to opposing forces, fire coordination requirements, and reduced communications capability. The GCE will frame a centralized plan of attack, fire support, and logistics support to conduct urban operations. However, the actual attack of a built-up area is a series of isolated, close-quarter battles carried out by small units. The execution of these attacks should be decentralized.

Communications planning procedures, as outlined in MCWP 3-40.3, *Communications and Information Systems* (under development), are applicable to the urban battle. Planners should be meticulous in their scrutiny of the environment and the effect it will have on communications. Manmade structures can create problems for single-channel radios. These structures inhibit LOS radio communications by absorbing or reflecting transmitted signals. However, the urban environment may have exploitable advantages such as the availability of electrical power, commercial telecommunications networks, and environmental control systems. Electrical power generation stations and other emergency power systems are normally found in protected structures and are probably usable. Due to their value to the enemy, these areas are more likely to be heavily defended. Enclosed areas offer excellent concealment and protection of communications and other command and control support equipment. Extensive commercial communications networks composed of miles of underground protected cable connecting central telephone exchanges are likely to be available, as well as a multitude of public service radio nets (police, fire, civil defense, taxi, etc.) complete with existing antenna systems and retransmission stations. To communicate effectively and continuously, commanders must minimize limitations imposed by the urban environment and maximize the advantage of existing civil communications.

b. Intelligence. Collection of intelligence information on urbanized terrain is difficult. Urbanized terrain provides overhead cover and concealment from observation by aircraft, reconnaissance personnel, and satellites.

(1) Information Requirements. In addition to information about the location of enemy units, nature of the terrain, trafficability of roads, etc., that all operations require, urban warfare requires detailed information on the following:

- Population density, characteristics, and culture
- Location and quality of water supply
- Width and construction of streets
- Layout of road networks
- Location, width, and load capacity of bridges
- Layout of underground networks such as subways, sewers, and utility ducts
- Types and construction of buildings
- Location of key communications and transportation facilities
- Location of canals and waterways.

Information on port facilities should include details on harbor sites, pier networks, unloading capabilities, warehouse/dockside facilities, harbor currents, and sandbars or other midstream obstructions.

(2) Collection. Accurate, detailed, and timely information is vital to the success of any urban operation. Collection operations conducted by national intelligence organizations may provide valuable information on the layout of the city, recent changes to transportation routes, political affiliations of various groups, and locations of military units. This information needs to be current and must be verified and updated as attacking forces close on the city. Reconnaissance activities such as the study of large-scale maps, aerial photographs, and background intelligence reports keep the commander's information current. The latest information on the enemy and the terrain should be continuously fed to the intelligence section using long-range reconnaissance patrols, human intelligence (HUMINT) reports, unmanned aerial vehicle (UAV) missions, and electronic intelligence. Use of reconnaissance assets should be tempered with a risk-versus-gain judgment. Reconnaissance assets should not be risked indiscriminately if the information can be collected by other means.

(3) Maps. Tactical maps should be supplemented by city street maps that contain street names and information on the location of important buildings and transportation terminals. These city street maps should be distributed to all units operating within the city. Although tactical maps are key in controlling fire support missions, simple street maps assist ground troops in maintaining their orientation within the city and in tracking buildings and areas that have been cleared.

c. Maneuver. Maneuver is the employment of forces on the urban battlespace through movement in combination with fire or fire potential to achieve a position of advantage (see MCWP 3-1). Maneuver is an essential element of combat power. Maneuver in itself cannot produce decisive results. Combined with mass, offensive action, economy of force, and surprise, maneuver provides favorable conditions for closing with the enemy. Maneuver contributes significantly to sustaining the initiative, exploiting success, preserving freedom of action, and reducing vulnerability. It is through maneuver that an inferior force can achieve decisive superiority at the necessary time and place. In many cases, maneuver is made possible only through the control of tempo and effective employment of firepower. The commander integrates supporting fires with the scheme of maneuver to create a dilemma for the enemy. Likewise, movement without fires exposes the force to effective enemy counteraction and risks losing the initiative and momentum. Maneuver that does not include violent action against the enemy will not be decisive. At all levels, successful application of this principle in the urban environment requires flexible and innovative plans and actions.

(1) Infantry/Armor Movement. Armored vehicles are vulnerable in built-up areas where streets and alleys provide ready-made fire lanes for defenders. Motorized traffic is greatly restricted, canalized, and vulnerable to ambush and close-range fire. Tanks are at a further disadvantage because their main guns cannot be depressed sufficiently to fire into basements or be elevated to fire into upper floors of buildings at close range (Figure 2-11). Whenever tanks are used in the direct-fire role, infantry must provide security against enemy ground attack. Movement of armor down narrow streets or down narrow paths through debris requires close coordination with infantry. Infantry and armor may use bounding overwatch techniques during movement (See MCWP 3-12, *Marine Tank Employment* [under development]). With this technique, infantry moves ahead of armored vehicles to clear the buildings on each side. Lead infantry units should not get so far ahead of armor that mutual support is lost. Conversely, if armor gets too far ahead of infantry, it may be quickly damaged or destroyed. Coordinated movement also contains a rear security element to protect armor from behind.

During movement down a street tanks should deploy with one tank on each side of the street focusing on the street and lower levels of buildings. *In MOUT, tanks are best deployed in pairs.* Other tanks of the tank platoon or section should move behind the infantry and fire at targets in the upper stories of buildings. When space is inadequate to deploy abreast, tanks may deploy in single file. Movement across open areas should be well coordinated to ensure mutual support. Suppression fire and smoke are used to cover force movements.

Armor can also be used in the reduction of obstacles. For example, grappling hooks and ropes attached to a tank can be used to pull down wire obstacles.

(2) Cover and Concealment. Walls and buildings can provide cover, concealment, and protection from enemy ATGM or heavy machine gun fire. Before armor enters a building, infantry should clear the building and check the ground floors to ensure that there is no basement into which an armored vehicle could fall and become trapped.

Tank and AAV generated smoke may be useful in concealing the location and movement of assaulting forces. Additionally, the thermal sights of the tank are very effective as they can use heat differentials to detect the movement or presence of concealed enemy soldiers, weapons, and vehicles.

(3) Tactical Considerations for Helicopterborne Movement. Helicopterborne operations are planned and conducted using the same combat fundamentals that apply to ground operations. However, there are considerations that apply only to helicopterborne operations. See MCWP 3-24, *Assault Support* (under development) for a detailed discussion on helicopterborne operations. These considerations include:

(a) Helicopterborne forces, once they have landed, lack vehicular mobility and heavy weapons. For this reason, it is important that the forces land on or near the objective. Such a force may be isolated for a time pending linkup with ground forces.

(b) Helicopters may be used for tactical deception. Helicopters may be used to make demonstration landings in several different zones to deceive the enemy as to the true objective of an operation.

(c) Helicopterborne assaults are normally conducted against undefended or lightly defended objectives. When attacking a well-defended objective, an LZ should be selected which provides cover and concealment for the assault force and is close to the objective.

(d) Helicopterborne forces are vulnerable to attack helicopters, fixed-wing aircraft, and surface-to-air missiles. To counter this vulnerability, friendly attack helicopters

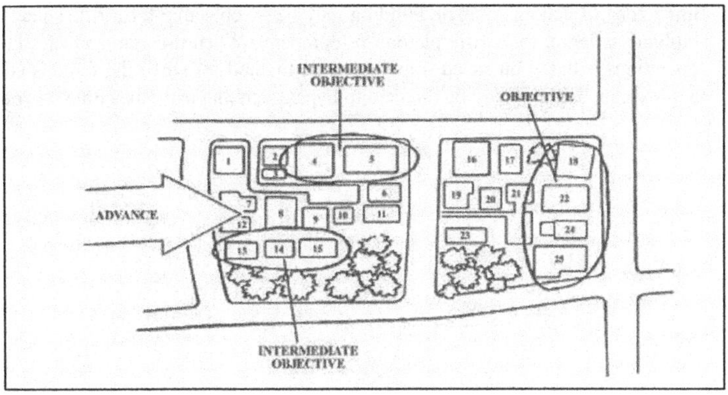

escort the helicopterborne forces during air movement, and indirect fires are used to suppress or neutralize enemy air defenses.

(4) Control Measures. The following control measures assist in the manuever of forces in MOUT:

(a) Objectives. When attacking to seize a foothold, the infantry unit may assign subordinate units the first block of buildings as their first objective. When an objective extends to a street, only the near side of the street is included. The final objective may be buildings or key terrain at the far edge of the built-up area. Key buildings, or groups

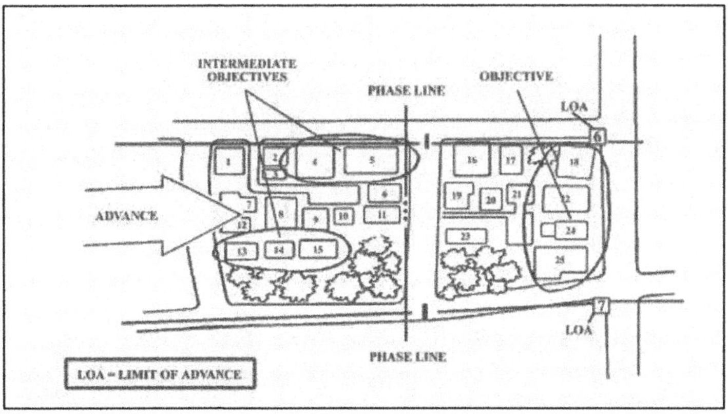

of buildings, may also be assigned as intermediate objectives. Buildings along the route of attack should be identified by numbers for clarification (Figure 2-6 on page 2-18).

During the seizing phase, bypassing buildings may increase the risk of attack from the rear or flank. It may be necessary for the unit to enter, search, and clear each building in its zone of action. A single building may be an objective for a rifle squad or, if the building is large, for a rifle platoon or company. When the commander's concept of operations is based on speed (rapid advance method) or when the force is conducting a hasty attack, a unit may be directed to bypass certain positions within its zone.

Figure 2-6. Objectives and Numbering System

(b) Phase Lines. Phase lines are control measures used to report progress or to control the advance of attacking units. Phase lines should be oriented on readily identifiable terrain features such as principal streets, rivers, and railroad lines. They should be placed on the near side of a street or open area. In an attack, a unit may have the mission to seize positions in its zone of action up to a particular phase line. (Figure 2-7).

Figure 2-7. Phase Lines

(c) Boundaries. Unit boundaries are used to define zones of action and are usually set within blocks so that a street is included in the zone. Both sides of a street should be included within the same unit's boundries.

(d) Checkpoints and Contact Points. Checkpoints aid in reporting locations and controlling movement. Contact points are used to designate specific points where units make physical contact. Checkpoints and contact points are designated by each unit as appropriate for command and control at street corners, buildings, railway crossings, bridges, or any other easily identifiable feature.

(e) Attack Position and Line of Departure (LD). A designated attack position may be occupied by forward units for last-minute preparation and coordination. The attack position is often behind or inside the last large building before crossing the LD. The LD should be located on the near side of an open area running perpendicular to the direction of attack, such as a street or rail line.

(5) Time of Attack. Ideally, the time of attack should be conducted during hours of darkness or limited visibility. Marines can exploit the poor visibility to cross open areas, to gain access to rooftops, to infiltrate enemy areas, and to gain a foothold. When attacking in unrestricted visibility, units should use smoke to conceal movement.

(6) Reserve. Reserves should be mobile and usually follow in trace of forward units. Battalion reserves normally follow one to two blocks to the rear of the lead company. If a company reserve is available, it follows within the same block so that it can immediately influence the attack. A unit with a reserve mission can be called upon to:

- Exploit an enemy weakness or friendly success
- Attack from another direction
- Attrit and/or clear bypassed enemy positions (normally requires designation of a new reserve)
- Secure the rear or a flank
- Maintain contact with adjacent units
- Counterattack

(7) Security. Each unit should ensure security of its flanks and rear (Figure 2-8 on page 2-20). Fighting in a built-up area is unique in that attacks can come from all three dimensions. Units assigned a mission to isolate can also provide security. Small units conduct security patrols and man OPs to supplement higher level reconnaissance and security operations.

Figure 2-8. Enemy Firing From Flank

(8) Engineers. Leading units should have engineers attached for immediate support. Tasks commonly assigned to engineers include:

- Preparing and using explosives to breach walls and obstacles
- Finding and exploding mines in place or helping to remove them

- Cratering roads and performing other countermobility measures
- Clearing rubble and obstacles

Requirements for engineer support in MOUT normally exceed capabilities. To compensate, all Marines should receive basic demolition training prior to conducting urban operations.

d. Fires. The attack of a built-up area may require extensive air and artillery preparation prior to the ground attack. Supporting fires suppress the defender's fire, restricts his movement, and possibly destroys his position. Consideration should be given to the rubbling effect produced by aerial and artillery bombardment. The assault should closely follow air and artillery fire to exploit its immediate effect on the defender. Maneuver units move near the final coordination line while the enemy is engaged by supporting fires. As the attacking force assaults, supporting fires lift and/or shift to block enemy withdrawal or to prevent the enemy from reinforcing their position. (See Chapter 4 for detailed discussion on air and artillery support during urban warfare.) Fire can be categorized into indirect and direct fires.

(1) Indirect Fire

(a) Artillery. Indirect artillery fire is planned to isolate objectives, to prevent reinforcement and resupply, to neutralize known and suspected CPs and OPs, and to suppress enemy defenders. Urbanized terrain requires that most indirect artillery fire use a high-angle trajectory.

(b) Mortars. Mortars are the most responsive indirect-fire weapon for the urban environment. They can be employed to hit targets of opportunity at the close ranges that are typical of combat in built-up areas.

(c) Other. Urban conflict is conducive to innovation. Many direct fire weapons can be employed in the indirect fire mode. Indirect fire weapons may also be effective at clearing the tops of buildings using variable-timed fuses.

(2) Direct Fire.

(a) Armor. Tanks are normally task organized with mechanized units for protection and security operations. The commander can employ tanks to take advantage of their long-range lethality, high-speed mobility, and survivability. They can be used outside the built-up area to cover high-speed armor avenues of approach. Tanks are normally employed in this role during the isolation phase. Tanks also support by fire the assault to seize a foothold and may support attacks to seize objectives within the built-up area. In MOUT, tanks can also be organized into special assault teams. The tank cannon's direct fire is very effective against structures. (Figure 2-9 on page 2-22 and Appendix B)

(b) Howitzers. Artillery rounds delivered by direct fire can be very effective for destroying targets in buildings. When used in this manner, the mobility, flexibility, responsiveness, and survivability of gun crews may be reduced. Artillery used in the direct fire role normally use HE shells.

(c) Infantry. Direct-fire systems, mainly machine guns, antitank guided missiles (ATGMs), shoulder- launched multipurpose assault weapons (SMAWs), and AT4s—are initially employed to support the seizure of a foothold. The positioning of antitank weapons in buildings must allow enough space for the effects of backblasts. Antitank weapons have limited as effectiveness when neutralizing targets behind walls. Designed primarily as armor penetrators, they neutralize a target only if that

Figure 2-9. Tank Direct-Fire Support Within the Built-Up Area

target is located directly behind the point of impact. Consideration must be given to weapons with minimum arming distances. Many of these factors can combine to limit the effects of weapons in the urban environment.

(d) Aviation. The vertical nature of large urban areas may block the pilot's line of sight with the target unless careful consideration is given to inbound headings and altitudes. Plan attack headings to take advantage of large open areas and street axis. Friendly ground forces must be clearly distinguishable from targets. The use of smoke, visible or infrared strobes, or other visual devices must be effectively employed by ground forces and clearly understood by friendly aircraft.

e. Logistics. Logistics support in urban operations should be timely, flexible, and responsive. Logistics planners, both in the supported units and the CSSE, should be fully integrated in the planning process. The final logistics plan must be thoroughly coordinated with the scheme of maneuver. Units may need special equipment such as ropes, grappling hooks, ladders, sandbags, and wire; medical supplies must be readied in anticipation to support increased casualties; and additional supplies must be prestocked. (See Chapter 5 for a detailed discussion on logistics and urban warfare, and MCWP 4-1, *Tactical Logistics*, for additional planning considerations.)

f. Force Protection. Force protection encompasses all those measures that preserve the fighting potential of the command. It is integrated throughout all the warfighting functions. It does not entail security operations alone. It involves the continuous implementation of active and passive measures by all units during all phases of an operation.

Some particular active and passive measures in MOUT include:

-Manmade structures in the urban environment provide cover and concealment for both the attacker and defender.

-Lines of communication can be more easily interrupted due to multiple avenues of approach.

-All units must be prepared to implement force protection in a 3-dimensional environment.

-Proximity of civilian population makes identification of non-combatants more difficult.

-ROE may be more restrictive

-Snipers and man-portable shoulder lauched air defense missiles are usually more difficult to detect.

Section II

Infantry Battalion Operations

2201. Overview. This section addresses some of the operations that a task-organized and/or reinforced infantry battalion could conduct in MOUT. These operations are seldom conducted independently and are usually part of a larger joint operation. In the example given below, the battalion has seized a foothold and is operating in its sector of a town of 100,000 people. Concurrent operations are on-going through other parts of the town. The threat in the battalion sector is a company (-) with a capability to reinforce.

2202. Seizure of a Key Objective. The tactical situation may require the immediate seizure of a building, industrial complex, or transportation site that is deemed decisive terrain. Examples of such objectives may include key bridges, intersections, railheads, or facilities that constitue the center of gravity for the enemy. As an example, a battalion may be assigned the mission of seizing the only bridge that is capable of supporting armored vehicles and resupply by heavy transports along a river bisecting a town. The intent is to isolate enemy forces and secure the bridge for follow-on forces. Intelligence indicates that the bridge is lightly defended at this time. In this case, speed of attack and surprise is critical to capture the objective. Some key intelligence information is required prior to the attack. This would include the number of enemy defending the bridge, their weapons, disposition, their ability to reinforce their defense, location of LZs near the bridge, whether the bridge has been set for demolition, and the location of enemy indirect fire assets that can range the objective.

The commander decides to conduct a penetration using a combination of helicopterborne attack to rapidly seize the bridge followed by a mechanized ground attack to link-up with the helicopterborne forces (Figure 2-10). The helicopterborne force can achieve surprise and speed of movement required while the mechanized force can thrust quickly along high-speed avenues to minimize the time the helicopterborne force must hold the objective against enemy counterattack.

For the ground attack, the battalion commander makes a determination that speed of movement outweighs the risks of rapidly extending his lines of communication. He directs, on the basis of his knowledge of the enemy situation and his mission, that his maneuver forces avoid decisive engagement with the enemy along the route to minimize the time it takes to link up with the helicopterborne force. The commander provides guidance on those enemy forces that should be bypassed and those that should be isolated and reduced.

The commander task organizes his battalion for rapid movement on two axes to allow flexibility in reacting to unexpected enemy contact. Lead and rear security is conducted throughout the operation. As lead units locate enemy positions, they fix them by fire, and quickly reduce or bypass them. They also report the contact so that follow-on forces can then deal with the threat.

As the ground attack force reaches the edge of the built-up area, Marines remain mounted so as not to slow the advance. All vehicle machine guns fire to flanks and into doors, windows, and

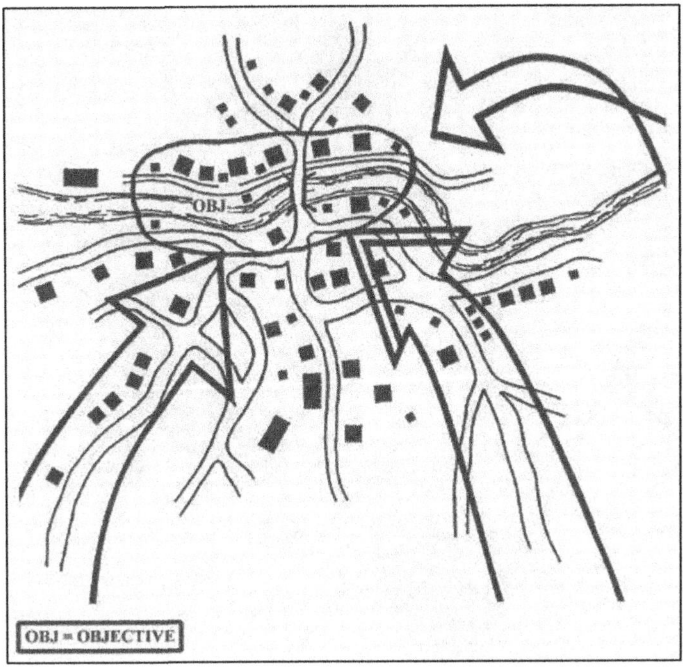

Figure 2-10. Seizure of a Key Objective

alleys to provide suppressing fires (Figure 2-11 on page 2-26). If heavy resistance is met and the column is halted, Marines dismount and the enemy and/or obstacles are cleared as quickly as possible. Once resistance and/or obstacles are cleared, the attack is immediately resumed. If necessary, units are dropped off to assume blocking positions and await follow-on clearing forces.

Once the objective is seized or linkup is accomplished, the battalion establishes a perimeter defense. The companies seize buildings around the objective and expand the size of the perimeter until it is large enough to secure the bridge. Attached engineers can then examine the bridge and clear it of any explosives.

2203. Infiltration. Infiltration is *"the movement through or into an area or territory occupied by either friendly or enemy troops or organizations. The movement is made, either by small groups or by individuals, at extended or irregular intervals. When the term is used in connection with the enemy, it infers that contact is avoided"* (Joint Pub 1-02).

Figure 2-11. Rapid Advance Movement

.3

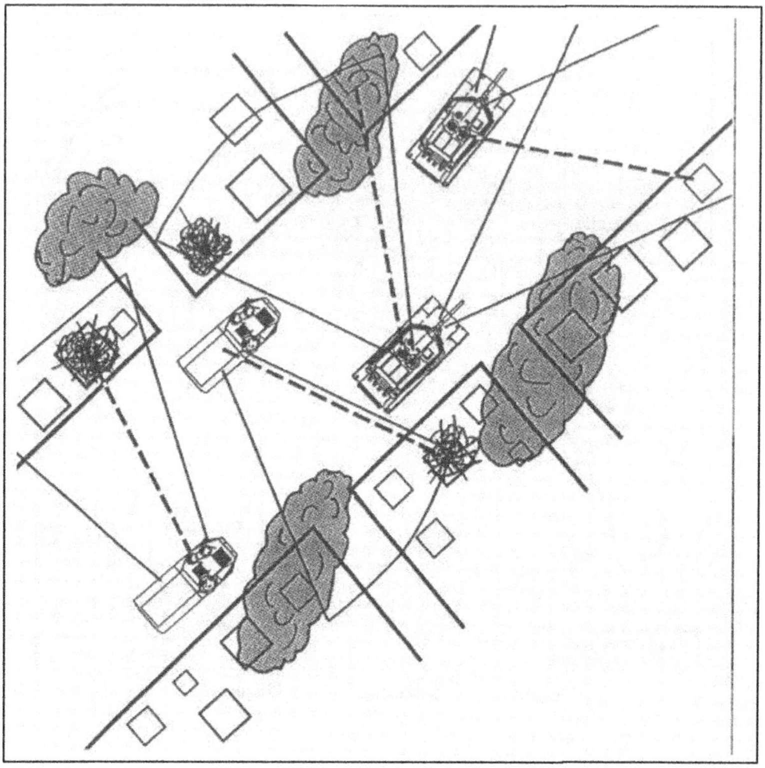

The outskirts of a town may not be strongly defended. The town's defenders may have only a series of antiarmor positions, security elements on the principal approach, or positions blocking the approaches to key features in the town. The enemy strongpoints and reserves may be positioned deeper in the city. In such a situation, an infantry battalion may be able to seize a part of the town by infiltrating companies and platoons between the enemy positions on the outskirts (Figure 2-12).

Moving by stealth on secondary streets, using the cover and concealment of alleys and buildings, the battalion may be able to seize key street junctions or terrain features, isolate enemy positions, and help following units move into the built-up area. Such an infiltration should be performed when visibility is poor.

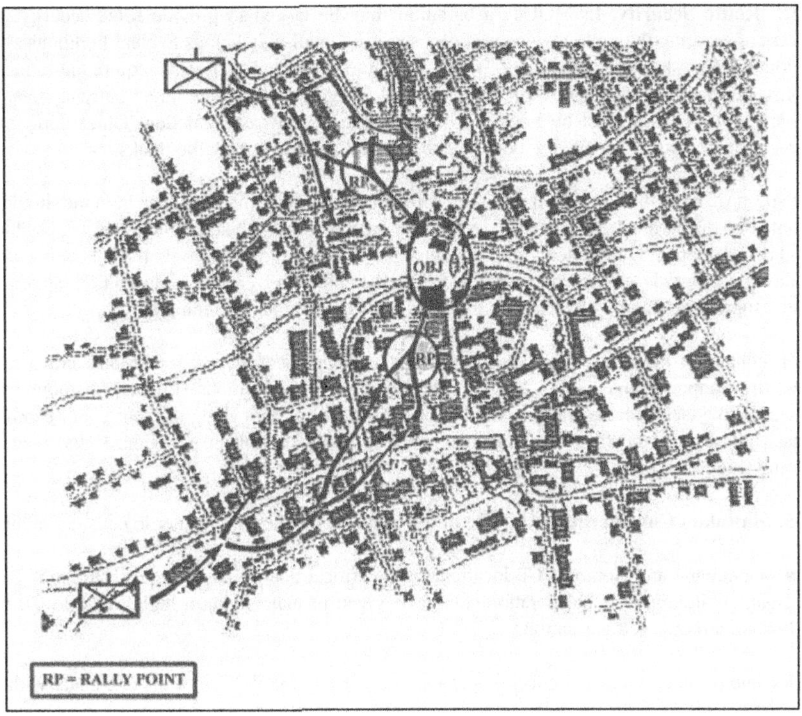

Figure 2-12. Infiltration

The infantry battalion is organized into infiltration companies with engineers attached and a designated reserve. Each company is assigned an infiltration lane that allows for infiltration by company-sized or smaller units. Depending on the construction of the built-up areas and streets, the infiltration lane may be 500 to 1,500 meters wide.

Mortar and artillery fire can be used to divert the enemy's attention and cover the sound of infiltrating forces. Armored vehicles and antitank weapons should be positioned to cover likely avenues of approach for enemy armored vehicles. Designated unit(s) provide isolation and security for infiltrating rifle companies. As the companies move into the built-up area, they secure their own flanks. Security elements may be dropped off along the route to warn of a flank attack. Engineers assist in breaching or bypassing minefields or other obstacles. Enemy positions are avoided but reported. The infiltrating companies proceed until they reach their rally points. At that time, they reorganize and make final preparations for attack of the objective.

2204. Route Security. In MOUT, a battalion may be tasked to provide route security. The process of securing the route may consist of a series of small-unit actions to clear buildings along the route. Fires should be planned to isolate the route as well. It may also require the reduction of obstacles and the emplacement of flank security along the entire route. In an outlying area, the forward units may proceed by bounds from road junction to road junction. Other units could augment flank security by moving down parallel streets and by probing the flanks.

The speed of movement required and the enemy situation determine whether the unit should be mounted or dismounted. For mechanized forces, narrow streets should be avoided if possible to avoid being trapped. Designated armored vehicles overwatch the squad to its front, keeping watch on the opposite side of the street. While armored vehicle movements and positioning provide mutual support, armored vehicles should be protected by dismounted infantry.

When contact is made with the enemy, armored vehicles seek covered positions and provide direct-fire support. Supporting fires fix and isolate enemy positions. Infantry dismount then maneuver to attack. Phase lines are used to coordinate movement along the route. For example, at each phase line, the forward companies might reestablish contact, reorganize, and continue clearing (Figure 2-13).

2205. Battalion Considerations. Some of the concerns for the commander to consider include:

a. Command and Control. CP location can be critical due to line of sight restrictions in the urban environment. Consideration should be given to maintaining a highly mobile CP with links to airborne retrans stations.

b. Intelligence. Organic collection capability resides primarily from units in contact with the enemy. The Scout Sniper Platoon also provides information but they may have obstructed visibility in the urban environment. Commanders may chose to employ them more often in a sniper role. Links to higher command intelligence sources may be key to maintaining current information on the enemy. The same line of sight restrictions encountered with command and contol will also restrict intelligence dissemination. Commanders may have to rely on messengers.

c. Manuever. Maneuver of mechanized forces in an urban environment may be restricted to the size, width, and load capabilities of the streets. Consideration must be given to plan for alternate routes for any attack. Significant manmade obstacles can be encountered throughout the battlespace. Additionally, the reserve is normally kept closer to the assault force. Movement of forces by aviation is less restrictive, but may also be vulnerable to surface to air missiles and small arms threat. Availability of LZs may also be a limiting factor.

d. Fires. A key consideration on the use of fires will be the ROE and the collateral damage that can result by their use. Rubbling of buildings can complicate an already restricted maneuver capability. Weapons effects can be more pronounced and can create a hazard for friendly forces. IPB determines potential firing positions for indirect fire assets that accounts

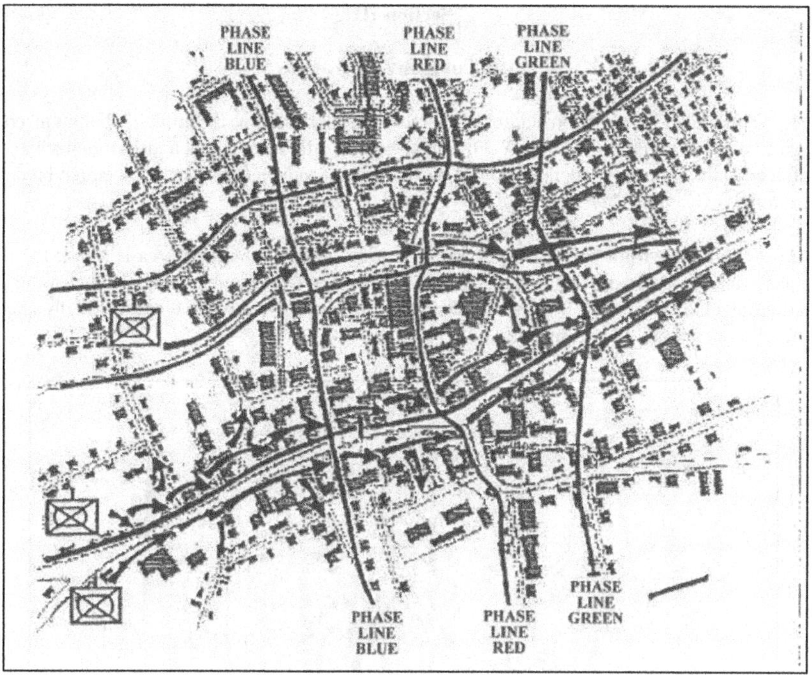

Figure 2-13. Clearing Along a Route

for the overhead masking by buildings and the ability to seat baseplates. Traditional indirect fire assets, such as artillery, may be used in the direct fire role.

e. Logistics. Historically in MOUT, the utilization of class I-V and VIII supplies is much greater. More vehicular assets should be allocated to the small unit level to transport ammunition and support the increased casualty evacuation requirements.

f. Force Protection. Lines of communication require extensive flank security in the urban environment. The commander must determine where his critical security requirements are and concentrate his force protection efforts in those areas while economizing in other areas. Manpower may further be depleted by security requirements for refugees and EPWs.

Section III

Rifle Company Operations

2301. Overview. This section addresses some of the operations the infantry battalion could assign to the rifle company in MOUT. For our focus, the rifle company is a task-organized and/or reinforced unit. Company operations will normally be conducted as part of a larger battalion operation.

2302. Attacking within a Built-up Area. Attacking within a built-up area is a task the rifle company may be assigned (Figure 2-14). This task involves the isolation, attack, and clearing of a specified block of structures. It is usually conducted in conjunction with adjacent friendly attacks or security operations. (See Appendix A.)

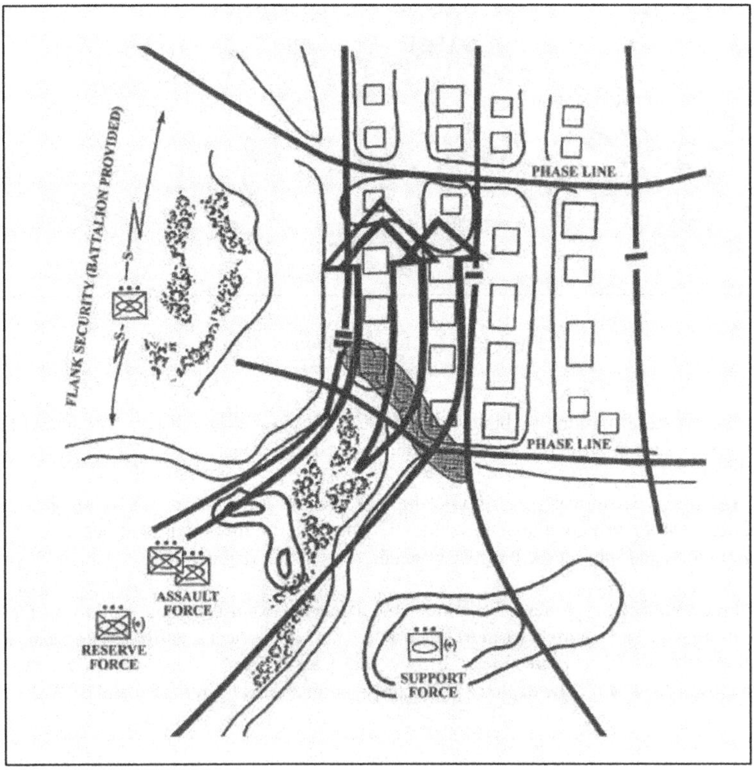

Figure 2-14. Company Attack of a Block

Before an assault, the company commander employs smoke to conceal the movements of assaulting platoons. He provides suppression fires to their flanks with direct-fire weapons. The battalion may provide additional security forces for flank protection.

(1) Concealed by smoke and supported by direct-fire weapons, an assaulting platoon attacks the first isolated building. The platoon must close on the building quickly while the enemy is still stunned by supporting fire. The company commander must closely coordinate the assault with supporting fire so that the fire is shifted at the last possible moment.

(2) The platoons continue to seize each building until the block is cleared. (See Appendix C for a detailed discussion on clearing of buildings.) After seizing the block, the company consolidates and reorganizes to repel a counterattack or to continue the attack.

The company commander directs platoon attacks on a designated front supported by direct and indirect fires. Success depends on isolating the enemy positions (which often become platoon objectives), suppressing enemy weapons, securing a foothold in the block, and systematically clearing the block's buildings room by room.

During the attack of a block, the company is usually task organized usually into four elements: *assault*, *support*, *security,* and *reserve*. The assault force is organized to conduct the actual assault of the objective. The support force provides direct- and/or indirect-fire support. The security force isolates the objective area and provides security. The reserve is organized, equipped, and positioned to exploit success. For example:

(1) The rifle company fighting in the outskirts of a city might organize as follows:

- Assault force: two platoons reinforced with the assault section
- Support force: armored vehicles, mechanized squad, and other direct- and indirect-fire weapons
- Security force: one reinforced squad.
- Reserve: mechanized platoon (-) reinforced.

(2) In a core or core periphery area (See Chapter 1), that same company might be organized as follows:

- Assault force: one rifle platoon, reinforced with engineers, SMAWs, Dragons, and a tank section (two tanks) per platoon
- Support force: One squad with direct- and indirect-fire weapons

.3

- Security force: one rifle platoon (-)
- Reserve: one rifle platoon, mechanized.

c. Armored vehicles, machine guns, and other direct-fire weapons under the company commander's control are assigned to isolate the objective and provide suppressive fires from covered positions. These weapons should not be fired for prolonged periods from one position. The gunners should use a series of positions and displace from one to another to gain better fields of fire and to avoid being targeted by the enemy. Direct-fire support tasks are assigned as follows:

- Machine guns fire along streets and into windows, doors, alleys, etc.; however, innovative use of direct fire weapons can be key in MOUT
- Tanks, ATGMs, and AT4s fire at enemy tanks and other armored vehicles.
- SMAWs can be used to make entrances in buildings and destroy identified bunker/fighting positions.
- Tanks can fire at targets protected by walls, make entrances in buildings, and engage enemy armored vehicles with their main gun and other targets with their machine guns.
- Riflemen engage targets of opportunity.

2303. Attack of an Enemy Outpost. A motorized or mechanized force may encounter enemy outposts, checkpoints, or strongpoints. In a hasty attack on an enemy outpost, the company commander makes a quick assessment based on the mission and situation, and then reacts in accordance with the commander's intent. If not able to bypass an enemy outpost, he may have to attack and reduce it.

The company occupies positions of advantage. The tanks or other armored vehicles support the dismounted infantry in the attack by assuming support-by-fire positions from which they can fire on the outpost, prevent the enemy from escaping, and destroy any reinforcements.

The infantry platoons dismount and move to engage the enemy (Figure 2-15). They should attack from a covered route to hit the outpost at a vulnerable point. As the platoons approach the outpost, smoke may be employed to screen their movement, and supporting fires are shifted. After platoons close on the outpost, they clear the buildings quickly and consolidate. The company is then ready to continue operations.

2304. Seizure of a Traffic Circle or Major Intersection. A company may have to seize a traffic circle or major road or traffic intersection (Figure 2-16 on page 2-34). This operation consists of seizing the buildings adjacent to the traffic circle or intersection while bringing direct-fire weapons into position to fire on likely avenues of approach.

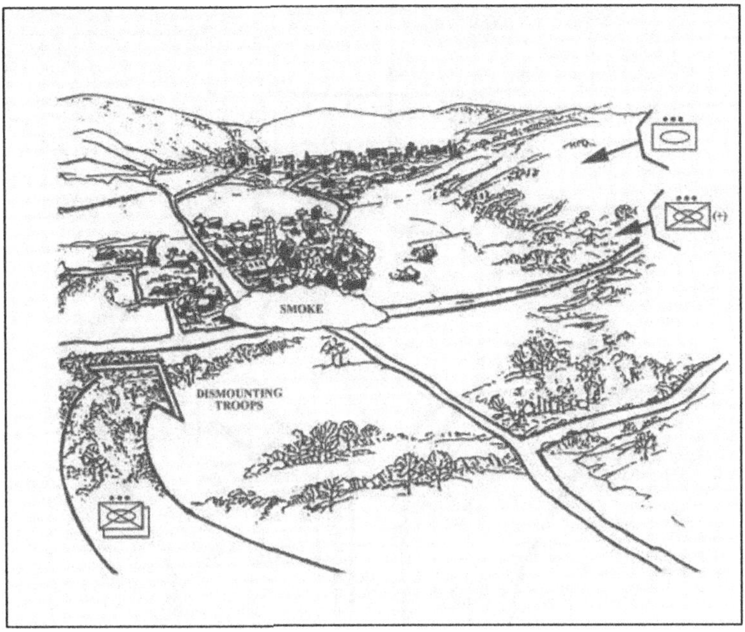

Figure 2-15. Hasty Attack of an Enemy Outpost

The company commander takes the following actions to seize a traffic circle or major intersection:

- Isolate the objective area.
- Seize and clear the buildings overlooking the traffic circle or major intersection under cover of direct-fire weapons.
- Secure the flanks against enemy attack.
- Consolidate and prepare for counterattack.

2305. Seizure of Key Terrain. The seizure of certain key terrain such as a bridge, intersection, facility, or structure may be critical to the success of an operation. Seizing key terrain for friendly use is a possible mission for a rifle company.

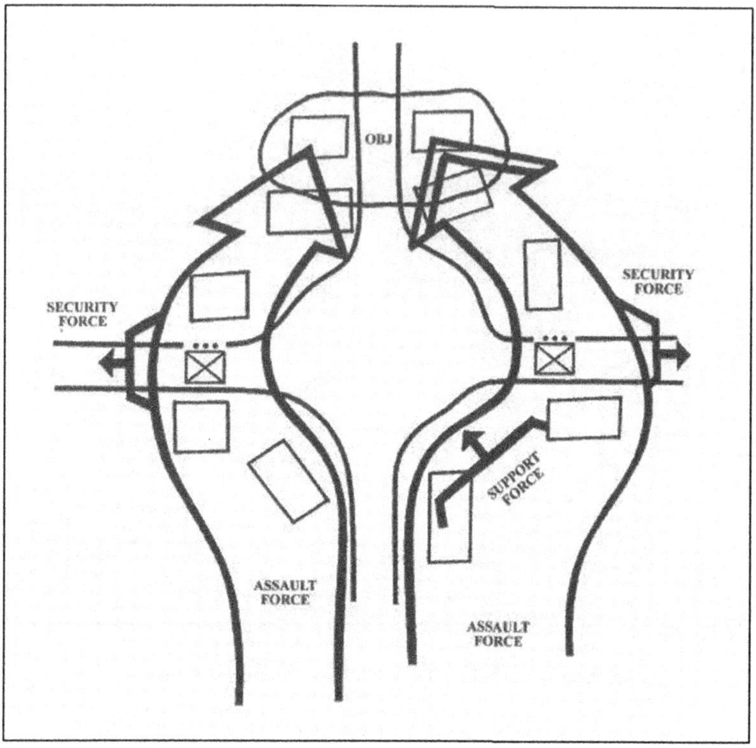

Figure 2-16. Seizure of a Traffic Circle or Major Intersection

To seize a bridge, a company should:

- Clear the buildings on the *near bank* that provide clear fields of fire for supporting weapons
- Suppress enemy weapons on the *far bank*
- Use screening smoke to obscure enemy observation
- Clear the buildings on the *far bank* that dominate the bridge
- Secure a perimeter around the bridge so that the engineers can clear obstacles or remove demolitions from the bridge.

The commander determines which buildings dominate the approaches to the bridge. Buildings that permit him to employ AT4s, ATGMs, machine guns, and riflemen are cleared. Supporting fire prevents the enemy from reinforcing units on the far bank. Enemy positions that can fire directly onto the bridge should be destroyed first. The company's forward observer (FO) should plan artillery and mortar fires to suppress the enemy's positions on the far bank.

The objectives of the assaulting platoons are buildings that dominate the far side of the bridge. One or two platoons are supported by the remainder of the company and assault across the bridge. Supporting fires are shifted and the infantry start clearing buildings. The assault continues until all the buildings in the objective area are cleared.

At this point, the engineers clear the bridge and its approaches of all mines, demolitions, and obstacles. The company commander may expand his perimeter to prepare for counterattack. Once

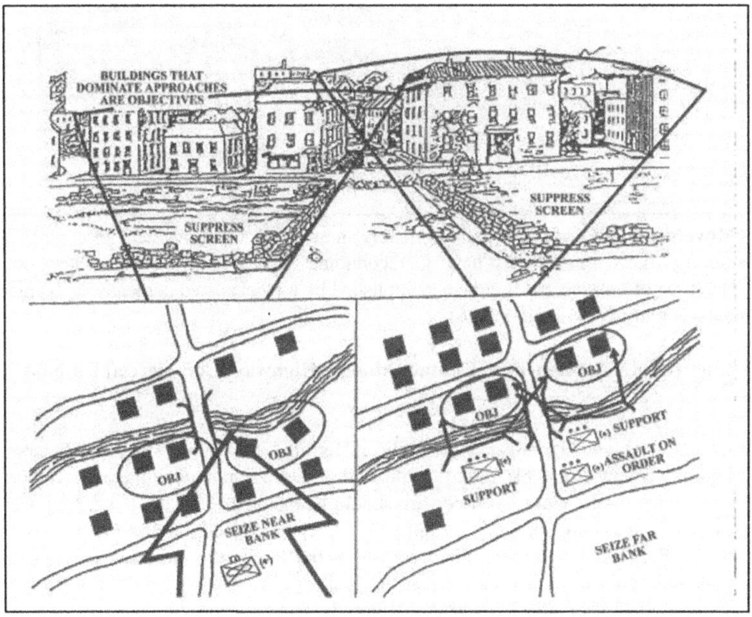

the bridge is cleared, the armored vehicles are brought across to the far bank (Figure 2-17).

Figure 2-17. Seizure of a Bridge

2306. Movement to Contact. A rifle company may conduct a movement to contact along a road. Similarly, the company may have to reconnoiter such a route to prepare for a battalion attack. This type of mission can best be accomplished by a mechanized company with appropriate attachments. (Figure 2-18 on page 2-36).

Figure 2-18. Movement to Contact Along a Highway (Commercial Ribbon)

The speed of movement depends on the mission, terrain, and enemy situation. In open areas where rapid movement is possible, tanks should lead. In close terrain, the infantry may lead while being overwatched by the tanks. Artillery fire should be planned along the route. Engineers may accompany the lead platoon on the main route to help clear obstacles and mines as they may be encountered. Between key terrain, when contact is not likely, the company remains on their vehicles. At key terrain or when enemy contact is likely, the company moves on foot to clear enemy positions or to secure the key terrain. Armored vehicles support the infantry.

Section IV

Rifle Platoon Operations

2401. Overview. This section discusses some of the various tactical missions that might be assigned to a rifle platoon.

2402. Attack of a Building. One offensive mission assigned to a rifle platoon in MOUT is the attack of a building. The attack requires isolating the building and entering the building at the least-defended point or through a breached hole (Figure 2-19 on page 2-38). When possible, clear the building from top to bottom. Close coordination between the assault and support forces is required.

When attacking a building, a platoon should be organized into an assault force, a support force, and a security force. The assault force provides the forces that actually conduct the assault. A support force provides supporting fires and efforts that assist the assault force. A security force covers the platoon's flanks and rear. (See Appendix A.) The platoon may be augmented by armored vehicles and from the company's weapons platoon or the battalions weapons company. If a reserve is required it should be provided by the company.

The assault consists of three steps:

(1) Isolate the building and deny the enemy the ability to maneuver or reinforce.

(2) Enter the building to secure a foothold for further action.

(3) Clear the building systematically.

Clearing is performed by individual squads and teams, which leapfrog as rooms and floors are secured. Platoons tasked to clear buildings may be reinforced with combat engineers.

2403. Movement Down a Street. When moving in urban terrain, a rifle platoon normally follows the same principles of movement as in other environments. However, some movement techniques must be modified to adjust to the unique dangers encountered in an urban environment. (See Appendix A, Section VII.) This section describes the movement down the street of the lead platoon of a company.

The speed of movement depends on the enemy situation. In outlying or lightly defended areas, a mechanized platoon may proceed along the street mounted but send dismounted infantry forward to reconnoiter danger areas. In the center of a built-up area the platoon dismounts to move on foot with two squads leading—one on each side of the road, using all available cover. They move through the buildings, if feasible, to avoid exposure on the streets. The squads coordinate to provide each other with all-around observation and mutual support.

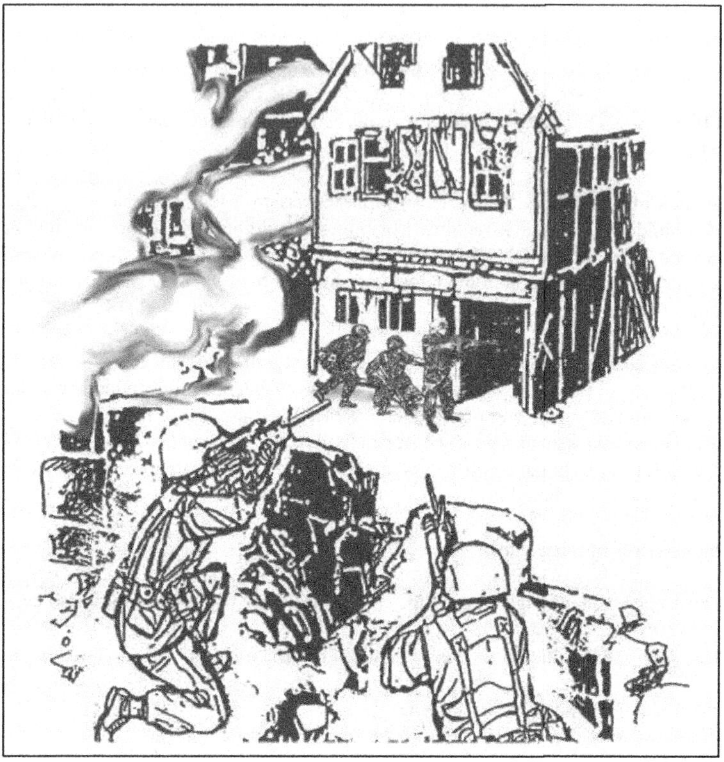

Figure 2-19. Attack of a Building

The platoon should organize and practice to move as two forces: a *maneuver* force and an overwatch force. The maneuver force (normally of one squad on narrow streets or two squads on wide streets) moves forward, scouts danger areas, and closes with the enemy. An *overwatch* force (the remainder of the platoon and its attached or supporting crew-served weapons), moves behind the maneuver force, secures the platoon's flanks and rear, and provides direct-fire support when needed. These two forces, can exchange roles from time to time to avoid overfatigue of the maneuver force (Figure 2-20).

2404. Utilization of the Reserve. In the offense, a rifle platoon may be assigned as a reserve. The reserve can be used to reinforce and attack or exploit success. Reserves are normally held at the company level.

Figure 2-20. Movement Down a Street

In outlying areas, where the terrain is relatively open, a tank platoon reinforced with mechanized assets (tank-heavy task force) may be tasked as the reserve. However, if inside the urban area, a mechanized platoon reinforced with tanks (mech-heavy task force) would most probably be directed to provide the reserve. Regardless of what unit is designated as the reserve, the following items must be considered for their employment in an urban environment:

- The reserve should plan to use a different axis of advance
- Route selection for the reserve may be more constructive since it may have to use building, side streets, sewers, etc.
- Reserves are kept closer to assault forces in MOUT than on open terrain
- Engineers should normally be attached
- Reserves should maintain a high degree of mobility

Chapter 3

Defensive Operations

The essence of defensive tactics is to place the enemy into a position that permits his destruction through the intelligent use of terrain and firepower.

3001. Introduction. In urban combat, the defender possesses key advantages over the attacker. The defender can shape the battlespace to his advantage by maximizing the natural restrictions and obstacles found in the urban environment. ROE for the attacker can add to the defender's advantage by placing restrictions on the application of force by the attacker. Knowledge of the terrain and time available for preparing defensive positions are advantages which may enable the defender to successfully resist a numerically superior force. A spirited and stubborn defense may persuade an attacker to abandon the attack. In some battles, urban defenders can be credited with repelling or decisively delaying an attacker which ultimately influenced the course of the war in their favor.

Marines may be called upon to defend a port city, an embassy located on or near a coastline, a transportation link, or the capital of a foreign government. They may also be tasked to defend from an urban area in an economy-of-force role in order to support operations elsewhere. In combat, offensive and defensive operations are inseparable. Marines need to be capable of conducting both offensive and defensive operations in an urban environment.

The fundamentals of defense do not change in an urban environment. The defenders of a city, however, usually have detailed knowledge of the terrain. This will allow them to establish an extensive defensive network that is designed to force an attacker to expend exorbitant amounts of time, supplies, equipment, and manpower. Commanders need to recognize both the advantages and disadvantages of defensive operations in an urban environment.

3002. Decision to Defend

a. Reasons for Defending Built-Up Areas. Historically, the following are some of the reasons for defending built-up areas:

(1) Denial of Important Strategy/Political Objectives. Capitals and cultural centers can be defended strictly for psychological or national morale purposes even if they do not offer a tactical advantage to the defender. Defending a city can cause an attacker to commit a significant amount of his forces which reduces his capability to attack elsewhere. The defense of a city can delay the overall offensive capability of the attacker.

(2) Retention of Key Economic Centers. In many countries, the entire nation's economic well-being may be tied to a few key cities. These key cities usually contain the country's primary industrial, transportation, and communications base. The capture of these key centers could result in the overthrow of the current government, or deny that government the ability to adequately support combat operations against enemy regular or insurgent forces.

(3) Control of Avenues of Approach. Most avenues of approach to large cities are straddled by small towns every few kilometers which must be controlled by defending forces. These areas can be used as battle positions or strongpoints to control the avenues of approach. For the attacker to utilize these avenues of approach he must sacrifice speed and expend resources. The defense of these cities or towns may constitute a large part of the overall defense of a city.

(4) Economy of Force. The tactical advantages provided by urbanized terrain allows the defender to engage a numerically superior force. The defender can conserve the bulk of his combat power so that it is available for use in other operations.

(5) Concealment of Forces. Reconnaissance and combat identification is more difficult in an urban environment. CPs, reserves, CSS complexes, and combat forces emplaced in built-up areas are much harder to detect.

3003. Reasons for Not Defending Built-Up Areas. The commander considers the following reasons for deciding not to defend built-up areas.

a. Unnecessary to the Defensive or Offensive Plan. If the built-up area is too far forward or back in a unit's defensive sector, is isolated, or is not astride an enemy's expected avenue of approach, the commander may choose not to defend it.

b. Bypassable. If the nature of nearby terrain allows the enemy to bypass the city, then it will not make a good blocking position. Some built-up areas, mainly smaller ones, are easily bypassed by existing main road and highway systems. A built-up area that can be easily bypassed normally will be, thereby effectively isolating that area's defenders from the remainder of their forces.

c. Inadequate Structures for Defense. Extensive areas of lightly built or flammable structures offer little protection to the defender. Built-up areas near flammable or hazardous industrial areas, such as refineries or chemical plants, also may not lend themselves to successful defense.

d. Adjacent Dominating Terrain. If the built-up area is small and dominated by close, prominent terrain, the commander may choose to defend on that terrain rather than in the built-up area.

e. Better Fields of Fire Elsewhere. The commander may choose to base all or part of his defense on the better fields of fire that exist outside built-up areas. An example would be an armor-heavy force defending in sectors from multiple, small, built-up areas surrounded by open or farm-type areas.

f. Open City. See chapter 2.

Section I

Defensive Planning

3101. Commander's Estimate. As in any other defensive operation, planning begins with the commander's estimate of the situation. The complxities involved with the defense of a built-up area require detailed and centralized planning. This section details the general procedures and principles for planning and organizing the defense of an urban area.

3102. METT-T

a. Mission. Mission analysis is the first step in conducting the estimate of the situation. When conducting mission analysis commanders and their staffs should analyze their assigned tasks and commander's intent to determine the scope of the defense. For example:

- Do I need to control every building or can I control the frontage from key buildings?

- Are certain parts of the urban area more significant? (i.e. main transportation centers, communication nodes government buildings)

b. Enemy. The commander must focus on the enemy and build his defense to concentrate his strength against enemy weakness. The IPB process is used to analyze the terrain and enemy capabilities. (See FM 34-130.) For example:

- If the attacker relies heavily on aviation for mobility, the defender may defend potential LZs and emplace manpad teams to counter the air threat

- If the enemy relies heavily on mechanized movement, the defender may employ additional countermobility assets.

c. Terrain and Weather

(1) Terrain. Analysis of all manmade and natural terrain features is critical for planning a defense in built-up terrain. A defending commander has the advantage of occupying the terrain on which the fight will occur, and therefore usually knows it better than the attacker.

The defender must make best use of the battlespace that he controls: ground level (streets and parks), above ground (buildings), and below ground (subways and sewers). Defenders must be aware that built-up areas are normally interdependent. A reservoir supplying water to a city may be many miles away, electricity can come from remote power stations, and the control points for these services and others, such as communications lines, may be vulnerable.

.3

(a) K: Key Terrain. Primary examples of key terrain in the urban environment are ports, airfields, power grids, communication nodes, bridges, government building complexes, or parks. We identify key terrain so that we can better select our defensive positions and also better determine the enemy's objectives.

1 **Villages.** Villages often dominate high-speed avenues of approach. If the buildings are constructed of stone, brick, or concrete, providing good protection against both direct and indirect fires, a formidable defense can be constructed

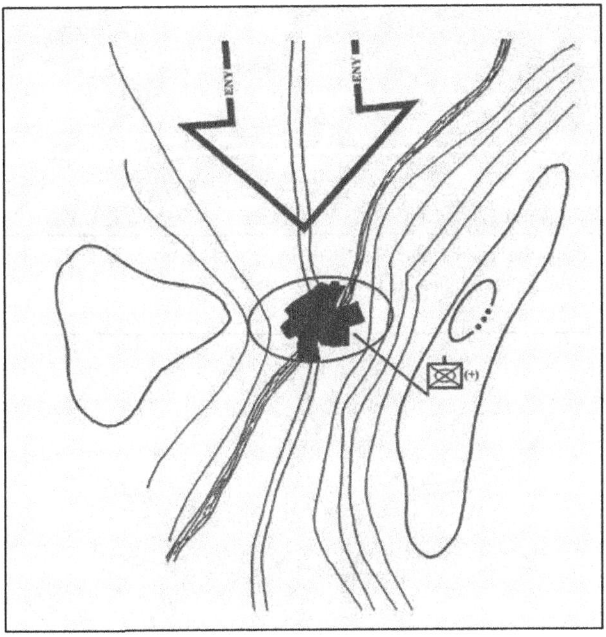

(Figure 3-1). Company-sized battle positions can be established in these small villages to block approaches into the larger urban area.

Figure 3-1. Village Defense (Chokepoint)

2 **Strip Areas.** Strip areas are built along roads or down valleys between towns and villages. They afford the defender the same advantages as villages. If effective fields of fire are available, a unit can occupy a few key positions within the strip to create an effective defense. (Figures 3-2A and 3-2B).

3 **Towns and Cities.** When a small town or city is positioned on a key enemy avenue of approach, a force can defend from the urban area while other forces

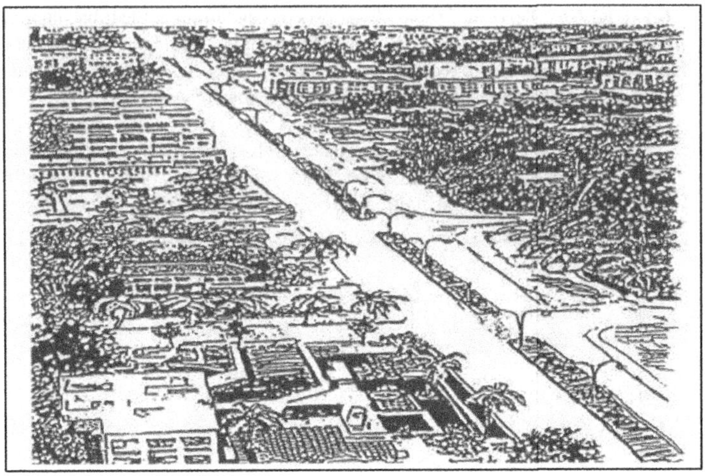

deny the enemy the ability to bypass. The town or city can be made into a strongpoint defense. (Figures 3-3A and 3-3B on page 3-6).

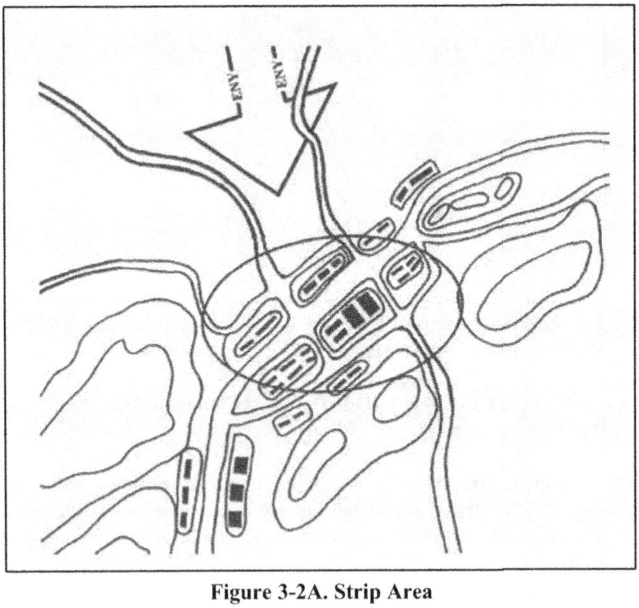

Figure 3-2A. Strip Area

Figure 3-2B. Strip Area Defense

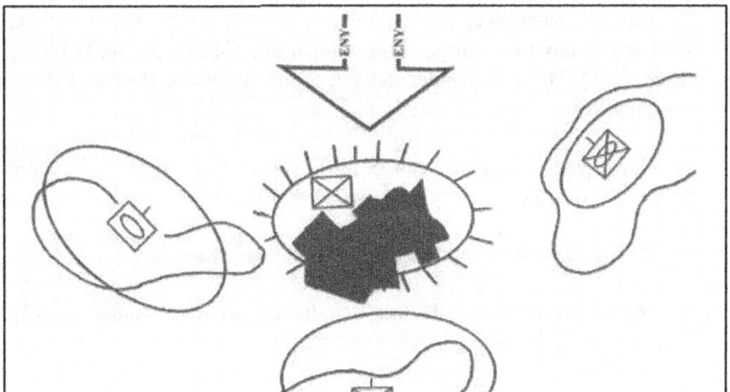

Figure 3-3A. Towns and Cities

Figure 3-3B. Defense of a Town or City

4 **Large Built-Up Areas.** Large buildings that are close together may require more forces and smaller defensive sectors than natural open terrain. The density of buildings, rubble, and street patterns will dictate the frontage of the unit.

(b) O: Observation and Fields of Fire. Attackers must generally advance by crossing streets and open areas between buildings where they are exposed to fires from concealed positions.

1 **Weapons and Range.** The defender must position weapons to obtain maximum effect and mutual support. FACs and FOs should be placed well above street level for increased observation. Fires and final protective fires should be preregistered on the most likely approaches.

2 **Limited Visibility.** The commander can expect the attacker to use periods of limited visibility to sustain or gain momentum. (See Appendix H for more detailed discussion.) The commander considers the following during periods of limited visibility:

- Unoccupied areas that can be observed and covered by fire during daylight may have to be occupied or patrolled at night.
- Remote sensors should be emplaced in dead space.
- Mines, noise-making devices, tanglefoot, and OPs should be positioned on avenues of approach to provide early warning.
- Artificial illumination to include street lamps, stadium lights, and flares should be integrated into the overall defense.
- When defenses are probed, indirect-fire weapons, grenade launchers, and hand grenades can be used to avoid disclosure of defensive position locations.

(c) C: Cover and Concealment. Although walls, floors, and ceilings can be used as protective cover, Marines should always improve these positions by using sandbags, rubble, etc. A defender can reduce his exposure by establishing routes between positions using:

- prepared breaches through buildings

.3

- reconnoitered and marked underground systems
- trenches and sewage systems
- concealment offered by smoke and darkness to cross open areas.

(d) O: Obstacles. A city is an obstacle because it canalizes and impedes an attack. Avenues of approach should be blocked by obstacles and covered by fire. (See Paragraphs 3404.d., 3404.f., 3405.e., and Appendix F.)

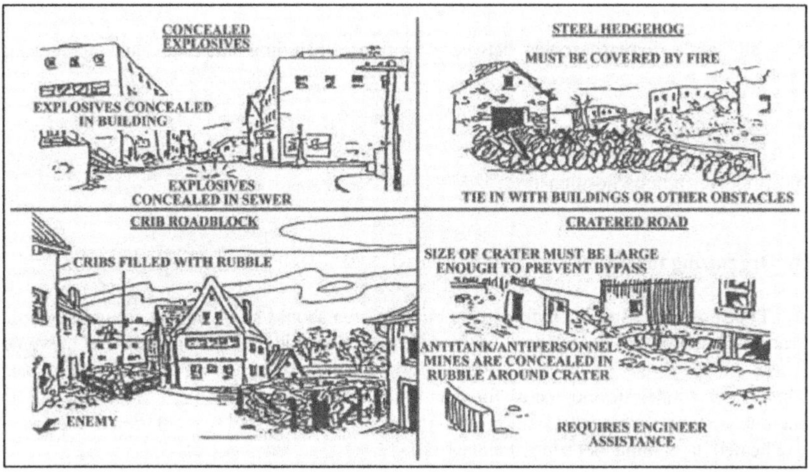

(e) A: Avenues of Approach. Avenues of approach (surface and subsurface) should be denied to the enemy in keeping with the overall defensive plan (Figure 3-4).

(2) Weather. The effects of weather as discussed in Chapter 2, Section II, pertain equally to defensive operations.

D. Troops and Support Available. The defender of a built-up area has a terrain advantage and can resist the attacker with much smaller forces. The defender stands a greater chance of success when he fights with combined arms and task organizes his forces. The discussion of troops and support available in Chapter 2, Section II, also pertains to defensive operations.

Figure 3-4. Denial of Avenues of Approach

e. Time Available. The commander establishes priorities of work, which depend on the time available. Many tasks can be accomplished at the same time, but priorities for work are accomplished according to the commander's order. An example priority-of-work sequence follows:

- Establish security and communications
- Assign sectors of responsibility, final protective fires
- Clear fields of fire
- Select and prepare initial fighting positions
- Establish and mark routes between positions (including alternate and supplementary routes)
- Emplace obstacles and mines
- Improve fighting positions.

3103. Organizing for a Defense

a. Types of Defense. The defense of a built-up area should be organized around key terrain features that preserve the integrity of the defense and provide the defender ease of movement. There are two types of defense: position and mobile. (See MCWP 3-1, *Ground Combat Operations* [under development] for details on defensive operations.) Most defenses will include a combination of position and mobile defense. The type of defense chosen is predicated on commander's intent and METT-T.

(1) Position Defense. The type of defense in which the bulk of the defending force is disposed in selected tactical localities where the decisive battle is to be fought. Principal reliance is placed on the ability of the forces in the defended localities to maintain their positions and to control the terrain between them. The reserve is used to add depth, to block, or restore the battle position by counterattack. (Joint Pub 1-02.) The position defense is usually characterized by defending key terrain. In the urban environment, this equates to the physical occupation of key public buildings, avenues of approach, transportation centers, industrial parks, etc. This defense focuses combat power to repel the attacker's advance and/or penetration while retaining the terrain. This type of defense is common for battalion-sized units and below.

The commander conducting a position defense organizes forces utilizes the same fundamentals that apply to any defense. Forces can be assigned sectors to defend and/or battle positions or strongpoints based on the commander's analysis of METT-T. Reserves are identified and located to respond quickly to exploit success.

(2) Mobile Defense. Defense of an area or position in which maneuver is used with organization of fire and utilization of terrain to seize the initiative from the enemy. (Joint Pub 1-02.) The mobile defense is characterized by its focus on the enemy, not on terrain. Terrain is still important. However, this defense is organized with a mobility capability to exploit enemy weakness. In order to conduct the mobile defense, the defender must have enough forces to defend the built-up area, and to rapidly conduct an enveloping or flanking attack.

As in the position defense, the sector, battle position, strongpoint, or any combinations thereof are also used in shaping the battlespace by canalizing the enemy into a designated location where he can be destroyed.

b. Defensive Options

(1) Defense Outside the City. The commander may defend from outside the city. This option is often selected when the terrain surrounding an urban area offers an advantage to a defender and when his defending force is of sufficient strength (See Cherbourg paragraph in Chapter 1). The presence of a large, friendly, civilian population, or buildings of historic significance or specific cultural value, would also contribute to deciding on an outside defense. A situation in which the built-up area has a preponderance of building structures that offer little protection to defenders or terrain that favors the attacker.

A commander organizes his defenses around strongpoints selected on the outer edges of the urban area when these may be suitable structures for siting weapons to gain maximum firepower range. The organization of an outer-city defense is also heavily influenced by the number and type of weapons with which the hostile force is armed. A defense against a predominantly infantry threat would differ from that planned to defeat an armored threat. The decision to defend or concentrate assets solely on the outer edge of a city are weighed against the enemy's ability to mass fires, infiltrate, or bypass strongpoints.

(2) Defense Inside the City. The commander may decide that his best defense is to defend within the city. This option is usually chosen when hostile forces possess strong, accurate, long-range fires and the defender wishes to minimize their effectiveness (See Quang Tri City example in Chapter 1). The strength of the defending force relative to the attacker may also lead to this option as an economy-of-force measure.

Structures on the outer edge of an urban area that obstruct fields of fire or that are likely to be used by an attacker should be prepared for demolition or demolished ahead of time. All defensive positions should be supplemented or covered by direct-fire weapons. Indirect fires should be planned for most likely avenues of approach. Additionally, close air support should be planned and integrated with the defense.

Security forces are placed outside the defended area to gain early enemy contact, inflict maximum casualties, and attempt to deceive the enemy as to the location of the friendly positions.

(3) Defense of a Key Sector. This option is usually employed when only a portion of an urban area be held or when strength is insufficient to defend an entire area (See the Japanese defense of Manila in Chapter 1). This option is often employed in situations where the mission requires clearing and holding a key sector, or sectors, in order to facilitate other operations.

(4) Entrapment and Ambush. This option is preferred when the defender lacks requisite strength to man organized, in-depth, defensive positions (See the highly successful Egyptian defense of Suez City example in Chapter 1). It requires the preparation of a series of blocking and/or ambush positions along major avenues of approach through an urban area. The defending force uses mines and obstacles to block alternate or possible detour routes in order to canalize the attack force into prepared ambush sites. The defender employs security forces forward of the main battle area and establishes inner or outer defensive strongpoints with extensive obstacles designed to assist in canalize the enemy into kill zones.

The entrapment and ambush option requires detailed planning and rehearsal by all echelons. Alternate plans should be prepared in the event the ambush is only partially successful or if the enemy fails to enter prepared ambush routes. Alternate ambush positions or counterattacks are planned.

Ambush planning in an urban environment does not differ significantly from planning in other environments. However, the urban environment offers an abundance of kill zones that afford the enemy little or no opportunity to escape. Ambush positions are established in selected buildings in the kill zone which offer optimum fields of fire for all weapons. Indirect fires and close air support should be planned to obtain maximum effect in the kill zone. The kill zone should also be covered with targets for indirect fire and CAS. Entrances into buildings not used as firing points should be blocked to entrap hostile forces in the kill zones.

(5) Defense in Depth. Defense in depth should be incorporated into any defensive plan. Units should be given enough terrain to allow for primary, supplementary, and alternate fighting positions as well as successive battle positions or strongpoints. This allows enough room for multiple ambush/entrapment kill zones and the positioning of a reserve.

c. Defensive Organization. The defensive battlespace is divided into three areas: the *security area*, *main battle area*, and *rear area*. Depending on the mission of the unit, forces defending a built-up area may have missions in any one or more of these areas. (Figure 3-5 on page 3-12).

.3

(1) Security Area. The defensive battle begins with a combined-arms force conducting security operations well forward of the main battle area. Security area missions can include screen , guard, and cover. Based on their assigned mission, security forces use all

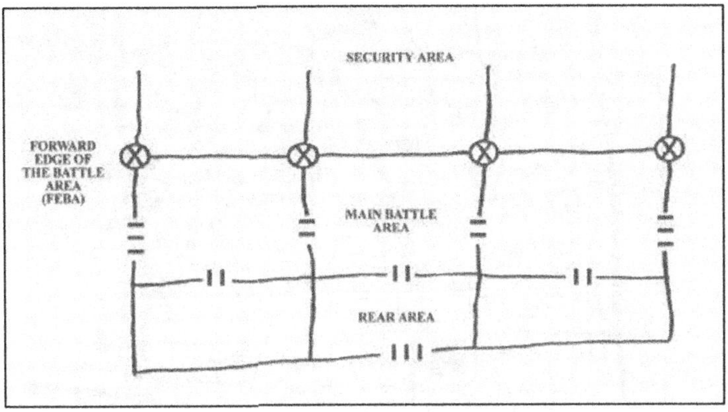

available combat power to gain time, destroy the enemy, or slow his momentum. The initial battle is fought using aviation assets and indirect fire weapons.

(a) Value of Security Area Operations. Security area operations complement to the defense by:

- Alerting the defense to the strength, location, and general direction of the enemy's main and supporting attacks
- Delaying enemy first-echelon detachments
- Initiating early engagement of enemy forces
- Deceiving the enemy as to the true location of the defender's main forces and main effort.

Figure 3-5. Organization of the Battlefield

(b) Withdrawal of Security Area Forces. The urban environment may complicate battle handover from the security force to the main battle area force. However, this transition must be accomplished smoothly to prevent the enemy from gaining momentum. The withdrawal of the security area forces must not result in an easing of pressure on the enemy.

(2) **Main Battle Area.** The decisive battle is fought in the main battle area. However, the

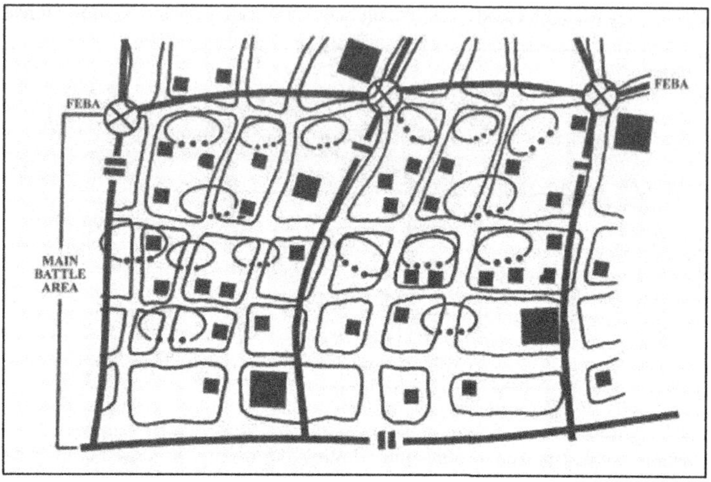

commander may deploy units on the forward edges of the city or in battle positions in depth. The defensive scheme should include forces to defend along the flanks. The commander normally employs security forces to the front to provide early warning and to deny the enemy intelligence on the unit's defensive dispositions (counterreconnaissance) (Figure 3-6).

(a) The size and location of battle positions depend on METT-T and the type of defense prepared. A sophisticated defense will often include a combination of both position and mobile defense but will always be planned in depth.

(b) A unit assigned battle positions on the forward edge of a city or town should:

- Provide early warning of the enemy's advance
- Engage the enemy at long range
- Deceive the enemy as to the true location of the defense.

Figure 3-6. Main Battle Area

(c) The defender employs all available fires to destroy and suppress the direct-fire weapons that support the ground attack. Depending on the concept of operations, weapons may engage targets at their maximum effective range or engage simultaneously for an ambush.

(d) As the attack develops, the actions of small-unit leaders become increasingly important. It is imperitive that all leaders understand their commander's intent.

(3) Rear Area. The rear area is located behind the main battle area and it contains the CSS and administrative support. Protection of these units is vital. Fortunately, the urban environment usually provides sufficient cover and concealment for CSS assets as well as covered and protected routes to the forward areas. Rear area operations are covered in detail in MCWP 3-41.1, *Rear Area Operations.*

3104. Warfighting Functions. The defensive planning sequence remains the same as in any other defensive operation. Defense planning considerations, based on METT-T, must take into account all the activities in each of the warfighting functions. The complexities involved in defense of a built-up area require detailed and centralized planning. Commanders and subordinate leaders must incorporate the following planning considerations for an urban environment when conducting a defensive operation.

a. Command and Control. The commander positions himself so that he can control the action. In an urban environment, this can be more difficult because of the close nature of fighting, prolific obstacles, poor visibility, and difficulty in communication.

Graphic control measures oriented on prominent terrain features are also used in planning and conducting combat in built-up areas. Prominent streets are ideal for use as phase lines and boundries. These and other control measures assist the commander in controlling manuever and fires throughout the battlespace.

COCs should be located in secure locations. (These can be inside or underground when possible.) (See Chapter 2, Section IV, Paragraph 2402.)

Commanders consider the effects of built-up areas on communications when they allocate time to establish communications. LOS limitations affect both visual and radio communications. Wire laid at street level is easily damaged by rubble and vehicle traffic. Also, the noise of urban combat is much louder than in other areas, making sound signals difficult to hear. Therefore, the time needed to establish an effective communications system may be greater than in more conventional terrain.

Wire is the primary means of communication for controlling the defense of a city and for enforcing security. However, wire can be compromised if interdicted by the enemy. Radio communication in built-up areas is normally degraded by structures and a high concentration of electrical power lines. Nonetheless, radio is an alternate means of communication. Messengers can be used as another means. Visual signals may also be used but are often not effective because of the screening effects of buildings, walls, and so forth. Signals must be planned, widely disseminated, and understood by all assigned and attached units. Commanders consider the following techniques when planning for communications:

(1) If possible, lay wire through buildings for maximum protection.

(2) Use existing telephone systems. (Remember, telephones are not secure.)

(3) Emplace radios and retransmission sites in buildings on top floors or atop buildings that are higher than surrounding structures. Ensure that antennas are placed on the side of buildings where the radio signal can be directed at friendly units. This will enhance LOS radio communications.

(4) Use messengers at all levels because they are the most reliable and secure means of communications.

b. Intelligence. The discussion in Chapter 2 pertains equally to a defensive operation. Intelligence gathering for defensive operations is not limited to studying only the enemy. Commanders must emphasize obtaining and using all intelligence. Commanders must emphasize the utility of the urban infrastructure to enhance his intelligence gathering and dissemination effort. The defender usually has the advantage of being familiar with the intricacies of the urban terrain that he is defending. For example, the:

- The sanitation or public works department will have knowledge of the subterranean sewer lines
- Civil engineers will have extensive knowledge of water and power distribution nodes
- Law enforcement representatives may have access to lists of people sympathetic to the attacker
- Local journalists may agree to use local sources to assist in intelligence gathering effort.

c. Manuever

(1) Positioning of Forces. Finding fighting positions in urban areas that provide both good fields of fire and cover is often difficult. The forward edges of a town usually offer the best fields of fire but can be easily targeted by enemy observation and supporting fire. These areas often contain residential buildings constructed of light material. Factories, civic buildings, and other heavy structures that provide adequate cover and are more suitable for a defense are more likely to be found deeper within the town, but they have limited fields of fire.

Because the forward edge of a town is the obvious position for the defender, it should generally be avoided. However, the defender may choose to use the edge of the town if

.3

the terrain limits the enemy's ability for engagement or if strongly constructed buildings that give defenders protection are available. The mission of such a force is to provide early warning of the enemy's advance, to engage the enemy at long range, and to deceive the enemy as to the true location of the defense. This force should withdraw in time to avoid decisive engagement. If there is limited observation from the forward edge, a force should be positioned on more favorable terrain to gain better observation and to engage the enemy at long range.

Reconnaissance by all defending elements should help select routes for use by defenders moving between positions. Movement is crucial in fighting in built-up areas. Early selection and marking of routes adds to the defender's advantages.

The commander covers probable landing and drop zones such as parks, stadiums, or large rooftops and heliports with obstacles or fire to prevent air assault. (See Appendix F.) In a large built-up area, a unit is given a sector to defend and normally establishes a series of defensive positions. Although mutual support between positions should be maintained, built-up terrain often allows the enemy to infiltrate between positions. Therefore, the defender must identify the following:

- Positions that allow suppressive fires on infiltrating routes
- Covered and concealed routes, such as subways and sewers, for friendly forces to use to move between positions
- Structures that dominate large areas
- Areas such as parks, boulevards, rivers, highways, and railroads where antiarmor weapons have longer fields of fire
- Firing positions for mortars
- Command locations that offer cover, concealment, and ease of command and control
- Protected storage areas for supplies.

Buildings that enhance the general plan of defense are selected. Mutual support between these positions is vital to prevent the attacker from maneuvering and outflanking the defensive position. Buildings chosen for occupation as defensive positions should:

- Offer good protection
- Have strong floors to keep the structure from collapsing under the weight of debris

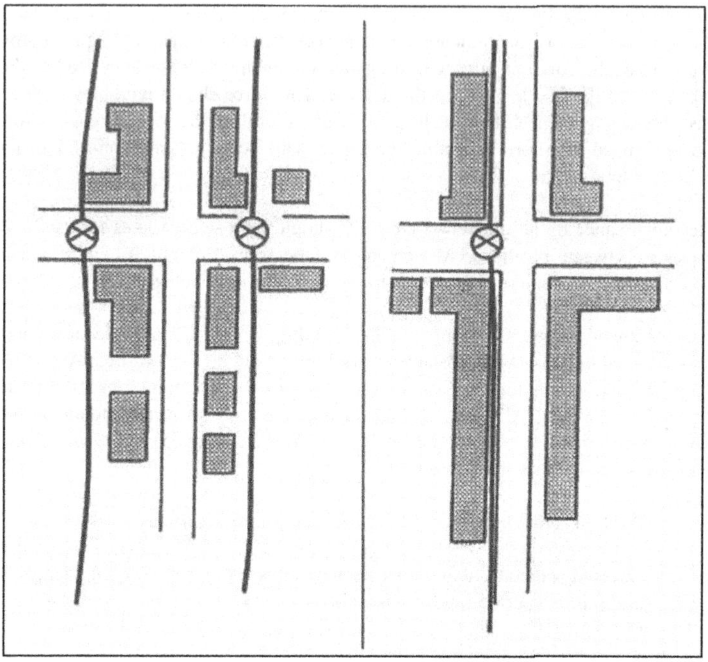

- Have thick walls
- Be constructed of nonflammable materials
- Be strategically located (corner buildings and prominent structures)
- Be adjacent to streets, alleys, vacant lots, and park sites (these buildings usually provide better fields of fire and are more easily tied in with other buildings)
- Provide for 360-degree protection and fields of fire
- Be stocked with adequate supplies as positions may become isolated and cut off for extended periods.

(2) Assign Sectors of Responsibility. Boundaries define sectors of responsibility. Sectors include areas where units may fire and maneuver without interference or coordination with other units. Responsibility for primary avenues of approach should never be split among units. In areas of semi-detached construction, where observation and movement are less restricted, boundaries should be established parallel to alleys or streets so that both sides of a street will be in a single sector. Where buildings present a solid

.3

front along streets, a boundary may have to be extended to one side of the street (Figure 3-7).

Figure 3-7. Sector Boundaries in Areas With Semi-detached and Solid-Front Construction

(3) Select and Prepare Initial Fighting Positions. Commanders select positions in depth. The unit should prepare positions as soon as they arrive and continue improving positions as long as they are occupied. Enemy infiltration or movement sometimes occurs between and behind friendly positions. Therefore, each position must be organized for all-around defense. The defender should also:

(a) Make minimum changes to the outside appearance of buildings where positions are located.

(b) Screen or block windows and other openings to keep the enemy from seeing in and tossing in hand grenades.

(c) Remove combustible material to limit the danger of fire. Fires are dangerous to defenders and create smoke that could conceal attacking forces. Defenders should store firefighting equipment (water, sand, etc.). The danger of fire also influences the type of ammunition used in the defense. Tracers or incendiary rounds should be avoided if the threat of fire exists.

(d) Turn off electricity and gas. Both propane and natural gas are explosive. Natural gas is also poisonous and is not filtered out by a protective mask. Propane gas, although not poisonous, is heavier than air. If it leaks into an enclosed area, it displaces the oxygen and can cause suffocation. Gas mains and electricity should be shut off at the main distribution nodes.

(e) Battle positions should be located so as not to establish a pattern. The unit should avoid obvious firing locations like church steeples.

(f) The unit should rehearse movements between positions.

(g) Camouflage positions.

(h) Reinforce positions with available materials, such as beds, furniture, etc.

(i) Block stairwells and doors with wire or other material to prevent or delay enemy movement. Create holes between floors and rooms to allow for covered movement within a building.

(j) Prepare range cards, fire plans, and sector sketches.

(k) Emplace machine guns in basements windows, where the guns can provide grazing fires. When basements are not used, access to them should be sealed to prevent enemy entry.

(l) Stockpile V (ammunition) and class VIII (medical supplies) items.

(4) Clear Fields of Fire. Commanders may need to rubble certain buildings and structures to provide greater protection and fields of fire. If the ceiling of a lower story room can support the weight of the rubble, collapsing the top floor of a building before the battle starts may afford better protection from indirect fires. Rubbling an entire building can increase the fields of fire and create an obstacle to enemy movement. Rubbling buildings too soon (or rubbling too many) may give away defensive locations and may destroy cover from direct fire. Rubbled buildings may also interfere with planned routes of withdrawal or counterattack.

(5) Improve Fighting Positions. When time permits, all positions, including supplementary and alternate positions, should be reinforced with sandbags and provided

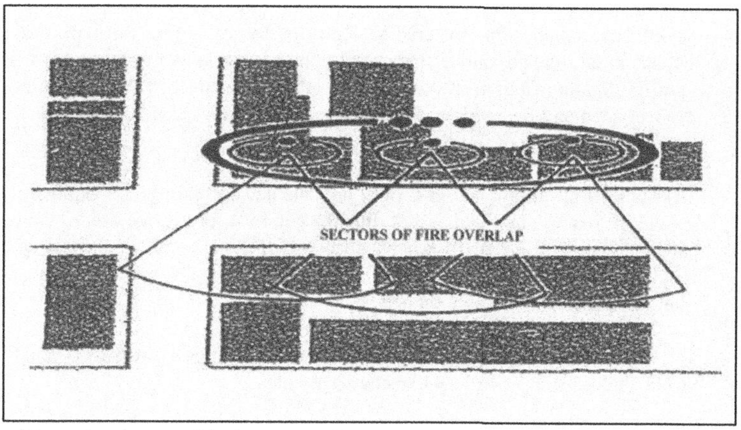

with overhead cover. Obtain support from attached engineers. (See Appendix D).

(6) Employment of Infantry

(a) The infantry battalion is normally assigned a sector to defend. However, depending on METT-T, it may be assigned to defend from a battle position or strongpoint. The battalion will usually assign its companies to either a battle position, a strongpoint, or a sector. These battle positions, strongpoints, or sectors are placed along avenues of approach to block or restrict the enemy's movement.

Depending on the type of built-up area, a company may be employed on the forward edge of the flanks of the area. This forces the enemy to deploy early without decisive engagement because it deceives the enemy as to the true location of the main defense. Other companies may then be employed in a series of battle positions and/or strongpoints in the center of the city or town. In all cases, mutual support between positions is vital. Companies and platoons should have designated alternate and supplementary positions.

(b) Once the rifle platoon commander has received his battle position or sector, he then selects the positions for his squads and crew-served weapons. Squad positions within the built-up area may be separated by rooms within buildings or be deployed in different buildings. Squad positions must be mutually supporting and allow for overlapping sectors of fire (Figure 3-8).

Figure 3-8. Sectors of Fire

(7) Employment of the Reserve. The commander's defensive plan always considers the employment of a reserve. The reserve force is organized and prepared to exploit success, to counterattack to regain key positions, to block enemy penetrations, to protect the flanks, or to assist by fire in the disengagement and withdrawal from endangered positions. For combat in a built-up area, a reserve force:

- Must be as mobile as possible
- May be a mechanized platoon, a platoon (-), or a squad at mechanized company level or a company, a company (-), or a platoon at battalion level
- May be supported by tanks or other armored vehicles
- Should be centrally located to facilitate manuever to threatened areas.

(8) Employment of Armored Vehicles. The commander plans to employ armored vehicles to take advantage of their long-range fires, heavy machine gun fires, self-generating smoke, and mobility. However, built-up areas restrict the mobility of armored vehicles and make them vulnerable to the antiarmor weapons of the enemy. Both the LAV and AAV are lightly armored and can be penetrated by heavy machine gun and antiarmor fires. Armored vehicles provide the commander with a mobile force to respond quickly to enemy threats on different avenues of approach.

(a) When armored vehicles are employed in the defense of a city, infantry should be positioned to provide them with security against close antitank fires and to detect targets for the armored vehicles. Armored vehicles should be assigned *primary*, *alternate*, and *supplementary* positions, as well as *primary* and *alternate* sectors.

(b) Armored vehicles should be located on likely avenues of approach to take advantage of their long-range fires. They may be:

- Positioned on the edge of the city in mutually supporting positions
- Positioned on key terrain on the flanks of towns and villages
- Used to cover barricades and obstacles by fire
- Placed with the reserve.

(c) Armored vehicles are normally employed as a platoon. However, sections of armored vehicles may be employed with rifle platoons or squads. This provides the armored vehicles with the close security of the infantry.

(9) Employment of Helicopterborne Forces

(a) Helicopterborne Task Force in the Defense. A helicopterborne force can defend against an infantry-heavy threat by utilizing its mobility to achieve a maneuver advantage. This force can be key in the execution of a mobile defense in both the security area and main battle area.

(b) Security Area. The helicopterborne force may be able to conduct security force operations for a larger force. Helicopters can position combat power and combat support quickly during rapidly changing situations.

(c) Main Battle Area. The mobility advantage that the helicopterborne force has over enemy infantry-heavy units may allow it to defend in greater depth in a large city. The helicopterborne force focuses on the destruction of advancing enemy forces and fights a series of battles in depth, it can attack the enemy from the front, flanks, and rear. Battle positions are selected and prepared throughout the main battle area along likely avenues of approach. Primary and alternate LZs and pick-up zones should be selected.

d. Fires. Fire support in the defense is as important as it is in the offense and has many of the same characteristics. The proximity of buildings to targets, minimum range restrictions, and repositioning requirements are all factors in the utilization of fire support in the defense. Additionally indirect fires are planned on top of and immediately around defensive positions for close support. Some of the indirect fire support considerations for the defense include:

(1) Historically, artillery fire has been used in some unique and innovative ways in an urban environment. Artillery fire support may be used in the direct- or indirect-fire mode. Artillery fire should be used to:

- Suppress and blind enemy overwatch elements

.3

- Disrupt or destroy an assault
- Provide counterfire
- Support counterattacks
- Provide direct fire when necessary
- Provide illumination in hours of darkness.

(2) Mortars at the infantry battalion and rifle company level are employed to maximize the effect of their high-angle fires. Mortars are one of the most effective indirect fire weapons in the urban environment. They should be used to engage:

- Enemy overwatch positions
- Enemy infantry before they seize a foothold
- Targets on rooftops
- Enemy reinforcements within range.

(3) Final protective fires are planned to stop dismounted assaults in front of the defensive positions. Fires within the city are planned along likely routes of advance to delay, disrupt, and destroy the enemy as he attempts to deepen a penetration.

(4) Commanders establish priorities of fire on the basis of enemy avenues of approach and threat systems that present the greatest danger to the defense. For example, during the attacker's initial advance, tanks, IFVs, and overwatching elements are the greatest threat to the defense. ATGMs should concentrate on destroying tanks first, then IFVs. Indirect fires should suppress and destroy enemy ATGMs and overwatch positions and/or elements. If enemy formations secure a foothold, priority is shifted to the destruction of enemy forces within the penetration.

(5) As the enemy attack progresses into the city, fires are increased to separate enemy infantry from its supporting armored vehicles. Friendly artillery concentrates on attacking infantry, conducting counterfire missions, and destroying reinforcements.

(6) Ample consideration must also be given to the rubbling effects of supporting arms. These effects often create massive rubbling of structures and buildings, making obstacles to both friendly and enemy movement.

(7) Mortars may be initially positioned forward in support of security forces. After their withdrawal, mortars are repositioned where they can support the in the main battle area.

(8) The commander assigns final protective fires and machine gun final protective lines (FPLs). Machine gun positions provide interlocking grazing fire and mutual support between adjacent units. FPLs are fired on planned signals. Proposed FPLs must be "walked out" to determine the extent of grazing fire available and to locate dead space. Dead space can be covered by:

- Rifle and light machine gun fire
- Grenade launchers
- Mines and boobytraps
- Indirect fires.

(9) Long-range antiarmor weapons are employed in the forward areas of the main battle area and the security area. Antiarmor weapons must be carefully positioned to prevent terrain and obstacles from reducing their range. Dismounted and manpacked antiarmor weapons should be positioned in buildings and along routes where engagement distances are reduced but where surprise and ambush are ideal.

e. Logistics. Just as in the offense, the defense utilizes a greater amount of class V and class VIII supplies in the urban environment. The defender needs to plan for multiple routes and means of supply. These can include:

- Using the city's existing transportation infrastructure to move supplies
- Prepositioning supplies in the defense in depth
- Using helicopter assets for rapid resupply
- Establishing and protecting supply routes
- Plan for utilization of existing repair and maintenance facilities
- Identify potable water sources

f. Force Protection

(1) Employment of Engineers. Engineers are employed at the battalion level or attached to companies and integrated down to squad level. Commanders must consider and assign the priority of work for engineer tasks that enhance survivability, mobility, and countermobility. Tasks that engineers can accomplish in the defense of a built-up area include:

- Constructing obstacles and rubbling for countermobility

- Clearing fields of fire
- Laying mines
- Preparing routes to the rear
- Preparing fighting positions.

(a) Employment of Obstacles. Obstacles are used to shape the battlespace. (See FM 5-102, *Countermobility*, and FM 20-32, *Mine/Countermine Operations*, for details.) The four types of obstacles are:

- **Disrupting.** Disrupting obstacles are used to disrupt the enemy's timetable and force him into conducting a time-costly bypass or breach, or cause his march columns to become staggered or "telescoped."
- **Turning.** Turning obstacles are used to canalize the enemy into engagement areas or kill zones where combined-arms fires can be concentrated and flanking shots are increased.
- **Fixing.** Fixing obstacles increase the time the enemy can be fired on in an engagement or target area and to cause increased confusion.
- **Blocking.** Blocking obstacles prevent or delay the enemy from moving past a certain point.

(b) Obstacle Belts. In defensive obstacle planning, obstacles may be emplaced in a series of belts. This enhances the overall effectiveness of the obstacle plan with successive belts designed for specific purposes. The following example lays out a three-belt obstacle plan.

1 **First Obstacle Belt.** The first obstacle belt is enplaced at the nearest buildings across from and parallel to the main defensive position. This belt consists of wire and improvised barriers (including inside buildings, in subterranean avenues of approach, and outside in open areas), danger areas, and dead space. These barriers and obstacles should be heavily boobytrapped and covered by fires. This belt is designed to impede enemy movement, breaks up and disorganize attack formations, and inflicts casualties.

2 **Second Obstacle Belt.** The second obstacle belt is placed between the first belt and the main defensive position buildings, but out of hand grenade range. It is designed to impede movement, canalize the enemy into the best fields of fire, break up attack formations, and inflict casualties. This belt is not meant to stop the enemy permanently. It consists mainly of wire obstacles, improvised barriers, road

craters, and minefields. It should be heavily boobytrapped. Triple-strand concertina wire is placed along the machine gun FPL to delay the enemy in the kill zone.

3 **Third Obstacle Belt.** The third obstacle belt is the defensive position's denial belt. It consists of wire obstacles placed around and through the defensive buildings, close-in minefields, and subterranean accesses. It is designed to impede the enemy's ability to gain a foothold in the defensive area. It should be heavily boobytrapped, and integrated with mines.

(c) Emplace Obstacles and Mines. While principles for employing mines and obstacles do not change in the defense of a urban area, techniques for their employment change. For example, burying and concealing mines in streets is usually not done. To save time and resources in preparing the defense, commanders must emphasize using all available materials (automobiles, railcars, rubble) to create obstacles. Civilian construction equipment and materials should be utilized. Obstacles must be integrated into buildings and rubble areas to increase their effectiveness and to canalize the enemy. The family of scatterable mines (FASCAM) may be effective on the outskirts of a city or in parks and open areas. (See Appendix F for further discussion on obstacles, mines, and demolitions.)

(2) Air Defense. Air defense (Stinger) weapons are positioned on rooftops and parking garages in order to obtain LOS coverage for the units they are protecting. Stinger teams can be assigned the mission of protecting specific positions or placed in general support (GS) of the battalion.

(3) Fire Hazards. All cities are vulnerable to destruction by fire, especially those with many wooden structures. The defender's detailed knowledge of the terrain permits him to avoid areas that are likely to be fire hazards. The defender can deliberately set fires:

- To disrupt and disorganize the attackers
- To canalize the attackers into more favorable engagement areas
- To obscure the attackers' observation.

Section II

Infantry Battalion Defense of a Built-Up Area

This section addresses some of the missions a task-organized and/or reinforced infantry battalion could be assigned. These missions are normally conducted as part of a larger joint operation.

3201. Defense in Sector. An infantry battalion is normally given the mission of defending a sector in a city (Figure 3-9). The interior buildings in the battalion sector form battle positions for the battalion's fixed defense. The size of a sector varies, but may cover the outlying structures and can be used to provide early warning and to delay the enemy.

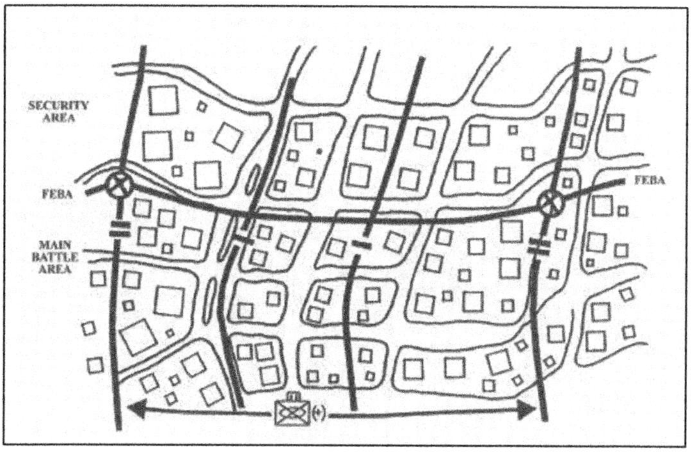

Figure 3-9. Defense of a Built-Up Sector

a. Defensive operations begin with the reconnoitering of the assigned sector to learn the terrain and to orient toward the enemy's expected advance. The main defensive positions should be formed on key terrain features, such as prominent buildings, road junctions, and positions which offer good fields of fire. Long-range antiarmor weapon systems should be deployed to engage the enemy as early as possible.

b. Depending on the situation, a reinforced rifle company may be used as a security force. The security force would operate forward of the FEBA and, depending on the mission, cause the enemy to deploy without without becoming decisively engaged. This can be done through maximum use of ambushes and obstacles while using covered and concealed routes. These routes are used for disengagement. Covered and concealed routes can be artificially made by making holes through buildings. The security force also attempts to detect the location of the enemy's main attack. Once the security force withdraws to the main battle area, it can be used as a reserve to reinforce or counterattack.

c. Based on METT-T considerations, the defense in sector (behind the FEBA in the main battle area) may consist of either company sectors, strongpoints, or battle positions. Strongpoints located on or covering decisive terrain are extremely effective in the defense.

d. Antiarmor vehicles and weapons should be used to engage enemy armored vehicles and/or cover obstacles with fire on likely avenues of approach.

e. Tanks are best used to engage enemy tanks, cover obstacles by fire, and engage in counterattacks. They should be employed in platoons where possible. However, in congested areas they may be employed in sections.

f. Artillery and mortar fire should be used to suppress and blind enemy overwatch elements, to engage enemy infantry on street approaches, provide counterbattery fire, and support counterattacks with direct and indirect fire.

g. Engineers are integrated throughout the defense. They provide support to forces in strongpoints to help prepare fighting positions. Engineers may also be attached to security forces to help in laying mines and constructing obstacles, clearing fields of fire, and preparing routes.

3202. Delay in a Built-Up Area. A delay is an operation in which a force under pressure trades space for time by slowing down the enemy's momentum and inflicting maximum damage on the enemy without becoming decisively engaged (MCWP 3-1). The delay can be oriented either on the enemy or on specified terrain such as a prominent building or industrial complex.

a. A delay in a built-up area may consist of a combination of ambushes and alternating or successive battle positions (Figures 3-10A and 3-10B on page 3-28).

(1) Ambushes are normally planned on observed obstacles. They require decentralized control and execution. Ambushes can be combined with limited-objective attacks on the enemy's flanks. They are most effective at the edge of open spaces, parks, wide streets, and so on. These attacks may be executed by dismounted infantry supported by tanks, LAVs/AAVs, or HMMWV-mounted machine guns and antitank vehicles.

(2) Battle positions should be placed where heavy weapons, antiarmor weapons, and machine guns will have the best fields of fire. Such locations are normally found at major street intersections, parks, and the edge of open residential areas. Battle positions should be carefully and deliberately prepared, reinforced by obstacles, and supported by indirect fire weapons. Battle positions are designed to inflict maximum casualties on the enemy and cause him to deploy for a deliberate attack.

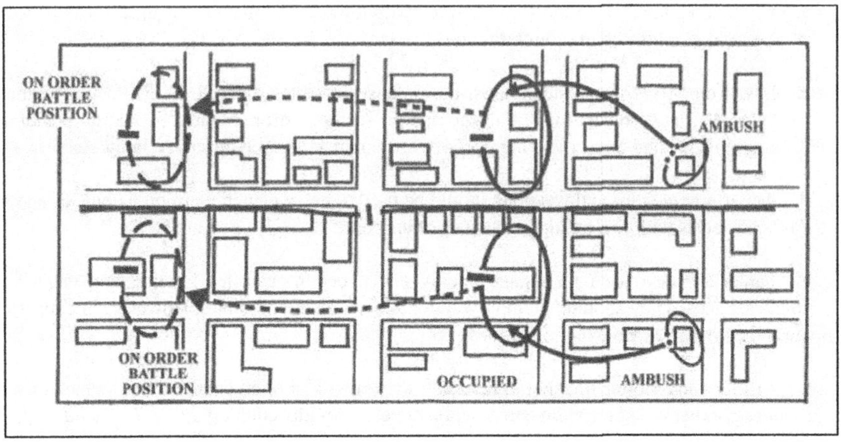

b. Tanks, LAVs/AAVs, and antiarmor weapons should prepare primary and alternate positions to improve their surivability and flexibility.

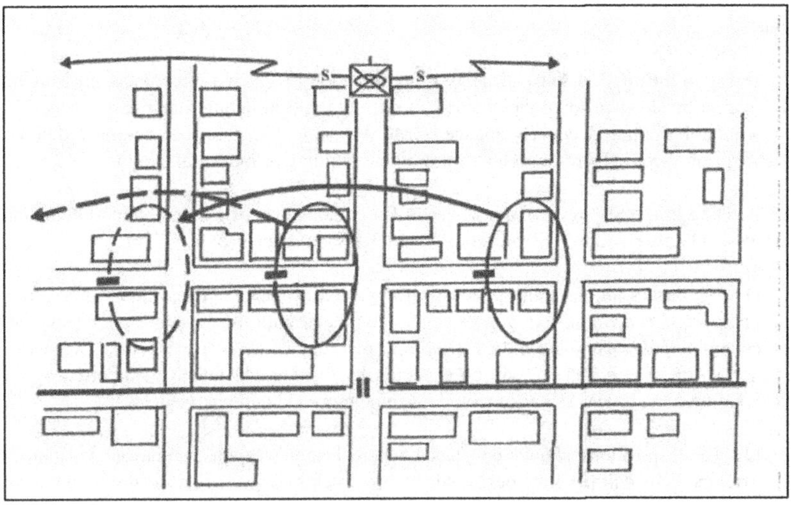

Figure 3-10A. Delay in a Built-Up Area (Successive Bounds)

Figure 3-10B. Delay in a Built-Up Area (Alternate Bounds)

c. The infantry battalion is most effective when deployed in successive delaying actions. Forces alternately engage the enemy to the maximum extent possible without becoming decisively engaged, then move on order in successive or alternate bounds. Smoke and demolitions are used to aid disengagement and cover the movement of the withdrawing force. Security units on the flank can be employed to prevent the enemy from outflanking the delaying unit. A small reserve can be used to react to unexpected enemy action and to conduct continued attacks on the enemy's flank.

d. The engineer effort should be centralized to support the preparation of battle positions, obstacle emplacement, and clearing routes.

e. The width of the infantry battalion sector depends on the nature of the buildings, obstacles along the street, and the time that the enemy must be delayed. Normally, narrow sectors supports delaying by alternate bounds while a wide sector better supports delaying by successive bounds.

3203. Defense of a Village. An infantry battalion assigned a defensive sector that includes a village may decide to incorporate the village as a strongpoint. This use is most common where the village stands astride a high-speed avenue of approach or lies between two difficult obstacles. To incorporate such an area into its defense, the infantry battalion must control the high ground on either side of the village to prevent the enemy from firing into the village. The battalion commander normally deploys one rifle company in the village to prepare a strongpoint defense

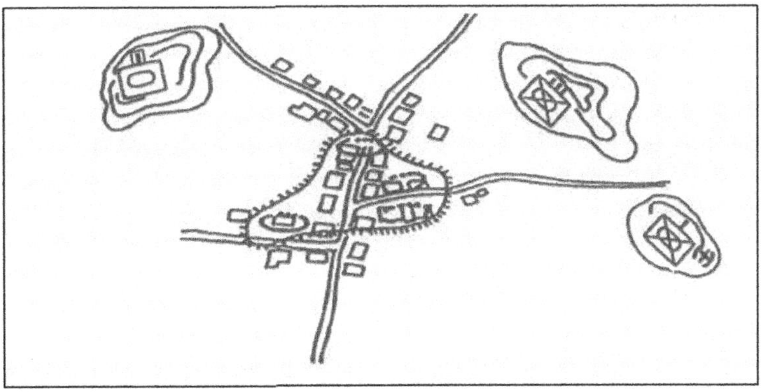

while the other rifle companies are overwatching the village, blocking bypass routes, and protecting the strongpoint's flanks. The strongpoint defense should provide a location where the enemy can be stopped and counterattacks can be launched (Figure 3-11).

Figure 3-11. Defense of a Village

a. Armored Vehicles and antiarmor weapons should be employed where the maneuver room is the greatest. This is usually on key terrain to the flanks of the village. As the security force withdraws and rifle companies become engaged, armored vehicles and weapons support by fire.

b. Although the infantry battalion's disposition should prevent large enemy forces from threatening the rear and flanks of the village, the danger of small-unit enemy infiltration requires the village be prepared for all-around defense.

c. Engineers should be assigned to the rifle company in the village strongpoint to provide continuous engineer support in preparation for the enemy's attack. The priority of engineer support should go to the strongpoint.

.3

Section III

Rifle Company Operations

This section addresses some of the missions the infantry battalion could assign to the rifle company. For the purpose of this publication, the rifle company is a task-organized and/or reinforced unit. Assigned missions are normally part of a larger battalion operation.

3301. Defense of a Village. Once the commander has completed his reconnaissance of the village and the surrounding terrain, he develops his plan for the defense (Figure 3-12). The commander must decide whether to defend on the leading edge of the village or within the village.

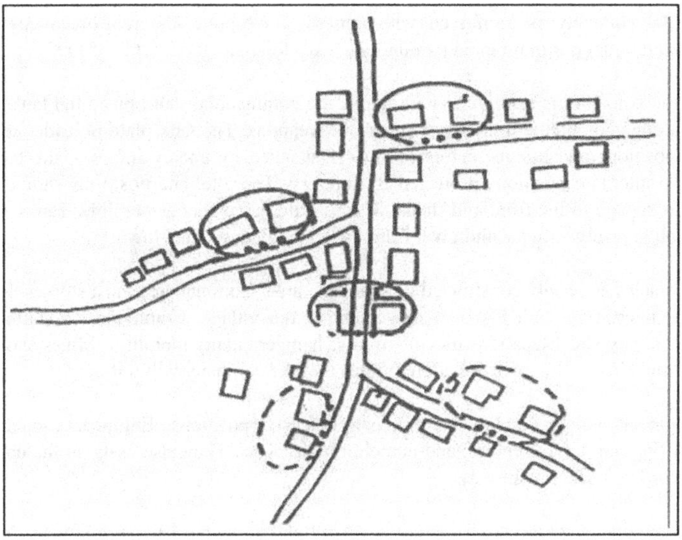

Figure 3-12. Defense of a Village

a. Several factors influence his decision. First, he must know the type of enemy that he will defend against. If the threat is mainly dismounted infantry, the greatest danger is in allowing them to gain a foothold in the town. If the threat is armor or motorized infantry, the greatest danger is that the company's defensive positions can be destroyed by direct fire. The commander must also consider the terrain forward and to the flanks of the village from which the enemy can direct fires against his positions.

b. The commander places platoons in a small group of buildings where each platoon leader establishes mutually supporting squad-sized positions. This positioning increases the area that the rifle platoon can control and hampers the enemy's ability to isolate or bypass a platoon. A rifle platoon can be responsible for the road through a village. The rest of the rifle company is positioned to provide all-around security and defense in depth.

c. Mortar positions must be protected from direct fire and allow for overhead clearance.

d. Combat vehicles should be assigned primary, alternate, and supplementary positions as well as primary and secondary sectors of fire. They should be positioned in defilade behind rubble and walls or inside buildings for cover, concealment, and movement into and out of the area.

e. If the company is reinforced with antiarmor weapons, the reinforcements should be positioned on high ground in and around the town to attain good fields of fire.

f. If the company is reinforced with tanks, the commander could place the tanks along the leading edge complement the other antiarmor weapons. The tank platoon leader selects exact firing positions and assign sectors of fire. If attacked by enemy infantry, the tanks should move to alternate positions protected by infantry. The alternate positions should allow the tanks to engage to the front and flanks with as little movement as possible. Tanks can be used as a mobile reserve after withdrawal from the leading edge of the town.

g. Obstacles are easily constructed in a built-up area. Commanders must stop enemy vehicles without interfering with his own movement in the village. Command detonating cratering charges at key street locations may be used to hamper enemy mobility. Mines should be laid on the outskirts of the town and along routes that the company will not use.

h. Engineers assist rifle platoons to prepare defensive positions. Engineers use explosives to make firing ports, mouseholes, and demolition obstacles. They also assist in the emplacement and recording of the minefields.

i. Ammunition expenditure is increased when fighting in an urban environment. Ammunition should be stockpiled at each squad position.

Other factors that the company commander may consider are:

- Resupply
- MEDEVAC
- Communications
- Firefighting

- Sleep and alert plans
- Security
- Limited visibility
- Control of noncombatants.

j. The company should install a wire net and develop a plan for pyrotechnic signals to ensure adequate communications. Backup wire should be laid in case primary lines are cut by vehicles, fires, or the enemy. The commander also plans for the use of messengers .

3302. Delay in a Built-Up Area. A rifle company may be assigned the mission to delay in a built-up area.(Figure 3-13). To perform this mission, a rifle company can destroy enemy reconnaissance elements forward of the outskirts of the town, prevent their penetration into the built-up areas, and gain and maintain contact with the enemy to determine the strength and

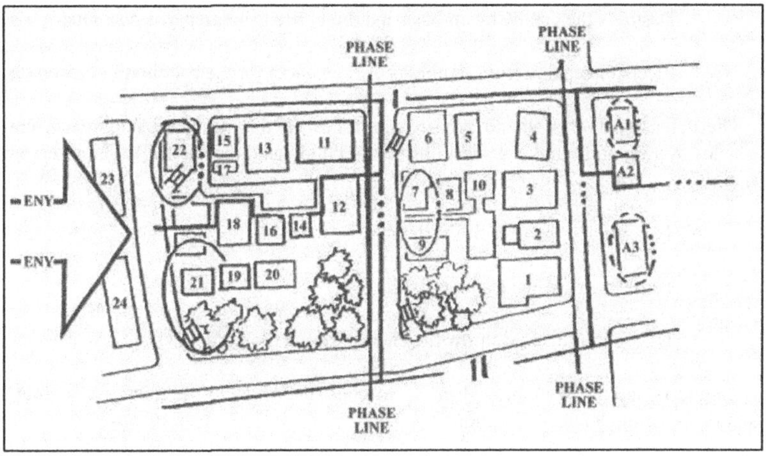

location of the main attack.

Figure 3-13. Delay in a Built-Up Area

a. Platoons delay by using patrols, OPs, and ambushes. Following each action, a platoon will disengage and withdraw.

b. The company's sector should be prepared with obstacles to enhance the delay. Engineers should prepare obstacles on anticipated enemy main routes of advance.

c. Antiarmor weapons may be positioned on the outskirts of the town to destroy the enemy at maximum range. They should be located in defilade positions or in prepared shelters. They fire at visible targets and then fall back or proceed to alternate positions.

d. Tanks in support may be used to engage enemy tanks, provide reinforcing fires, aid the disengagement of platoons, and cover obstacles by fire.

3303. Defense of a City Block. The rifle company may be assigned to defend a city block. The company actions should be coordinated with those of the security forces charged with delaying to the front of the company's position. The defense should take advantage of the protection of buildings that dominate the roads.

a. A well-organized company defense:

- Stops the attack of the enemy on the roads by using obstacles and enfilade fire
- Destroys the enemy by ambush and direct fire from prepared positions
- Ejects the enemy from footholds or fixes them for the battalion's counterattack.

b. The defensive forces should prepare ambushes on the avenues of approach, cover the obstacles by fire, and build a defense inside the buildings (Appendix D). The reserve can be tasked:

- To reinforce the fires of the defense
- To react to a danger on the flank
- To counterattack to throw the enemy from a foothold.

c. Engineers should be employed in GS to construct obstacles, prepare access routes, and assist in preparing defensive positions.

d. Tanks attached to the rifle company may be used to direct-fire support, engage enemy tanks, and support a counterattack.

3304. Defense of a Traffic Circle or Key Intersection. A rifle company may be assigned the mission to defend a key traffic circle or road junction (Figure 3-14).

a. The commander analyzes enemy avenues of approach and buildings that dominate those avenues. He plans to focus all possible firepower on the traffic circle and its approaches. He also plans for the all-around defense of the buildings that dominate the traffic circle to prevent encirclement. The commander prepares as many covered and concealed routes between these

.3

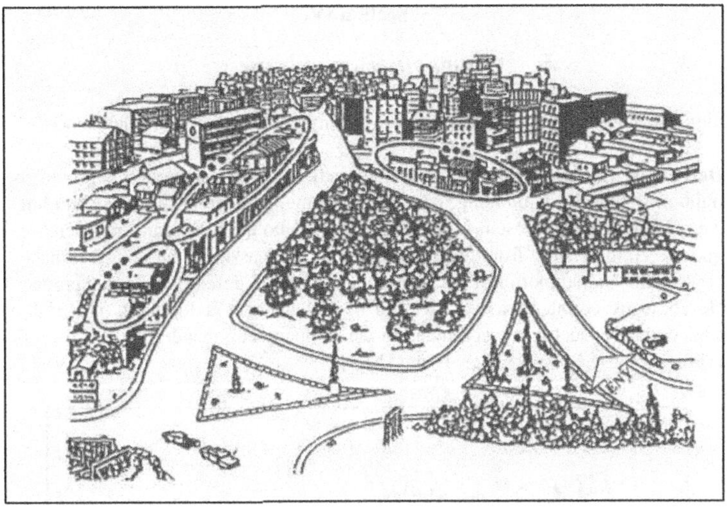

buildings as possible. These routes make it easier to mass or shift fires and to execute counterattacks.

Figure 3-14. Defense of a Traffic Circle

b. Obstacles can be used to deny the enemy the use of the traffic circle. In this case, obstacle planning should consider whether friendly forces require access to the traffic circle. Antiarmor weapons can be employed across the traffic circle if designated fields of fire exist. Tanks can be used to engage enemy tanks and provide direct-fire for counterattacks.

Section IV

Rifle Platoon Operations

This section discusses some missions the rifle company might assign to a rifle platoon.

3401. Defense of a Strongpoint. A rifle platoon may be given the mission to provide or conduct the strongpoint defense of a building, part of a building, or a group of small buildings (Figure 3-15). The rifle platoon's defense is normally integrated into the rifle company's defense. The rifle platoon must keep the enemy from gaining a foothold in its assigned buildings. It makes the best use of its weapons and supporting fires, organizes all-around defense, and counterattacks or calls for a rifle company counterattack to eject an enemy that has a foothold. The platoon leader analyzes his defensive sector to recommend to the company commander the best use of obstacles and supporting fires. (See Appendices B and D.)

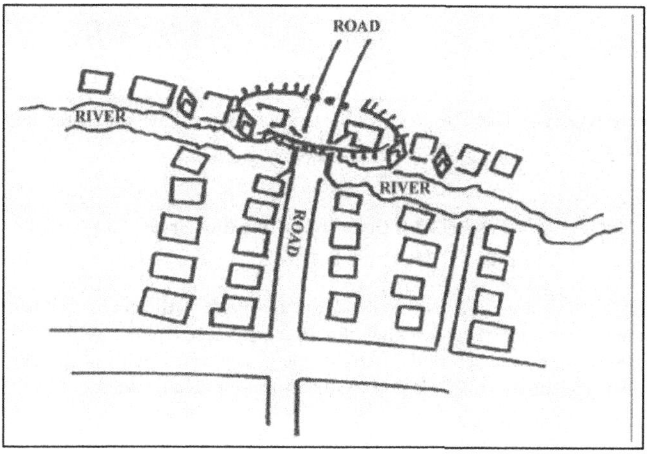

Figure 3-15. Defense of a Strongpoint

a. The rifle platoon should be organized into a series of firing positions located to cover avenues of approach, cover obstacles, and provide mutual support. Snipers may be located on the upper floors of the buildings. Unengaged units should be ready to counterattack, fight fires, or reinforce other elements of the rifle platoon.

b. Depending on the length of the mission, the rifle platoon should have available:

- Pioneer equipment (axes, shovels, hammers, picket pounders)

- Barrier material (barbed wire, sandbags)
- Munitions (especially grenades)
- Food and water
- Medical supplies
- Firefighting equipment.

3402. Defense Against Armor. Urbanized terrain is well suited to an infantry's defense against mechanized infantry and armored forces. Armored forces will usually avoid built-up areas but may be forced to pass through them.

a. Built-up areas have certain aspects that favor infantry antiarmor operations.

(1) Buildings provide cover and concealment for defending infantry.

(2) The streets restrict armor maneuver, fields of fire, and communications, thereby reducing the enemy's ability to reinforce.

(3) Rubble in the streets can be used to block enemy vehicles, conceal mines, and cover and conceal defending infantry.

(4) Rooftops, alleys, and upper floors provide good positions for flanking and overhead fire.

(5) Sewers, storm drains, and subways provide underground routes for infantry forces.

b. To prepare for antiarmor operations, a platoon should:

(1) Choose a Good Engagement Area. The best way for infantrymen to engage tanks is one at a time so that they can destroy one tank without being vulnerable to the fires of another. Good engagement areas include narrow streets, turns in the road, "T" intersections, bridges, tunnels, split-level roads, and rubbled areas. Less obvious locations can also be created by using demolitions or mines to create obstacles.

(2) Select Good Weapons Positions. The best weapons positions are places where the tank is weakest (tank deadspace) and the infantry is most protected. Forcing tanks to button up greatly reduces their ability to see and fire on close-in street and overhead targets. Figure 3-16 (on page 3-38) shows close-in tank weapon vision and engagement limitations (tank buttoned up) at street and overhead levels.

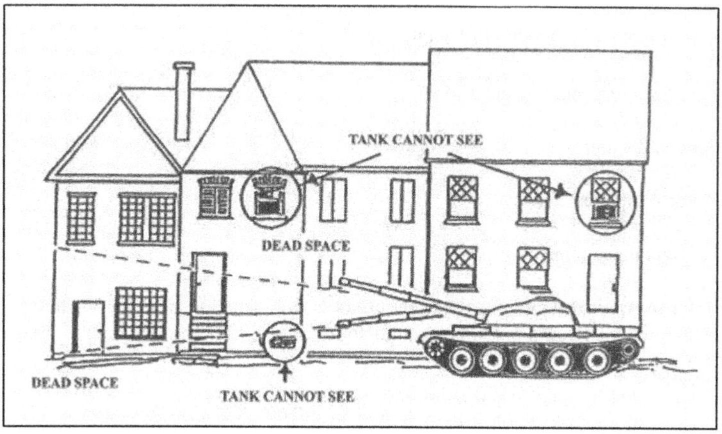

Figure 3-16. Close-in Tank Limitations at Street and Overhead Levels

(3) Assign Target Reference Points and Select Method of Engagement. After selecting the weapons positions, target reference points should be assigned to ensure adequate coverage and control fires. The target reference points should be clearly visible through the gunner's sights and should be relatively resistant to battle damage (for example, large buildings or bridge abutments, but not trees or cars). The three types of antiarmor engagement techniques are frontal, crossfire, and depth. Frontal fire is the least preferred because it exposes the gunner and it attacks the tank where the armor is the thickest. (For more information on target engagement techniques, see MCWP 3-15.5 *Antiarmor Operations*.)

Infantry forces should fire at tanks on the flanks and rear from ground level or at the top of tanks from an elevated position. A suitable antiarmor defense might be set up as shown in Figure 3-17. For a safe engagement from an elevated position, Marines should allow the tank to approach to a range three times the elevation of the weapons. However, engagement at longer ranges risks counterfire because the attacking weapon's position will not be in the tank's overhead dead space.

(4) Coordinate Target Engagement. Tanks are more vulnerable to infantry operations when they are buttoned up. The first task is to compel the tanks to button up by using direct and indirect fire. The next task is to coordinate the fires of the antitank weapons so that all targets in the engagement area are attacked simultaneously (Figure 3-18).

Figure 3-17. A Rifle Platoon's Antiarmor Defense

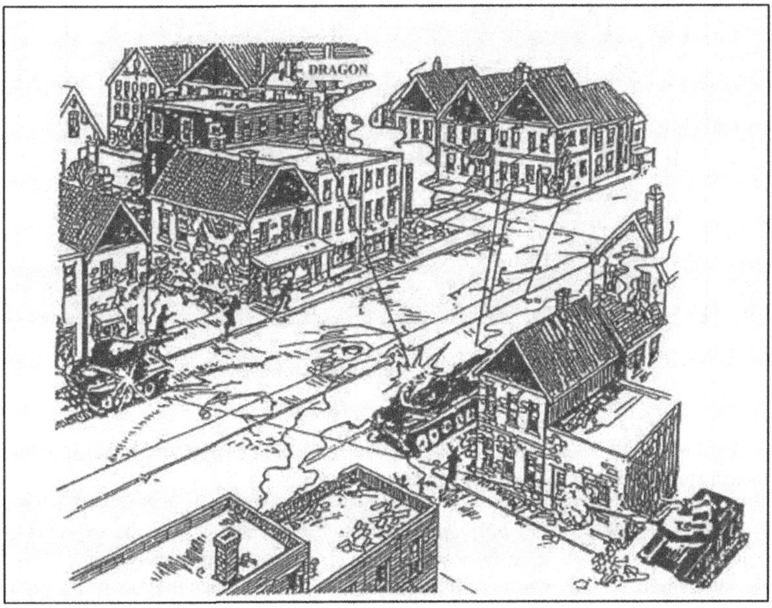

Figure 3-18. Coordinated Antiarmor Ambush

c. Infantry often accompany armored vehicles in built-up areas to provide them with close-in protection. Antiarmor weapons should be supported by an effective antipersonnel defense.

d. At a planned signal (for example, the detonation of a mine), all targets in the engagement area are attcked at the same time. If that is not possible, then the most dangerous target should be engaged first. Although tanks present the greatest threat, BMPs are also dangerous because their infantry can dismount and destroy friendly antiarmor positions. Rubble and mines should be used to reduce target mobility so that a target's time in the engagement area is increased.

3403. Conduct of an Armored Ambush. A rifle company can use an attached tank platoon to conduct an armored ambush (Figure 3-19). The tank platoon should be reinforced with infantry.

Figure 3-19. Armored Ambush

a. An armored ambush requires maneuver on obstacle-free roads. Obstacles outside the ambush area can be used to canalize and delay the enemy.

b. Ambushing tanks should hide approximately 1,000 meters from the expected enemy avenue of approach. A security post, located at a chokepoint, observes and reports the approach, speed, security posture, and activity of the enemy. When the enemy is reported at a trigger point or target reference point, the tank platoon leader orders the ambush.

c. The tanks move quickly from their hiding positions to firing positions, taking advantage of all available concealment. They attempt to fire flank shots on the approaching enemy at an average range of 300 to 400 meters. Firing at these ranges does not expose tanks to enemy infantry. After the engagement, the tank breaks contact and moves to a rally point with close security provided by infantry. They may then move to a new ambush site.

Chapter 4

Combat Support

Combat support units provide critical combat functions in conjunction with combat arms units to secure victory (MCRP 5-2A).

4001. Introduction. Success against a determined enemy is achieved in part by generating decisive combat power through the application of *combined arms*. This is especially true in MOUT. The MAGTF commander must effectively integrate all the capabilities of the command element (CE), GCE, ACE, and CSSE to win the battle in the shortest amount of time with the fewest casualties. The MOUT environment requires special considerations pertaining to combat support forces and capabilities.

Section I

Fire Support

4101. Indirect and Direct Fire Support

a. General. MOUT operations present unique fire support considerations. On urbanized terrain, buildings provide excellent cover and concealment to the enemy while limiting friendly observation and targeting efforts. Targets are generally exposed for brief periods of time and are often in close proximity to friendly forces. Observers will experience difficulty in finding OPs with adequate fields of observation. Terrain masking by tall buildings may restrict the delivery of indirect fires. Collateral damage and rubble effects must be considered during ammunition selection. The importance of effective communications, ROE, control measures, and procedures to prevent fratricide is magnified.

b. Offensive Support. In the offense, fire support plans should include fires to isolate the objective area, support the assault, and support the clearing action. Fires are delivered to isolate and fix the enemy and deny him the use of avenues of approach into and out of the built-up area. Most fires are normally planned and executed at the GCE or higher echelons. Fires are employed to rupture the enemy's established defenses and screen friendly maneuver, in order to maintain the momentum of the attack. Fire support is also allocated to units involved in clearing operations. Fire support plans should incorporate the employment of aerial observers and UAVs to compensate for restrictions to observation and to assist in the delivery of deep fires. Procedures for designating the forward line of own troops (FLOT), marking targets, shifting fires, and communicating in the urban environment should also be considered.

c. Defensive Support. In the defense, fire support plans address fires to disrupt and slow the enemy attack. Fires are delivered at maximum ranges along avenues of approach to separate armor and infantry forces, to canalize the enemy into killing zones, and to deny the enemy use of key facilities in the city, such as communications and transportation facilities. Most fires are planned and executed at the GCE or higher echelon. Defensive fires planned at lower echelons support fighting within the built-up area. Defensive fire support plans identify fire support coordination procedures necessary to execute the delivery of fire support.

4102. Artillery

a. Employment of Artillery. The mission of artillery does not change in MOUT. However, the method of employment is tailored to the unique requirements posed by the urban battlespace. MCWP 3-16.3, *Marine Artillery Support,* provides a detailed discussion on the employment of artillery. Considerations for MOUT include:

(1) Organization for Combat. Standard command relationships for tactical missions are used to organize for combat. During the initial phases of offensive MOUT, particularly isolation of the objective area and the initial assault to secure a foothold, control of artillery is more centralized than in other types of offensive operations. Centralized control provides better responsiveness in shifting and massing fires. As the attack progresses into the clearance phase, control of artillery becomes more decentralized with direct support (DS) artillery providing direct and indirect fires. In the defense, control of artillery should be sufficiently centralized to enable the rapid massing of fires.

(2) Movement and Positioning. Key considerations for the positioning of artillery in MOUT include:

- Mobility restrictions
- Limited availability of firing positions
- Masking of fires
- Security
- Enemy counterbattery fires

(a) Offense. Artillery is initially positioned outside or in the outskirts of the city. Artillery may displace to suitable firing positions within the city as the attack progresses. Positions should be selected that minimize masking, provide several routes for escape, and afford as much cover and concealment as possible. Potential firing positions can be found in parks, racetracks, sports fields, school yards, stadiums, parking lots, and similar areas. Special techniques for the emplacement of howitzers may be required if the ground is not suitable for normal emplacement. Some artillery positions may be maintained outside the city (see appendix B). Artillery generally makes few displacements in MOUT because it requires extensive route reconnaissance and security. If and when artillery is moved, it should be given priority of movement during displacements. Movement is conducted during night or periods of reduced visibility.

(b) Defense. Artillery is positioned on the friendly outskirts of the built-up area. Artillery displaces along predetermined routes to avoid traffic congestion.

(3) Delivery of Fires. The commander, in coordination with the fire support coordinator (FSC), considers the following when planning fire support in a built-up area:

(a) Target acquisition may be more difficult because of the increased cover and concealment afforded by the terrain. Ground observation is limited; therefore, FOs should be placed near tops of buildings.

(b) For targets in defilade, fire may be adjusted laterally so that the rounds impact on the street perpendicular to the observer's position. Airburst rounds are more suitable for this adjustment. Adjustments may have to be made by sound. When rounds impact on a perpendicular street, they are adjusted for range. When the range is correct, a lateral shift is made onto the target and the observer fires for effect.

(c) Special consideration must be given to shell and fuze combinations when effects of munitions are influenced by buildings. These include:

- Careful use of variable time is required to avoid premature arming.
- HE ammunition with impact or delay fuzes may create unwanted rubble.
- The close proximity of enemy and friendly forces requires careful coordination.
- White phosphorus (WP) may create unwanted fires and smoke.
- Fuze delay should be used to penetrate fortifications.
- Illumination rounds can be effective. However, friendly positions should remain in shadows and enemy positions should be highlighted. Tall buildings may mask the effects of illumination rounds.
- HE/variable time, HE/time fuze, and improved conventional munitions are effective for clearing enemy positions, observers, and antennas from rooftops.
- Swirling winds may degrade smoke operations.
- FASCAM may be used to impede enemy movements. FASCAM effectiveness is reduced when delivered on a hard surface.

(4) Targeting is difficult on urbanized terrain because the enemy has many covered and concealed positions and movement lanes. The enemy may be on rooftops and in buildings and may use sewer and subway systems. Targets should be planned on rooftops to clear enemy FOs, communications, and radar equipment. Targets should also be planned on major roads, at road intersections, and on known or likely enemy fortifications. Employing artillery in the direct-fire mode to destroy fortifications should be considered. Also, restrictive fire support coordination measures (such as a restrictive fire area or no fire area) may be imposed to protect civilians and critical installations.

(5) The 155-mm howitzer can be effective in neutralizing concrete targets by direct fire. A 155-mm round can penetrate 36 inches of concrete at ranges up to 2,200 meters. Restrictions may be placed on the type of ammunition used in order to reduce rubbling on friendly avenues of approach.

(6) FOs must be able to determine the location and size of dead space. Dead space is the area in which indirect fires cannot reach the street level because of buildings. This area is a safe haven for the enemy because he is protected from indirect fires. For low-angle artillery, the dead space is about five times the height of the building. For mortars and high-angle artillery, the dead space is about one-half the height of the building.

(7) Aerial observers and UAVs are effective in providing observation behind buildings immediately to the front of friendly forces. They are extremely helpful in adjusting fire. Aerial observers can also relay calls for fire when communications are degraded because of power lines or building masking.

(8) Weapons locating radars are effective in locating enemy artillery and mortar targets in an urban environment because of their high percentage of high-angle fires. However, LOS must be maintained in order to exploit this advantage. Radars should not be placed in the midst of a city due to masking. They are best located on the outskirts of the city.

(9) The use of airburst fires is an effective means of clearing snipers and manpads from rooftops. HE shells with delay fuzes may be effective against enemy troops in the upper floors of buildings, however, because of the overhead cover provided by the building, such shells have little effect on the enemy in the lower floors. (The planning and employment of artillery in offensive and defensive operations is also addressed in Chapters 2 and 3.)

b. Local Security. Artillery crews and howitzers require close security when used in the direct-fire mode. Artillery is particularly vulnerable when units are employed in isolated firing positions.

c. Communications and Information Systems (CIS). CIS must be able to satisfy the unique C2 requirements for artillery in MOUT. CIS should provide the commander with the ability to collect, process, analyze, and exchange information. Requirements such as massing fires and call for fire must be carefully planned and adapted to meet the requirements of

MOUT. For example, small units conducting independent operations must have the communications capabilities and maps to allow them to call for fire.

4103. Mortars

a. Employment of Mortars. Mortars are often the primary indirect-fire weapon for forward units in the assault or defense of a built-up area. Their high rate of fire, short minimum range, and high trajectory give mortar units the ability to fire in the tight confines of the city. Mortars can provide obscuration, neutralization, suppression, or illumination fires.

(1) Organization for Combat. Normally, 81-mm mortars are employed in GS of the battalion. The 60-mm mortar is normally used at the discretion of the rifle company commander.

(2) Movement and Positioning. 81-mm mortars employed in GS normally occupy a position within or near the infantry battalion reserve. Key considerations for the selection of positions include:

- The minimum range of the weapon
- Mask and overhead clearance
- Terrain suitable for setting base plates
- Dispersion, and accessibility.

Special techniques can be used to position and lay the 81-mm mortar. These techniques include placing buffers under base plates, using curbs and sandbags to anchor or brace mortars, and filling cans with dirt or using sandbags to emplace aiming posts. Fixed objects may be used as distant aiming points. When the depth of the defensive position is shallow, or when suitable firing positions are not available, mortars may have to be positioned behind the reserve. Positioning of mortars behind buildings greatly enhances survivability. Displacement is often executed by section. Mortars are usually not placed on top of buildings because the location lacks cover and concealment and recoil/concussion could collapse roofs.

(3) Delivery of Fires. Mortar fires are effective in the offense as well as in the defense. The indirect fires are extremely responsive. Mortars are well suited for combat in built-up areas because of their high rate of fire, steep angle of fall, and short minimum range. Mortar fires can be used to inhibit enemy fires and movement, allowing friendly forces to maneuver to a position of advantage. Effectively integrating mortar fires with dismounted maneuver is key to successful combat in a built-up area.

.3

(a) HE Ammunition. HE is used more than any other type of ammunition. Although mortar fires are often targeted against roads and other open areas, the natural dispersion of indirect fires will result in many hits on buildings. Leaders must use care when planning mortar fires to minimize collateral damage. Considerations when using HE include:

- HE ammunition is particularly effective when used on lightly built structures within cities. However, it does not perform well against the reinforced concrete found in larger urban areas.

- Point-detonating fuzes should be used. The use of proximity fuzes should be avoided because the nature of built-up areas causes proximity fuzes to function prematurely. However, proximity fuzes are useful in attacking targets such as OPs on tops of buildings.

- During World War II, the Middle East conflicts, and most recently in Bosnia, mortar HE fires have been used extensively to deny the use of streets, parks, and plazas to enemy and civilian personnel.

(b) Incendiary Fires. Employment of these munitions is planned in detail to prevent the burning of buildings and surrounding terrain that may create obstacles to friendly forces.

(c) Illumination. In the offense, illumination rounds should burst above the objective to put enemy troops in the light. Buildings reduce the effectiveness of the illumination by creating shadows. If the illumination is behind the objective, the enemy troops would be in the shadows rather than in the light. In the defense, illumination is planned to burst behind friendly forces to put them in the shadows and place the enemy forces in the light. Continuous illumination requires close coordination between the FO and the fire direction center (FDC) to produce the proper effect.

(d) Other Considerations. When planning the use of mortars, commanders must consider the following:

- FOs should be positioned on tops of buildings so target acquisition and adjustments in fire can best be accomplished.

- Commanders must understand ammunition effects to correctly estimate the number of volleys needed for the specific target coverage. Also, the effects of using white or red phosphorus may create unwanted smoke screens or limit visibility, which could interfere with the tactical plan.

- Mortar sections should plan to provide their own security.

- Commanders must give consideration to where, when, and how mortars are to displace in order to maintain immediate indirect fire support. Combat in built-up areas may adversely affects the ability of mortars to displace because of rubbling.

b. Communications. An increased use of wire, messenger, and visual signals will be required. However, wire should be the primary means of communication between the FOs and the FDC because elements are close to each other. Also, frequency-modulated (FM) radio transmissions in built-up areas are likely to be erratic. Remote siting of antennas to upper floors or roofs may improve communications and enhance operator survivability. The use of radio retransmissions can also be effective. Both in the offense and defense, commanders consider using existing civilian systems to supplement the unit's capability.

c. Magnetic Interference. Anything that uses or generates electricity generates its own magnetic field. In an urban environment, all magnetic instruments are affected by surrounding structural steel, electrical cables, and automobiles. Minimum distance restrictions for the use of the M-2 aiming circle will be difficult to overcome. However, an azimuth may be obtained from a distant aiming point.

4104. Naval Surface Fire Support (NSFS) and Naval Gunfire (NGF)

a. General. When NSFS ships are available, naval surface fires and NGF can provide some fire support.

b. Employment of Naval Surface Fire Support. The urban environment severely limits NSFS, and NGF in particular, due to its flat trajectory. NSFS can be used to supplement existing fire support but should not be planned as the primary fire support asset for MOUT. NSFS is most effective when firing on targets to isolate the battlefield and engage targets located on the seaward side of a city. Only when the ROE allows extensive collateral damage can the full effect of NSFS be brought to bear on units occupying a city.

(1) Organization for Combat. NSFS is best employed under centralized control because of the limited use of NSFS within the confines of the city.

(2) Positioning. Many of the fires delivered by NSFS ships require a careful orientation of the gun target line and may limit the size of a fire support area.

(3) Delivery of Fires. NGF can deliver a variety of ammunition in MOUT. NGF can deliver HE, incendiary, and illuminating projectiles. The NGF high-capacity projectile is very effective in producing fragmenting and penetration effects. Offensive and defensive fire support plans can incorporate NSFS fires on bridges, roads, reinforcements, artillery positions, and command and control centers. NSFS can be effectively delivered on targets along the outskirts of the city to rupture established defenses. These fires are particularly effective on buildings and on targets with a linear axis such as streets and airfields. Because of its flat trajectory, NGF employment is limited on targets within the city. Firing

.3

reduced charges at high angles does enable NGF to hit more targets but will reduce its range and accuracy.

4105. Aviation

a. Employment of Aviation. Both fixed- and rotary-wing aircraft can provide fire support in MOUT. The aviation combat element (ACE) provides the MAGTF commander with a variety of options in ordnance delivery. Aerial delivered ordnance is particularly effective against hard targets. Cluster bomb units, rockets, cannons, laser-guided bombs, and electro-optically guided missiles can be used against a variety of targets. However, the presence of civilians, the requirement to preserve key facilities, and the residual effects caused by rubble and collateral damage must be taken into consideration when planning for and using aviation in an urban environment. (See MCWP 3-24 *Assault Support Helicopter Tactical Manual* (under development), MCWP 3-23.1 *Close Air Support* (under development),and NWP 55-3-AH1, *AH1 Tactical Manual,* for greater detail on MOUT considerations for aviation.)

(1) Fixed-Wing Aircraft

(a) Offensive/Defensive Operations. During offensive operations, fixed-wing aircraft can isolate, interdict, and reduce enemy strongpoints throughout the battlespace. During defensive operations, they attrite enemy forces as they approach, attack forces attempting to bypass, or canalize the enemy to take avenues of approach that are advantageous to the defender.

(b) Other Considerations

1 Fixed-Wing Attack of a Built-Up Area. Air strikes can reduce or destroy vital supplies and facilities supporting the enemy. The resulting shock, concussion, and psychological effects of such attacks can reduce the efficiency and fighting spirit of enemy forces.

2 Rubble and Debris. The rubble and debris resulting from air attacks may increase the defender's cover and create obstacles to the movement of friendly forces.

3 Proximity of Friendly Forces. The proximity of opposing forces to friendly forces increases the potential for fratricide. This may require the use of precision guided munitions (PGMs) or the temporary disengagement of friendly forces in contact.

4 Locations of Indigenous Civilians or Key Facilities. The use of aerial delivered ordnance may be restricted by the ROE because of the presence of civilians or the requirement to preserve key facilities within a city.

5 **Limited Ground Observation.** Limited ground observation may require the use of forward air controllers (airborne) (FAC(A)).

(2) Rotary-Wing Aircraft. Attack helicopters provide a distinct advantage to the commander in MOUT. They can rapidly maneuver within the built-up area to attack a variety of targets. Urban structures offer ideal cover for attack helicopters. The ability to deliver PGMs, rockets, and 20 mm cannon fire makes attack helicopters valuable assets in reducing strongpoints, breaking up enemy concentrations, and providing suppressive fires for attacking forces. Attack helicopters also provide a highly manueverable antiarmor capability.

(a) Attack Helicopter Fire Support During Offensive/Defensive Operations

1 Provide close air support in support of the commander's scheme of maneuver

2 Provide suppression and/or destruction of enemy forces within buildings and strongpoints

3 Conduct counterattacks

4 Isolate the built-up area or objectives within the built-up area

5 Conduct armed reconnaissance and patrols.

6 Conduct rapid concentration of fires.

4106. Armored Vehicles

a. Employment of Armored Vehicles. Tanks, LAVs, and AAVs are primarily mobility assets that function best in an unrestricted, mobile environment. However, these vehicles can play an important role inside the city by using their firepower in support of offensive and defensive operations. LAVs and AAVs highly are susceptible to armor-piercing rounds and manpacked antiarmor weapons. All vehicles are susceptible to top-attack munitions. Some of the fire support roles armored vehicles may provide include:

- Suppression and/or destruction of enemy forces within buildings and strongpoints (tanks/LAVs/AAVs)
- Provide machinegun suppression fires
- Reserve or counterattack suppression fires (tanks and/or AAVs)
- Creating entry/exit points in buildings
- Isolation of the built-up area or objectives within the built-up area (tanks/LAVs/AAVs)

- Artillery-towing vehicle (LAVs/AAVs)
- Antiarmor fires
- Breaching obstacles in a direct fire mode.

b. Special Consideration. The existence of waterways, canals, and rivers in and around urban areas can provide avenues of ingress into designated objective areas. If used in conjunction with supporting fires, smoke, and/or deception operations, AAVs can capitalize on their amphibious capability and make use of waterways previously viewed as obstacles.

4107. Antiarmor Weapons. Antiarmor weapons have more utility in the defense than in the offense. Their primary task is defeating enemy armor. Secondary tasks may involve blasting holes in structures and fortified positions. As with other weapons systems, antiarmor weapons are normally employed in a decentralized fashion at the small unit level. Enemy armor should be engaged in predetermined locations that afford the most advantage to the defender. Weapon positions should be selected that provide maximum range fires. TOWs and Dragons may be fired from an enclosed area if it is a sturdy ventilated structure of appropriate size. For a more detailed discussion of antiarmor weapons, see MCWP 3-15.5, *Antiarmor Operations*.

4108. Employment of Snipers

a. General. Snipers are highly trained in field skills and marksmanship and can deliver long-range precision fire at selected targets from concealed positions. The method by which snipers are employed in urban warfare will be governed by the nature of the terrain/weather, number of snipers available, and whether the enemy employs snipers. A sniper can play an important role in MOUT. Employment of snipers influences the action by:

- Creating adverse psychological affects on the enemy by negating feelings of security
- Inflicting casualties on enemy leadership
- Enabling the infantry to seem to cover a large area, regardless of whether the terrain is physically occupied or not
- Detailing enemy positions and activity through the use of advanced optical equipment and observation techniques
- Controlling fire support
- Conducting reconnaissance in conjunction with their sniper role
- Disrupting enemy movement, observation, and infiltration, and negating the possibility of surprise.

b. Offensive Employment. Snipers should be used to gain and maintain contact with the enemy. This enhances security and prevents surprise by keeping constant and unrelenting pressure on the enemy. Some of the considerations for sniper employment include:

(1) As the "eyes" of a commander, a sniper increases the commander's flexibility by gathering and transmitting information on fire support targets.

(2) Snipers attack those targets that affect the enemy's ability to wage battle and those that will cause the maximum amount of confusion on the battlefield. Potential targets for a sniper are:

- Officers and noncommissioned officers
- Enemy snipers and marksmen
- Engineer personnel
- Personnel manning crew-served weapons
- Communications personnel and equipment
- Fire support observers and/or equipment
- Commanders of armored vehicles.

c. Combat on Urbanized Terrain

(1) Snipers can be employed to operate for extended periods of time from hidden positions to dominate and establish a "no-man's land," screen flanks, protect the rear, and deny the enemy access to certain areas or avenues of approach. Snipers can operate with the covering element to deliver accurate fires in support of search elements. Optical devices enhances their ability to detect movement and engage it instantly.

(2) Snipers can also support (by fire) infantry movement across streets. They can provide immediate precision fire on enemy machine gun nests and enemy snipers. The best countersniper weapon is the sniper. Snipers are assigned the following supporting missions:

- Delaying and inflicting casualties on the advancing enemy
- Observing and reporting potential targets
- Covering (by fire) avenues of approach and obstacles

.3

- Harassing the enemy and causing him to deploy prematurely

- Directing supporting arms fire.

The urban environment provides an abundance of secure firing positions that are not highly accessible to countersniping.

d. Defensive Employment. In defensive operations, the sniper is best employed forward of the FEBA to provide early warning of enemy approach, disrupt it, and if possible, cause the enemy to deploy prematurely. Snipers are employed to:

- Prevent enemy infiltration

- Detect and destroy infiltrators

- Protect unit at the FEBA from surprise attacks

- Protect patrols from ambush

- Screen the flanks and rear of the defensive positions.

More detailed information regarding the employment of snipers can be found in Marine Corps Warfighting Publication (MCWP) 3-15.3, *Scout Sniping* (under development).

Section II

Other Combat Support

4201. Engineers

a. General. The nature of the urban battlespace requires extensive use of engineers. In both the offense and defense, the proper employment of engineers can be decisive. Engineers must also be prepared to repair and maintain the internal support facilities of the city in order to support further operations. MAGTF commanders plan to use all available engineer assets, including engineers in the GCE, ACE, and CSSE, as well as any naval construction forces (NCFs).

b. Engineer Organization. Engineers are almost always task organized. The compartmentalized, closed-in nature of the city, which limits observation, mobility, and communications, will dictate the best organization for the mission. Teams should be formed to perform specific tasks for small units and must be positioned for quick employment. In MOUT, engineer support is key to maintaining the mometum of the attack. Engineer teams will usually move with the units they are supporting. Two-man demolition teams may be assigned to support a rifle platoon or squad to breach obstacles or to neutralize booby traps and improvised explosive devices (IEDs).

Commanders must carefully analyze mission requirements, the alignment of enemy forces, the condition of the battlespace, and available assets, to ensure an organization that will provide optimum engineer support. Most of the actual engineer work is done at the small unit level.

c. Engineer Tasks

(1) Offense. Urban combat is by nature a very slow, deliberate affair. Even if the enemy has not prepared the area for a defense, it will take time to secure the city. A well-constructed urban defense will slow or stop any attack unless impediments to movement can be quickly removed or bypassed. Properly organized and positioned engineers can reduce delays by performing a variety of tasks which include:

(a) Engineer reconnaissance to determine the level of effort required to penetrate outer and inner defenses, clear obstacles, and identify bypass routes.

(b) Bridge repair and construction on critical avenues of approach. Bridge construction may also open additional avenues of approach that the enemy is not prepared to defend.

(c) Breaching obstacles both outside and inside the city. This includes breaching minefields and neutralizing booby traps and IEDs.

(d) Assault breaching into buildings.

(e) Reducing enemy strongpoints using demolitions and heavy equipment.

(f) Preparing helicopter landing zones (HLZs).

(g) Assist in the preparation of defensive positions in the event of a counterattack.

(2) Defense. An urban area can be made nearly impregnable with proper organization of the battlespace and skillful preparation by engineers. Obstacles in outlying areas can delay/disrupt the enemy and canalize him into engagement areas. The streets and alleys inside a city can also be use to canalize enemy forces into engagement areas and kill zones. In preparing the defense of an urban area, engineers perform several critical tasks to include:

(a) Engineer preparation of the urban battlespace. This should begin well outside the city with complex obstacles in support of security forces. Selective demolition of bridges and roads and other countermobility measures can force the enemy into taking avenues of approach that favor the defense.

(b) Constructing obstacles forward of and between defensive positions to block, disrupt, fix, and turn enemy attacks. A majority of the obstacles will be directed at interfering with the movement of personnel and vehicles, but some effort must be directed at preventing vertical and subterranean movement. Material designed especially for obstacles and barricades, such as barbed wire and tetrahedron(s), is necessary, but improvised barriers can be just as effective and are readily available. Vehicles and building rubble are the most common improvised barriers.

(c) Enhancing mobility by opening and/or marking preplanned routes into attack positions for counterattacks or alternate positions to protect a flank.

(d) Providing technical advice on which buildings are more suitable for use as strongpoints or which rooftops are strong enough for helicopter landings can be vital. Any reinforcement of building roofs should be supervised by the engineers.

(e) Preparing tank positions that offer them cover and concealment.

(f) Assisting infantry units in clearing clear fields of fire, especially for longer range weapons like the TOW.

(g) Playing a potential role in the maintenance of critical urban services such as power, water, and communications. This function may also be performed by naval construction units.

(h) Assisting in the preparation and maintenance of railroads, airfields, and ports. This function may also be performed by naval construction units.

4202. Assault Support Helicopters

a. Combat Assault Transportation. The mobility provided by helicopters can be a significant advantage in the urban environment. In many modern cities, office buildings may have helipads on their roofs. Other buildings, such as parking garages, are usually strong enough to support the weight of a helicopter. The delivery of Marines onto a building can also be accomplished by rappelling, fastroping, or jumping out of the helicopter as it hovers just above the roof. The rooftops should be inspected to ensure that no obstacles exist, such as electrical wires, telephone poles, antennas, or mines and wire. The use of helicopters in this role is usually predicated on the antiair threat. Helicopterborne forces are highly vulnerable to small arms fire, AAA and manpads.

b. Small-Scale Helicopterborne Assaults. Small units may be helo-lifted onto the rooftop of a key building. Depending on the construction of the roof, rappelling or fastroping Marines from the helicopter may be more advantageous than landing them on the rooftop. Fastroping provides a rapid and safer means of insertion than jumping from a low hover. With practice, Marines can accomplish a fastrope insertion with minimum exposure.

c. Large-Scale Helicopterborne Assaults. For large-scale assaults, rooftop landings are not practical; however, open spaces (parks, parking lots, sports arenas) within the built-up area may be used. Depending on METT-T, helicopterborne forces may assault into an LZ in one wave or scheduled waves.

d. Mobility of Forces and Supplies. In the urban environment, movement of forces by helicopters may become a major requirement. Units engaged in house-to-house fighting normally suffer more casualties than units fighting in open terrain. These casualties must be evacuated and replaced quickly with new forces. At the same time, roads are likely to be crowded with resupply and evacuation vehicles and may also be blocked with craters or rubble. Helicopters provide a means of overcoming these obstacles.

e. Helicopterborne Assault Techniques. Helicopter assaults into enemy-held territory are extremely difficult (Figure 4-1 on page 4-16). One technique is to fly nap-of-the-earth down a broad street or commercial ribbon while attack helicopters and assault helicopter door gunners suppress buildings on either side of the street. Artillery preparations can be incorporated into the helicopter assault plan to suppress rooftop and building antiair threat. Feints and demonstrations in the form of false insertions can deceive the enemy as to the intended objective. Helicopters often deliver forces to the last covered position, short of the fighting, then return without exposure to enemy direct fire. Similar flight techniques can be used for transportation of supplies and casualties.

Figure 4-1. Helicopterborne Assault of a Built-Up Area

f. Combat Roles. Below are some of the roles that assault support helicopters may play in urban warfare:

- Conduct insertion and extraction of forces (See MCRP 3-11.7A, *Helicopter insertions and Extractions* [under development])
- Provide command/control and aerial retransmission
- Support CSS operations
- Provide air reconnaissance.

4203. Employment of Reconnaissance Forces. Reconnaissance units traditionally make maximum use of rural camouflage and terrain. Area, zone, and route reconnaissance are conducted from well-concealed vantage points through clandestine insertion by foot, vehicle, rubber boat, or helicopter. These methods of insertion may still apply in village and strip areas for collection of information, sensor employment, conducting OP radio relays, or a combination of missions. In major urban areas, undetected insertions are difficult to accomplish because of the high density civilian population.

The use of reconnaissance assets must be used wisely. Reconnaissance assets must not be risked indiscriminately if the information can be collected by other means. Consideration should be given to flank and rearward employment of ground reconnaissance units, in addition to close (0 to about 2,000 meters), distant (approximately 2,000 - 20,000 meters), and deep (from about 20,000 meters and beyond, plus preassault) missions. Reconnaissance of bypassed pockets of resistance, avenues of approach, canals, bridges, subways, and sewer systems (possibly in conjunction with combat engineers) is feasible. Observation, sniper, and listening posts manned by reconnaissance personnel are also missions that reconnaisance personnel perform. The limited combatant diving capability of division reconnaissance may be employed to advantage in many coastal cities.

4204. Military Police (MP). MPs can play a significant role in the urban environment. They may conduct battlefield circulation control, area security, enemy prisoner of war (EPW) and law and order operations. These operations require continuous coordination with MAGTF units as well as with host nation military and civilian police.

a. MP units support area damage control operations. With the increased possibility of rubbling, MP units report on and block off affected areas and reroute movement to alternate road networks.

b. When conducting EPW operations, MPs collect prisoners as far forward as possible and transport them to the rear. They operate temporary collecting points and holding areas to briefly retain EPWs and civilian internees. EPW operations are important in built-up areas because the rate of capture can be higher than normal.

c. Commanders must realize that MP support may not always be available and that infantry units may have to assume certain MP functions. These functions include:

(1) Providing convoy escort and security for lines of communications.

(2) Controling roads, waterways, and railroad terminals that are critical chokepoints in the main supply routes.

(3) Providing security of critical sites and facilities, including communications centers, government buildings, water and electrical supply facilities, command and control nodes, airfields, storage facilities, and other essential areas.

(4) Provide refugee control in close cooperation with host nation civil authorities.

(5) Collect and escort EPWs.

4205. Communications

a. General. Manmade structures create many challenges for CIS in an urban environment. Other CIS challenges include extensive commercial and communications networks and public service radio nets. Commercial communication networks are commposed of miles of underground protected cable which connect and control telephones and computer networks. Public service radio nets, (including police, fire, civil defense, and taxi) compete with existing communication systems. To communicate on an effective, continuous basis in the urban environment, commanders must minimize communication limitations imposed by this environment and maximize the advantage of existing civil communications capabilities.

b. Radio. Manmade structures will decrease the effectiveness of VHF radio communications. The physical locations of radios are critical for effective communications. Some specific locations to avoid include powerlines, underground areas, and steel bridges.

Very high frequency (VHF) radios provide the primary communication capability at the battalion level. It is imperative that these radios be employed as effectively as possible. Communicators can use the following techniques to enhance VHF communications equipment:

- Relocation of VHF radios to overcome phase cancellation due to numerous dead spots
- Use of the higher end (50 - 75 MHz) of the available frequency spectrum
- Use of retransmission facilities to improve the reliability of communications
- Use of radio remotes.
- Use of field expedient antennas.

In setting up retransmission stations, antennas should be camouflaged to blend in with urban surroundings and to prevent them from being targeted by enemy fires. Antennas can be concealed by such objects as water towers and steeples.

In larger cities with many tall buildings, it may be necessary to construct a terrain profile of the area of operations. A communications plateau can be deduced by calculating the average height of the structures in a zone. Operation of antennas above the median line of this plateau will result in better wave propagation and aid in retransmission.

c. Wire Communications. Wire may be used extensively in urban environments. It offers two primary advantages over radio communications:

- It is not susceptible to interference and absorption suffered by radio systems.
- Security of the overall system is greater because wire systems are essentially point-to-point.

Maximum use should be made of existing facilities for concealment. Sewers, subway tunnels, cable stairwells, elevator shafts, intact buildings, and pole lines offer ideal concealment for military wire systems. Precautions must be taken to lessen the possibility of vehicles tearing lines and disrupting communications. Existing telephone poles and tunnels can be used to keep wire above or below the street level.

Maximum use of existing commercial facilities should be made to supplement or replace military CIS. Telephone exchanges are linked together by underground cable systems and constitute a key resource in the urban communications system. These facilities are normally windowless buildings constructed of steel-reinforced concrete and provide ideal CPs and communications centers. The physical location of these exchanges, along with circuit diagrams and telephone directories, should be determined as soon as possible. Action should be initiated to secure these facilities early. Other facilities that contain extensive internal/external telephone systems may also be used as command centers. These include most municipal buildings, department stores, local police stations, and large business offices.

d. Messengers. Messengers provide a secure and flexible means of communication, particularly in an electronic warfare (EW) environment. However, careful selection of messenger routes is necessary to avoid pockets of enemy resistance. Disorientation in an urban environment occurs more often and more easily than in a rural environment. Marking facilities, transportation networks, and city streets on overlays often facilitates messenger service. Selected primary and alternate routes can often interface with designated logistics/resupply routes. Early restoration of street signs and recognizable landmarks are is essential.

e. Telephone Networks. Captured civil or military communications facilities can be utilized to augment organic communications capabilities. A civilian phone system can provide a reliable means of communication. If a telephone encryption device is not available, codes and authentication tables should be used.

Telephone technology differs from place to place, but there are universal concepts and practices in their operation. The most important concept to the military planner is network hierarchy.

Network hierarchy is a concept of organization by which a telephone company can rapidly and expeditiously allocate calls through a particular group of offices. As an ancillary function, it represents a control feature by which traffic can be rerouted to avoid congested areas or damaged facilities. This network hierarchy could play an important role in the planning and execution of the scheme of maneuver and fire support.

4206. Employment of Sensors

a. Sensor systems (seismic, infrared, acoustic, or magnetic) are normally employed in a rural context but also have traditional applications in village and strip area terrain. Use of sensors in urban areas is largely undocumented. In major urban areas, the capability for emplacement by air over select areas is reduced. Emplacement in, under, or around urban structures may also reduce microphone efficiency. Tactical employment will be degraded by such obstacles as concrete and steel structures, high-density population, and traffic.

b. Training in camouflage techniques for microphones and sensor terminals in an urban environment should be conducted. Operators should receive sufficient training to distinguish background-generated clutter from enemy foot, vehicular, and mechanized traffic. Use of and reliance on sensors may be seriously reduced during offensive operations in major urban areas. However, greater use of sensors will occur in subterranean defensive operations. Sensors should continue to be employed as far forward as possible and be centrally controlled.

c. Sensor employment should be coordinated with counterintelligence (CI) efforts for placement, readouts, and recovery. Sensors can also play a role in rear areas to augment rear security operations. Sensors, in themselves, are only component elements of an integrated intelligence system. Their employment is limited only by the imagination.

4207. Employment of Air Defense. Basic air defense doctrine does not change when units operate on urbanized terrain. The employment principles of mix, mass, mobility, and integration all apply to the employment of air defense assets. The commander must consider the following when developing his air defense plan:

a. Enemy air targets such as principal lines of communication and road and rail networks and bridges are often found in and around built-up areas.

b. Good firing positions may be difficult to find and occupy for long-range air defense missile systems in the built-up areas.

c. Movement between positions is normally restricted in built-up areas.

d. Long-range systems can provide air defense cover from positions on or outside of the edge of the city.

e. Radar masking and degraded communications reduce air defense warning time for all units. Air defense control measures must be adjusted to permit responsive air defense within this reduced warning environment.

f. Positioning of air defense weapons in built-up areas may be limited to more open areas without masking, such as parks, fields, and rail yards.

g. Stinger teams provide protection for infantry battalions the same as in any operation. When employed within the built-up area, rooftops normally offer the best firing positions.

h. Heavy machine guns emplaced on rooftops can also provide additional air defense.

Chapter 5

Combat Service Support and Legal Aspects of Combat

During combat in built-up areas, the terrain and the nature of operations create unique demands on the CSS system. Meeting these demands will require innovative techniques and in-depth planning.

Section I

Combat Service Support

5101. Introduction. CSS is "the essential capabilities, functions, activities, and tasks necessary to sustain all elements of operating forces in theater at all levels of war" (Joint Pub 1-02). The CSS organization throughout the MAGTF, down to battalion level, must provide a responsive CSS system that can perform all functions and tasks associated with meeting identified CSS requirements. This responsibility does not change during urban operations. Forward support for combat forces continues to be the basic concept governing CSS operations. No significant changes in doctrine or organization are required. However, the characteristics and nature of urban combat do affect how CSS is provided.

5102. Combat Service Support Resupply, Maintenance/Repair, and Replacement

a. Resupply of Ammunition. Combat in built-up areas is characterized by extremely high ammunition expenditure rates. Not only do individual Marines expend more ammunition, but they also use greater quantities of munitions such as smoke, concussion, and fragmentation grenades; AT4s; claymore mines; demolitions; and so on. The ammunition consumption rate for the first day of combat in a built-up area can be up to four times the normal rate. Even though this rate decreases during succeeding days, consumption remains high. Leaders should plan to meet these high consumption rates. The plan must include how ammunition and demolitions are to be moved forward to the companies. AAVs should be designated for the movement of ammunition if rubble or glass prevents wheeled-vehicle traffic. Marines may need to organize into carrying elements if streets are blocked by rubble.

b. Resupply of Fuel. The amount of bulk fuel needed by forces during combat in built-up areas is greatly reduced. Combat vehicles normally use less fuel in built-up areas because they travel shorter distances and perform less cross-country traveling. Engineer equipment and power generation equipment may use more fuel, but requirements are small. A unit may not use much fuel daily, but when it does need fuel, a problem exists in delivering bulk fuel to the vehicle. In open terrain, a vehicle that has run out of fuel can be recovered later. In built-up areas, the same vehicle is probably going to be lost quickly. Leaders should plan for and provide the means of moving limited amounts of bulk fuel forward to combat units.

c. Maintenance and Repair. Maintenance contact teams must operate well forward to support units fighting in built-up areas. Although some maintenance operations may be consolidated in civilian facilities, many vehicles will have to be repaired near fighting positions. Battle damage assessment (BDA) and repair procedures allow mechanics to analyze, repair, and return damaged vehicles to a serviceable condition. Other considerations:

(1) Combat in built-up areas generates a high demand for tires.

(2) The dust and rough handling that are characteristic of combat in built-up areas also place great strains on communications and night vision devices.

(3) The unit armorers and their small-arms repair kits provide only limited maintenance. S-4s should plan for increased weapons maintenance demands.

d. Replacements. Units conducting combat in built-up areas must expect high casualty rates. Casualty reports must be prepared scrupulously and forwarded via the S-1 to the battalion personnel officer located at the consolidated administration center (CONAD). Other considerations:

(1) The S-4s must plan to expedite the evacuation of wounded from the built-up area. Location of battalion aid stations and evacuation routes must be planned and disseminated to the lowest level. Higher casualty rates should be expected and may require the stockpiling of medical supplies and augmentation of additional HSS personnel.

(2) The personnel officer quickly processes replacements and coordinates their movement forward via the S-1 officer.

(3) Proper accountability of personnel at all levels is required. Timely and accurate personnel accountability and strength reporting support decisionmaking and initiate the replacement cycle. Leaders maintain accountability through the use of battle rosters and by establishing procedures for periodic reporting of numbers to higher commands. During combat, strength reports are provided on request or as significant changes occur.

5103. Critical Classes of Supply. Requirements for supply and resupply will vary from those encountered during operations in other tactical environments. The critical classes of supply listed below will assume greater importance during combat inside a built-up area.

a. Class I (Rations). The process of ordering and moving rations to forward units is complicated by the dispersed nature of combat in built-up areas and its increased caloric demands on Marines. Hot meals should be provided when practical.

(1) Combat in built-up areas not only causes great stress on Marines, but also requires great physical exertion. This combination of stress and exertion quickly causes dehydration. Unless potable water is continuously provided, Marines will seek local sources, which are usually contaminated by petroleum, oils, and lubricants (POL) runoff, sewage, bacteria, or unburied corpses. Marines who are not provided sufficient quantities of potable water become casualties as a result of drinking from contaminated sources or from dehydration. Waterborne contaminants can quickly render entire units combat ineffective.

(2) Water and other liquid supplements such as coffee, tea, or soup that must be forwarded to exposed positions may need to be backpacked at night.

b. Class II (General Supplies). Combat in built-up areas increases wear and tear on combat uniforms and footgear. Supply officers should increase on-hand stocks of uniforms, boots, and individual combat equipment such as protective masks and armored vests. Nuclear, biological, and chemical (NBC) protective suits become damaged quickly when worn in the urban environment. Extra stocks of these and protective mask filters should be kept on hand. (For further information on NBC considerations, see Appendix G.) Limited amounts of other Class II and IV (barrier materials) items may be available locally. These should be gathered and used if authorized and practical. Local shops may provide such items as hand tools, nails, bolts, chains, and light construction equipment, which are useful in preparing a defense or reducing enemy-held positions. The unit's organic wire communications net may be augmented with locally obtained telephone wire and electrical wire.

c. Class III (POL). Bulk fuel may have to be brought forward from fuel tankers by using 5-gallon cans. One man can carry a fuel can long distances, even over rubble, if it is lashed to a pack frame. Supplies of bulk Class III items and some prepackaged POL may be available at local gas stations and garages. These may be contaminated or of poor quality. The S-4s should coordinate to have fuel tests performed.

d. Class IV (Barrier Materials). If a unit is defending a built-up area, the requirements for Class IV materials are less than in other areas. This class of supply is probably the most available locally. After coordinating the effort with higher headquarters, S-4s and supporting engineer officers gather materials for use in strengthening the defense. Cargo trucks, wreckers, or recovery vehicles from maintenance platoons or engineer units can be used to load and move barrier materials. The defense of a built-up area may require concertina wire and/or barbed wire to restrict the enemy infantry's movements. Barriers can be built of abandoned cars and buses, which are dragged into position, turned on their sides, and chained together through the axles.

e. Class V (Munitions). Combat in built-up areas causes ammunition to be expended at extremely high rates. Commanders should plan for early resupply of explosives, grenades, and ammunition for small arms, direct fire, and indirect fire.

(1) In the defense, the S-4 should prestage as much ammunition as practical in dispersed storage areas. These storage areas should be protected, and they should be easily accessible to the forward defensive positions. In the offense, attacking forces should not be overburdened with excessive ammunition. Mobile distribution points may be set up as low as at company level.

(2) Leaders should plan to continuously deliver ammunition to the leading units as they advance. Ammunition may be carried by armored vehicles close behind the advancing forces or by designated carrying elements. Modern ammunition, particularly missiles, is characterized by extensive amounts of packing material. S-4s must remove the ammunition depot overpack before the ammunition is transported forward. Resupply by helicopter (prepackaged slingloads) may be feasible.

(3) Removing the overpack from large amounts of ammunition can be a time-consuming process. It may require being augmented by available Marines. If carrying elements are used to move ammunition forward, a Marine can carry about 75 to 90 pounds by using a pack frame. Bulky and heavier loads can be carried by lashing them to litters and using teams of two to four men. Loads of up to 400 pounds can be carried moderate distances using four-man teams.

Note: DO NOT use medical corpsmen to carry ammunition forward as described above—it is a violation of the Geneva Accords.

f. Class VIII (Medical Supplies). Because of the decentralized nature of combat in built-up areas, medical supplies should be dispersed throughout the unit, not just consolidated with the aid station and the individual corpsman. Marines should carry additional bandages, and units should have additional splints and stretchers.

5104. Health Service Support. The regimental and battalion surgeons are responsible for planning and executing HSS within their respective units. The most critical functions during combat in built-up areas include preventive medicine, trauma treatment, and evacuation. In addition, there should be a plan for the treatment and evacuation of NBC-related casualties that could occur in combat in built-up areas.

a. Combat in built-up areas exposes Marines not only to combat wounds, but also to the diseases endemic to the AO. Commanders must enforce preventive measures against the spread of infectious diseases. The unit surgeon advises the commander on how best to implement the use of prophylactics.

b. Corpsmen should be placed at strongpoints, at battle positions, and in units likely to be fighting in somewhat isolated positions.

c. Battalion aid stations must be placed farther forward than in operations in more open terrain. Protection offered by urban structures will permit this forward location, but the limited range of observation will require that personnel involved in evacuation be aware of their surroundings and the threat.

d. Corpsmen attached to rifle platoons are trained in the treatment of traumatic injuries; however, they can quickly become overwhelmed by the number of casualties needing care. If time and training are available before the conduct of urban operations, the commander should increase first aid training. Immediate first aid support by a buddy to an injured Marine may be the difference between life and death and will greatly assist the corpsmen. The aid station should plan to care for the increased casualties that are inherent to combat in built-up areas, as the incidence of crushing injuries, eye injuries, burns, shrapnel wounds, and fractures increases.

e. The difficulties encountered when evacuating casualties from urbanized terrain are many and require innovative techniques and procedures. The planning for evacuation on urbanized terrain must include special equipment requirements, use of litter teams, use of air MEDEVAC, use of an ambulance shuttle system, communications requirements, and techniques for locating casualties.

(1) Special equipment requirements include ropes, pulleys, skid litters, axes, crowbars, and other tools used to break through barriers.

(2) Although litter teams are labor intensive, they are required for evacuation from buildings, where casualties can occur on any level. Also, rubble in the streets, barricades, and demolition of roads will require a heavy reliance on litter teams.

(3) Communications presents one of the biggest obstacles to casualty evacuation. Urbanized terrain renders LOS radios ineffective. Also, individual Marines normally do not have access to radios. Therefore, a wounded Marine within a building may be difficult to find and evacuate. The unit SOP should contain alternate forms of communication such as colored panels or other forms of markers that can be displayed to hasten rescue when the battle is over. Also, a systematic search of the area after the battle may be required to recover casualties.

f. The use of local medical facilities, hospitals, professional medical help, and medical supplies may be available during combat in large, built-up areas. The commander must adhere to the guidelines established within the theater as to when and how these facilities can be used. If civilians are wounded in the unit's area, the commander is responsible for providing them with aid and protection without disrupting military operations. A commander cannot confiscate civilian medical supplies unless he makes provisions to provide adequate replacements if civilians are wounded.

g. The commander is responsible for the evacuation of deceased personnel to the nearest mortuary affairs collection point, whether they are U.S., allied, or enemy personnel, or civilians. (See Joint Pub 4-06, *JTTP for Mortuary Affairs in Joint Operations*.) Some general considerations for the handling of deceased personnel include:

(1) The theater commander is the approval authority for hasty burial.

(2) The deceased person's personal effects must remain with the body to assist in the identification of the body and to facilitate shipment of personal effects to the next of kin.

.3

Retention of personal items is considered looting and is, therefore, punishable by the Uniform Code of Military Justice (UCMJ).

(3) When operating under NBC conditions, the bodies of deceased personnel should be decontaminated before removal from contaminated areas to prevent further contamination and casualties.

(4) Care must be exercised when handling deceased personnel. Improper handling of deceased personnel can result in a significant decrease in unit and civilian morale.

5105. Personnel Services. Timely and accurate personnel services are just as important during combat in built-up areas as in any other operation.

a. The S-1 plans for all personnel services that support and sustain the morale and fighting spirit of the battalion. Among the most important of these services are:

- Replacement operations
- Strength accounting
- Casualty reporting.

These functions allow the commander to maintain accountability of his forces and to redistribute personnel or change taskings for subordinate units to maintain combat power and tempo.

Chapter 6

Noncombatant Considerations in Urban Operations

Noncombatants can have a significant impact on the conduct of military operations.

Section I

6101. Introduction. Commanders must be well educated in the legal aspects of military operations involving noncombatants. The urban environment contains numerous legal pitfalls in dealing with the civilian population. Adherence to the ROE and the standards of the Geneva Accords will keep the commander on firm legal ground.

6102. Noncombatant Impact. Large concentrations of civilians can greatly impede tactical operations. Noncombatants may have the following effects on military operations:

a. Mobility. Noncombatants civilians, attempting to escape the battlespace, can block military movement. Commanders plan routes to be used by civilians and seek the assistance of the civil police in refugee control.

b. Firepower. The presence of noncombatants can restrict the use of firepower. Areas may be designated no-fire areas to prevent noncombatant casualties. Other areas may be limited to small-arms fire and grenades. The control of fire missions may be complicated by the requirement for positive target identification. Detailed guidance on the use of fires in the presence of noncombatants is promulgated by the MAGTF commander. In the absence of guidance, the general rules of the law of land warfare apply.

c. Security. The presence of noncombatants increases security requirements in an urban environment to preclude:

- Noncombatants entering defensive areas
- Pilferage of equipment
- Sabotage
- Terrorism

d. Obstacle Employment. The presence and movement of Noncombatants will influence the MAGTFs commander's obstacle plan. Minefields may not be allowed on designated refugee routes or, if allowed, must be guarded until the passage of refugees is completed. The use of boobytraps may be curtailed until noncombatants have been evacuated. Commanders need to review and understand current national and international treaties concerning the employment of mines and boobytraps.

6103. Command Authority. The limits of command authority over government officials and noncombatants must be established. The host government's responsibility for its populace and territory can affect the commander's authority in civil-military matters. In less secure areas, where the host government may be only partly effective, the commander may have to assume greater responsibility for noncombatants.

6104. Source Utilization. Operations in highly populated areas may require the diversion of men, time, equipment, and supplies to accomplish humanitarian tasks. If host government agencies collapse, the impact on military resources could be substantial.

6105. Health and Welfare. The disruption of noncombatant health and sanitary services may sharply increases the risk of disease among both noncombatant and military personnel.

6106. Law and Order. Marines may have to augment civilian forces to protect life and property and to restore order. Marines may also have to secure vital government facilities for the host nation. (For more information on how to control civilians violating civil law, see FM 9-15, *Civil Disturbances*.)

6107. Public Affairs Officer and Media Relations. Free media access to units in the field is desirable; however, operational security, existing guidelines, and ROE considerations have priority. All members of the media should have an escort officer. Escort officers may be detailed from line units when public affairs personnel are unavailable. Enforcing established guidelines helps prevent negative publicity which could jeopardize the operation or national and strategic objectives.

6108. Civil Affairs Activities and Psychological Operations (PSYOP).

a. General. CA and PSYOP are mutually supporting and may have essential roles in MOUT. They can be critical force multipliers. CA and PSYOP may offer the possibility of victory without the destruction, suffering, and horror of protracted urban conflict.

(1) CA activities are activities performed by commanders, staffs, department of defense elements and units, and foreign military forces that (1) embrace the relationship between military forces and civil authorities and populations in areas where military forces are present; and (2) involve application of civil affairs functional specialty skills, in areas normally the responsibility of civilian government, which enhance conduct of civil-military operations (Joint Pub 1-02). CA activities are inherently civil-military in nature. They assist in coordinating military efforts to strengthen host government legitimacy, and help prevent or reduce violence by bridging critical gaps between civil and military sectors. MAGTF responsibilities for CA activities are determined during planning. U.S. Army CA units can be requested to support the MAGTF. The Marine Corps Civil Affairs Group (CAG), located in the Marine Corps Forces Reserve, may be activated and assigned to the MAGTF.

(a) CA activities are characterized by applications of functional specialties in areas normally the responsibility of indigenous government or civil authority. CA activities may extend to assumption of government functions required in an occupied territory during or immediately subsequent to hostilities. Refer to Joint Pub 3-57, *Doctrine for Joint Civil Affairs*.

(b) CA support to a MAGTF would normally include procuring local goods and services for military use, reducing civil interference in military operations, and assisting a MAGTF commander in fulfilling moral and legal obligations to the local government and noncombatant populace. See MCWP 3-33.1, *MAGTF Civil Affairs*, for more detailed information.

(2) Psychological operations are planned operations to convey selected information and indicators to foreign audiences to influence their emotions, motives, objective reasoning, and ultimately, their behavior. The purpose of psychological operations is to induce or reinforce foreign attitudes and behavior favorable to the originator's objectives (Joint Pub 1-02).

PSYOPs are planned and conducted by the MAGTF. PSYOPs are integrated and coordinated down to the lowest tactical level. PSYOPs can weaken enemy resistance and reduce its combat effectiveness. PSYOPs can also help prevent civilian interference with military operations.

PSYOPs may include the use of television, radio, posters, leaflets, and loud speakers, to disseminate propaganda and information. Television, including video tapes, is one of the most effective mediums for persuasion. In areas where television is not common, televisions may be distributed to public facilities and selected individuals. See MCWP 3-36.2/FM 33-1, *Psychological Operations*, for more detail.

6109. Refugee Control. The control of refugee movements in combat operations is primarily a host nation function.

The most important principle in refugee control is reliance on host government forces (military/civilian) whenever possible. U.S. forces will have limited refugee control capabilities. If a host government is incapable of performing the function, MAGTFs and MPs may be required to conduct any or all of the following refugee control measures:

- Enforce standfast orders or prohibit other unauthorized movements

- Screen and take custody of enemy agents or other hostile personnel discovered among the refugees

- Screen refugees for suspicious activities (including detection of infiltration by guerrillas, enemy agents, and escaping members of hostile forces)

.3

- Perform traffic control on designated refugee routes
- Escort refugee columns
- Direct or divert refugee columns to secondary roads and through areas not used by military elements
- Perform police duties such as traffic control or security at refugee assembly areas, collecting points, and centers or camps
- Enforce prescribed restrictive measures such as curfews, identification and pass systems, restricted area restraining lines, no-passage lines, checkpoints and roadblocks, travel permits or manifests, prohibition on the use of vehicles, priorities of movement, and limits on personal belongings.

Section II

6201. Commander's Legal Authority and Responsibilities

a. General. Leaders at all levels of command are responsible for protecting noncombatants and minimizing collateral damage to their property. Looting, vandalism, and brutal treatment of civilians are strictly prohibited by the Geneva Accords. Individuals who commit such acts may be severely punished. Civilians, and their religions and customs, must be treated with respect. Women are especially susceptible to many forms of abuse, and special care must be taken to protect them. Urban warfare offers many opportunities for abuses of the civilian population. This section focuses on those civilian-military confrontations that are most common in the urban environment.

b. Control Measures. Commanders may enforce control measures to conduct operations, maintain security, or ensure the safety and well-being of the civilians.

(1) Curfew. A commander with the mission of defending a town could establish a curfew to maintain security or to aid in control of military traffic. However, a curfew would not be legal if imposed strictly as punishment.

(2) Evacuation. A commander can require civilians to evacuate towns or buildings if the purpose of the evacuation is to use the town or building for essential military purposes, to enhance security, or to safeguard those civilians. The commander must specify and provide security for those being evacuated. Food, clothing, and sanitary facilities should be provided at the destination.

(3) Forced Labor. The Geneva Accords prohibit the use of civilians in combat operations. However, they may be required to perform some forms of forced labor prior to and after the conduct of combat operations. Guidelines for use of civilian labor should be published. The commander may force civilians over 18 years of age to work if the work does not oblige them to take part in military operations. An example of forced labor that is permitted would be maintenance of public utilities as long as those utilities are not used in the general conduct of the war. Jobs can also include services to the local population such as care of the wounded and burial. Civilians can also be forced to help evacuate and care for military wounded, as long as doing so does not involve any physical danger. Prohibited forced labor include digging entrenchments, constructing fortifications, transporting supplies or ammunition, or acting as guards. Only civilians who are volunteers can be employed in such work.

c. Civilian Resistance Groups. Another situation that commanders might encounter is combat with a civilian resistance group. The combatant status of these groups varies with their organization, methods, and adherance to the Geneva Accords. For example:

(1) Civilians accompanying their armed forces (with an identity card authorizing them to do so) are treated as EPWs when captured. Examples would include civilian members of military aircraft crews, war correspondents, supply contractors, and members of labor units or of service organizations responsible for the welfare of the armed forces.

(2) Captured civilians of a nonoccupying territory who have taken up arms against an invading enemy without time to form regular armed forces are considered EPWs if they wear a distinctive insignia that can be seen at a distance, carry their weapons openly, and operate according to the rules and customs of warfare. Other civilians who provide assistance to such groups may not be entitled to status as combatants, depending on whether they are actually members of the resistance group. They are normally best treated as combatants until a higher authority determines their status.

(3) Armed civilian groups that do not meet the criteria of a legal resistance group or individuals caught in the act of sabotage, terrorism, or espionage are not considered legal combatants. If captured, they will be considered criminals under the provisions of the law of land warfare. They should be detained in a facility separate from EPWs and should be quickly transferred to the MPs. Reprisals, mass punishments, hostage taking, corporal punishment, pillaging, or indiscriminate destruction of property are prohibited.

(4) The law of land warfare lets the commander control the civil population under the conditions already described using his own organic resources. However, language and cultural differences between U.S. and foreign personnel make it good practice to use local authorities such as the police for such purposes. Use of the police does not relieve a commander of his responsibility to safeguard civilians in his area.

d. Protection of Property. Like civilian personnel, civilian buildings and towns normally have a protected status. As such, they are not legal targets. They lose their protected status if authorities determine that the enemy is using them for military purposes. All doubt as to whether a town or building is a military target must be resolved before it can be targeted. For example:

(1) If the enemy is using a building or a portion of the town for military purposes—such as a supply point or a strongpoint—that building or that portion of the town is a legal target. However, engagement of the target should be limited to the buildings or portion of the town being occupied by the enemy.

(2) Normally, religious, historical, and cultural objects and buildings are not legal targets. They are sometimes marked with symbols to signify their status. Medical facilities are identified under the internationally recognized Red Cross, Red Crescent, Red Lion, or Red Star of David symbols. The fact that such symbols are absent does not relieve a commander of his responsibility to protect objects he recognizes as having religious, cultural, medical, or historical value.

(3) The misuse of such objects by the enemy is grounds to disregard their protected status. A demand should be made for the enemy to stop his misuse of the protected object before any action is taken. However, if an enemy FO or sniper uses a church for an OP, a commander would be justified in destroying it immediately because of the immediate danger posed to his troops. Another example is a religious shrine that is used as a telephone switchboard; in this case a warning would be appropriate because it would take some time to dismantle the system. Attacks on these types of targets should be limited to the least amount necessary to neutralize the enemy installations.

(4) The destruction, demolition, or military use of other buildings is permitted under the law of land warfare if required by clear military necessity. Destroying a house to obtain a better field of fire would be a legal act. But destroying it as a reprisal would not. However, firing on any houses that are occupied or defended by an enemy force is usually legal.

Chapter 7

The Urban Environment and Restrictions to Operations

"The responsibility of officers . . . and the training necessary are of a very different order from their responsibilities and training in ordinary military duties. Instead of striving to generate the maximum power with the forces available, the goal is to gain decisive results with the least application of force and the consequent minimum loss of life."

USMC, 1940 (FMFRP 12-15, *Small Wars Manual*)

Section I

Military Operations Other Than War

7101. Introduction. One of the most likely missions that U.S. Marines will undertake abroad will be military operations other than war (MOOTW). These missions typically will take place in the Third World and are normally associated with urban areas. MOOTW focuses on deterring war, resolving conflict, promoting peace, and supporting civil authorities in response to domestic crises. It is important to remember that political considerations permeate at all levels. Although some types of MOOTW, such as peace enforcement, may have the same characteristics of war, they normally have more restrictive rules of engagement (ROE).

Urban environments are inherently complex, and sometimes unpredictable. Local governments may have limited functions. Services may be overburdened, leading to allocation on the basis of personal connections and corruption. Services may be provided by a mixture of formal, high-technology and informal, labor-intensive systems. Large portions of the population may be poor, young, and underemployed. Nevertheless, there may be a high degree of social organization and interaction. Word of mouth and informal organizations may be as important as mass media and formal organizations to the communication structure of the area. Indigenous populations may have experienced centuries of foreign interventions and become quite adept at innovative ways of communicating. These factors and others often combine to make MOOTW a challenging and frustrating endeavor.

7102. Types of MOOTW. MOOTW encompasses a wide variety of military operations, that may take place in an urban environment. Forward deployed MAGTFs are ideal forces to respond to contingency operations involving MOOTW. These operations include:

- Combatting Terrorism
- DOD support to counterdrug operations
- Enforcement of Sanctions/MIO
- Enforcing Exclusion Zones

- Humanitarian Assistance
- Military Support to Civil Authorities
- Nation Assistance/Support to Counterinsurgency
- NEO
- Peace Operations
- Recovery Operations
- Show of Force Operations
- Strikes and Raids.

7103. Recent Experience and Lessons Learned. Recent Marine Corps urban experiences— Lebanon (1982 - 1984); Grenada (1983); Panama (1989); and Somalia (1992 - 1994)— are examples of MOOTW. Although each operation was unique, the lessons learned are similar. For detailed information, commanders may obtain after action reports through the Marine Corps Lessons Learned System (MCLLS). Some of the lessons learned include:

(1) Demonstrate a powerful military presence. Where and when possible, use tanks, armored vehicles, heavy automatic weapons, fixed-wing aircraft, and attack helicopters.

(2) PSYOP, CI, HUMINT teams, and linguists, are essential to communications and intelligence gathering.

(3) Topographic detachment is useful to provide necessary sketches and maps.

(4) UAVs can be used for security, reconnaissance, and surveillance.

(5) Whenever entering an AO, isolate the area, patrol, and use intelligence to evaluate specific buildings or groups of buildings for search-and-clear operations. PSYOP, CI, HUMINT teams, and linguists can be used, as appropriate, to support patrols.

(6) Take advantage of the night by using night vision devices.

(7) The most effective means of dealing with snipers is with counter snipering operations.

(8) Snipers are also a tremendous asset to overwatch patrol movements, provide intelligence, and direct fire support. (See MCWP 3-15.3.)

(9) Random patrolling, traffic control points, random checkpoints/roadblocks, and ambushes disrupt enemy movements, reduce looting, seize curfew violators, and convince the population that order has been restored.

(10) Use armored vehicles to move quickly to new checkpoints and set up roadblocks or ambushes. They also increase protection for infantry. Their night vision sights also provide increased security and recognition during periods of limited visibility.

(11) Extensive patrolling provides current combat information used to maintain situational awareness. It helps to familiarize units with the AO and presents a strong military presence. (See Appendix A, Section VII)

(12) Tanks, armored vehicles, or attack helicopters increase military presence and can provide immediate, accurate gunfire.

(13) Before search and clear of a building, use loudspeakers to explain your intentions to noncombatants and enlist their cooperation.

(14) Use a minimum amount of force when dealing with noncombatants and their property. This will assist you in obtaining them to cooperate and may prevent turning them into active combatants. Assign engineers to search-and-clear forces, as necessary.

(15) MPs can instruct small unit leaders on the proper method of conducting vehicle, individual, and room search techniques. In some cases, it may be appropriate to attach MPs to search-and-clear forces. (See MCWP 3-34.1, *Military Police in Combat.*)

(16) Maintain a strong, mobile reserve.

(17) Avoid becoming predictable.

7104. Urban Terrorism. Terrorism is "the calculated use of violence or threat of violence to inculcate fear; intended to coerce or to intimidate governments or societies in the pursuit of goals that are generally political, religious, or ideological" (Joint Pub 1-02). Terrorism is a criminal act that is symbolic in nature. Its intent is to influence an audience other than the victim.

The likelihood of Marine involvement with terrorism increases in the urban environment. Terrorism in an urban environment can occur across the spectrum of conflict. Terrorists are an integral element in an insurgency and also play a major role in unconventional warfare. Terrorists can disrupt economic functions, expose a government's incompetence, eliminate opposition leaders, and elevate social anxiety. (See MCRP 3-02D, *Combating Terrorism*; and MCRP 3-02E, *The Individual's Guide for Understanding and Surviving Terrorism.*)

In countering this threat, Marines should determine whether it is internally or externally directed terrorism. Terrorism rooted externally must be severed from its roots. Against internal terrorism,

.3

Marines should attempt to penetrate the infrastructure and destroy the leadership of the terrorist groups.

The keys to defeating terrorists are awareness, education, and intelligence in order to deny, deter, delay, and detect terrorist acts. Rapid coordination among military units, military services, local police, and other government agencies is essential in denying the terrorists targets and refuge.

a. Nature of Urban Terrorism. Terrorism is generally an urban phenomenon. Incidents of terrorism are increasing in the international community, and in the United States, and pose a significant threat for U.S. military forces. Military personnel in noncombatant roles are considered terroist targets. For example, the attack on the Marine compound in Lebanon can be considered a terrorist attack instead of a military action because the Marines were officially in a noncombat activity (a multinational peacekeeping activity).

b. Terrorists in the Urban Area. Cities are target rich environments for terrorists. They offer ample cover and concealment. They contain large amounts of resources, transportation, communications, and potential recruits or sympathizers to their cause. Cities also provide the media attention essential for terrorists to accomplish their mission.

c. Combating Terrorism

(1) Counterterrorism. Counterterrorism includes the full range of offensive measures to prevent, deter, and respond to terrorism (Joint Pub 1-02). Population protection and control to combat the terrorist threat may be a major effort within the urban environment. The ability to protect the general population from attack and intimidation is essential.

(2) Antiterrorism. Antiterrorism includes defensive measures used to reduce the vulnerability of individuals and property to terrorist acts, to include limited response and containment by local military forces (Joint Pub 1-02). The extent of these measures varies based on assessments of the local threat. These include personnel awareness and knowledge of personal protection techniques. They also include crime prevention and physical security programs to "harden" the target, making installations and personnel less appealing as terrorist targets.

Section II

Restrictive Conditions

7201. Introduction. Commanders can expect to encounter restrictions on the use of force in MOOTW, especially in the urban environment. While we tend to focus on a traditional image of urban combat, MOUT extends beyond the house-to-house, close-combat effort to seize a city. Recent history provides many examples of the need for precise application of force to minimize collateral damage. MOOTW generally requires the application of force in varying degrees. Future missions in urban areas will likely follow this same pattern. While Marine forces must be prepared for more intensive urban combat, they must also train to apply only the level of force necessary to accomplish the mission. *Our tactics may have to rely more on innovation than firepower.*

Basic tenets of manuever warfare remain the same in MOOTW, but the tactics, techniques, and procedures may have to be modified to stay within established ROE and to avoid unnecessary collateral damage.

7202. Rules of Engagement. ROE are "directives issued by competent military authority which delineate the circumstances and limitations under which United States forces will initiate and/or continue combat engagement with other forces encountered" (Joint Pub 1-02). ROE for Marine forces are usually set by the JTF commander. They are based on National Command Authorities (NCA) guidance, the mission, the threat, the laws of war, and host nation constraints on force deployment.

MOOTW focuses on deterring war, resolving conflict, promoting peace, and supporting civil authorities in response to domestic crises. It is important to remember that political considerations permeate at all levels. Although some types of MOOTW, such as peace enforcement, may have the same characteristics of war, they normally have more restrictive rules of engagement (ROE). Restraint requires the careful balancing of the need for security, the conduct of operations, and the political objective. Excessive force antagonizes those parties involved, thereby damaging the legitimacy of the organization that uses it while possibly enhancing the legitimacy of the opposing party. Ensure all personnel understand the ROE and incorporate the practice into training exercises at all levels.

ROE must be fully integrated in the MAGTFs plans. As an example, table 7-1, presents the ROE used in Operation United Shield:

ROE Used for Operation United Shield
Nothing in these Rules of Engagement limits your right to take appropriate action to defend yourself and your unit.
a. You have the right to use deadly force in response to a hostile act or when there is a clear indication of hostile intent.
b. Hostile fire may be returned effectively and promptly to stop a hostile act.
c. When US forces are attacked by unarmed hostile elements, mobs and/or rioters, US forces should use the minimum force necessary under the circumstances and proportional to the threat.
d. Inside designated security zones, once a hostile act or hostile intent is demonstrated, you have the right to use minimum force to prevent armed individuals/crew-served weapons from endangering US/UNOSOM II forces. This includes deadly force.
e. Detention of civilians is authorized for security reasons or in self-defense.
Remember: 1. The United States is not at war. 2. Treat all persons with dignity and respect. 3. Use minimum force to carry out mission. 4. Always be prepared to act in self-defense.

Table 7-1. ROE Used in Operation United Shield

7203. Planning. Commanders and planners thoroughly study each situation to identify the restrictive measures necessary to prevent needless civilian casualties and damage while also protecting the lives of the Marines who will execute the mission.

a. Mission/Command Guidance. The mission statement itself may determine much about the required level of force for a given operation. A humanitarian relief operation may require only individual weapons for security and personal protection. A raid on an insurgent communication center may need more. The requirements for an NEO conducted in a permissive environment would be different from those in a nonpermissive one. The key is to conduct a thorough analysis of the mission and determine the necessary combat power and how it must be applied to accomplish the mission. Many of these missions will be accompanied by a great deal of additional guidance concerning operational restrictions. These restrictions may come from the NCA, State Department, Department of Defense (DoD), or the joint task force (JTF) commander, and other sources. Such guidance must be carefully studied and incorporated into the plan. The commander can impose restrictions on his own forces that are more stringent then those dictated by higher authority. A careful analysis of the ROE will provide key insights into other restraints that the commander may wish to impose on his own forces.

b. Intelligence/Threat. Obtaining detailed and specific intelligence on the target and the area is essential to minimize collateral damage. Information on the physical layout of an area is often as important as information about the enemy. This is especially true for an NEO where avoiding conflict is a primary concern.

c. Location. The geographic area into which operations are to be conducted will normally dictate some restrictions on operations. In an urban environment, where the city is located in a desert, a common restriction is placed on the destruction of water producing facilities. If the operation is conducted where there is a large Hindu population, special protection may be given to cattle. If the operation is on an island nation, restrictions may be placed on destroying port or airfield facilities. In an urban environment with a large amount of shanty towns, restrictions may be placed on the use of incendiary devices.

d. Restrictions on Force Employment. The *right of self-defense is inherent in all military operations*. Marines should not be placed in a situation where they are not provided the means and authority to protect themselves or others. Beyond that, the restrictions placed on the forces conducting a particular operation depend on a variety of factors. These include:

- NCA guidance
- Type of mission
- Nature of the threat
- Political environment
- Host nation attitude.

e. Lethal Fires. Most restrictions in MOOTW are normally concerned with the application of fires. For instance, a NEO in a permissive environment may not require the landing of fire support assets However, a raid may require supporting arms to be landed with detailed instructions on their use. Aviation in MOOTW is commonly restricted in the amount and type of ordnance it can deliver because of the potential for collateral damage. However, in both a permissive and nonpermissive environment, supporting arms still should be planned. The most common types of fire support assets that will be restricted in the urban environment include artillery mortars, tanks, armored vehicles, and fixed-wing attack aircraft.

f. Search and Clear Buildings. Marine forces organize to search and clear a building under restrictive conditions in the same ways as discussed in Appendix C. However, because noncombatants may be occupying rooms within the building, clearing procedures must be modified as follows:

(1) Squad organization remains the same and is reinforced as appropriate (see Appendix A), except assault elements become search elements. Search elements enter and search rooms in the same manner as assault elements enter and clear rooms, as discussed in Appendix A.

(2) ROE must be identified and known by all personnel before entering or clearing a room or building.

(3) PSYOP or CA teams can help remove noncombatants before a search or battle starts.

(4) The building or area of search should be isolated with security forces and the intent to search the building should be communicated via loudspeakers.

(5) Assault forces move to the building and begin systematically to search room by room and floor by floor. Marines should carry ammunition and other equipment as in normal urban operations. However, Marines will be restricted in the use of their weapons and other equipment as dictated by the ROE. If permitted, a riot control agent (such as CS gas) can be used to assist in the clearing of a building. Additionally, shotguns and stun grenades are more appropriate than rifles and machine guns or fragmentation grenades for this type of restricted, close-combat operation.

g. Communications and Information Systems (CIS).

(1) CIS must be able to satisfy the C2 requirements of MOOTW. CIS systems must provide the commander with the ability to rapidly collect, process, analyze, and exchange information in order to keep the potential for violence to a minimum. These systems must make available the information needed, when it is needed, whenever it is needed. When supporting arms are on standby, fire support coordination and air traffic control nets are particularly critical.

(2) CIS officers (G/S-6) must carefully analyze the mission, situation, threat, and specific characteristics of the urban environment to design reliable, flexible, and responsive CIS. These systems should include single and multichannel radios, tactical telephones, and tactical satellite communications. A detailed backup CIS plan should also be prepared. The success of CIS in MOOTW depends on designing, planning, and employing CIS that satisfies the commander's C2 requirements.

Appendix A

Organization for Urban Combat and Fundamental Combat Skills

Successful combat operations in built-up areas depend on training in fundamental skills and proper employment techniques and procedures.

Before conducting urban operations, Marines must be trained in fundamental combat skills to conduct urban operations. Training for this unique environment will reduce casualties because properly prepared Marines will be able to fight more effectively. This appendix supersedes all other fundamental combat skills taught for fighting in urbanized terrain. Urban combat will require Marines to use their best initiative and judgment in applying the techniques and procedures described in this appendix.

Section I

Organization

1. Structure. Basic infantry organization does not change when the battlefield is moved into an urban area. However, infantry units may be task organized or reinforced down to squad level in order for small units to fight and win in this violent, three-dimensional battlespace. Several new terms that address organization and employment have been added to clearly define mission assignments. Figures A-1 through A-3 show platoon through fire team organization breakdown.

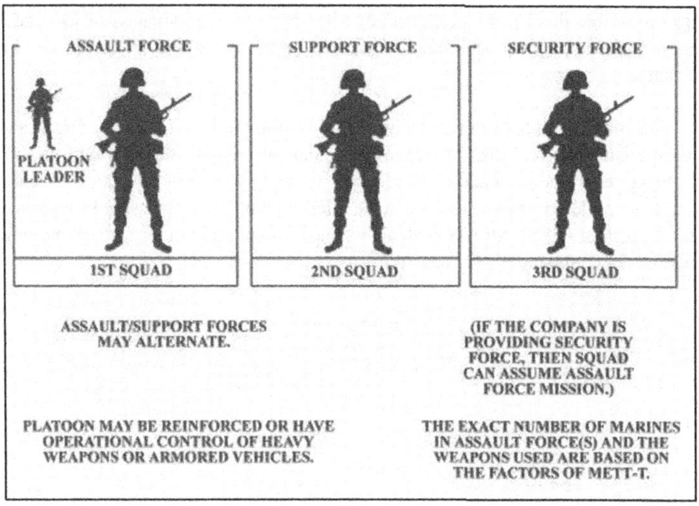

Figure A-1. Task-Organized Marine Platoon

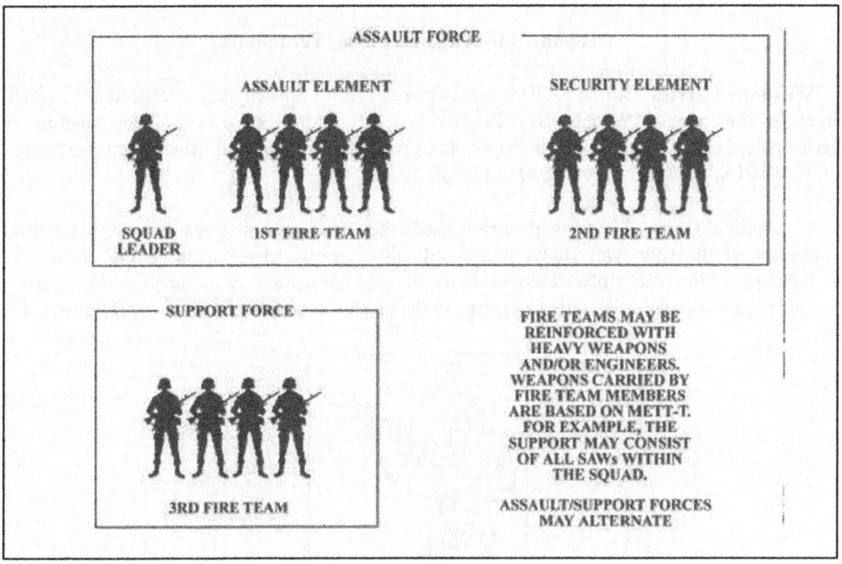

Figure A-2. Task-Organized Marine Squad

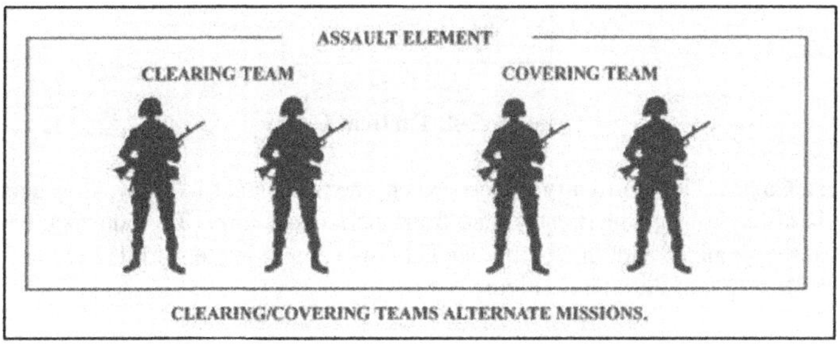

Figure A-3. Task-Organized Marine Fire Team

Section II

Weapons Handling and Firing Techniques

2. Weapons Carries. During MOUT, a Marine often finds himself very unexpectedly engaging targets at close ranges. Weapons carries provide a safe and effective method for handling the service rifle. For additional in-depth discussion of weapons handling and firing techniques, refer to MCWP 3-01X, *Field Firing of the M16A2 Rifle.*

a. Tactical Carry. The tactical carry is used when no immediate threat is present. It permits control of the rifle while moving and still allows quick engagement of the enemy. The buttstock of the rifle is placed alongside the body at approximately hip level, and the barrel is angled upward approximately 45 degrees in the general direction of the enemy (Figure A-4).

Figure A-4. Tactical Carry

b. Alert Carry. The alert carry is used when enemy contact is likely. Engagement of the enemy is faster from the alert carry than from the tactical carry. The buttstock of the rifle is placed in the shoulder with the muzzle angled down approximately 45 degrees and pointed in the likely direction of the enemy (Figure A-5).

In the alert carry, the Marine keeps both eyes open and scans for any threat with the weapon always in line with the Marine's LOS (this is called "guns 'n' eyeballs"). When a threat is spotted, the Marine quickly snaps his weapon into a firing position, looks over the rear sight aperture, and aligns the tip of the front sight post center mass on the target. Once the threat has been eliminated, the Marine returns to the alert carry and continues to scan for targets. Accuracy is the primary goal.

Figure A-5. Alert Carry

c. Ready Carry. The ready carry is employed when contact with the enemy is imminent. It allows for immediate target engagement. The buttstock of the rifle is in the shoulder with the muzzle of the rifle pointed in the direction of the enemy (Figure A-6). As in the alert carry, sight alignment and sight picture are achieved as the shot is fired. Marines should always strive for a clear tip of the front sight post and center mass hold.

Figure A-6. Ready Carry

d. Short Stocking. Rifles may be "short stocked" to increase their maneuverability in an enclosed area and to reduce possible "target indicators" for the enemy (muzzles sticking around corners, for example). The weapon is held in such a manner as to reduce the length of the weapon without sacrificing too much accuracy (Figure A-7). The stock is positioned so that the pistol grip is behind the Marine's head. The Marine may use his index finger or thumb to manipulate the trigger. The handguard of the weapon is placed against the Marine's cheek, and a firm stockweld is attained. The Marine should use the tip of the front sight post. Short stocking may be conducted from either side. However, care should be taken not to mask the ejection port cover. The short stocking technique may be used in the tactical, alert, or ready carry.

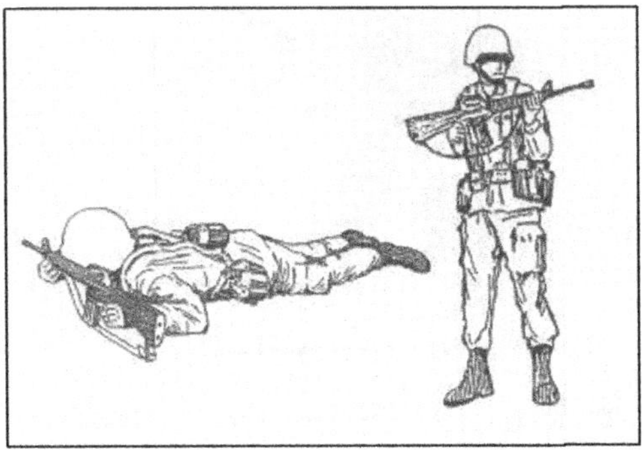

Figure A-7. Short Stocking

e. Weak-Hand Carries. Marines may carry their weapon on the weak-hand side to obtain maximum cover and fire from the weak-hand position to avoid exposing themselves. However, firing from the weak-hand position may reduce a Marine's ability to fire accurately. This may be acceptable when Marines are providing supporting fires, but in close kill-or-be-killed engagements, rapid, accurate firing from the strong-hand position is preferred.

3. Firing Techniques. The speed and uncertainty of combat require Marines to act without hesitation while achieving accurate target engagement. Several firing techniques are listed below. For an in-depth discussion of firing techniques that can be applied to MOUT operations, see MCWP 3-01X.

a. Pieing. Pieing is an effective technique for clearing dead space inside rooms and buildings to gain security of hallways, stairwells, mouseholes, and so on. It is conducted by using the ready carry position or short stocking technique. The weapon is aimed at a sector of a window, doorway, corner, or hallway and slowly moved at different angles, sectoring off the window, doorway, corner, or hallway until each sector is cleared of any threat. As soon as a hostile threat is seen, immediate, accurate fire can be placed on the threat. Figures A-11, A-34, A-35, A-38, A-47, and A-49 show examples of the pieing technique.

b. Aimed Quick Fire. The Marine's initial focus is on the target. As the rifle is brought up, the firing eye looks through or just over the rear sight aperture, and the front sight post is used to aim at the target. The Marine fires two quick shots. Focus remains on the front sight post throughout the aiming process.

Immediately after target engagement, the Marine scans for additional targets. To scan, perform the following steps:

(1) Lower the weapon to look over the sights.

(2) Place the trigger finger straight along the receiver.

(3) Scan the area for targets and assess the situation. Wherever the head moves, the muzzle moves (eyes, muzzle, target). Keep both eyes open to increase the field of view.

c. Pointing Quick Fire. The pointing system is based on the phenomenon that when a person looks at an object and simultaneously points a finger at it, the finger aligns itself on the point of focus of the eyes with no conscious effort on the part of the individual. When a Marine looks at an object and simultaneously brings his rifle to his shoulder, the rifle in effect becomes an extension of the pointed finger. Consequently, it aligns itself naturally with the object on which the shooter is focusing.

When a target appears, the Marine will keep both eyes open, concentrating intensely on a small, specific focal point near the base of the target mass. The rifle is brought simultaneously to the hollow of the shoulder. The head is held high, stock welded to the jaw. The eyes are 2 - 3 inches over the top of the sights, staring intently at the target. As soon as the rifle is brought to the shoulder, two quick shots are fired. Focus on the target is not broken during the interval between initially seeing the target and discharging the rifle.

d. Instinctive Shooting. There may be situations in which a Marine is surprised and may need to react immediately. If possible, the Marine should engage the threat by using the tip of the front sight post. However, *speed* may be more important. The Marine's weapon and body are quickly "pointed," and the target is engaged. It is important that the body be turned with the weapon in order to achieve a natural point of aim. Simply pointing the weapon will usually result in a miss. Once the first two shots have been fired and the Marine regains the initiative, the weapon should be quickly moved to the Marine's shoulder and the tip of the front post used for sighting subsequent shots.

Section III

Movement

Movement techniques within built-up areas are the first fundamental skills that Marines should master. These movement techniques should be practiced until they become habitual. To reduce exposure to enemy fire, Marines should avoid silhouetting themselves, avoid open areas, and select their next covered position before moving.

4. Crossing a Wall. Each Marine should learn the correct method of crossing a wall (Figure A-8). After reconnoitering the other side, the Marine quickly rolls over the wall, keeping a low silhouette. The speed of his movement and a low silhouette deny the enemy a good target.

Figure A-8. Crossing a Wall

5. Observation Around Corners. The area around a corner should be observed before the Marine moves beyond it. The most common mistake the Marine makes at a corner is allowing his weapon to extend beyond the corner before observing, thereby exposing his position. Using the short stocking technique (Paragraph 2.d.) reduces exposure. Additionally, Marines can use a technique called "popping the corner" to reduce exposure time (Figure A-9). The Marine will get into a prone position near the corner of a building or obstacle around which he needs to observe. The weapon is short stocked, and the muzzle is pointed in the direction the Marine is looking. This allows the Marine to engage a target, if necessary, when he observes around a corner. The Marine will crawl to the corner but not expose himself. He will raise his upper body onto his elbows. The Marine will then push his body forward with his feet and legs without moving his elbows. His upper body, with the weapon ready, will move forward. The final position will expose the weapon, the Marine's helmet, and a minimal amount of the Marine's face. The Marine's forearms will come to rest on the deck giving him a low profile, the ability to observe around the corner, and the immediate capability to engage targets with his weapon.

Figure A-9. Correct Technique for Observing Around a Corner

6. **Movement Past Windows.** Windows present another hazard to Marines. The most common mistake made when passing a window is exposing the head to enemy observation from inside the room. If a Marine shows his head, an enemy gunner inside the building could engage him through the window without exposing himself to return fires.

a. The correct technique for passing a window is for the Marine to stay below the window level. He makes sure he does not silhouette himself in the window; he "hugs" the side of the building. An enemy gunner inside the building would have to expose himself to covering fires to engage the Marine (Figure A-10A).

Figure A-10A. Moving Past Windows

b. The most common mistake Marines make when passing a basement window is not being aware of it. A Marine should not walk or run past a basement window because he presents a good target to an enemy gunner inside the building. When using the correct procedure for negotiating a basement window, the Marine stays close to the wall of the building and steps or jumps past the window without exposing his legs (Figure A10-B).

c. An alternate method is for the pointman to pie off the window (or other opening) as he approaches it. As the pointman moves toward the window, the second Marine (behind the pointman) steps out to provide security to the direct front. The pointman continues to cover the opening by pieing, all the while using the near edge for cover. The pointman covers and sectors off that part of the interior that can be viewed from the outside. The second Marine moves with the pointman and maintains security to the front, eventually moving past the pointman as the pointman clears the inside corners of the opening. Once the opening is cleared, the second man may assume the duties of point, or when subsequent Marines move up to secure the opening, the original pointman may reassume those duties. Figure A-11 illustrates this alternate method.

Figure A-10B. Marine Passing Basement Windows

7. Use of Doorways. Doorways should not be used as entrances or exits because they are normally covered by enemy fire. If a Marine must use a doorway as an exit, he should move quickly through it to his next position, staying as low as possible to avoid silhouetting himself (Figure A-12). Preselection of positions, speed, a low silhouette, and the use of covering fires should be emphasized when exiting doorways.

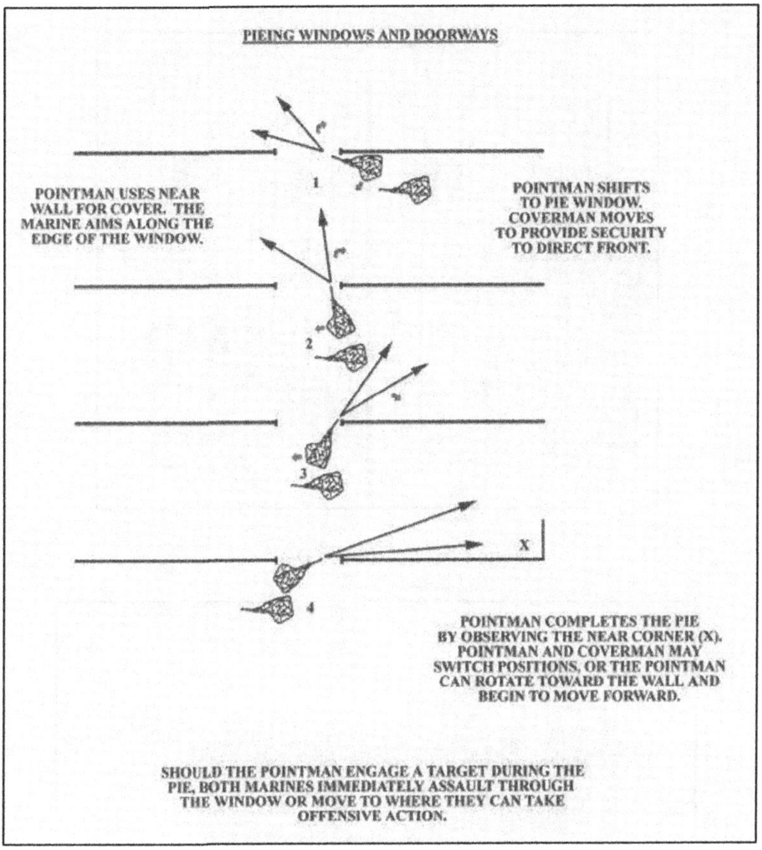

Figure A-11. Alternate Technique for Moving Past Windows

8. Movement Parallel to Buildings. Marines and small units may not always be able to use the inside of buildings as a route of advance. In such cases, they should move on the outside of the buildings (Figure A-13). Smoke, covering fires, and cover and concealment should be used to hide movement. When correctly moving along the outside of a building, the Marine hugs the side of the building, stays in the shadow, presents a low silhouette, and moves rapidly to his next position (Figure A-14). If an enemy gunner inside the building fires on a Marine, he exposes himself to fire from other squad/fire team members who are covering him. Furthermore, an enemy gunner farther down the street may have difficulty detecting and engaging the Marine.

Figure A-12. Exiting a Doorway

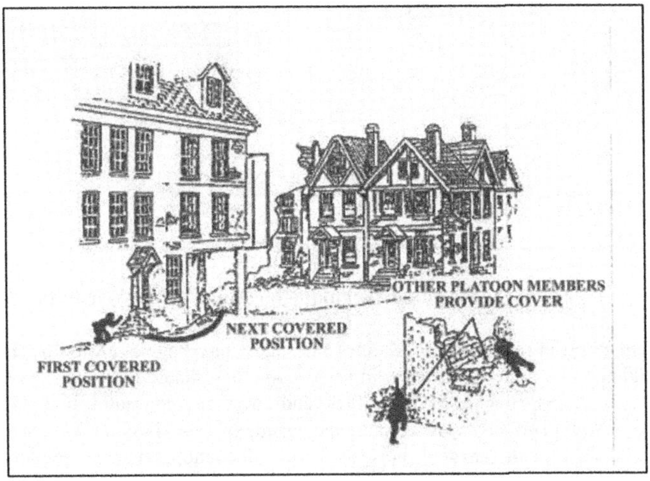

Figure A-13. Moving Outside a Building

Figure A-14. Selection of the Next Position

9. Crossing Open Areas. Open areas such as streets, alleys, and parks should be avoided. They are natural kill zones for enemy crew-served weapons. They can be crossed safely if certain fundamentals are applied by Marines and small-unit leaders.

a. When using the correct procedure for crossing an open area, the Marine develops a plan for his own movement. (Smoke from hand grenades or smoke pots should be used to conceal the movement of all Marines.) The Marine runs the shortest distance between buildings and moves along the far building to the next position. By doing so, he reduces the amount of time during which he is exposed to enemy fire.

b. Before moving to another position, the Marine should make a visual reconnaissance and select the position that offers the best cover and concealment. At the same time, he should select the route that he will take to get to that position.

c. When moving from position to position, each Marine should be careful not to mask his supporting fires. When he reaches his next position, he should be prepared to cover the movement of other members of his assault force or element.

10. Assault Element Employment. Moving as an assault element from building to building or between buildings presents a problem because an assault element presents a large target to enemy fire. When moving from the corner of one building to another, the assault element should move across the open area in groups of two, maintaining their integrity within their respective clearing or covering team (Figure A-15). The assault element leader (fire team leader) is responsible for the secure movement of his unit. Smoke should be used to screen the assault element's movement. The covering team will move first to establish security on the far side. Their movement is protected by the fire of the clearing team. Once the covering team is in position, they provide cover as the clearing team moves across and prepares, then enters, the structure. Moving from the side of one building to the side of another presents a similar problem, and the technique of movement employed is the same (Figure A-16).

There may be situations in which it is best to move in larger groups. An example is that Marines crossing in pairs may key enemy snipers in to the movement, setting them up for good shots at the next pair to cross. Leaders should remain aware of the situation and threat and plan movements accordingly.

Figure A-15. Assault Element Movement

11. Movement Inside a Building. When moving within a building that is under attack (Figure A-17), Marines should avoid silhouetting themselves in doors and windows. If forced to use a hallway (Figure A-18), Marines should move by hugging the wall to avoid presenting targets to the enemy.

Figure A-16. Movement to an Adjacent Building

Figure A-17. Movement Within a Building Under Attack

Figure A-18. Moving Down a Hallway

Section IV

Entry Techniques

Marines should enter a building with minimum exposure. They should select an entry point before moving toward the building; avoid windows and doors; use smoke to conceal their advance to the building; use demolitions, tank rounds, and so on to make new entrances; precede the entry of the clearing team with a grenade; enter immediately after the grenade explodes; and be covered by the covering team.

12. Upper Building Levels. Clearing a building from the top down is the preferred method. Clearing or defending a building is easier from an upper story. Gravity and the building's floor plan become assets when throwing hand grenades and moving from floor to floor.

a. An enemy who is forced to the top of a building may be cornered and fight desperately or escape over the roof, but an enemy who is forced down to the ground level may withdraw from the building, thus exposing himself to friendly fires from the outside.

b. Various means such as ladders, drainpipes, vines, helicopters, or the roofs and windows of adjoining buildings may be used to reach the top floor or roof of a building. In some cases, one Marine can climb onto the shoulders of another and reach high enough to pull himself up. Another method is to attach a grappling hook to the end of a scaling rope so that a Marine can scale a wall, spring from one building to another, or gain entrance through an upstairs window.

13. Use of Ladders. Ladders offer the quickest method to gain access to the upper levels of a building (Figure A-19). Units can get ladders from local civilians or stores or obtain material to build ladders through supply channels. If required, ladders can be built with resources that are available throughout the urban area; for example, lumber can be taken from inside the walls of buildings (Figure A-20). Although ladders may not permit access to the top of some buildings, they may offer security and safety through speed.

14. Use of Grappling Hook. A suitable grappling hook and rope are selected to meet the need. The grappling hook should be sturdy, portable, easily thrown, and equipped with hooks that can hold inside a window. The scaling rope should be 5/8 of an inch to 1 inch in diameter and long enough to reach the objective window. Knots are tied in the rope at 1-foot intervals to make climbing and holding easier. Marines should follow the procedures outlined below.

a. When throwing the grappling hook, the Marine should stand as close to the building as possible (Figure A-21). If the Marine stands close to the building, he is less exposed to enemy fires and he does not have to throw the hook as far.

Figure A-19. Using Ladders To Get to Upper Levels

Figure A-20. Getting Lumber From Inside the Walls

Figure A-21. Grappling Hook Thrown at Close Range

b. Making sure there is enough rope to reach the target, the Marine should hold the hook and a few coils of rope in the throwing hand. The remainder of the rope, in loose coils, should be in the other hand. The rope should be allowed to play out freely. The throw should be a gentle, even, upward lob of the hook, with the other hand releasing the rope as it plays out.

c. Once the grappling hook is inside the window (or on the roof), the Marine should pull on the rope to obtain a good hold before beginning to climb. When using a window, the hook should be pulled to one corner to ensure a good "bite" and to reduce the Marine's exposure to lower windows during the climb.

d. The use of grappling hooks is the least preferred method for gaining entry to upper levels of buildings. Grappling hooks should be used only as a last resort and away from potential enemy positions. This method may potentially be used on adjacent buildings that offer concealed locations and a roof that connects to enemy positions.

15. Scaling Walls. When forced to scale a wall during exposure to enemy fire, all available means of concealment should be used. Smoke and diversionary measures improve the chances of a successful exposed movement. When using smoke for concealment, Marines should plan for wind direction. They should use fire, shouts, and fake movement to distract the enemy.

a. A Marine scaling an outside wall is vulnerable to enemy sniper fire. Marines moving from building to building and climbing buildings should be covered with friendly fire. Areas between buildings offer good fields of fire to the enemy. Properly positioned friendly weapons can suppress and eliminate enemy fire. The M203 grenade launcher is effective in clearing the enemy from rooms inside buildings (Figure A-22).

Figure A-22. Employment of the M203 Grenade Launcher

b. A Marine scaling a wall with a rope should avoid silhouetting himself in windows of uncleared rooms or exposing himself to enemy fires from lower windows. He should climb with his weapon slung over the firing shoulder so that he may quickly bring it to a firing position. He should clear the lower room with a hand grenade before going outside the window. Before doing this, the Marine first loosens the safety pin so that he needs only one hand to throw the grenade. The objective upper story window should not be entered before a hand grenade has been thrown in.

c. The Marine enters the objective window with a low silhouette (Figure A-23). Entry can be head first; however, a preferred method is to hook a leg over the windowsill and enter sideways, straddling the ledge.

16. Rappelling. Rappelling (Figure A-24) is an entry technique that Marines can use to descend from the rooftop of a tall building into a window. (See Army manual TC 21-24 for more information on rappelling.)

17. Entry at Lower Levels. Buildings should be cleared from the top down. However, it may be impossible to enter a building at the top; therefore, entry at the bottom or lower level may be the only course of action. When entering a building at the lower level, Marines avoid entry through windows and doors because both can be easily boobytrapped and are usually covered by enemy fire.

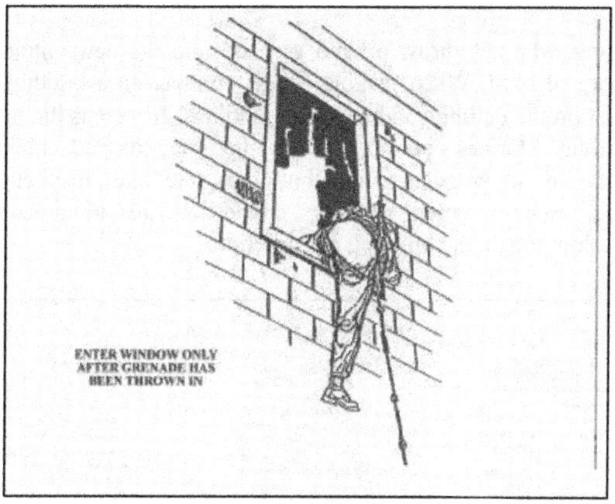

Figure A-23. Entering the Objective Window

Figure A-24. Rappelling

a. Ideally, when entering at lower levels, demolitions, artillery, tank fire, antiarmor weapons fire, or similar means are used to create a new entrance to avoid boobytraps. Quick entry is required to follow through on the effects of the blast and concussion.

b. When the only entry to a building is through a window or door, supporting fire should be directed at that location. (Specific lower level entry techniques are shown in Figure A-25.) If no supporting fire is available, AT4s can be employed instead.

c. Before entering, Marines throw a hand grenade into the new entrance to reinforce the effects of the original blast. When making a new entrance in a building, they consider the effects of the blast on the building and adjacent buildings. If there is the possibility of a fire in an adjacent building, Marines coordinate with adjacent units and obtain permission before starting the operation. In wooden-framed buildings, the blast may cause the building to collapse. In stone, brick, or cement buildings, supporting fires are aimed at the corner of the building or at weak points in the building construction.

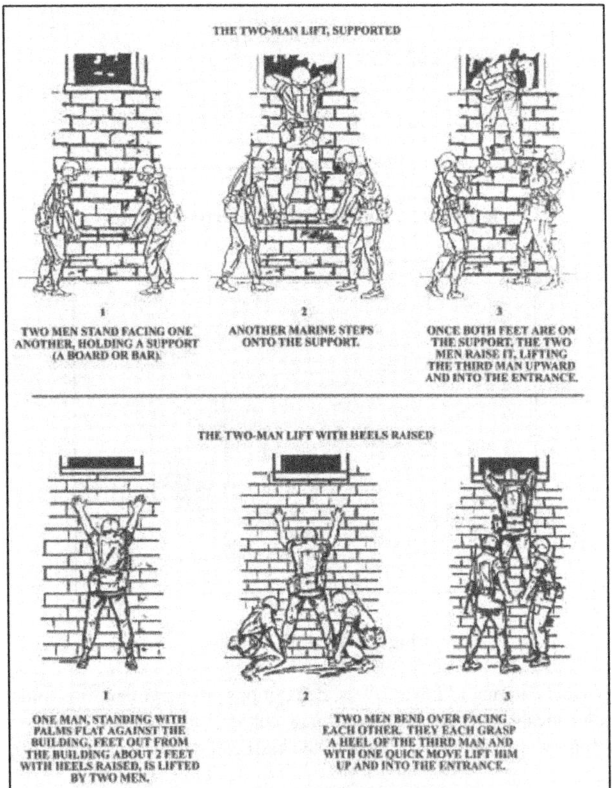

Figure A-25. Lower Level Entry Techniques

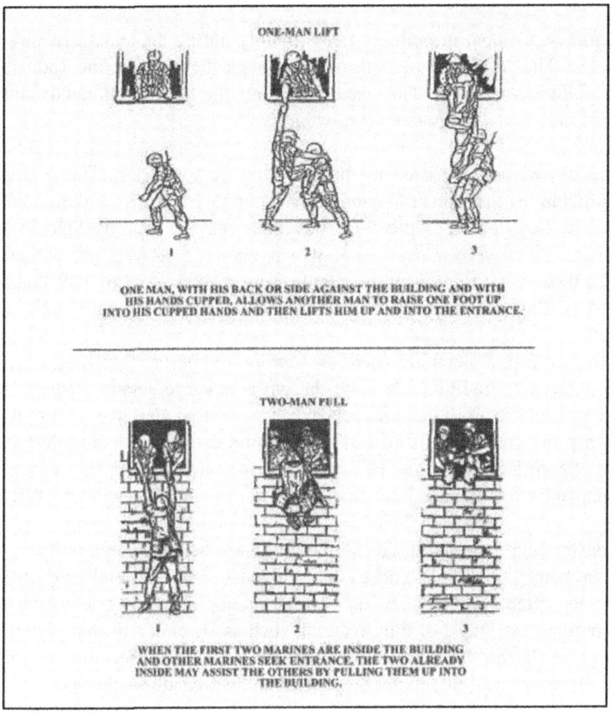

Figure A-25. Lower Level Entry Techniques (Continued)

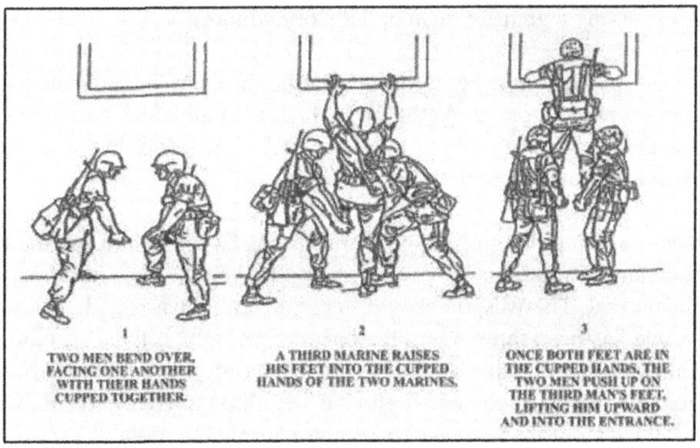

Figure A-25. Lower Level Entry Techniques (Continued)

18. Hand Grenades. Combat in built-up areas (mainly during the attack) requires extensive use of hand grenades. The Marine should throw a grenade before entering rooms or negotiating staircases, mouseholes, and so on. This usually requires the use of both hands and employs both the overhand and underhand methods of throwing.

a. Two techniques may be used in the throwing of a grenade. The *preferred technique* involves throwing the grenade hard enough that it bounces or skips around, making it difficult to pick up. The hard-throw, skip/bounce technique may be used by Marines in training and combat. The *least preferred technique*, which involves cooking-off the grenade for two seconds then throwing it to prevent the enemy from grabbing the grenade and tossing it back, may be used by Marines, as appropriate, during actual combat but not in training.

b. Nonverbal or verbal alerts are used, as appropriate, before throwing a grenade. A visual showing (preferred) of the grenade to be thrown is made to assault element members, and a visual acknowledgment from them is received. A nonverbal alert may ensure that the enemy is surprised when the grenade is thrown. If the situation demands, a voice alert can be used, but the element of surprise may be lost. When or if a voice alert is used, the voice alert is "**FRAG OUT**;" when an enemy grenade has been identified, friendly forces shout, "**GRENADE**."

c. The construction material used in the building being cleared influences the use of grenades. In some situations, concussion grenades may be preferred over fragmentary grenades during offensive operations or when defending from hasty defensive positions. If the walls of a building are made of thin material, such as sheetrock or thin plyboard, the Marine should either lie flat on the floor with his helmet pointing toward the area of detonation or move away from any wall that might be penetrated by grenade fragments.

d. Marines should throw grenades into an opening before entering a building to eliminate any enemy that might be near the entrance (Figure A-26). Using the M203 grenade launcher is the best method for putting a grenade in an upper story window.

e. When a hand grenade must be used, the Marine throwing the grenade should stand close to the building, using it for cover. At the same time, the individual and the rest of the element should have a planned area to which they can move for safety if the grenade does not go through the window, but falls back to the ground.

f. The Marine throwing the grenade should step out far enough to lob the grenade into the upper story opening. The Marine's weapon should be kept in the nonthrowing hand so that it can be used if needed. The weapon should never be laid down outside or inside the building. Once the grenade has been thrown into the opening, assaulting forces should move swiftly to enter the building. This technique should be employed only when the window has already been broken. Otherwise, the chances are high that the thrown grenade will bounce off of the window and fall back onto the ground without going into the room.

Figure A-26. Hand Grenade Thrown Through Window

WARNING

After throwing the grenade, the Marine should immediately announce, "FRAG OUT," to indicate that a grenade has been thrown. He then takes cover because the grenade may bounce back or be thrown back, or the enemy may fire at him.

g. If Marines must enter the building by using the stairs, they first look for boobytraps. Then they throw a grenade through the stairwell door, let it detonate, and move quickly inside. They can use the staircase for cover.

h. The best way to enter a building is to breach the exterior wall. Again, a grenade should be thrown through the hole while using all available cover, such as the lower corner of the building (Figure A-27).

i. Another way to enter a room is to blast mouseholes with demolitions. In moving from room to room through mouseholes, Marines should use grenades to clear the room as they do in moving through open doorways.

j. Although buildings are best cleared from the top down, this is not always possible. While clearing the bottom floor of a building, Marines may encounter stairs that must also be cleared. Once again, grenades play an important role. Before climbing the stairs from the bottom floor, Marines should first inspect for boobytraps then toss a grenade to the head of the stairs, if appropriate (Figure A-28). The use of a grenade in these situations is based on stair/building construction and building layout. *Throwing grenades up staircases is extremely* *dangerous, and extreme caution and common sense should be used.* In most situations, throwing a grenade up a staircase is not recommended. (Again, the grenade is being used in the initial building entry or clearing of the bottom floor of a building having a staircase.) If a grenade is thrown too hard or the staircase is too steep, the grenade can roll back down the staircase. For this reason, if a grenade is thrown, it should be thrown underhand to reduce the risk of it bouncing back and rolling down the stairs. A nonverbal or, as appropriate, voice alert is used when throwing the grenade. Once the first grenade has detonated, another grenade should be thrown over and behind the staircase banister and into the hallway, destroying any enemy hiding to the rear.

Figure A-27. Entering Through a Breached Hole

Figure A-28. Tossing a Grenade Up a Stairway

k. After the stairs have been cleared, assaulting forces move to the top floor and clear it. Upon clearing the top floor, forces move downstairs to clear the center and bottom floors and to continue with the mission.

Note: Because large quantities of hand grenades are used when clearing buildings, a continuous supply must be available to forces having this mission within a built-up area.

c. Before entering, Marines throw a hand grenade into the new entrance to reinforce the effects of the original blast. When making a new entrance in a building, they consider the effects of the blast on the building and adjacent buildings. If there is the possibility of a fire in an adjacent building, Marines coordinate with adjacent units and obtain permission before starting the operation. In wooden-framed buildings, the blast may cause the building to collapse. In stone, brick, or cement buildings, supporting fires are aimed at the corner of the building or at weak points in the building construction.

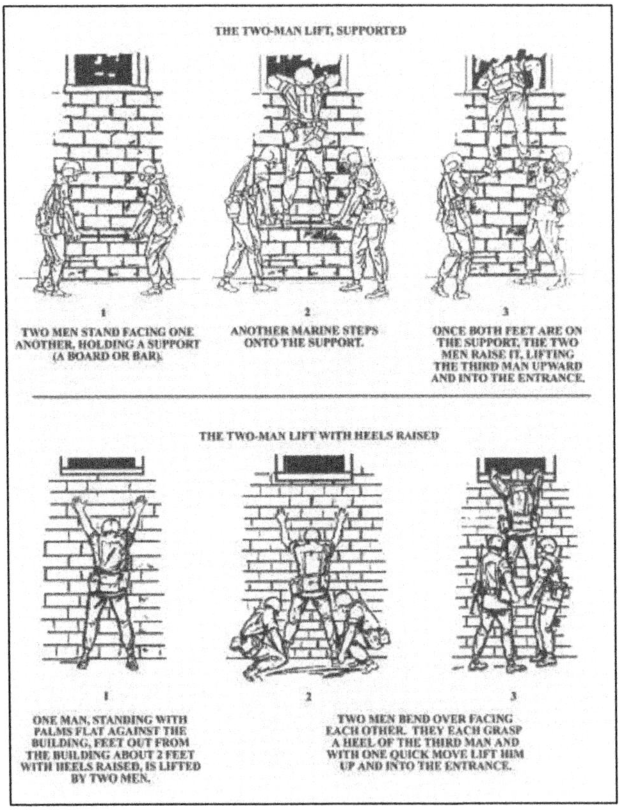

Figure A-25. Lower Level Entry Techniques

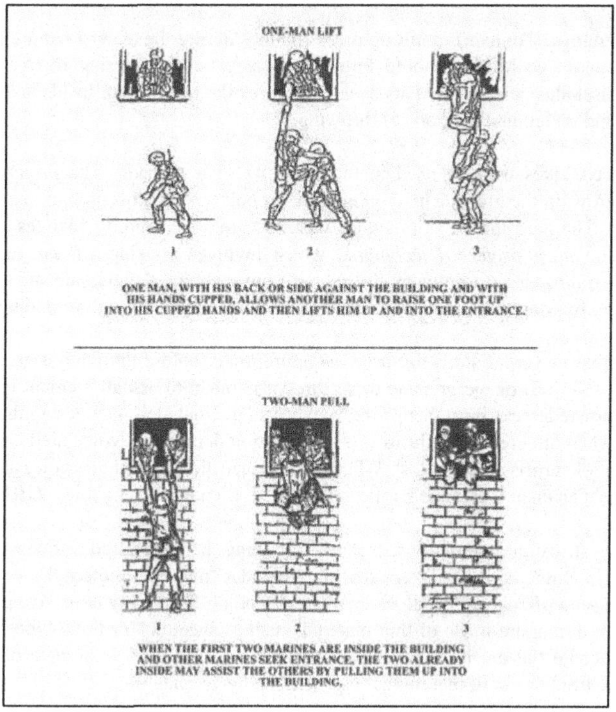

Figure A-25. Lower Level Entry Techniques (Continued)

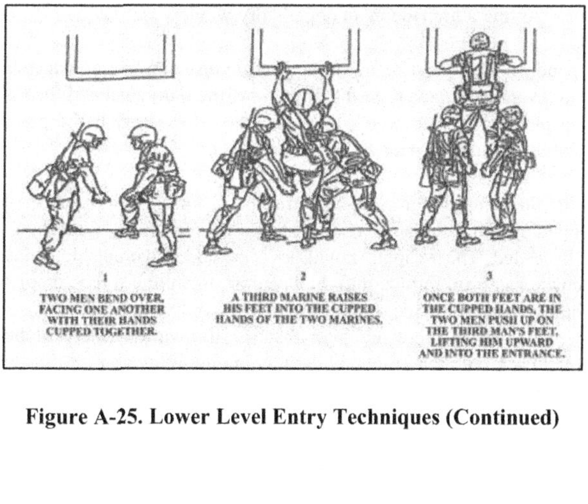

Figure A-25. Lower Level Entry Techniques (Continued)

18. Hand Grenades. Combat in built-up areas (mainly during the attack) requires extensive use of hand grenades. The Marine should throw a grenade before entering rooms or negotiating staircases, mouseholes, and so on. This usually requires the use of both hands and employs both the overhand and underhand methods of throwing.

a. Two techniques may be used in the throwing of a grenade. The *preferred technique* involves throwing the grenade hard enough that it bounces or skips around, making it difficult to pick up. The hard-throw, skip/bounce technique may be used by Marines in training and combat. The *least preferred technique*, which involves cooking-off the grenade for two seconds then throwing it to prevent the enemy from grabbing the grenade and tossing it back, may be used by Marines, as appropriate, during actual combat but not in training.

b. Nonverbal or verbal alerts are used, as appropriate, before throwing a grenade. A visual showing (preferred) of the grenade to be thrown is made to assault element members, and a visual acknowledgment from them is received. A nonverbal alert may ensure that the enemy is surprised when the grenade is thrown. If the situation demands, a voice alert can be used, but the element of surprise may be lost. When or if a voice alert is used, the voice alert is "**FRAG OUT**;" when an enemy grenade has been identified, friendly forces shout, "**GRENADE**."

c. The construction material used in the building being cleared influences the use of grenades. In some situations, concussion grenades may be preferred over fragmentary grenades during offensive operations or when defending from hasty defensive positions. If the walls of a building are made of thin material, such as sheetrock or thin plyboard, the Marine should either lie flat on the floor with his helmet pointing toward the area of detonation or move away from any wall that might be penetrated by grenade fragments.

d. Marines should throw grenades into an opening before entering a building to eliminate any enemy that might be near the entrance (Figure A-26). Using the M203 grenade launcher is the best method for putting a grenade in an upper story window.

e. When a hand grenade must be used, the Marine throwing the grenade should stand close to the building, using it for cover. At the same time, the individual and the rest of the element should have a planned area to which they can move for safety if the grenade does not go through the window, but falls back to the ground.

f. The Marine throwing the grenade should step out far enough to lob the grenade into the upper story opening. The Marine's weapon should be kept in the nonthrowing hand so that it can be used if needed. The weapon should never be laid down outside or inside the building. Once the grenade has been thrown into the opening, assaulting forces should move swiftly to enter the building. This technique should be employed only when the window has already been broken. Otherwise, the chances are high that the thrown grenade will bounce off of the window and fall back onto the ground without going into the room.

Figure A-26. Hand Grenade Thrown Through Window

WARNING

After throwing the grenade, the Marine should immediately announce, "FRAG OUT," to indicate that a grenade has been thrown. He then takes cover because the grenade may bounce back or be thrown back, or the enemy may fire at him.

g. If Marines must enter the building by using the stairs, they first look for boobytraps. Then they throw a grenade through the stairwell door, let it detonate, and move quickly inside. They can use the staircase for cover.

h. The best way to enter a building is to breach the exterior wall. Again, a grenade should be thrown through the hole while using all available cover, such as the lower corner of the building (Figure A-27).

i. Another way to enter a room is to blast mouseholes with demolitions. In moving from room to room through mouseholes, Marines should use grenades to clear the room as they do in moving through open doorways.

j. Although buildings are best cleared from the top down, this is not always possible. While clearing the bottom floor of a building, Marines may encounter stairs that must also be cleared. Once again, grenades play an important role. Before climbing the stairs from the bottom floor, Marines should first inspect for boobytraps then toss a grenade to the head of the stairs, if appropriate (Figure A-28). The use of a grenade in these situations is based on stair/building construction and building layout. *Throwing grenades up staircases is extremely* *dangerous, and extreme caution and common sense should be used.* In most situations, throwing a grenade up a staircase is not recommended. (Again, the grenade is being used in the initial building entry or clearing of the bottom floor of a building having a staircase.) If a grenade is thrown too hard or the staircase is too steep, the grenade can roll back down the staircase. For this reason, if a grenade is thrown, it should be thrown underhand to reduce the risk of it bouncing back and rolling down the stairs. A nonverbal or, as appropriate, voice alert is used when throwing the grenade. Once the first grenade has detonated, another grenade should be thrown over and behind the staircase banister and into the hallway, destroying any enemy hiding to the rear.

Figure A-27. Entering Through a Breached Hole

Figure A-28. Tossing a Grenade Up a Stairway

k. After the stairs have been cleared, assaulting forces move to the top floor and clear it. Upon clearing the top floor, forces move downstairs to clear the center and bottom floors and to continue with the mission.

Note: Because large quantities of hand grenades are used when clearing buildings, a continuous supply must be available to forces having this mission within a built-up area.

Section V

Clearing Techniques

This section will discuss basic clearing techniques; however, the fundamentals may have to be modified depending on the various situations and building/room layouts that Marines may encounter. Caution should be taken when using a stacked type of position as personnel may be concentrating in a danger area where they could be exposed to fires penetrating walls or explosive effects from boobytraps or grenades thrown by the enemy. *Marines should always remain aware of the protective, or lack of protective, capabilities of the materials comprising the walls, floors, ceilings, and doors within the areas they are clearing.*

19. Clearing a Room, Door Closed or Open, From Stacked Positions. The following discussion and series of figures will focus on the two-man clearing team in the conduct of clearing a room. The covering team is positioned by the assault element leader to provide security as the clearing team begins its procedures to enter and clear a room.

a. Shooter Number Two is behind Shooter Number One (stacked position). This positioning is the same if the door is already open or nonexistent (Figure A-29).

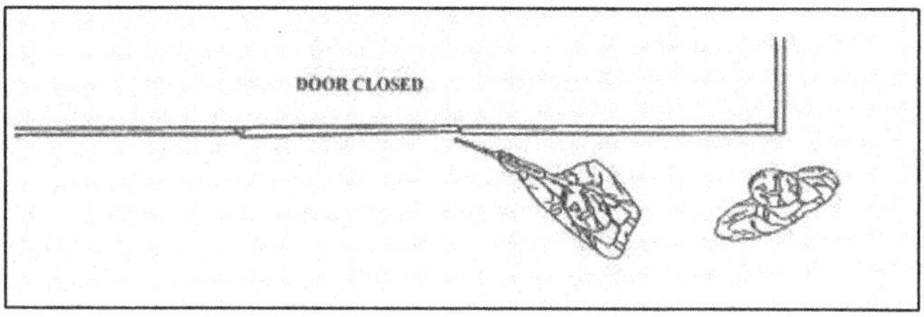

Figure A-29. Clearing a Room, Door Closed

b. Shooter Number One determines the direction in which the door opens then quickly informs Shooter Number Two and forces the door open.

When a door is the only means of entering a room, Marines should beware of fire from enemy soldiers within the room and of boobytraps. When opening a door, Marines should not expose themselves to fires through the door. The two-man clearing team forces the door open by directing a short burst of automatic fire through the door around the latch (Figure A-30) and then kicking it open. Another method of opening the door is to use a shotgun with 12-gauge slugs to shoot the hinges and/or the latch. When the doorknob is locked and/or deadbolt locks are present, shooting the hinges of the door is recommended. The door may then be opened by kicking or using a battering ram.

Figure A-30. Shooting the Door Open

Marines can also use an ax or demolitions if they are available. As a last resort, Marines can kick the door open with a front kick (Figure A-31) or mule kick. This is the least preferred technique because it is difficult and tiring to the Marine. It also rarely works the first time, thereby giving any enemy units within the room ample warning (and giving the enemy time to shoot through the door). In situations where restrictions on firing and limiting structural damage have been imposed and/or boobytraps are not expected, using the door handle to open the door may be appropriate (apply METT-T).

Figure A-31. Opening a Door With a Front Kick

c. Once the door is open, Shooter Number Two immediately throws a hand grenade into the room. Shooter Number One maintains cover for Shooter Number Two as he prepares to throw the grenade into the room. Shooter Number Two then returns to the original stacked position behind Shooter Number One (Figure A-32). If Marines have no grenades, the procedures remain the same except for throwing the grenade. *Note: When the door is already open or nonexistent, the stacked position is always used.*

WARNING

Because fragments from M67 fragmentation grenades may injure Marines outside the room, they should not be used. Marines should use MK3A2 offensive hand grenades instead. Also, the technique of cooking-off hand grenades can be deadly unless properly performed.

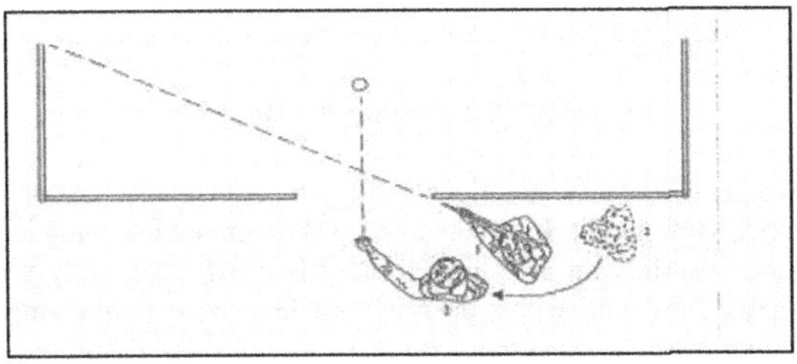

Figure A-32. Throwing a Grenade Into a Room

(1) Throwing Grenades. Two techniques are available for using grenades in room clearing. The *preferred technique* for Marines is to throw a concussion hand grenade into the room so hard that it skips and bounces, making it difficult for the enemy to pick up and throw back. The skip/bounce technique should be used by Marines during training and combat. The *least preferred technique* is to cook-off a concussion hand grenade by removing the grenade's safety pin, releasing the safety lever, counting off two seconds by thousands (one thousand and one, one thousand and two), and then throwing the grenade into the room. Cooking-off the grenade will be used only as appropriate during combat.

(2) Nonverbal and Verbal Alerts. To alert all that a grenade will be thrown, a visual showing of the grenade is made to assault element members, and a visual acknowledgment from them is received. A nonverbal alert may ensure that the enemy is surprised when the grenade is thrown. If the situation demands, a voice alert can be used, but the element of surprise may be lost. When or if a voice alert is used, the voice alert is "**FRAG OUT**;" when an enemy grenade has been identified, friendly forces shout, "**GRENADE**." This allows Marines to distinguish between warnings for outgoing and incoming grenades.

d. After the grenade explodes, Shooter Number One steps across the threshold and clears his immediate area (Figure A-33). He engages targets from the ready carry position. Shooter Number Two follows immediately behind Shooter Number One, buttonhooks, and clears his area.

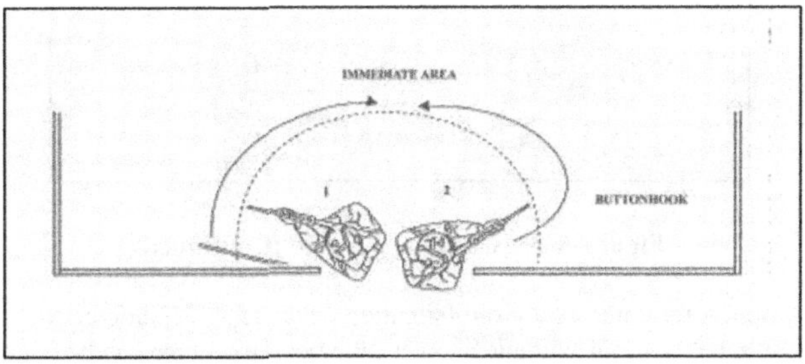

Figure A-33. Shooters Enter the Room

e. Both shooters clear the immediate area and along their respective walls, starting from the nearest respective corner and continuing to the farthest respective corner (Figure A-34). Shooters use the pieing technique to systematically clear the room by sector. (See Paragraph 3.a. of this appendix.)

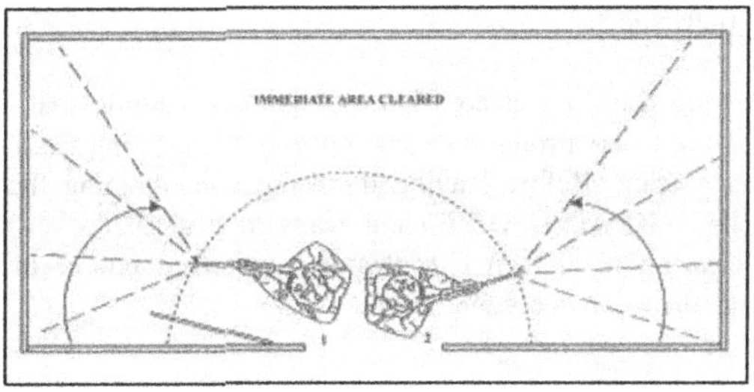

Figure A-34. Clearing a Room

f. Both shooters then establish a dominant position in the room one step away from the wall and two steps into the room and clear the room by sector, pieing to the opposite side of the room (Figure A-35). Meanwhile, the covering team (two-man team) in position outside the room being cleared provides security.

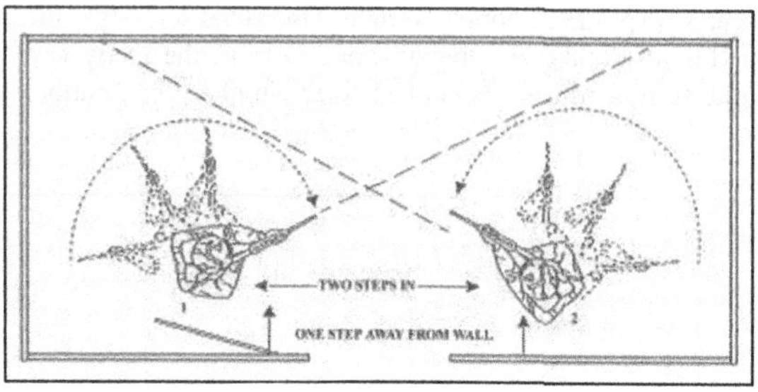

Figure A-35. Clearing a Room (Continued)

g. Another option for entering a room (situation dependent): If Shooter Number One of the clearing team enters a room without Shooter Number Two immediately following him (this could be due to room size, entering the room through a narrow opening, or a number of factors caused by the situation), then Shooter Number One will position himself inside the room to the left or right of the door and quickly scan the entire room. He then gives the command to Shooter Number Two, "**NEXT MAN IN, LEFT (RIGHT).**" Shooter Number Two shouts, "**COMING IN, LEFT (RIGHT),**" enters the room, positions himself up against the wall to the left (right) of the entrance as designated, and scans the room. Once in position, the clearing team leader can call in additional members with the "**NEXT MAN IN**" command, as the situation dictates. It is critical that all assault element members tell each other where they are to avoid fratricide.

h. When the clearing team has cleared the room and is ready to exit the room, it uses the following voice alerts. Once a room has been cleared, the clearing team yells, "**CLEAR,**" to inform the covering team. Before leaving the room and rejoining the covering team, the clearing team yells, "**COMING OUT**" and waits to hear the covering team yell, "**ALL CLEAR.**" The clearing team then executes its movement out of the room. The assault element then marks the room according to unit SOP.

When moving up or down a staircase, the appropriate team yells, "**COMING UP**" or "**COMING DOWN.**" As before, it waits for the response "**ALL CLEAR**" before executing movement and rejoining its element. When exiting a building, the clearing team or assault element yells, "**COMING OUT**" and waits for the support force or covering team to respond with "**ALL CLEAR.**"

20. Clearing a Room, Door Closed, Split Positions

a. Before opening a door, Shooter Number Two of the clearing team positions himself opposite Shooter Number One on the other side of the door and away from the wall in a safe position that allows Shooter Number One to shoot the door-opening mechanism (Figure A-36). This positioning allows Shooter Number Two to be in position to immediately move to a wall position opposite Shooter Number One. No matter what method is used to open the door, Shooter Number Two should get in a position on the side of the door opposite Shooter Number One. If the door is already open, the stacked positions in Paragraph 19.a. of this section are used.

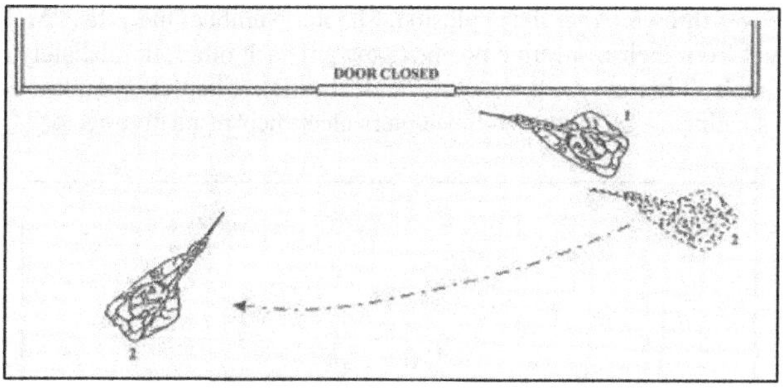

Figure A-36. Positioning To Open a Door

b. After Shooter Number One shoots the door open, Shooter Number Two immediately moves to a kneeling position against the wall (Figure A-37). Shooter Number One kicks the door open from a standing position. If the Marines have grenades, then Shooter Number Two immediately throws a grenade into the room as discussed in Paragraph 19.c. of this section. If the Marines do not have any grenades, the stand/kneel positioning of shooters places the shooters' weapons at different levels, thus allowing each shooter to engage targets inside the room while staying out of the other shooter's line of fire. The shooters scan the room for targets.

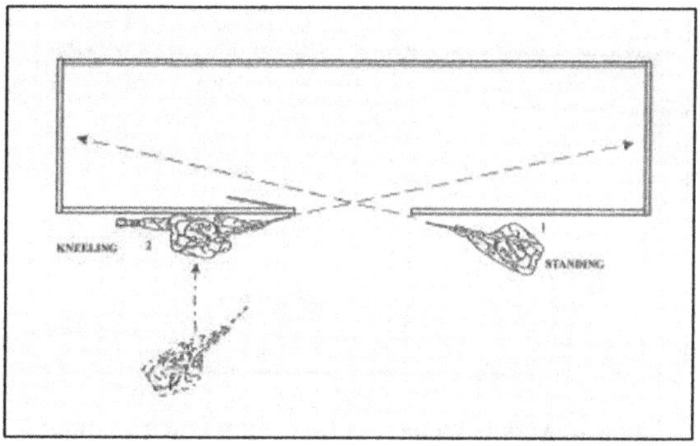

Figure A-37. Positioning To Enter a Room

c. If no grenade was thrown: Once the initial scan of the room is complete, Shooter Number One yells, "**MOVE**," and both shooters pivot from their respective positions toward each other, pieing (see Paragraph 3.a. of this appendix) while maintaining eye-muzzle-target contact. Shooter Number Two is kneeling, and Shooter Number One is still standing (Figure A-38). Both shooters clear their respective sector of fire.

If a grenade was thrown: After the explosion, Shooter Number One yells, "**MOVE**," and both shooters pivot from their respective positions toward each other, immediately enter the room, and buttonhook (Figures A-38, A-39A, and A-39B). Shooter Number Two goes from kneeling to standing as he pivots. Both shooters clear their respective sector of fire.

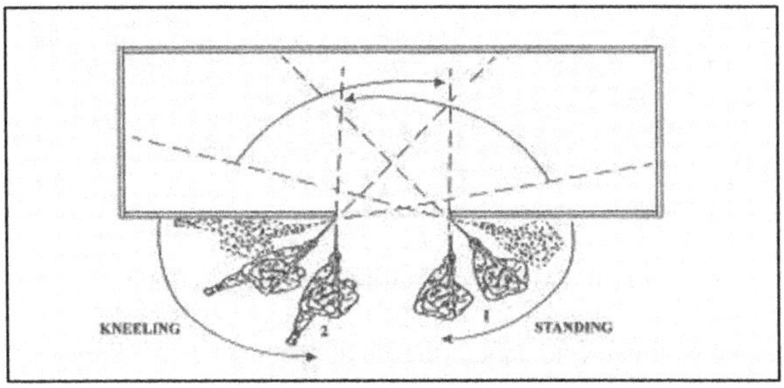

Figure A-38. Shooters Clear Sectors of Fire From Doorway

d. If no grenade was thrown: After scanning is completed, Shooter Number Two stands up and yells, "**READY**," and Shooter Number One Yells, "**MOVE**." Both shooters enter the room together (Figure A-39A).

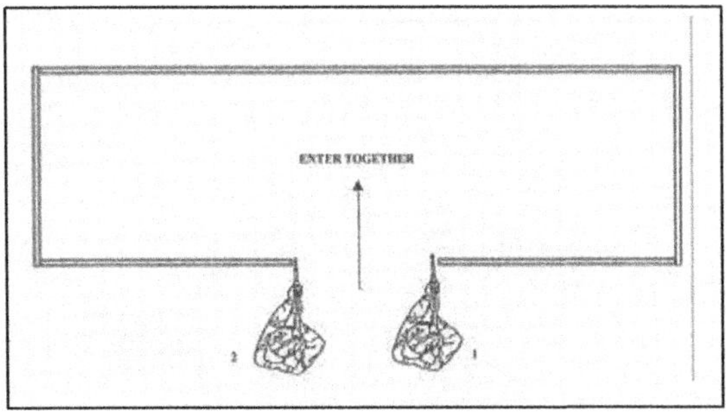

Figure A-39A. Shooters Enter the Room Together

e. Both shooters pass through the doorway together and immediately face respective corners, using the adjacent wall for cover. Both shooters buttonhook their respective areas (Figure A-39B). At this point, clearing the room is the same as discussed in Paragraphs 19.e. - 19.h.

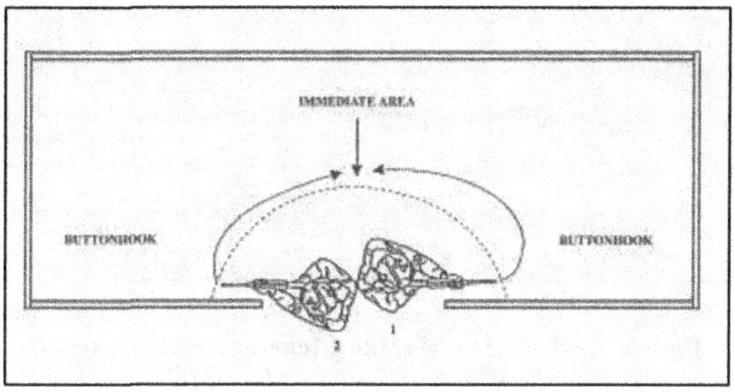

Figure A-39B. Shooters Clear Immediate Area

f. Cross Method. When employing the cross method, two Marines position themselves on either side of the entryway. Each Marine faces into the room covering the corner of the room opposite his position. On a prearranged signal, each Marine alternately enters the room. Each Marine crosses quickly to the opposite corner while covering the half of the room toward which he is moving. Once in the near corner, he assumes an outboard kneeling position to reduce his silhouette and continues to maintain coverage of his half of the room. He may change the position of his weapon to best cover the room. (Figure A-40)

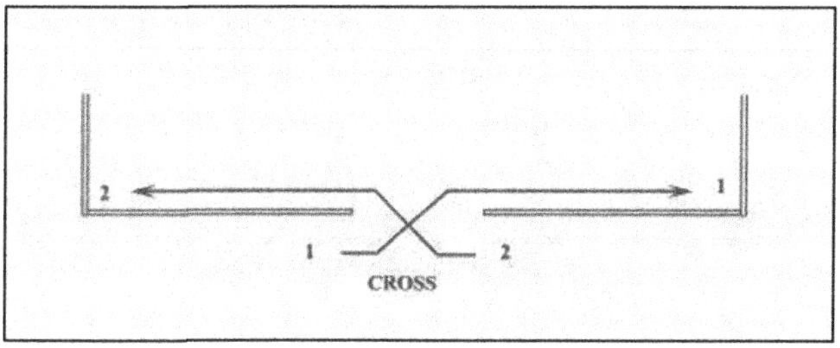

Figure A-40. Cross Method of Entering and Clearing Rooms

21. Covering Team Members Used To Support Clearing of a Room

a. Three Marines Clearing a Room. Clearing a room with three men involves the same procedures used by the aforementioned two-man clearing team, with a third Marine (from the covering team) following immediately (stacked) behind Shooters Number One and Two. Shooter Number Three moves to one side of the door and establishes a center sector of fire coverage (Figure A-41). The remaining covering team Marine provides outside security. If the split position is used, Shooters Number Two and Three are placed on the opposite side of the door.

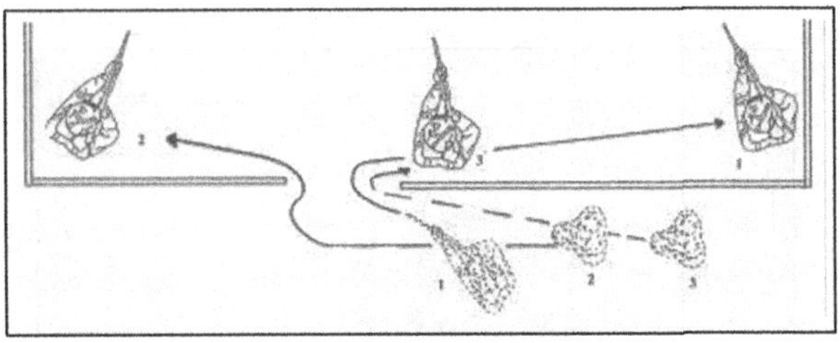

Figure A-41. Three-Marine Clearing Team (Stacked)

b. Four Marines Clearing a Room. When four Marines (entire assault element) are used to clear a room, outside security should be established by another assault element. (If outside security cannot be established, then the two- or three-Marine clearing team should be used.) The two-Marine clearing team procedures are used for initial entry into the room. Shooters Number Three and Four (split position example) stack up opposite Shooters Number One and Two. Shooters Number Three and Four follow Shooter Number Two into the room. Shooter Number Three crosses the threshold and establishes a center sector of fire coverage. Shooter Number Four follows Shooter Number Three and establishes a center sector of fire coverage opposite Shooter Number Three (Figure A-42). If the stacked position is used to enter a room, all shooters stack up on the same side of the door and follow each other in with the same positioning.

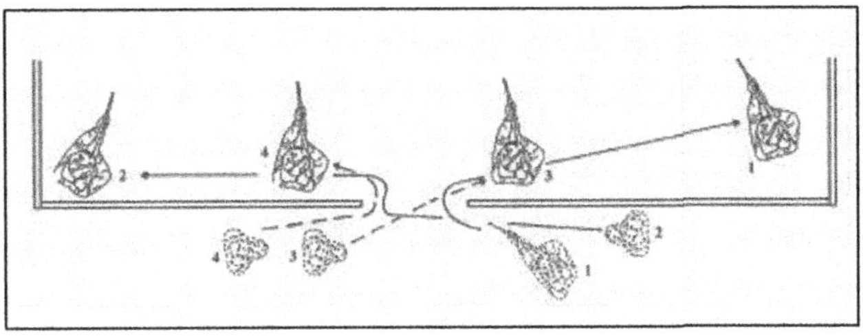

Figure A-42. Four-Marine Clearing Team (Split)

c. Single Marine Clearing a Room. This is the *least preferred* method of clearing a room. However, there may be times when this is the only method available, for example, when gaining access to an upper story window. When this is the case, the Marine gains entry to the room by shooting, kicking, ramming, or breaking a window. He then throws a grenade into the room if one is available and no restrictions apply and takes cover. Immediately after the grenade explodes, he crosses the threshold using the buttonhook technique depicted in earlier figures, scans the room for targets, and engages targets with two shots as they appear.

22. Clearing a Room, Entering Through Mousehole. Mouseholes measure about 2 feet wide and are blown or cut through a wall so that Marines can enter a room (Figure A-43). Mouseholes are safer entrances than doors, which can be easily boobytrapped and should be avoided. Procedures discussed in Paragraph 19 are used. The assault element leader positions the covering team to provide security.

a. In this situation, Shooter Number One would enter the room through a mousehole after throwing a grenade and waiting for the detonation.

b. Shooter Number One will position himself inside the room to the left or right of the mousehole and quickly scan the entire room. He then gives the command to Shooter Number Two, "**NEXT MAN IN, LEFT (RIGHT).**" Shooter Number Two shouts, "**COMING IN, LEFT (RIGHT),**" enters the room, positions himself against the wall to the left (right) of the entrance, and scans the room.

c. The clearing team continues clearing the room as discussed in Paragraph 19.

Figure A-43. Entering Through a Mousehole

23. Clearing an L-Shaped Hallway. After entering a room or building, the clearing team may encounter an L-shaped hallway that should be cleared. The following procedures are used:

a. The clearing team finds an L-shaped hallway.

b. Each shooter takes a dominant position.

c. Shooter Number Two aims his weapon to cover the dead space (Figure A-44).

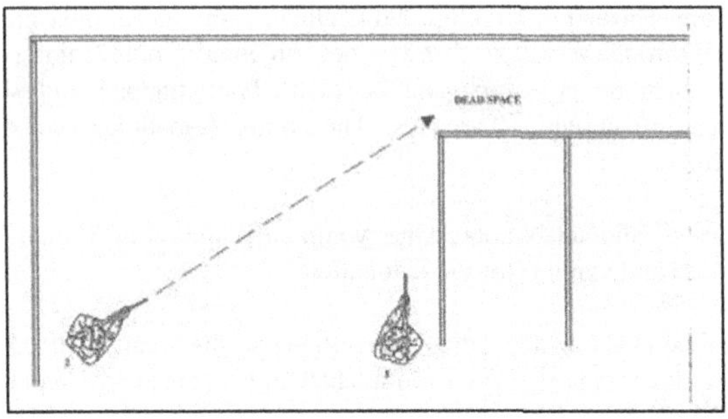

Figure A-44. Positioning To Clear an L-Shaped Hallway

d. Shooter Number Two, with his weapon covering the dead space, pies as far as possible and moves forward to increase his angle of fire further into the dead space. Shooter Number One moves accordingly with Shooter Number Two (Figure A-45).

e. Shooter Number Two stops pieing just before he reaches the position where he cannot cover the dead space. Shooter Number One stops short of the hallway and goes to a kneeling position.

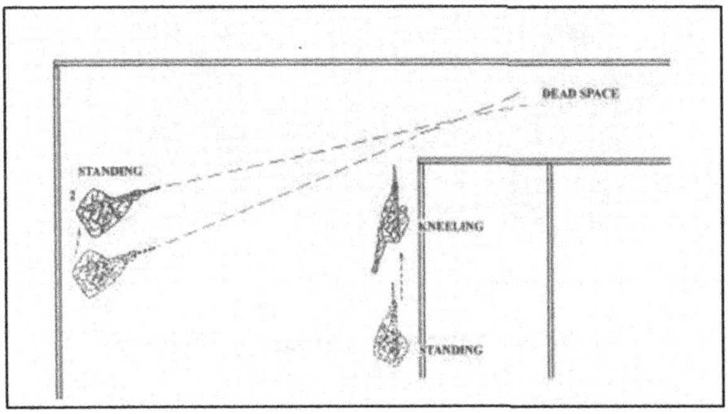

Figure A-45. Clearing an L-Shaped Hallway

f. Shooter Number One says, "**READY!**"

g. Shooter Number Two then says, "**MOVE!**" and both shooters move simultaneously to clear the hallway (Figure A-46). Shooter Number Two remains standing, and Shooter Number One remains kneeling as he pivots around the corner of the wall into the hallway.

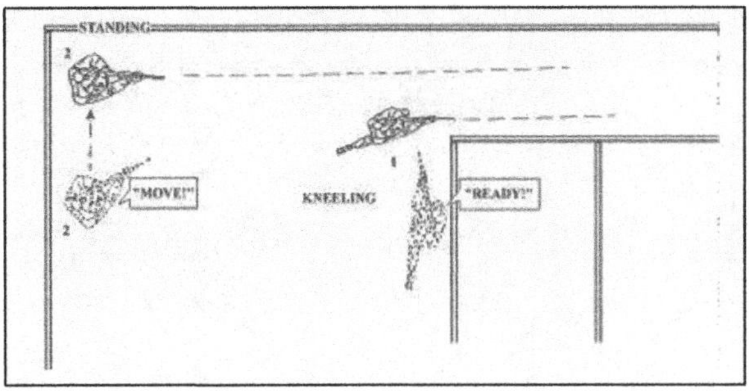

Figure A-46. Clearing an L-Shaped Hallway (Continued)

24. Clearing a T-Shaped Hallway. After entering a room or building, the clearing team may find a T-shaped hallway that should be cleared. The following procedures are used:

a. Both shooters approach the T-shaped intersection together, pieing their individual sectors while maintaining eye-muzzle-target contact (Figure A-47).

b. Both shooters move to dominant positions without entering the intersection.

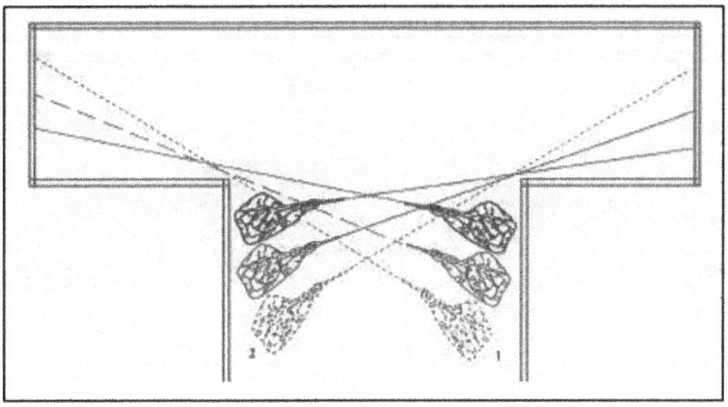

Figure A-47. Positioning To Clear a T-Shaped Hallway

c. Shooter Number One clears his sector of fire and then says, "**READY!**" (Figure A-48).

d. Shooter Number Two clears his sector of fire and then says, "**MOVE!**"

e. Both shooters buttonhook into their next positions.

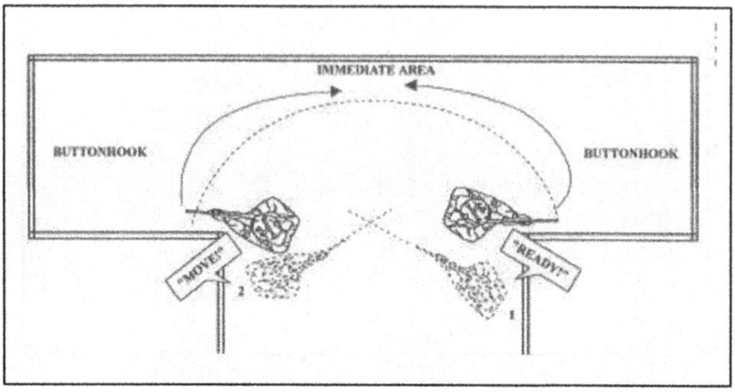

Figure A-48. Clearing a T-Shaped Hallway

f. Both shooters clear their immediate areas along their respective walls, starting from the nearest respective corner and continuing to the farthest respective corner (Figure A-49).

g. Both shooters establish dominant positions that give them control of the hallway and the doorways leading into hallways. The covering team can then be called forward to provide security while the clearing team clears one end of the hallway and associated rooms. The clearing team then clears the other end of the hallway.

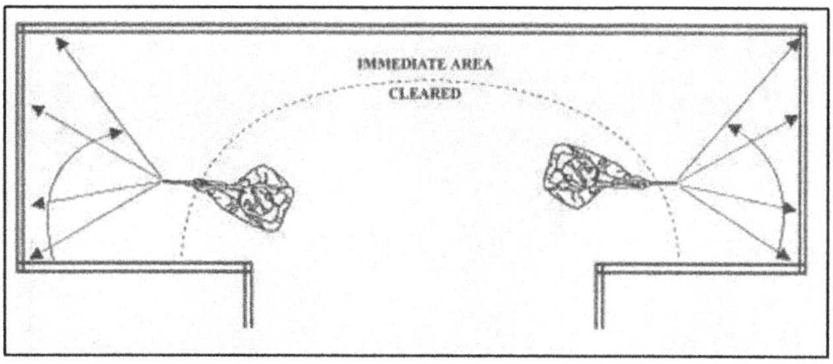

Figure A-49. Clearing a T-Shaped Hallway (Continued)

25. Clearing a Stairwell. After entering a building, the assault element may encounter stairwells. The following procedures are used:

a. Shooter Number One leads upstairs, one step ahead of Shooter Number Two (Figure A-50). Upon reaching the point just before he can be engaged from above, Shooter Number One turns around and covers overhead. From this point, Shooter Number One ascends the stairs, moving backward while covering behind and above.

b. Shooter Number Two follows Shooter Number One upstairs, one step behind and to the side of Shooter Number One. When Shooter Number One turns to cover overhead, Shooter Number Two remains oriented to the front, covering directly up the stairwell.

The clearing team's speed of movement is determined by Shooter Number One. Marines pie as much of an area as possible before ascending each step.

Figure A-50. Clearing a Stairwell

Section VI

Firing Positions

Whether a unit is attacking, defending, or conducting retrograde operations, its success or failure depends on the ability of the individual Marine to place accurate fire on the enemy while providing the least exposure to return fire. Consequently, Marines should immediately seek and properly use firing positions.

26. Hasty Firing Position. A hasty firing position is one that is normally occupied in the attack or the early stages of the defense. It is a position from which a Marine can place fire on the enemy while using available cover for protection from return fire. The Marine may occupy this position voluntarily, or he may be forced to occupy it because of enemy fire. In either case, the position lacks preparation before occupation. Some of the more common hasty firing positions in a built-up area involve firing around corners of buildings, firing from behind walls, firing from windows, firing from unprepared loopholes, and firing from the peak of a roof.

a. Corners of Buildings. The corner of a building provides cover for a hasty firing position if used properly.

(1) The shooter should be capable of both right-handed and left-handed firing of his weapon (using the "short stocking" technique) to be effective around corners. Short stocking the weapon will prevent the muzzle from protruding and keep the weapon ready to fire the instant visual contact with the enemy is made. Furthermore, it reduces the Marine's exposure as a target. (Figure A-51)

(2) A common mistake made when firing around corners is firing from the standing position. The shooter exposes himself at the height the enemy would expect a target to appear and risks exposing the entire length of his body as a target for the enemy.

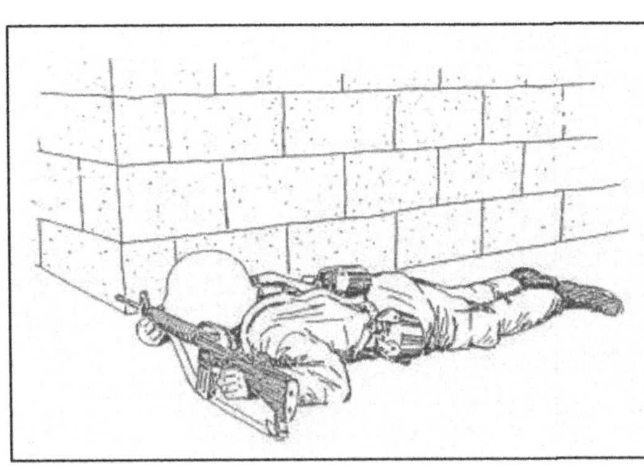

Figure A-51. Firing, Using Either Hand, Around the Corner of a Building

b. Walls. When firing from behind walls, a Marine should fire around cover—not over it (Figure A-52). Marines in this situation should be able to fire from the right or left shoulder. When firing from behind cover, a Marine should remain far enough back from that cover so that the weapon does not extend beyond it. This aids in weapon retention by ensuring that no enemy on the other side of the wall can get control of the barrel.

Figure A-52. Firing Around Cover

c. Windows. In a built-up area, windows provide convenient firing ports. A Marine should avoid firing from the standing position because it exposes most of his body to return fire from the enemy and could silhouette him against a light-colored interior beyond the window. This is an obvious indicator of the firer's position, especially at night when the muzzle flash can easily be observed. When using the proper method of firing from a window (Figure A-53), the Marine is well back into the room to prevent the muzzle flash from being seen, and he is kneeling to limit exposure and avoid silhouetting himself.

Figure A-53. Firing From a Window

d. Loopholes. A Marine may fire through a hole torn in the wall and avoid windows (Figure A-54). He stays well back from the loophole so the muzzle of the weapon does not protrude beyond the wall and the muzzle flash is concealed.

Figure A-54. Firing From a Loophole

e. Roof. The peak of a roof provides a vantage point for snipers that increases their field of vision and the ranges at which they can engage targets (Figure A-55). A chimney, a smokestack, or any other object protruding from the roof of a building can reduce the size of the exposed target and should be used.

Figure A-55. Firing From the Peak of a Roof

f. No Position Available. When a Marine is subjected to enemy fire and none of the positions mentioned are available, he should try to expose as little of himself as possible. When a Marine is in an open area between buildings (a street or alley) with no cover available and he is fired upon by an enemy from one of the buildings to his front, he should lie prone as close as possible to a building on the same side of the open area as the enemy. To engage the Marine, the enemy must then lean out of the window and expose himself to return fire.

g. No Cover Available. When no cover is available, target exposure can be reduced by firing from the prone position, by firing from shadows, and by presenting no silhouette against buildings.

27. Prepared Firing Position. A prepared firing position is one built or improved to allow the shooter to engage a particular area, avenue of approach, or enemy position, while reducing his exposure to return fire. Examples of prepared positions include: barricaded windows, fortified loopholes, sniper positions, antiarmor positions, and machine gun positions.

a. The natural firing port provided by windows can be improved by barricading the window and leaving a small hole for the firer's use (Figure A-56). The barricading may be accomplished by using materials torn from the interior walls of the building or any other available material. When barricading windows, a Marine should avoid:

(1) Barricading only the windows that will be used as firing ports. The enemy will soon determine that the barricaded windows are firing positions.

(2) Neat square or rectangular holes that are easily identified by the enemy. A barricaded window should not have a neat, regular firing port. The window should keep its original shape so that the position of the firer is hard to detect. Firing from the bottom of the window gives the firer the cover advantage of the wall because the firing port is less obvious to the enemy. Sandbags are used to reinforce the wall below the window and to increase protection for the firer. All glass should be removed from the window to prevent injury to the firer. Lace curtains permit the firer to see out and prevent the enemy from seeing in. Wet blankets should be placed under weapons to reduce dust that may create a signature. Wire mesh placed over the window keeps the enemy from throwing in hand grenades.

b. Although windows are usually good firing positions, they do not always allow the firer to engage targets throughout his sector.

(1) To avoid establishing a pattern of always firing from windows, an alternate position is required such as the prepared loophole (Figure A-57). This involves cutting or blowing a small hole in the wall to allow the shooter to observe and engage targets in his sector.

(2) Sandbags are used to reinforce the walls below, around, and above the loophole. If the position is on the second floor or higher, two layers of sandbags are placed on the floor under the firer to protect him from an explosion on a lower floor. A wall of sandbags, rubble, furniture, and so on should be constructed to the rear of the position to protect the firer from explosions in the room.

Figure A-56. Window Firing Position

(3) A table, a bedstead, or other available material provides overhead cover for the position. This prevents injury to the shooter from falling debris or explosions above his position.

(4) The position should be camouflaged by knocking other holes in the wall. This makes it difficult for the enemy to determine which hole the fire is coming from. Siding material should be removed from the building in several places to make loopholes less noticeable.

Figure A-57. Prepared Loopholes

c. A chimney or other protruding structure provides a base from which a sniper position can be prepared. Part of the roofing material is removed to allow the sniper to fire around the chimney. He should stand inside the building on the beams or on a platform with only his head and shoulders above the roof (behind the chimney). Sandbags placed on the sides of the position protect the sniper's flanks.

d. When the roof has no protruding structure to provide protection, the sniper position should be prepared from underneath on the enemy side of the roof (Figure A-58). The position is reinforced with sandbags, and a small piece of roofing material should be removed to allow the sniper to engage targets in his sector. The missing piece of roofing material should be the only sign that a position exists. Other pieces of roofing should be removed to deceive the enemy as to the true sniper position. The sniper should be invisible from outside the building, and the muzzle flash should be hidden from view.

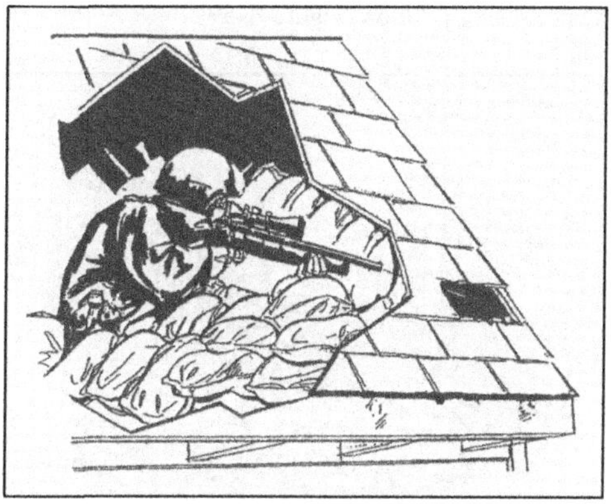

Figure A-58. Sniper Position

e. Some rules and considerations for selecting and occupying individual firing positions are:

(1) Make maximum use of available cover and concealment.

(2) Avoid firing over cover; when possible, fire around it.

(3) Avoid silhouetting against light-colored buildings, the skyline, and so on.

(4) Carefully select a new firing position before leaving an old one.

(5) Avoid setting a pattern; fire from both barricaded and unbarricaded windows.

(6) Keep exposure time to a minimum.

(7) Begin improving a hasty position immediately after occupation.

(8) Use construction material that is readily available in the built-up area for prepared positions.

(9) Remember that positions that provide cover at ground level may not provide cover on higher floors.

f. In attacking a built-up area, recoilless weapon and ATGM crews may be hampered in choosing firing positions by the backblast of their weapons. They may not have enough time to knock out walls in buildings and clear backblast areas. They should select positions that allow the backblast to escape, such as corner windows where the round fired goes out one window and the backblast escapes from another. The corner of a building can be improved with sandbags to create a firing position (Figure A-59).

Figure A-59. Corner Firing Position

g. The rifle squad is often reinforced with attached antitank weapons during an attack on and in defense of a built-up area. Therefore, the rifle squad leader should be able to choose good firing positions for the antitank weapons under his control.

h. Various principles for employing antitank weapons have universal applications, such as making maximum use of available cover, trying to achieve mutual support, and allowing for the backblast when positioning recoilless weapons, TOWs, Dragons, and AT4s.

i. Operating in a built-up area presents new considerations. Marines should select numerous alternate positions, particularly when the structure does not provide cover from small-arms fire. They should position their weapons in the shadows and within the building.

j. Recoilless weapons and ATGMs firing from the top of a building can use the chimney for cover (Figure A-60). The rear of this position should be reinforced with sandbags.

Figure A-60. Antitank Weapon Firing From a Rooftop

k. When selecting firing positions for recoilless weapons and ATGMs, Marines should make maximum use of rubble, corners of buildings, and destroyed vehicles to provide cover for the crew. Recoilless weapons and ATGMs can also be moved along rooftops to obtain a better firing angle from which to engage enemy armor. When buildings are elevated off the ground, positions can be prepared using a building for overhead cover (Figure A-61). The backblast under the building must not damage or collapse the building or injure the crew.

Note: When firing from a slope, ensure that the angle of the launcher relative to the ground or firing platform is not greater than 20 degrees. When firing within a building, ensure that the enclosure is at least 10 feet by 15 feet, is clear of debris and loose objects, and has windows, doors, or holes in the walls to allow the backblast to escape.

l. The machine gun can be emplaced almost anywhere. In the attack, windows and doors offer ready-made firing ports (Figure A-62). For this reason, the enemy normally has windows and doors under observation and fire; they should therefore be avoided. Any openings in walls that were created during the fighting may be used. When other holes are not present, small explosive charges can create loopholes (Figure A-63). Regardless of what openings are used, machine guns should be sited within the building and in the shadows.

Figure A-61. Prepared Positions Using a Building for Overhead Cover

Figure A-62. Emplacement of a Machine Gun in a Doorway

m. Upon occupying a building, Marines board up all windows and doors. By leaving small gaps between the slots, Marines can use windows and doors as good alternate firing positions.

n. Loopholes should be used extensively in the defense. They should not be constructed in any logical pattern, nor should they all be at floor or tabletop level. Varying their height and location makes them hard to pinpoint and identify. Dummy loopholes, knocked-off shingles, or holes that are cut but not intended to be used as firing positions aid in deception. Loopholes located behind shrubbery, under doorjambs, and under the eaves of a building are hard to detect.

Figure A-63. Use of a Loophole With a Machine Gun

o. Increased fields of fire can be obtained by locating the machine gun in the corner of the building or sandbagging it under a building (Figure A-64). Available materials, such as desks, overstuffed chairs, couches, and other items of furniture, should be integrated into the construction of bunkers to add both cover and concealment (Figure A-65).

Figure A-64. Sandbagged Machine Gun Emplacement Under a Building

p. Although grazing fire is desirable when employing the machine gun, it may not always be practical or possible. Where destroyed vehicles, rubble, and other obstructions restrict the fields of grazing fire, the gun can be elevated to a position from which it can deliver plunging fire over obstacles. Firing from loopholes on the second or third story may be necessary. A firing platform can be built under the roof (Figure A-66) and a loophole constructed. Again, the exact location of the position should be concealed by knocking off shingles in isolated patches over the entire roof.

Figure A-65. Corner Machine Gun Bunker

Figure A-66. Firing Platform Built Under a Roof

28. Target Acquisition. Built-up areas provide unique challenges to acquiring targets. Buildings mask movement and the effects of direct and indirect fires. Also, the rubble from destroyed buildings provides concealment and protection for attackers and defenders alike, making target acquisition difficult.

a. The techniques of patrolling and using OPs apply in the city as well as in wooded terrain. These techniques enable units to locate the enemy, to develop targets for direct and indirect fires in the defense, and to find uncovered avenues of approach in the offense.

b. Most weapons and vehicles produce recognizable signatures. These come from design features or from the environment in which the equipment is used. For example, firing a tank main gun in dry, dusty, and debris-covered streets raises a dust cloud; a tank being driven in built-up areas produces more noise than one moving through an open field; people moving through rubble on a street or in the halls of a damaged building create more noise than in a wooded area. Marines should learn to recognize signatures so they can locate and identify targets. Seeing and hearing assist in detecting and identifying signatures that lead to target location, identification, and rapid engagement. Marines should look for targets in areas where they are most likely to be employed.

c. Target acquisition should be continuous, whether halted or moving. Built-up areas provide both the attacker and defender with good cover and concealment but usually favor the defender. This makes rapid and accurate target acquisition extremely important.

d. When a unit is moving and enemy contact is likely, the unit should employ an overwatching element. This principle applies in built-up areas as it does in other kinds of terrain except that in built-up areas the overwatching element should observe both the upper floors of buildings and the street level.

e. Stealth should be used when moving in built-up areas because little distance separates attackers from defenders. Only hand and arm signals should be used until contact is made. The unit should stop periodically to listen and observe to ensure that it is not being followed or that the enemy is not moving parallel to the unit's flank for an ambush. Routes should be carefully chosen so that buildings and piles of rubble can be used to mask the unit's movement.

f. Observation duties should be clearly issued to squad members to ensure all-around security as they move. This security continues at the halt. All of the senses should be used to acquire targets, especially hearing and smelling. Marines soon recognize the sounds of vehicles and people moving through streets that are littered with rubble. The smell of fuel, cologne, or food cooking can disclose enemy positions.

g. OPs are "positions from which military observations are made, or fire directed and adjusted, and which possess appropriate communications" (Joint Pub 1-02). They are positions from which Marines can watch and listen to enemy activity in a specific sector. They warn the unit of enemy approach and should be positioned in the upper floors of buildings to give Marines a better vantage point than they would have at street level.

h. In the defense, a platoon leader positions OPs for local security as ordered by the company commander. The platoon leader selects the general location, but the squad leader establishes the OP (Figure A-67). Normally, there is at least one OP per platoon. An OP consists of two to four men and is within small-arms supporting range of the platoon. Leaders look for positions that allow good observation of the target sector. Ideally, an OP has a field of observation that overlaps those of adjacent OPs. The position selected for the OP should provide cover and concealment for units moving to and from the OP. The upper floors of houses or other buildings should be used. The squad leader should not routinely select obvious positions such as water towers or church steeples because those positions naturally attract the enemy's attention.

A SQUAD LEADER IS GIVEN THE GENERAL LOCATION OF THE OP BY THE PLATOON LEADER. THE SQUAD LEADER SELECTS THE EXACT POSITION.

LEADERS LOOK FOR POSITIONS THAT:

- ALLOW GOOD OBSERVATION OF THE DESIRED AREA OR SECTOR. (IDEALLY, AN OP HAS A FIELD OF OBSERVATION THAT OVERLAPS THOSE OF ADJACENT OPs.)
- PROVIDE COVER AND CONCEALMENT. (GOOD OBSERVATION OF THE SECTOR MAY REQUIRE THE OP TO ACCEPT LESS COVER AND CONCEALMENT AND REQUIRE TROOPS TO SELECTIVELY CLEAR FIELDS OF OBSERVATION.)

Figure A-67. Selection of an OP Location

i. The Marine should be taught how to scan a target area from OPs or from fighting positions. Use of proper scanning techniques enables observers to quickly locate and identify targets. The Marine searches quickly without optics for obvious targets, using all of his senses to detect target signatures. If no targets are found and time permits, he makes a more detailed search (using binoculars, if available) of the terrain in the assigned sector using the 50-meter method. First, he searches a strip 50 meters deep from right to left. He then searches a strip from left to right that is farther out, overlapping the first strip. This process is continued until the entire sector has been searched. In the city core or core periphery where the observer is faced with multistory buildings, the overlapping sectors may extend up rather than out.

j. Marines who man OPs and other positions should employ target acquisition devices. These devices include binoculars, image intensification devices, and thermal sights. All of these devices can enhance the unit's ability to detect and engage targets, especially at night or during periods of reduced visibility.

k. Target acquisition techniques used at night are similar to those used during the day. At night, whether using daylight optics or the unaided eye, the observer should not look directly at an object but a few degrees off to the side. The side of the eye is more sensitive to dim light. When scanning with off-center vision, the Marine moves his eyes in short, abrupt, irregular movements. At each likely target area, he pauses for a few seconds to detect any motion.

l. At night, sounds and smells can aid in acquiring targets. They transmit better in the cooler, damper night air. Running engines, vehicles, and people moving through rubble- covered streets can be heard for great distances. Odors from diesel fuel, gasoline, cooking food, burning tobacco and aftershave lotion can reveal enemy and friendly locations.

29. Firefighting Planning and Operations. Incendiary ammunition, special weapons, and the ease with which incendiary devices can be constructed from gasoline and other flammables make fire a true threat in operations in built-up areas. During defensive operations, firefighting should be a primary concern. Proper steps should be taken to reduce the risk of a fire that could make a chosen position indefensible.

a. Marines choose or create positions that do not have large openings. These positions provide as much built-in cover as possible to prevent penetration by incendiary ammunition. All unnecessary flammable materials are removed, including ammunition boxes, furniture, rugs, newspapers, curtains, and so on. The electricity and gas coming into the building should be shut off.

b. A building of concrete block construction with concrete floors and a tin roof is an ideal place for a fighting position. However, most buildings have wooden floors or subfloors, wooden rafters, and wooden inner walls that may require improvement. Inner walls are removed and replaced with blankets to resemble walls from the outside. Sand is spread 2 inches deep on floors and in attics to retard fire.

c. All available firefighting gear is prepositioned so that it can be used during combat. For the individual Marine, such gear includes entrenching tools, helmets, sand, and blankets. These items are supplemented with fire extinguishers.

d. Fire is so destructive that it can easily overwhelm personnel regardless of extraordinary precautions. Marines should plan routes of withdrawal so that evacuation can be accomplished from their fighting positions. This allows Marines to exit through areas that are free of combustible material but that still provide cover from enemy direct fire.

e. The confined space and large amounts of combustible material found in built-up areas may influence the enemy to use incendiary devices. First-aid problems that are expected include burns and smoke/flame inhalation. These can easily occur in buildings, thereby rendering the victims combat ineffective. Although there is little defense against flame inhalation and lack of oxygen, smoke inhalation may be greatly reduced by wearing the individual protective mask. Regardless of the fire hazard, defensive planning for combat in built-up areas should include location of corpsmen. Corpsmen must reach victims and their equipment and should stock extra supplies for the treatment of burns and inhalation injuries.

f. Offensive operations also require plans for firefighting because the success of the mission can easily be threatened by fire. Poorly planned use of incendiary munitions can make fires so extensive that they become obstacles to offensive operations. The enemy may use fire to cover his withdrawal and to create obstacles and barriers to the attacker.

g. Every Marine participating in an attack should be ready to deal with the problems of fire. The normal firefighting equipment available includes the entrenching tool, helmet (for carrying sand or water), and blankets (for snuffing out small fires). Fire extinguishers are normally available on vehicles.

30. Employment of Snipers. The value of the sniper to a unit operating in a built-up area depends on several factors. These factors include the type of operation, the level of conflict, and the ROE. Where ROE allow destruction, snipers may not be needed because other weapons systems available to the infantry commander have a greater destructive effect. However, snipers can contribute to the fight. Where the ROE prohibit collateral damage, snipers may be the most valuable tool the commander has. (See MCWP 3-15.3 (under development).)

a. Sniper effectiveness depends in part on the terrain. Control is decentralized by the characteristics of an urban area. To provide timely and effective support, the sniper should have a clear picture of the commander's mission and intent.

b. Where possible, snipers should be positioned in masonry buildings. These buildings should also offer long-range fields of fire and all-around observation. The sniper has an advantage because he does not have to move with or be positioned with lead elements. He may occupy a higher position to the rear or flanks of, and some distance away from, the element that he is supporting. By operating far from the other elements, a sniper avoids decisive engagement but remains close enough to kill distant targets that threaten the unit. Snipers should not be placed in obvious positions such as church steeples and rooftops because the enemy often watches these and targets them for destruction. Also, snipers should not be positioned where there is heavy traffic; these areas also invite enemy observation.

c. Snipers should operate throughout the AO, moving with and supporting the companies as necessary. Some teams may be designated to operate independently of other forces. They may be tasked to search for targets of opportunity, especially for enemy snipers. The sniper team may occupy multiple positions. A single position may not afford adequate observation for the entire team without increasing the risk of detection by the enemy. Separate positions should maintain mutual support. Alternate and supplementary positions should also be established in urban areas.

d. Snipers may be assigned the following tasks:

- Killing enemy snipers (countersniper fire)

- Killing targets of opportunity (These targets should be prioritized by the commander. Types of targets might include enemy snipers, leaders, vehicle commanders, radio men, sappers, and machine gun crews.)

- Denying enemy access to certain areas or avenues of approach (controlling key terrain)

- Providing fire support for covering barricades and other obstacles

- Maintaining surveillance of flank and rear avenues of approach (screening)

- Supporting local counterattacks with precision fire.

Section VII

Advancing/Patrolling Along City Streets

31. Advancing/Patrolling Along City Streets. During MOUT operations, skills for advancing/patrolling along city streets will be required. The platoon, squad, or fire team mission may be such that building-to-building clearing is neither required nor desired until contact with the enemy has been made. This section will discuss advancing/patrolling along urban streets. The standard tactics, techniques, and procedures for the conduct of a patrol, as outlined in MCWP 3-11.6 *Scouting and Patrolling* (draft), apply except as otherwise discussed within this section.

a. Organization. The platoon's three-squad organization or the squad's three-fire team organization is already structured to conduct an advance/patrol down an urban street. The typical patrol is organized into assault, support, security, and command and control elements. Security is divided into front, flank, and rear security responsibilities. The platoon or squad can move down the street in a double or single column.

b. Formation and Movement

(1) Double Column. The double column is the *preferred method* for moving along an urban street (Figures A-68 and A-69). The double column provides 360 degrees of security and allows for mutual support and interlocking sectors of fire. The double column is flexible but can make control more difficult. There should be at least two squads or fire teams to conduct a double column. The two squads or fire teams move down the sides of the street using the walls of the buildings for cover and concealment. The pointmen for the two columns should always stay abreast of each other and not move in front of or behind the opposite column. The Marines following in trace of the pointmen stagger themselves accordingly. The platoon or squad leaders move where they can best control their units. If a squad is conducting the advance/patrol along the street, the trail fire team is never split.

(a) Sectors of Responsibility. Every Marine moving in a column has a particular sector for whose security he is responsible. The pointmen are responsible for security to the direct front and for pieing off windows and doorways. They are also responsible for staying in line with one another.

The Marines immediately behind the pointmen are designated as the covermen. They are responsible for providing guidance to their respective pointmen from the leader of the unit and for covering the 45-degree oblique angle across from one another. The second coverman is also responsible for covering the direct front when the pointman is pieing and checking for the enemy. The third and fourth covermen in the column are responsible for the far-side flank security; one covers the upper level and one covers the lower level building windows, doors, and stairwells. Marines will alternate providing security on upper level and lower level windows, doors, and stairwells. The fourth Marine also provides rear security if he is the last man in the column or interlocks his sector of fire with that of the lead Marine in the next fire team in the column. Should the second coverman have to take the point while the pointman pies a window or door, the third coverman would automatically shift his sector of fire to the 45-degree oblique angle. If the second coverman has to assist the pointman in pieing a large opening, the third coverman would provide security to the direct front while the fourth coverman would take the 45-degree oblique angle. Marines following in the column would adjust their sectors of fire accordingly.

The two columns provide mutual support for each other and the greatest all-around security. Teamwork and coordination between the two columns are vital.

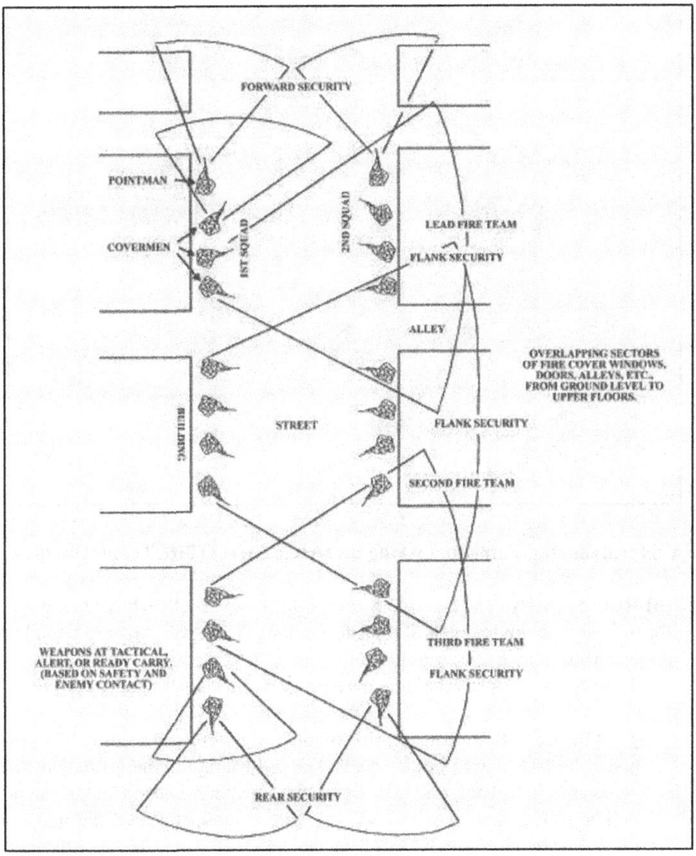

Figure A-68. Advancing/Patrolling Along an Urban Street (Squad, Double Column)

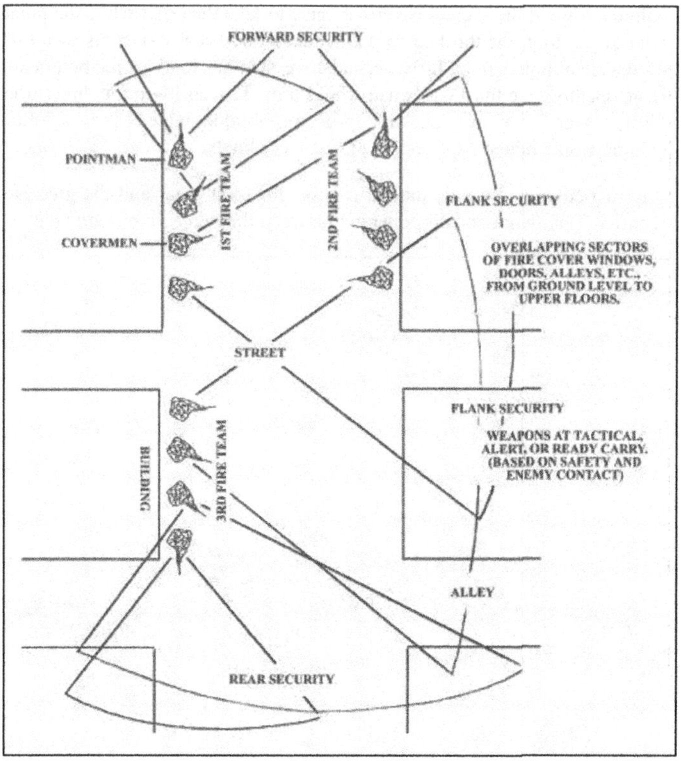

Figure A-69. Advancing/Patrolling Along an Urban Street (Fire Team, Double Column)

(b) Rear Security. The last two Marines in the squad of each column are responsible for the rear security of each of their formations. They will move in bounds in order to allow a Marine to always provide active security to the squad's rear (Figures A-70 and A-71). Two methods of bounding may be used: *alternate* or *successive bounding*.

1 **Alternate Bounding.** In this method, the two rear-security Marines alternate being the last Marine in the squad. The last Marine in the formation will have his weapon at the ready and will be in the kneeling position along the wall of the building. The second to last Marine will move in trace of the unit until he has moved approximately 10 to 15 meters. At this point, he will step out one pace from the wall. He will then assume a good kneeling firing position and announce, "**SET**." The last Marine will rise from his position, turn around, and move next to the wall to catch up with the column. It is important that the Marine moving back

does not step out and mask the other Marine's field of fire. At this point, the Marine covering the rearward movement of the last Marine can assume the rear guard position and move back to the wall as the Marine passes by his position (Figure A-70). The process is repeated continuously. The rear fire team leaders coordinate the movements of the rear guard to keep them abreast of each other and ensure that the column does not move too fast and leave the rear guard behind.

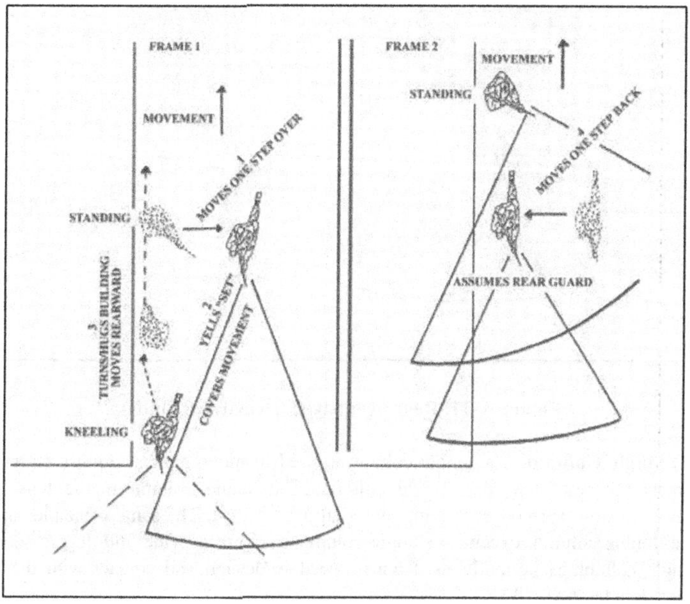

Figure A-70. Rear Security (Alternate Bounds)

2 **Successive Bounding.** In this method, the last Marine in the squad continues to be the rear guard after bounding. The same procedures are employed as in alternate bounding; however, the rear-guard Marine bounds abreast of the Marine providing cover, turns around, drops to a good kneeling position, provides rear-guard security, and announces, "**SET**." The covering Marine moves behind the rear-guard Marine and continues to move 10 to 15 meters before repeating the process (Figure A-71). The rear fire team leaders coordinate the movements of the rear guard as in alternate bounding.

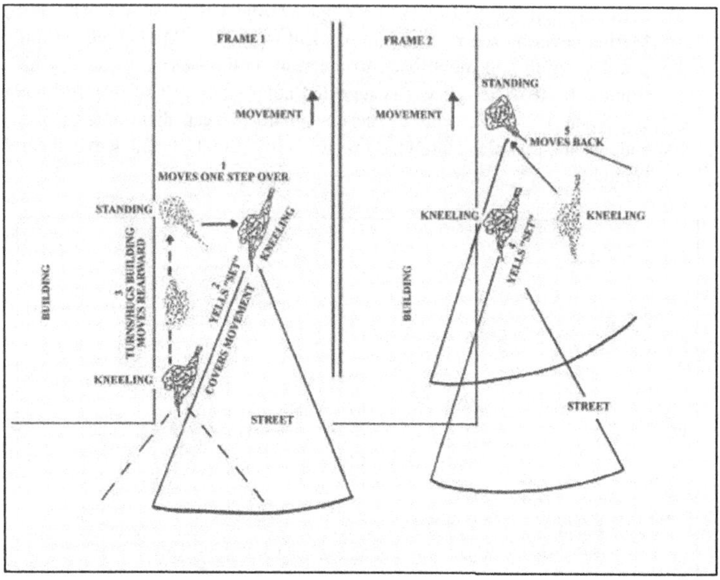

Figure A-71. Rear Security (Successive Bounds)

(2) Single Column. The single column is used to move rapidly along a street or when streets are restricted to a single column. The single column is the least preferred advance/patrol method of moving along an urban street. The same principles apply as in the double column, except the single column does not provide 360-degree security. The single column is normally used when speed is desired and contact with the enemy is considered to be unlikely.

32. Crossing a Street Intersection. Four-way intersections are very dangerous areas that should be crossed quickly. Security at an intersection is extremely important for the protection of crossing units. When moving, the pointmen notify the unit leaders, through the covermen, of an approaching intersection. Crossing the intersection will force the three elements of a unit moving along a city street to rotate positions and assume appropriate responsibilities.

a. First Fire Team To Cross. The lead (1st) fire teams in the two columns will simultaneously set security around the corners of the building. They will provide mutually supporting security to the direct front, security to the building on the far side of the intersection, and intersection security (Figure A-72). The middle (2nd) fire teams provide mutually supporting flank and rear security. On order, the rear (3rd) fire teams move forward, cross the intersection, and assume the lead (1st) fire teams' forward security mission.

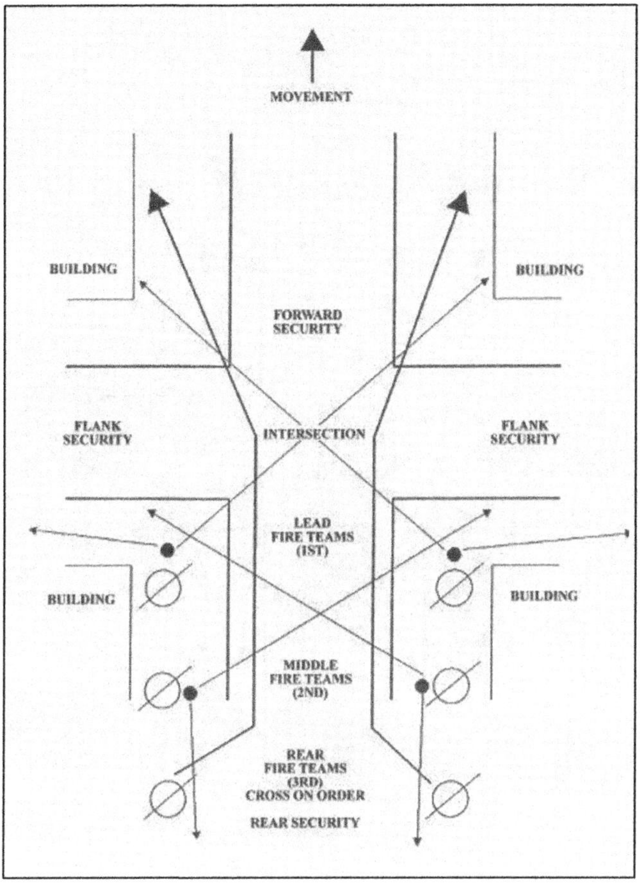

Figure A-72. Crossing an Intersection (Frame 1)

b. Second Fire Team to Cross. The 1st fire teams (now the middle fire teams) continue to provide intersection security and assume rear area security. The 2nd fire teams (now rear fire teams) cross the intersection on order. The second fire teams position themselves to provide rear and intersection security to support the covering of the last fire teams (Figure A-73).

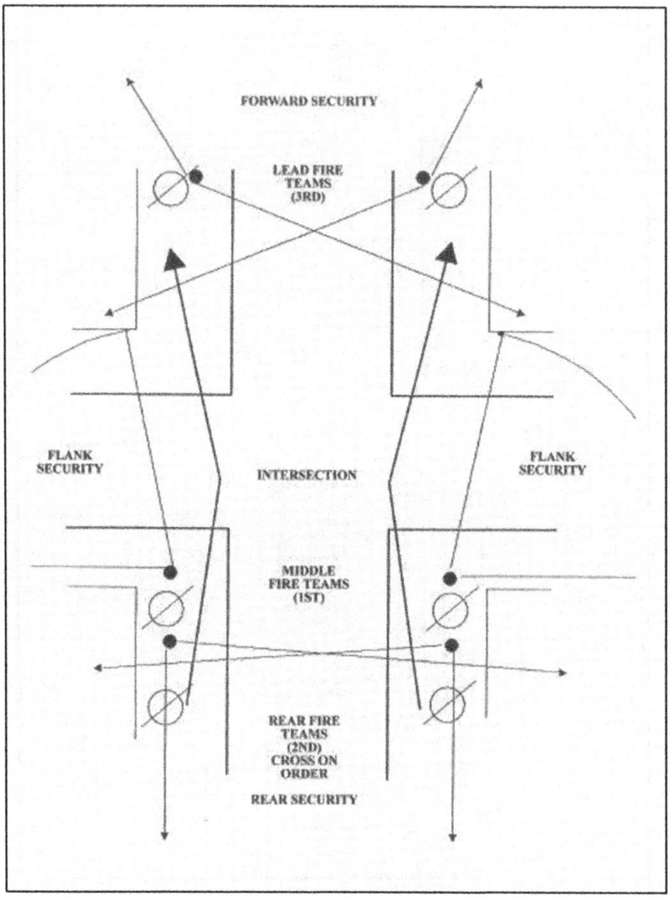

Figure A-73. Crossing an Intersection (Frame 2)

c. Third Fire Team To Cross. The 1st fire teams (now the rear fire teams) provide rear security and cross the intersection on order. The 1st fire teams bypass the 2nd fire teams to become the middle fire teams and provide mutually supporting forward and flank security. The 2nd fire teams become the rear security (Figure A-74).

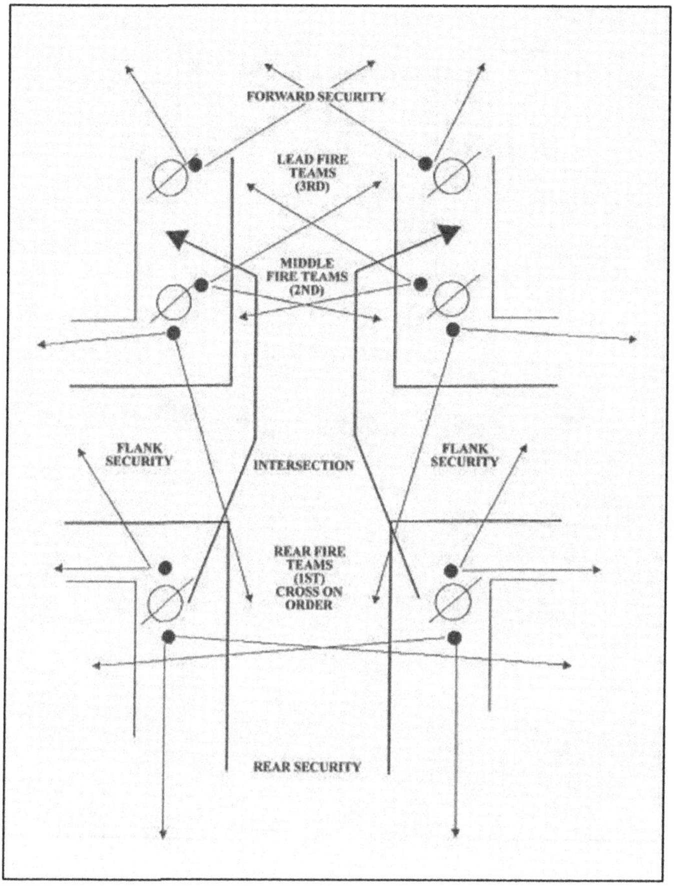

Figure A-74. Crossing an Intersection (Frame 3)

d. The squads and three fire teams have now crossed the intersection, and the fire teams have rotated (Figure A-75). This rotation is convenient because it alternates the lead units, who may quickly become fatigued during MOUT operations. The same procedures are used to cross an intersection with a single column.

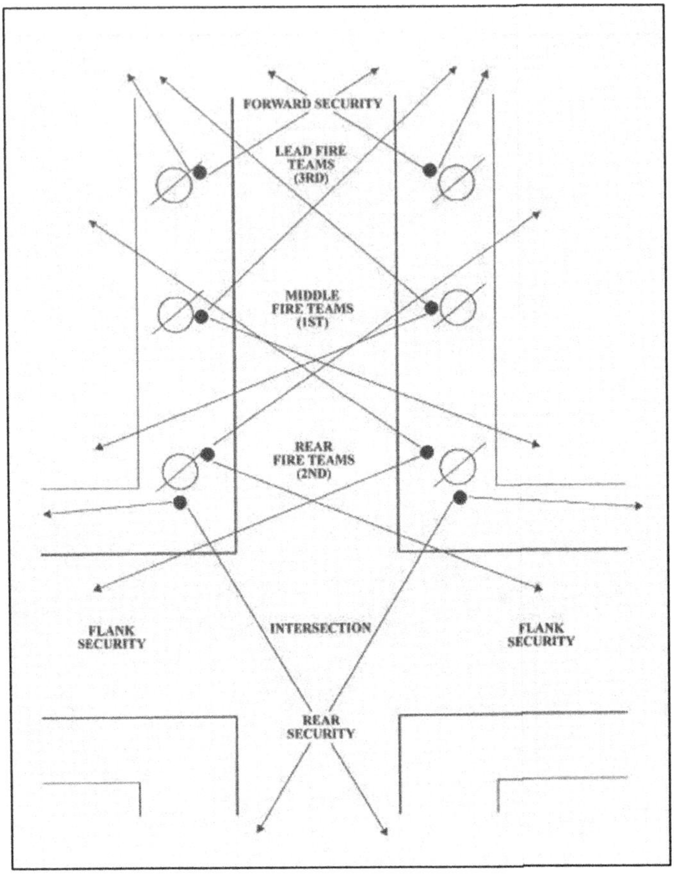

Figure A-75. Crossing an Intersection (Frame 4)

33. Reacting to Enemy Contact. When a platoon or squad is moving along a city street, contact with the enemy could happen at any time, even if the area has been declared secure. If a unit engages or is engaged by the enemy, there are two basic options that Marines may follow. The first and most preferred option is to immediately return fire and conduct a hasty clearing of a structure to seek cover inside of the building to which the unit was adjacent. The second and least preferred method is to remain outside of the building(s) and fight from the street.

a. Hasty Clearing. To conduct a hasty clearing of a structure in order to seek cover, Marines immediately return fire and enter buildings as quickly as possible. The basic principles of entry are applied as discussed earlier in the appendix; however, speed of action is essential. Once inside, Marines quickly scan the area and engage any threat. If no threat is present, the Marines acquire the enemy's positions and deliver carefully aimed shots to achieve fire superiority. At this point, the situation is developed as rapidly as possible and an attack to clear the enemy may be ordered.

b. Remain Outside and Fight. This option should not be conducted unless strict ROE do not allow the occupation of buildings without being fired on from them first or unless buildings cannot be entered because of obstruction, obstacles, or boobytraps. In such cases, the Marines should quickly acquire the target(s) and return fire to gain fire superiority. They should also seek whatever cover may exist. Once fire superiority is achieved, an assault to clear the enemy may be conducted.

Section VIII

Navigation in Built-Up Areas

Navigation in built-up areas present a unique set of challenges. Deep in the city core, the normal terrain features depicted on maps may not apply—buildings become the major terrain features, and units become tied to streets. Fighting in the city destroys buildings, whose rubble then blocks streets. Street and road signs are destroyed during the fighting if they are not switched or removed by the defenders. Operations in subways and sewers present other unique challenges. However, maps and photographs are available to help units overcome these problems. Additionally, the global positioning system (GPS) can help supplement navigational abilities in built-up areas.

34. Military Maps. The military city map is a topographical map of a city that delineates streets and shows street names, important buildings, and other urban elements. The scale of a city map can vary from 1:25,000 to 1:50,000 to 1:100,000 depending on the importance and size of the city, the density of detail, and intelligence information available.

a. Special maps prepared by supporting topographical engineers can assist units in navigating in built-up areas. These maps are designed or modified to provide information that is not covered on a standard military map. This may include maps of road and bridge networks, railroads, built-up areas, and electric power fields. Special maps can be used to supplement military city maps and topographical maps.

b. Once in the built-up area, Marines may use street intersections as reference points as they would use hills and streams in rural terrain. City maps supplement or replace topographical maps as the basis of navigation. These maps enable units moving in the built-up area to know where they are and to move to new locations even though streets have been blocked or a key building has been destroyed.

c. The techniques of compass reading and pace counting can still be used, especially in a blacked-out city where street signs and buildings are not visible. The presence of steel and iron in the MOUT environment may cause inaccurate compass readings. Sewers may be navigated in much the same way. Maps providing the basic layout of the sewer system are maintained by city sewer departments. This information includes directions in which the sewer lines run and distances between manhole covers. Along with basic compass and pace-count techniques, such information enables a unit to move through the city sewers with accuracy.

d. Operations in a built-up area adversely affect the performance of some types of communications-electronic devices such as the GPS, the Position Location Reporting System (PLRS), or other data-distribution systems. These systems function the same as some communications equipment—by LOS. They cannot determine underground locations or positions within a building. These systems should be employed on the tops of buildings, in open areas, and down streets where obstacles will not affect LOS readings.

e. City utility workers are assets to units fighting in built-up areas. They can provide maps of sewers and electrical fields and information about the city. This is especially important with regard to the use of the sewers. Sewers can contain pockets of methane gas that are highly toxic to humans. City sewer workers know the locations of these danger areas and can advise a unit on how to avoid them.

35. Aerial Photographs. Current aerial photographs are also excellent supplements to military city maps and can be substituted for a map. A topographical map or military city map could be obsolete if compiled many years ago. A recent aerial photograph shows changes that have taken place since the map was made. This could include destroyed buildings and streets that have been blocked by rubble as well as enemy defensive preparations. More information can be gained by using aerial photographs and maps together than by using either one alone.

Section IX

Camouflage

To survive and win in combat in built-up areas, a unit should supplement cover and concealment with camouflage. To properly camouflage men, carriers, and equipment, Marines should study the surrounding area and make their fighting positions blend with the local terrain.

36. Application. Only the proper amount of material needed for camouflaging a position should be used. Excess material could reveal the position. Material should be obtained from a wide area around the position. For example, Marines defending a cinder-block building do not strip the front, sides, or rear of the building to camouflage a position.

a. Buildings provide numerous opportunities for establishing concealed positions. Armored vehicles can often find prospective positions under archways or inside small industrial or commercial structures. Thick masonry, stone, or brick walls offer excellent protection from direct fire and may provide concealed routes into or out of the position.

b. After camouflage is completed, the Marine inspects his position from the enemy's viewpoint. He makes routine checks to see whether the camouflage remains natural looking and actually conceals the position. If it does not look natural, the Marine should rearrange or replace it until a natural look is achieved.

c. Positions should be progressively camouflaged as they are prepared. Work should continue until all camouflage is complete. If the enemy has air superiority, work may be possible only at night. Shiny or light-colored objects that attract attention from the air should be hidden.

d. Camouflage face paint is issued in three standard, two-tone sticks. When issue-type face-paint sticks are not available, burnt cork, charcoal, or lampblack can be used to tone down exposed skin. Mud may be used as a last resort because it dries and may peel off, leaving the skin exposed, and may also contain harmful bacteria.

37. Use of Shadows. Buildings throw sharp shadows that can be used to conceal vehicles and equipment (Figure A-76). Marines should avoid movement or positions in areas that are not in shadows. Vehicles may have to be moved periodically as shadows shift during the day. Emplacements inside buildings provide better concealment and cover as well.

38. Color and Texture. Standard camouflage-pattern painting of equipment is not as effective in built-up areas as a solid, dull, dark color hidden in shadows. Because repainting vehicles before entering a built-up area is not always practical, the lighter sand-colored patterns should be subdued with mud or dirt.

Figure A-76. Use of Shadows for Concealment

a. The need to break up the silhouette of helmets and individual equipment exists in built-up areas the same as it does elsewhere. However, burlap or canvas strips are a more effective camouflage than foliage (Figure A-77). Predominant colors are normally browns, tans, and sometimes grays rather than greens, but each camouflage location should be evaluated to determine ideal colors and patterns.

Figure A-77. Helmet Camouflaged With Burlap Strips

b. CPs and logistics emplacements are easier to camouflage and better protected if located underground. Antennas can be remoted to upper stories or to higher buildings based on remote capabilities. Field telephone wire should be laid in conduits, in sewers, or through buildings.

c. Marines should consider the background to ensure that they are not silhouetted or skylined, but rather blend into their surroundings. To defeat enemy urban camouflage, Marines should be alert for common camouflage errors such as:

- Tracks or other evidence of activity
- Shine or shadows
- Unnatural color or texture
- Muzzle flash, smoke, or dust
- Unnatural sounds and smells
- Movement.

d. Dummy positions can be used effectively to distract the enemy and make him reveal his position by firing on them.

e. Built-up areas afford cover, resources for camouflage, and locations for concealment. Following are the basic rules of cover, camouflage, and concealment to which Marines should adhere:

(1) Use the terrain and alter camouflage habits to suit your surroundings.

(2) Employ deceptive camouflage of buildings.

(3) Continue to improve positions. Reinforce fighting positions with sandbags or other fragment- and blast-absorbent material.

(4) Maintain the natural look of the area.

(5) Keep positions hidden by clearing away minimal debris for fields of fire.

(6) Choose firing ports in inconspicuous spots when available.

Note: Remember that a force that *covers* and *conceals* itself has a significant advantage over a force that does not.

Appendix B

Employment and Effects of Weapons

Success against the enemy in urban operations to a large extent depends on the effective employment of weapons. Marines at all levels must understand the effects that their weapons will produce in urban combat.

1. Effectiveness of Weapons and Demolitions. The characteristics and nature of combat in built-up areas affect the results and employment of weapons. Leaders at all levels must consider the following factors in various combinations when choosing weapons to employ.

a. Hard, smooth, flat surfaces are characteristic of urban targets. Rounds rarely impact perpendicular to these flat surfaces, but rather tend to impact at some angle of obliquity. This reduces the effect of a round and increases the threat of ricochets. The tendency of rounds to strike glancing blows against hard surfaces means that up to 25 percent of impact-fuzed explosive rounds may not detonate when fired into rubbled areas.

b. Engagement ranges in MOUT are close. Studies and historical analyses have shown that only 5 percent of all targets are more than 100 meters away. About 90 percent of all targets are located 50 meters or less from the identifying Marine. Few personnel targets will be visible beyond 50 meters, and most occur at 35 meters or less. Minimum arming ranges and unit safety from backblast or fragmentation effects must be considered.

c. Engagement times are short because of the close fight. Enemy personnel present only fleeting targets. Enemy-held buildings or structures are normally covered by mutually supporting fires and often cannot be engaged with deliberate, well-aimed shots without prolonging exposure to their observation and weapons.

d. Depression and elevation limits for some weapons create dead space. Tall buildings form deep canyons that are often safe from indirect fires. Some weapons, such as the M203, can fire rounds to ricochet behind cover and inflict casualties. Target engagement from oblique angles, both horizontal and vertical, demands superior marksmanship skills.

e. Smoke from burning buildings, dust from explosions, shadows from tall buildings, and the lack of light penetrating inner rooms all combine to reduce visibility and to increase a sense of isolation. Added to this is the masking of fires caused by rubble and manmade structures. Targets, even at close range, tend to be indistinct.

f. Urban fighting often leads to confused melees with several small units attacking on converging axes. The risks from friendly fires, ricochets, and fratricide must be considered during the planning phase of operations, and control measures must be continually adjusted to lower these risks. Marines and leaders must maintain a sense of situational awareness and clearly mark their progress in accordance with unit SOPs to avoid fratricide.

g. The firer and target may each be inside or outside the same or separate buildings. The enclosed nature of combat in built-up areas means that a weapon's signature effect, such as a muzzle blast or backblast, must be considered, as well as the round's impact on the target.

h. Usually the manmade structure that is protecting the enemy must be attacked before enemy personnel inside can be engaged. Therefore, weapons and demolitions can be chosen for employment based on their effects against masonry and concrete rather than against enemy personnel.

i. Modern engineering and design improvements mean that most large buildings constructed since World War II are resilient to the blast effects of bomb and artillery attack. Even though modern buildings may burn easily, they often retain their structural integrity and remain standing. Once high-rise buildings burn out, they are still useful to the military and are almost impossible to damage further. A large structure can take 24 to 48 hours to burn out and become cool enough for people to enter.

j. The most common worldwide building type is the 12- to 24-inch brick building. Table B-1 lists the frequency of occurrence of building types worldwide.

Type of Building	Frequency of Occurrence (Percentage)
30-inch stone	1
8- to 10-inch reinforced concrete	6
12- to 24-inch brick	63
6-inch wood	16
14-inch steel and concrete (heavy clad)	2
7-inch steel and concrete (light clad)	12

Table B-1. Types of Buildings and Frequency of Occurrence

2. M16 Rifle and M249 Squad Automatic Weapon. The M16A2 rifle and the M249 squad automatic weapon are the most common weapons used in combat in built-up areas. The M16A2 rifle and the M249 are used to kill enemy personnel, to suppress enemy fire and observation, and to penetrate light cover. Leaders can use 5.56-mm tracer fire to designate targets for other weapons.

a. Employment. Close combat is the predominant characteristic of urban engagements. Marine riflemen must be able to hit small, fleeting targets from bunker apertures, windows, and loopholes. This requires pinpoint accuracy with weapons fired in the semiautomatic mode. Killing an enemy through an 8-inch loophole at a range of 50 meters is a challenge, but one that may be common in combat in built-up areas.

(1) When fighting and engaging targets inside buildings, the weapons handling and firing techniques discussed in Appendix A are used.

(2) Within built-up areas, burning debris, reduced ambient light, strong shadow patterns of varying density, and smoke all limit the effectiveness of night vision and sighting devices. Aiming stakes and noise/light boobytraps coupled with night vision goggles and firing techniques can be used in the defense. During the offense, night vision goggles and illumination munitions can be used. Any Marine using night vision goggles should be teamed with at least one Marine not wearing them.

b. Weapon Penetration. The penetration that can be achieved with a 5.56-mm round depends on the range to the target and the type of material being fired against. Single 5.56-mm rounds are not effective against structural materials (as opposed to partitions) when fired at close range—the closer the range, the less the penetration.

(1) For the 5.56-mm round, maximum penetration occurs at 200 meters. At ranges of less than 25 meters, penetration is greatly reduced. At 10 meters, penetration by the M16 round is poor as a result of the tremendous stress placed on this high-speed round, which causes it to yaw upon striking a target. Stress causes the projectile to break up, and the resulting fragments are often too small to penetrate.

(2) Even with reduced penetration at short ranges, interior walls made of thin wood paneling, sheetrock, or plaster offer no protection against 5.56-mm rounds. Common office furniture such as desks and chairs cannot stop these rounds, but a layer of books 18 to 24 inches thick can.

(3) Wooden-framed buildings and single cinder-block walls offer little protection from 5.56-mm rounds. When clearing such structures, Marines must ensure that friendly casualties do not result from rounds passing through walls, floors, or ceilings.

(4) Armor-piercing rounds are slightly more effective than ball ammunition in penetrating urban targets at all ranges. They are also more likely to ricochet than ball ammunition, especially when the target presents a high degree of obliquity.

c. Protection. The following common barriers in built-up areas stop a 5.56-mm round fired at less than 50 meters:

- One thickness of sandbags
- A 2-inch concrete wall (unreinforced)
- A 55-gallon drum filled with water or sand
- A small ammunition can filled with sand

- A cinder block filled with sand (block will probably shatter)
- A plate-glass windowpane at a 45-degree angle (glass fragments will be thrown behind the glass)
- A brick veneer
- A car body (5.56-mm round will penetrate but normally not exit).

d. Wall Penetration. Although most structural materials repel single 5.56-mm rounds, continued and concentrated firing can breach some typical urban structures (See Table B-2).

Type	Penetration	Rounds Required
8-inch reinforced concrete	Initial Loophole	35 250
14-inch triple brick	Initial Loophole	90 160
12-inch cinder block with single-brick veneer	Loophole Breach hole	60 250
9-inch double brick	Initial Loophole	70 120
16-inch tree trunk or log wall	Initial[1]	1 to 3
12-inch cinder block (filled w/sand)	Loophole	35
24-inch double sandbag wall	Initial[1]	220
3/8-inch mild steel door	Initial[1]	1

[1]Penetration only, no loophole

Table B-2. Structure Penetration Capabilities of the 5.56-mm Round Against Typical Urban Targets (Range 25 to 100 Meters)

(1) The best method for breaching a masonry wall is by firing short bursts (3 - 5 rounds) in a U-shaped pattern. The distance from the gunner to the wall should be minimized for best results—ranges as close as 25 meters are relatively safe from ricochet. Ballistic eye protection, a protective vest, and a helmet should be worn.

(2) Ball ammunition and armor-piercing rounds produce almost the same results, but armor-piercing rounds are more likely to fly back at the firer. The 5.56-mm round can be used to create either a loophole (about 7 inches in diameter) or a breach hole (large enough for a man to enter). When used against reinforced concrete, the M16 rifle and M249 cannot cut the reinforcing bars.

3. Medium/Heavy Machine Guns and Sniper Rifles (7.62 mm and .50 caliber)

a. Machine Guns. In the urban environment, the .50-caliber machine gun and the 7.62-mm M240G machine gun provide high-volume, long-range, automatic fires for the suppression or destruction of targets. They provide final protective fires along fixed lines and can be used to penetrate light structures; the .50-caliber machine gun is most effective in this role. Tracers from both machine guns are likely to start fires, but the .50-caliber tracer is more apt to do so.

(1) Employment. The primary consideration affecting the employment of machine guns within built-up areas is the limited availability of long-range fields of fire. Although machine guns should be emplaced at the lowest level possible, grazing fire at ground level is often obstructed by rubble.

(a) The .50-caliber machine gun is often employed on its vehicular mount during both offensive and defensive operations. If necessary, it can be mounted on the M3 tripod for use in the ground role or in the upper level of buildings. When mounted on a tripod, the .50-caliber machine gun can be used as an accurate, long-range weapon and can supplement sniper fires.

(b) The M240G machine gun is cumbersome, making it difficult to use inside while clearing a building. However, it is useful outside to suppress and isolate enemy defenders. The M240G can be fired from either the shoulder or the hip to provide a high volume of assault and suppressive fires. The use of the long sling to support the weapon and ammunition is preferred.

(c) Because of their reduced penetration power, M240G machine guns are less effective against masonry targets than .50-caliber machine guns. However, their availability and light weight make them well suited to augment heavy machine gun fire, to be used in areas where .50-caliber machine guns cannot be positioned, or as a substitute when heavy machine guns are not available. The M240G machine gun can be employed on its tripod to deliver accurate fire along fixed lines and can then quickly be converted to bipod fire to cover alternate fields of fire.

(2) Penetration. The ability of the 7.62-mm and .50-caliber rounds to penetrate is also affected by the range to the target and type of material against which the rounds are fired. Close ranges affect penetration of the 7.62-mm round less than the 5.56-mm round; the .50-caliber round's penetration is reduced the least.

(a) At 50 meters, the 7.62-mm ball round cannot penetrate a single layer of sandbags. It can penetrate a single layer at 200 meters, but not a double layer. The armor-piercing round does only slightly better against sandbags. It cannot penetrate a double layer but can penetrate up to 10 inches of sand at 600 meters.

(b) The penetration of the 7.62-mm round is best at 600 meters, but most urban targets are closer. The longest engagement range is usually 200 meters or less. Table B-3 lists the penetration capabilities of a single 7.62-mm (ball) round at closer ranges.

Range (meters)	Pine Board (inches)	Dry, Loose Sand (inches)	Cinder Block (inches)	Concrete (inches)
25	13	5	8	2
100	18	4.5	10	2
200	41	7	8	2

Table B-3. Penetration Capabilities of a Single 7.62-mm (Ball) Round

(c) The .50-caliber round is also optimized for penetration at long ranges (about 800 meters). For hard targets, .50-caliber penetration is affected by obliquity and range. Both armor-piercing and ball ammunition penetrate 14 inches of sand or 28 inches of packed earth at 200 meters if the rounds impact perpendicular to the flat face of the target. Table B-4 depicts the effect of a 25-degree obliquity on a .50-caliber penetration.

Thickness (feet)	100 Meters (rounds)	200 Meters (rounds)
2	300	1,200
3	450	1,800
4	600	2,400

Table B-4. Number of Rounds Needed To Penetrate a Reinforced Concrete Wall at a 25-Degree Obliquity

(3) Protection. Barriers that offer protection against 5.56-mm rounds are also effective against 7.62-mm rounds with some exceptions. The 7.62-mm round can penetrate a windowpane at a 45-degree obliquity, a hollow cinder block, or both sides of a car body. It can also easily penetrate wooden-framed buildings. The .50-caliber round can penetrate all of the commonly found urban barriers except a sand-filled 55-gallon drum.

(4) Wall Penetration. Continued and concentrated machine gun fire can breach most typical urban walls. Such fire cannot breach thick reinforced concrete structures or dense natural stone walls. Internal walls, partitions, plaster, floors, ceilings, common office furniture, home appliances, and bedding can be easily penetrated by both 7.62-mm and .50-caliber rounds (Tables B-5 and B-6).

(a) The M240G machine gun can be hard to hold steady enough to repeatedly hit the same point on a wall. The dust created by the bullet strikes also makes precise aiming difficult. Firing with a tripod is usually more effective than without, especially if sandbags are used to steady the weapon. Short bursts of three to five rounds fired in a U-type pattern are best.

(b) Breaching a brick veneer over cinder block presents a special problem for the M240G machine gun. Rounds penetrate the cinder block but leave a net-like structure of unbroken block. Excessive ammunition is required to destroy the net because most rounds only pass through a previously eroded hole. One or two minutes of work with an entrenching tool, crowbar, or ax can remove this web and allow entry through the breach hole.

(c) The .50-caliber machine gun can be fired accurately from the tripod using the single-shot mode. This is the most efficient method for producing a loophole. Automatic fire in three- to five-round bursts in a U-type pattern is more effective in producing a breach.

Type	Thickness (Inches)	Hole Diameter (Inches)	Rounds Required
Reinforced concrete	8	7	100
Triple brick wall	14	7	170
Concrete block with single brick veneer	12	6 and 24	30 and 200
Cinder block (filled)	12	N/A[1]	18
Double brick wall	9	N/A[1]	45
Double sandbag wall	24	N/A[1]	110
Log wall	16	N/A[1]	1
Mild steel door	3/8	N/A[1]	1

[1] Penetration only, no loophole

Table B-5. Structure Penetrating Capabilities of 7.62-mm Round (NATO) Against Typical Urban Targets (Range 25 Meters)

Type	Thickness (Inches)	Hole Diameter (Inches)	Rounds Required
Reinforced concrete	10	12	50
		24	100
	18	7	140
Triple brick wall	12	8 and 26	15 and 50
Concrete block with single brick veneer	12	10 and 33	25 and 45
Armor plate	1	N/A[1]	1
Double sandbag wall	24	N/A[1]	5
Log wall	16	N/A[1]	1

[1]Penetration only, no loophole

Table B-6. Structure Penetrating Capabilities of .50-caliber Ball Against Typical Urban Targets (Range 35 Meters)

b. Sniper Rifles. The 7.62-mm match ball ammunition fired from the M-40A1 sniper rifle will perform similarly to the 7.62-mm ball round. The M-82A1A special application sniper rifle (SASR) can fire the .50-caliber armor-piercing incendiary (API) ammunition, which will have the same performance as when fired from the M-2 HB. The M-82A1A also fires the .50-caliber RAUFOSS ammunition, which contains a tungsten penetrator and a more powerful explosive charge than the API ammunition. When the RAUFOSS penetrator punches through the metal, the explosive detonates inside, acting as an incendiary round. Although the SASR's exact penetration capabilities have yet to be defined, it has penetrated an inch of steel at 2000 yards. Both sniper rifles are organic to the Marine infantry battalion.

4. Grenade Launchers, 40-mm (M203 and MK19). Both the M203 dual-purpose weapon and the MK19 grenade machine gun fire 40-mm HE and HE dual-purpose (HEDP) ammunition. Ammunition for these weapons is not interchangeable, but the grenade and fuze assembly that actually hits the target is identical. Both weapons provide point and area destructive fires as well as suppression. The MK19 has a much higher rate of fire and a longer range. The M203 is much lighter and more maneuverable.

a. Employment. The main consideration affecting the employment of 40-mm grenades within built-up areas is the typically short engagement range. The 40-mm grenade has a *minimum arming range of 14 to 28 meters*. If the round strikes an object before it is armed, it will not detonate. Both the HE and HEDP rounds have 5-meter burst radii against exposed forces, which means that the *minimum safe firing range for combat is 33 meters*. The 40-mm grenades can be used to suppress the enemy in a building or inflict casualties by firing through apertures or windows. The MK19 can use its high rate of fire to concentrate rounds against light structures. This concentrated fire can create extensive damage. The 40-mm HEDP round can penetrate the armor on the flank, rear, and top of Soviet-made BMPs and BTRs. Marines can use the M203 from upper stories to deliver accurate fire against the top decks of armored vehicles. Multiple hits are normally required to achieve a kill.

b. Weapon Penetration. The 40-mm HEDP grenade has a small shaped charge that penetrates better than the HE round. It also has a thin wire wrapping that bursts into a dense fragmentation pattern, creating casualties out to 5 meters. Because they explode on contact, 40-mm rounds achieve the same penetration regardless of range. Table B-7 lists the penetration capabilities of the HEDP round.

Target	Penetration (Inches)
Sandbags	20
Sand-filled cinder block	16
Pine logs	12
Armor plate	2

Table B-7. Penetration Capabilities of the HEDP Round

(1) If projected into an interior room, the 40-mm HEDP can penetrate all interior partition-type walls. It splinters plywood and plaster walls, making a hole large enough to fire a rifle through. It is better to have HEDP rounds pass into a room and explode on a far wall, even though much of the round's energy is wasted penetrating the back wall (Figure B-1). The fragmentation produced in the room causes more casualties than the high-explosive jet formed by the shaped charge.

(2) The fragments from the HEDP round do not reliably penetrate interior walls. They are also stopped by office furniture, sandbags, helmets, and protective vests (flak jackets). The M203 dual-purpose weapon has the inherent accuracy to place grenades into windows at 125 meters and bunker apertures at 50 meters. These ranges are significantly reduced as the angle of obliquity increases. Combat experience shows that M203 gunners cannot consistently hit windows at 50 meters when forced to aim and fire quickly.

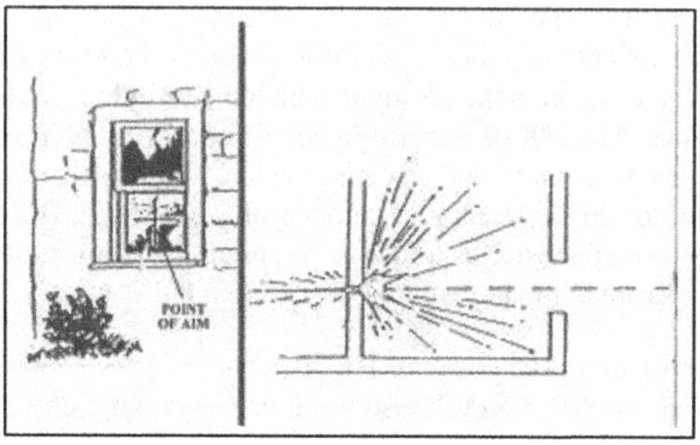

Figure B-1. Aim Point for 40-mm HEDP

c. Wall Penetration. The M203 cannot reasonably deliver the rounds needed to breach a typical exterior wall. The MK19 can concentrate its fire and achieve wall penetration. Firing from a tripod using a locked-down traversing and elevating mechanism is best for this role. Brick, cinder block, and concrete can be breached by using the MK19; individual HEDP rounds can penetrate 6 to 8 inches of brick. The only material that has proven resistant to concentrated 40-mm fire is dense stone such as that used in some European building construction. No precise data exist as to the number of rounds required to produce loopholes or breach holes with the MK19. However, the rounds' explosive effects are dramatic and should exceed the performance of the .50-caliber machine gun.

5. Light and Medium Recoilless Weapons. Light and medium recoilless weapons are used to attack enemy personnel, field fortifications, and LAVs. They have limited capability against main battle tanks, especially those equipped with reactive armor, except when attacking from the top, flanks, or rear. This category of weapons includes the AT4, the M47 Dragon, and the SMAW.

a. Employment. Other than defeating LAVs, the most common task for which light recoilless weapons are used is to neutralize fortified firing positions. Because of the design of the warhead and the narrow blast effect, these weapons are not as effective in this role as heavier weapons such as a tank main gun round. Their light weight allows Marines to carry several AT4s. Light recoilless weapons can be fired from the tops of buildings or from areas with extensive ventilation.

(1) Light and medium recoilless weapons with the exception of the SMAW employ shaped-charge warheads. As a result, the hole they punch in walls is often too small to use as a loophole. The fragmentation and spall that these weapons produce are limited. Normally, shaped-charge warheads do not neutralize enemy forces behind walls unless they are located directly in line with the point of impact.

(2) Against structures, shaped-charge weapons should be aimed about 6 inches below or to the side of a firing aperture (Figure B-2). This enhances the probability of killing the enemy behind the wall. A round that passes through a window wastes much of its energy on the back wall. Because these shaped-charge rounds lack the wire wrapping of the 40-mm HEDP, they burst into few fragments and are often ineffective casualty producers.

(3) Sandbagged emplacements present a different problem (Figure B-3). Because sandbags absorb much of the energy from a shaped charge, the rounds should be aimed at the center of the firing aperture. Even if the round misses the aperture, the bunker wall area near it is usually easiest to penetrate.

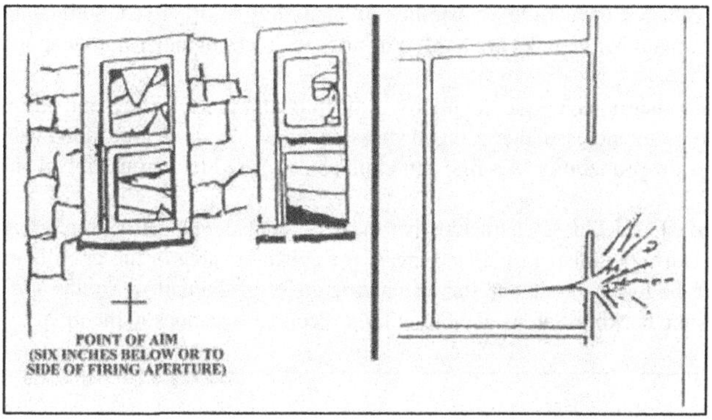

Figure B-2. Point of Aim for a Shaped-Charge Weapon Against a Masonry Structure

Figure B-3. Point of Aim for Sandbagged Emplacement

(4) Light and medium recoilless weapons obtain their most effective short-range antiarmor shots by firing from upper stories or from the flanks and rear. When firing at main battle tanks, these weapons should always be employed against these weaker areas in volley or paired firing. They normally require multiple hits to achieve a kill on a tank. Flank, top, and rear shots hit the most vulnerable parts of armored vehicles. Firing from upper stories protects the firer from tank main gun and coaxial machine gun fire because tanks cannot sharply elevate their cannons. The BMP-2 *can* elevate its 30-mm cannon to engage targets in upper stories. The BTR-series armored vehicles *can* also fire into upper stories with their heavy machine gun.

(5) Modern IFVs such as the BMP-2 and the BTR-80 have significantly improved frontal protection against shaped-charge weapons. Many main battle tanks have some form of reactive armor in addition to their thick armor plate. Head-on, ground-level shots against these vehicles have little probability of obtaining a kill. Even without reactive armor, modern main battle tanks are hard to destroy with a light antiarmor weapon.

(6) The easiest technique to improve the probability of hitting and killing an armored vehicle is to increase the firing depression angle. A 45-degree downward firing angle doubles the probability of a first-round hit as compared to a ground-level shot.

b. Backblast. Backblast characteristics must be considered when employing all recoilless weapons. During combat in built-up areas, the backblast area in the open is more hazardous because of the loose rubble and the channeling effect of the narrow streets and alleys. Figure B-4 shows the backblast areas of Marine light recoilless weapons in the open.

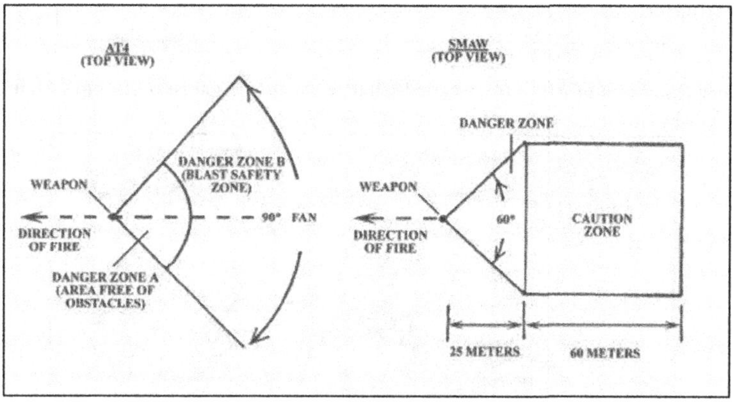

Figure B-4. Backblast Areas of Light Recoilless Weapons in the Open

(1) When firing recoilless weapons in the open, Marines should protect themselves from blast and burn injuries caused by the backblast. All personnel should be out of the danger zone. Anyone not able to vacate the caution zone should be behind cover. Marines in the caution zone should wear helmets, protective vests, and eye protection. The firer and all Marines in the area should wear earplugs.

(2) Extensive testing on the effects of firing recoilless weapons from within enclosures has been conducted since 1948. These tests showed that firing these weapons from enclosures presented no serious hazards, even when the overpressure was enough to produce structural damage to the building. The following were other findings of these tests.

(a) Little hazard exists to the gunnery or crew from any type of flying debris. Loose items were not hurled around the room.

(b) No substantial degradation occurs to the operator's tracking performance as a result of obscuration or blast overpressure.

(c) The most serious hazard that can be expected is hearing loss. This must be evaluated against the advantage gained in combat from firing from cover. To place this hazard in perspective, a gunner wearing earplugs and firing the loudest combination (the Dragon from within a masonry building) is exposed to less noise hazard than if he fired a LAW or AT4 in the open without earplugs.

(d) The safest place for other Marines in the room with the firer is against the wall from which the weapon is fired. Plastic ignition plugs are a hazard to anyone standing directly behind a LAW or TOW when it is fired.

(e) Firers should take advantage of all available sources of ventilation by opening doors and windows. Ventilation does not reduce the noise hazard, but it helps clear the room of smoke and dust and reduces the effective duration of the overpressure.

(f) The only difference between firing these weapons from enclosures and firing them in the open is the duration of the pressure fluctuation.

(g) Frame buildings, especially small ones, can suffer structural damage to the rear walls, windows, and doors. Large rooms suffer slight damage, if any.

(3) Recoilless weapons fired from within enclosures create some obscuration inside the room but almost none from the gunner's position looking out. Inside the room, obscuration can be intense, but the room remains inhabitable. Table B-8 shows the effects of smoke and obscuration.

Building	Weapon	From Gunner's Position Looking Out	Inside the Room	From Outside at a Distance
Masonry	LAW	None	Moderate	Slight smoke
	Dragon	Slight	Moderate	Small flash

Bunker	Dragon TOW	None None	Slight Slight	Moderate flash Moderate smoke
Small frame	LAW Dragon	None None	Moderate Severe	Moderate Moderate
Medium frame	LAW Dragon	None None	Slight Severe	Moderate Slight flash
Large frame	LAW Dragon TOW	None Slight None	Slight Severe Severe	None Slight flash Slight smoke

Table B-8. Smoke and Obscuration

(4) The Dragon causes the most structural damage, but only in frame buildings. There does not seem to be any threat of injury to the gunner because the damage is usually to the walls away from the gunner. The most damage and debris is from flying plaster chips and pieces of wood trim. Large chunks of plasterboard can be dislodged from ceilings. The backblast from a LAW, AT4, Dragon, or TOW rarely displaces furniture. Table B-9 shows the test results of structural damage and debris.

		Damage		
Building	**Weapon**	**Structure**	**Wall Covering**	**Debris Movement**
Masonry	LAW Dragon	None None	Slight Slight	Slight Slight
Bunker	Dragon TOW	None None	None None	None Leaves and dust disturbed
Small frame	LAW Dragon	None Severe	Slight Severe	None None
Medium frame	LAW Dragon	None Slight	None Slight	Slight Lamp and chair overturned
Large frame	LAW Dragon TOW	None Slight Slight	Slight Moderate Severe	Slight None None

Table B-9. Structural Damage and Debris Movement

(5) To fire a LAW, AT4, or SMAW from inside a room, the following safety precautions must be taken (Figure B-5).

(a) The building should be of a sturdy construction.

(b) The ceiling should be at least 7 feet high with loose plaster or ceiling boards removed.

(c) The floor size should be at least 15 feet by 12 feet. (The larger the room, the better.)

(d) At least 20 square feet of ventilation (room openings) should exist to the rear or side of the weapon. An open 7- by 3-foot door would provide minimum ventilation.

(e) All glass should be removed from windows, and small, loose objects should be removed from the room.

(f) Floors should be wet to prevent dust and dirt from blowing around and obscuring the gunner's vision.

(g) All personnel in the room should be forward of the rear of the weapon.

(h) All personnel in the room should wear helmets, protective vests, ballistic eye protection, and earplugs.

(i) If the gunner is firing from the prone position, his lower body must be perpendicular to the bore of the weapon, or the blast could cause injury to his legs.

Figure B-5. Firing an AT4 From Inside a Building

c. Weapon Penetration. The most important tasks to be performed against structures are the neutralization of fortified firing positions, personnel, and weapons behind barriers. Recoilless weapons can be used in this role; none, however, are as effective as heavy, direct-fire weapons or standard demolitions. Each recoilless weapon has different penetrating ability against various targets. Penetration does not always mean the destruction of the integrity of a position. Usually only those enemy soldiers directly in the path of the spall from a high-explosive antitank (HEAT) round become casualties. Other soldiers inside a fortification could be deafened, dazed, or shocked but could eventually return to action.

(1) AT4. The AT4 is heavier than the LAW and has a diameter of 84 millimeters, which gives the warhead much greater penetration. The AT4 can penetrate more than 17.5 inches (450 mm) of armor plate. Its warhead produces highly destructive results behind the armor. Tests against typical urban targets are still ongoing, but the AT4 should penetrate at least as well as the 90-mm recoilless rifle, if not better. The AT4 has a *minimum arming distance of 10 meters*, which allows it to be fired successfully against close targets. Firers should be well covered and protected when firing at close targets.

(2) MK153 83-mm SMAW. The SMAW is a lightweight assault weapon that is easily carried and placed into action by one man. It is used against fortified positions, but it is also effective against LAVs. The SMAW has a 9-mm spotting rifle and a 3.8-power telescope that ensure accuracy over ranges common to combat in built-up areas. The SMAW has excellent incapacitating effects for enemies behind walls and inside bunkers and can arm within 10 meters. Its fuze has the ability to distinguish between armor and soft earth, maximizing its capabilities against buildings, bunkers, or light armor. The warhead detonates immediately against hard targets but delays detonation against soft targets and burrows in to explode inside. The SMAW can destroy most bunkers with a single hit. The 83-mm HE warhead can create an 8-inch mousehole in reinforced concrete but will not cut the steel reinforcing bar. The SMAW is an outstanding urban support weapon that can be decisive during an urban assault. The SMAW allows for quick reduction of obstacles and bunkers as well as creation of entry points. SMAWs fired in a volley can be devastating to a building.

d. Wall Breaching. Wall breaching is a common combat task in built-up areas for which light recoilless weapons can be used. Breaching operations improve mobility by providing access to building interiors without using existing doors or windows. Breaching techniques can also be used to create loopholes for weapons positions or to allow hand grenades to be thrown into defended structures. Breaching holes for unit mobility should be about 24 inches (60 centimeters) in diameter. Loopholes should be about 8 inches (20 centimeters) in diameter (Figure B-6). Only the SMAW provides a one-shot wall breaching ability. However, in breaching walls a number of shots should be planned.

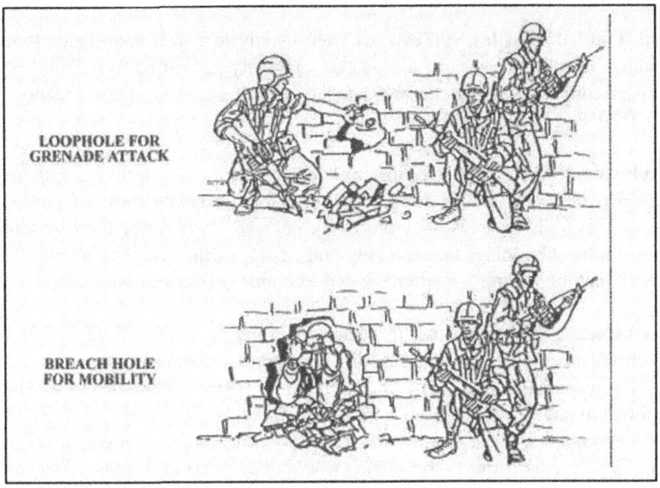

Figure B-6. Tactical Use of Holes in Masonry Walls

(1) Of all the common building materials, heavy stone is the most difficult to penetrate. The AT4 normally will not penetrate a heavy European-style stone wall. Surface cratering is usually the only effect it will have.

(2) Layered brick walls are also difficult to breach with light recoilless weapons. Some brick walls can be penetrated by multiple firings, especially if they are less than three bricks thick. Multiple AT4 rounds fired at the same spot on an 8-inch (double-brick) wall normally produce a loophole. The SMAW produces a hole in brick walls that is often large enough to be a breach hole.

(3) Wooden structural walls offer little resistance to light recoilless weapons. Even heavy timbered walls are penetrated and splintered. Multiple AT4 rounds fired at the same area of a wood-framed wall usually produce a man-sized hole. The SMAW has a devastating effect against wood-framed walls. A single round produces a breach hole as well as significant spall.

(4) Because of its high velocity, the AT4 may penetrate a soft target such as a car body or frame building before exploding.

(5) None of the light recoilless weapons are as effective against structural walls as demolitions or heavier weapons such as tank main guns or field artillery. Of all the light recoilless weapons, the SMAW is the most effective.

6. Antitank Guided Missiles. ATGMs are used mainly to defeat main battle tanks and other armored combat vehicles. They have a moderate capability against bunkers, buildings, and other fortified targets commonly found during combat in built-up areas. This category of weapons includes the TOW and Dragon missiles.

a. Employment. TOWs and Dragons provide overwatch antitank fires during the attack of a built-up area and an extended range capability for the engagement of armor during the defense. Within built-up areas, they are best employed along major thoroughfares and from the upper stories of buildings to attain long-range fields of fire. Their *minimum firing range of 65 meters* could limit firing opportunities in the confines of densely built-up areas.

(1) Obstacles. Rubble or other obstacles could interfere with the flight of missiles when fired from street level. At least 3.5 feet (1 meter) of vertical clearance over such obstacles must be maintained. Figure B-7 shows the most common obstacles to ATGM flight that are found in built-up areas. *Power lines are a special obstacle and present a unique threat to ATGM gunners. If the power in the lines has not been interrupted, the ATGM guidance wires could create a short circuit. This would allow extremely high voltage to pass to the gunner in the brief period before the guidance wires melted. This voltage could either damage the sight and guidance system or injure the gunner. Before any ATGM is fired over a power line, an attempt must be made to determine whether or not the power to that line has been interrupted.*

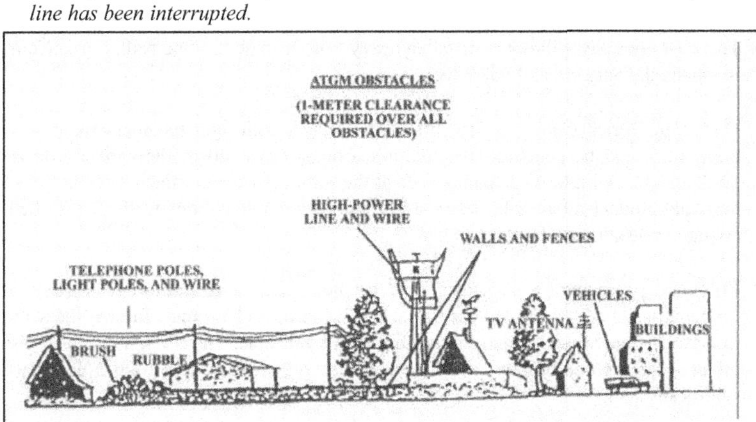

Figure B-7. Common Obstacles to ATGM Flights

(2) Dead Space. Three aspects of dead space that affect ATGM fires are arming distance, maximum depression, and maximum elevation.

(a) Both the Dragon and TOW missiles have a minimum arming distance of 65 meters, which severely limits their use in built-up areas. Few areas in the inner city permit fires much beyond the minimum arming distance. However, ground-level, long-range fires down streets or rail lines and across parks or plazas are possible. ATGMs may be used effectively from upper stories or roofs of buildings to fire into other buildings.

(b) The TOW is limited much more than the Dragon by its maximum depression and elevation. The maximum depression and elevation limits of the TOW mount could result in dead space and preclude the engagement of close targets (Figure B-8). A target located at the minimum arming range (65 meters) cannot be engaged by a TOW crew located any higher than the sixth floor of a building due to maximum depression limits. At 100 meters, the TOW crew can be located as high as the ninth floor and still engage the target.

(3) Backblast. Backblast for ATGMs is of more concern during combat in built-up areas than in open country. Any loose rubble in the caution zone could be picked up and thrown by the backblast, and the channeling effect of walls and narrow streets is even more pronounced due to the greater backblast. If the ATGM backblast strikes a wall at an angle, it can pick up debris or be deflected and cause injury to unprotected personnel (Figure B-9). Both types of ATGM can be fired from inside some buildings. In addition to the helmet and protective vest, eye protection and earplugs should be worn by all personnel in the room.

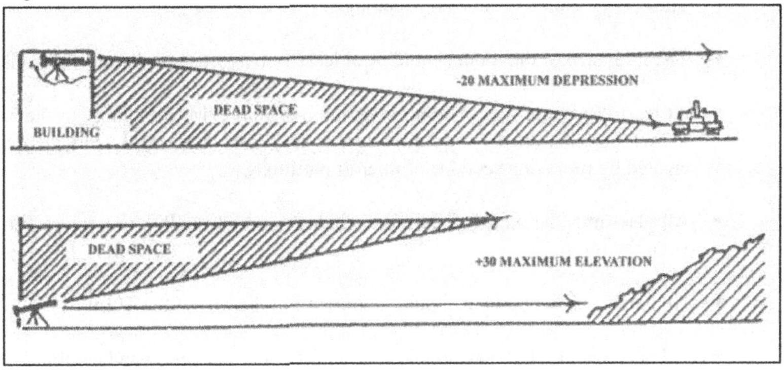

Figure B-8. TOW Maximum Elevation and Depression Limitations

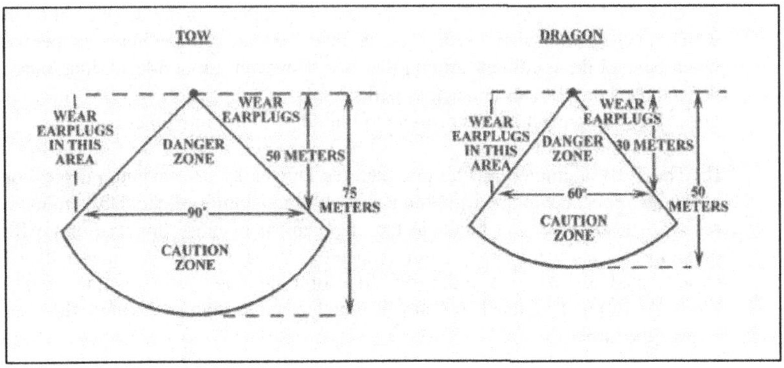

Figure B-9. ATGM Backblast in an Open Street

(a) To fire a TOW from inside a room, the following safety precautions must be taken (Figure B-10):

- The building must be of sturdy construction.
- The ceiling should be at least 7 feet high.
- The floor size of the room should be at least 15 by 15 feet, or larger if possible.
- At least 20 square feet of room ventilation should exist, preferably to the rear of the weapon. An open 7- by 3-foot door is sufficient. Additional ventilation can be created by removing sections of interior partitions.
- All glass must be removed from the windows and all small, loose objects removed from the room.
- All personnel in the room should be forward of the rear of the TOW.
- All personnel in the room should wear ballistic eye protection and earplugs.
- A clearance of 9 inches (23 centimeters) must exist between the launch tube and the aperture from which it is fired. (See FM 23-34, *TOW Weapon System*, for more detailed safety information.)

Figure B-10. TOW Fired From Inside a Room

(b) To fire a Dragon from inside a room, the following safety precautions must be taken.

- The building must be of sturdy construction.
- The ceiling should be at least 7 feet high.
- The floor size should be at least 15 by 15 feet, or larger if possible.
- At least 20 square feet of ventilation should exist (room openings), preferably to the rear of the weapon. An open 7- by 3-foot door would provide minimum ventilation.
- All glass should be removed from windows, and small, loose objects should be removed from the room.
- The room should be clean or the floors must be wet to prevent dust and dirt (kicked up by the backblast) from obscuring the vision of other soldiers in the room.
- All personnel in the room must be forward of the rear of the weapon.
- All personnel in the room must wear ballistic eye protection and earplugs.
- At least a 6-inch clearance must exist between the launch tube and the aperture from which it is fired.

b. Weapon Penetration. ATGMs can penetrate and destroy heavily armored tanks. They have large warheads that contain shaped charges. Because of their size, these warheads can achieve significant penetration against typical urban targets. Penetration, however, does not mean a concurrent destruction of the structural integrity of a position. The shaped-charge warhead produces relatively little spall. Enemy personnel not standing directly behind or near the point of impact of an ATGM may escape injury.

(1) Standard TOW Missiles. The basic TOW missile can penetrate 8 feet of packed earth, 4 feet of reinforced concrete, or 16 inches of steel plate. The improved TOW (ITOW), the TOW 2, and the TOW 2A have all been modified to improve their penetration. They all penetrate better than the basic TOW. All TOW missiles can defeat triple sandbag walls, double layers of earth-filled 55-gallon drums, and 18-inch log walls.

(2) TOW 2B. The TOW 2B uses a different method of defeating enemy armor. It flies over the target and fires an explosively formed penetrator down onto the top armor, which is thinner. *Because of this design feature, the TOW 2B missile cannot be used to attack nonmetallic structural targets. When using the TOW 2B missile against enemy armor, gunners must avoid firing directly over other friendly vehicles, disabled vehicles, or large metal objects such as water or oil tanks.*

(3) Dragon Missile. The Dragon missile can penetrate 8 feet of packed earth, 4 feet of concrete, or 13 inches of steel plate. It can attain effective short-range fire from upper stories or from the rear or flanks of a vehicle. These engagements are targeted against the most vulnerable parts of tanks and can entrap tanks in situations where they are unable to counterfire. Elevated firing positions increase the first-round hit probability. Firing down at an angle of 20 degrees increases the chance of a hit by 67 percent at 200 meters. A 45-degree downward angle doubles the first-round hit probability, compared to a ground-level shot.

c. Breaching Structural Walls. Firing ATGMs is the least efficient means to defeat structures. Because of their small basic load and high cost, ATGMs are better used against tanks or enemy-fortified firing positions. They can be effective against bunkers or other identified enemy firing positions.

7. Flame Weapons. Flame weapons are characterized by both physical and psychological casualty-producing abilities. Flame does not need to be applied with pinpoint accuracy, but it also must not spread to structures needed by friendly forces. Large fires in built-up areas are catastrophic. If they burn out of control, fires can create an impenetrable barrier for hours. The Marine Corps does not possess flame-generating equipment. However, the M34 WP grenade can be used to ignite and destroy flammable objects, especially wooden structures. The Army's M2A1-7 portable flamethrower is stored in war reserve status as a standard "C" item. Its availability is limited; however, Marines could be issued flame-support weapons for urban operations depending on the status of international agreements on the use of flame weapons that are in effect at the time.

a. Employment. Flame weapons used against fortified positions should be aimed directly at the aperture. Even if the round or burst misses, enough flaming material enters the position to cause casualties and disrupt the enemy occupants. The M34 WP grenade is difficult to throw far or into a small opening such as a bunker aperture. However, its effects are dramatic when thrown into a room or building.

b. Effects. The M34 is used to ignite and destroy flammable objects, especially wooden structures. It is also used to create an immediate smoke cloud to conceal movement across a narrow open space such as a street. Its smoke is not toxic but can cause choking in heavy concentrations.

(1) The grenade's explosion, bright flash, smoke, and burning WP particles all combine to make the M34 one of the most effective psychological weapons available. The M34 hand grenade throws WP fragments up to 35 meters from the point of detonation. These fragments can attach to clothing or skin and continue burning. Because of its weight, most infantrymen can throw this grenade only 30 to 40 meters.

(2) The Marine must avoid injury from friendly use of the M34. The M34 can ignite if the WP inside is exposed to the air. Bullets and shell fragments have been known to strike and rupture M34 grenades; therefore, grenades must be protected from enemy fire.

(3) The M34 WP grenade is an effective weapon against enemy armored vehicles when used in the close confines of combat in built-up areas. It can be thrown or dropped from upper stories onto enemy vehicles. The M34 can be combined with flammable liquids, detonating cords, blasting caps, and fuze igniters to create the eagle fireball, a field-expedient antiarmor device. (See FM 21-75, *Combat Skills of the Soldier*, Appendix H.)

(4) The M34 is also excellent as a screening device. A grenade can be thrown from behind cover into an open street or plaza. When it explodes, the enemy's observation is temporarily obscured. Thus, friendly forces can quickly cross the open area—if the enemy fires, it is unaimed and presents less of a danger. If screening smoke is used to cover a squad's movement across short open areas, it will reduce expected casualties from small-arms fire by about 90 percent.

8. Hand Grenades. Hand grenades are used extensively during combat in built-up areas. Smoke grenades are used for screening and signaling. Riot control grenades are used to drive the enemy out of deep fortifications. Fragmentation and concussion grenades are used to clear the enemy out of rooms and basements. Hand grenades are the most-used explosive munition during intense combat in built-up areas. In World War II, it was common for a battalion fighting in a city to use more than 500 hand grenades each day.

a. Employment. Smoke and riot-control grenades have similar employment techniques. Fragmentation grenades are used to produce enemy casualties.

(1) The AN-M8 HC grenade produces a dense white or grey smoke. It burns intensely and cannot be easily extinguished once it ignites. The smoke can be dangerous in heavy concentrations because it makes breathing difficult and causes choking. The M8 grenade is normally used for screening. It produces a slowly building screen of longer duration than the M34 WP grenade without the problem of collateral damage caused by scattered burning particles.

(2) The M18-series smoke grenades produce several different colors of smoke that are used for signaling. Yellow smoke is sometimes difficult to see in built-up areas.

(3) The M7A3 CS riot control grenade can be used to drive enemy troops out of fortifications when civilian casualties or collateral damage constraints are considerations. Built-up areas often create variable and shifting wind patterns. When using CS grenades, Marines must prevent the irritating smoke from affecting friendly forces. The CS grenade burns intensely and can ignite flammable structures. Enemy troops wearing even rudimentary chemical protective masks can withstand intense concentrations of CS gas.

(4) The MK3A2 offensive hand grenade, commonly referred to as the concussion grenade, produces casualties during close combat while minimizing the danger to friendly personnel. For this reason, it is the preferred hand grenade for use during offensive operations in a MOUT environment. The grenade produces severe concussion effects in enclosed areas. It can be used for light blasting and demolitions and for creating breach holes in interior walls. The concussion produced by the MK3A2 is much greater than that of the fragmentation grenade. It is very effective against enemy soldiers within bunkers, buildings, and underground passages.

(5) The M67 fragmentation grenade is the most commonly available grenade during combat in built-up areas. It provides suppression during room-to-room or house-to-house fighting, and it is used while clearing rooms of enemy personnel. The M67 grenade has a 3- to 5-second delay fuze. When used at close ranges, it can be cooked-off for two seconds to deny the enemy the time to throw it back. The fragmentation grenade can be rolled, bounced, or ricocheted into areas that cannot be reached by 40-mm grenade launchers.

b. Effects. Each type of hand grenade has its own specific effect during combat in built-up areas.

(1) Both HC smoke and CS gas *displace* oxygen from enclosed areas. Gas masks will not prevent this from occurring because they only filter air, not provide oxygen. Smoke grenades produce dense clouds of colored or white smoke that remain stationary in the surrounding area. They can also cause fires if used indiscriminately. If trapped and concentrated within a small space, smoke grenades and CS gas *can suffocate Marines*.

(2) The fragmentation grenade has more varied effects in combat in built-up areas. It produces a large amount of small, high-velocity fragments that can penetrate sheetrock partitions and are lethal at short ranges (15 to 20 meters). Fragments lose their velocity quickly and are less effective beyond 25 meters.

The fragments from a fragmentation grenade cannot penetrate a single layer of sandbags, a cinder block, or a brick building, but they can perforate wood-framed and tin buildings if exploded close to their walls.

(3) Inside rooms, fragmentation barriers consisting of common office furniture, mattresses, doors, or books can be effective against the fragmentation grenade. For this reason, a room should never be considered safe just because one or two grenades have been detonated inside. Fragmentation grenades detonated on the floor not only throw fragments laterally, but also send fragments and spall downward to lower floors. Predicting how much spall will occur is difficult because flooring material varies, but wooden floors are usually affected the most.

(4) Some foreign grenades throw fragments that are much larger than those of the U.S.-made M67. Light barriers and interior walls would probably be less effective against these grenades than against the M67. A major problem with the U.S.-made fragmentation grenade is its tendency to bounce back off of hard targets. Grenades are often directed at window openings on the ground floor or second floor. At ranges as close as 20 meters, a thrower's chances of missing a standard 1-meter by 1-meter window are high. The fragmentation grenade normally breaks through standard window glass and enters a room. If the grenade strikes at a sharp angle or the glass is thick plate, the grenade could be deflected without penetrating.

(5) Hand grenades are difficult weapons to use. They involve a high risk of fratricide. Commanders should conduct precombat training with hand grenades as part of normal preparations. Marines must be very careful when throwing hand grenades up stairs to avoid having the grenade roll back down on them.

(6) Once pulled, the pins of both fragmentation and concussion grenades can be replaced if the thrower decides not to use the weapon. This pin replacement must be done carefully (see FM 23-30, *Grenades and Pyrotechnic Signals*) and prior to the spoon being released or detaching.

(7) An estimate of the situation (METT-T) and existing ROE will dictate what type of grenade will be used to clear each room. Because of the high rate of expenditure of grenades, units should use assault packs to carry additional grenades of all types. Additional grenades can also be carried in empty ammunition or canteen pouches.

9. 25-mm Automatic Gun. The 25-mm automatic gun mounted on the LAV-25 is an effective weapon during combat in built-up areas. The primary roles of LAV-25s during combat in built-up areas are to provide suppressive fire and to breach exterior walls and fortifications.

a. Obliquity. The 25-mm gun produces its best urban target results when fired perpendicular to the hard surface (zero obliquity). In combat in built-up areas, however, finding a covered firing position that permits low-obliquity firing is unlikely unless the streets and gaps between buildings are wide. Most shots impact the target at an angle, which normally reduces penetration. With the armor-piercing, discarding, sabot with tracer (APDS-T) round, an angle of obliquity of up to 20 degrees can actually improve breaching. The rounds tend to dislodge more wall material for each shot but do not penetrate as deeply into the structure.

b. Target Types. The 25-mm gun has different effects when fired against different urban targets.

(1) Reinforced Concrete. Reinforced concrete walls, which are 12 to 20 inches thick, present problems for the 25-mm gun when trying to create breach holes. It is relatively easy to penetrate, fracture, and clear away the concrete, but the reinforcing rods remain in place. These create a "jail window" effect by preventing entry but allowing grenades or rifle fire to be placed behind the wall. Steel reinforcing rods are normally 3/4 of an inch thick and 6 to 8 inches apart—there is no quick way of cutting these rods. They can be cut with demolition charges, cutting torches, or special power saws. Firing with either APDS-T or high-explosive, incendiary with tracer (HEI-T) rounds from the 25-mm gun will not always cut these rods.

(2) Brick Walls. Brick walls are more easily defeated by the 25-mm gun regardless of their thickness, and they produce the most spall.

(3) Bunker Walls. The 25-mm gun is devastating when fired against sandbag bunker walls. Obliquity has the least effect on the penetration of bunker walls. Bunkers with earth walls up to 36 inches thick are easily penetrated. At short ranges typical of combat in built-up areas, defeating a bunker should be easy, especially if the 25-mm gun can fire at an aperture.

c. Burst Fire. The 25-mm gun's impact on typical urban targets seems to be magnified if the firing is in short bursts. At close ranges, the gunner might need to shift his point of aim in a spiral pattern to ensure that the second and third bursts enlarge the hole. Even without burst fire, sustained 25-mm gunfire can defeat almost all urban targets.

d. Weapon Penetration. The penetration achieved by the two combat rounds (HEI-T and APDS-T) differs slightly, but both are effective.

(1) APDS-T. The APDS-T round penetrates urban targets by retaining its kinetic energy and blasting a small hole deep into the target. The APDS-T round gives the best effects behind the wall, and the armor-piercing core often breaks into two or three fragments that can create multiple enemy casualties. The APDS-T needs as few as four rounds to achieve lethal results behind walls. Table B-10 lists the number of APDS-T rounds needed to create different-sized holes in common urban walls.

Target	Loophole	Breach Hole
3-inch brick wall at 0-degree obliquity	22 rounds	75 rounds
3-inch brick wall at 45-degree obliquity	22 rounds	35[1] rounds
5-inch brick wall at 0-degree obliquity	32 rounds	50[1] rounds
8-inch reinforced concrete at 0-degree obliquity	22 rounds	75 rounds (Note: Reinforcing rods still in place)
8-inch reinforced concrete at 45-degree obliquity	22 rounds	40[1] rounds (Note: Reinforcing rods still in place)

[1] Obliquity and depth tend to increase the amount of wall material removed.

Table B-10. Breaching Effects of APDS-T Rounds

(a) When firing single rounds, the APDS-T round provides the greatest capability for behind-the-wall incapacitation. The APDS-T round can penetrate more than 16 inches of reinforced concrete with enough energy left to cause enemy casualties. It penetrates through both sides of a wood-framed or brick veneer building. Field fortifications are easily penetrated by APDS-T rounds. Table B-11 lists the number of APDS-T rounds needed to create different-sized holes in commonly found bunkers.

(b) The APDS-T round creates a hazardous situation for exposed personnel because of the pieces of sabot that are thrown off of the round.

Type Bunker	Obliquity	Penetration	Loophole	Small Breach Hole
36-inch sand/timber	0 degree	1 round	25 rounds	40 rounds
36-inch sand/6-inch concrete	0 degree	6 rounds	6 rounds	20 rounds

Table B-11. Number of APDS-T Rounds Needed To Create Different-Sized Holes in Bunkers

WARNING

Personnel not under cover forward of the 25-mm gun's muzzle and within the danger zone could be injured or killed by these sabots, even if the penetrator passes overhead to hit the target.

The danger zone extends at an angle of about 10 degrees below the muzzle level, out to at least 100 meters, and about 17 degrees left and right of the muzzle. Figure B-11 shows the hazard area of the APDS-T round.

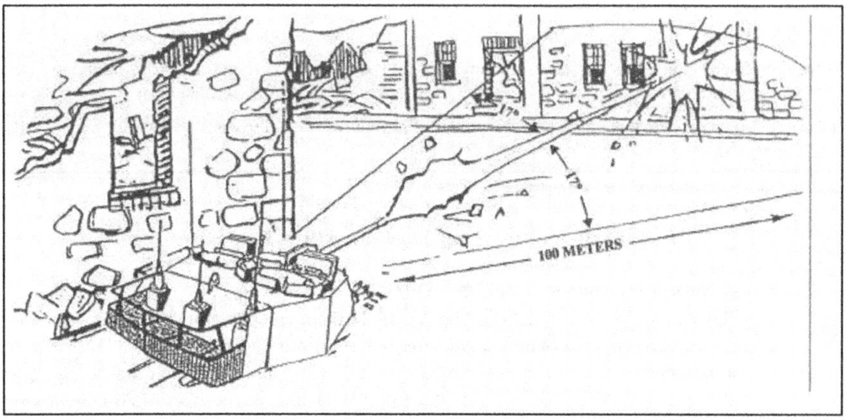

Figure B-11. APDS-T Danger Zone

(2) HEI-T. The HEI-T round penetrates urban targets by blasting away chunks of material.

(a) The HEI-T round does not penetrate an urban target as well as the APDS-T, but it does strip away a greater amount of material for each round. The HEI-T does more damage to an urban target when fired in multiple short bursts because the cumulative impact of multiple rounds is greater than the sum of individual rounds. Table B-12 lists the number of HEI-T rounds needed to create different-sized holes.

Target	Loophole	Breach Hole
3-inch brick wall at 0-degree obliquity	10 rounds	20 rounds
3-inch brick wall at 45-degree obliquity	20 rounds	25 rounds
5-inch brick wall at 0-degree obliquity	30 rounds	60 rounds

8-inch reinforced concrete at 0-degree obliquity	15 rounds	25 rounds
8-inch reinforced concrete at 45-degree obliquity	15 rounds	30 rounds

Table B-12. Number of HEI-T Rounds Needed To Create Different-Sized Holes in Bunkers

(b) The HEI-T round does not provide single-round perforation or incapacitating fragments when used against any external masonry structural wall. It can create first-round fragments behind wood-framed and brick veneer walls. HEI-T rounds cannot penetrate a bunker as quickly as APDS-T rounds, but they can create more damage inside the bunker once the external earth has been stripped away. Against a heavy bunker, about 40 rounds of HEI-T ammunition are needed to strip away the external earth shielding and breach the inner lining of concrete or timber. The HEI-T round is also used for suppression against known or suspected firing ports such as doors, windows, and loopholes.

10. Tank Cannon. The powerful, high-velocity cannon mounted on the M1A1 tank provides Marines with a key requirement for victory in built-up areas—heavy direct-fire support. Although the infantry assumes the lead role during combat in built-up areas, tanks and infantry work as a close team. Tanks move down streets after the infantry has cleared them of any suspected ATGM positions and, in turn, support the infantry with fire. The tank is one of the most effective weapons for heavy fire against structures. The primary role of the tank cannon during combat in built-up areas is to provide heavy direct fire against buildings and strongpoints that are identified as targets by the infantry. The wall and fortification breaching effects of the 120-mm tank cannon are major assets to Marines fighting in built-up areas.

a. Obliquity. The tank cannon produces its best urban target effects when fired perpendicular to the hard surface (zero obliquity). During combat in built-up areas, however, finding a covered firing position that permits low-obliquity firing is unlikely. Most shots strike the target at an angle that would normally reduce penetration. With tank cannon APDS rounds, obliquity angles of up to 25 degrees have little effect, but angles greater than 45 degrees greatly reduce penetration.

b. Ammunition. Armor-piercing, fin-stabilized, discarding sabot (APFSDS) rounds are the most commonly carried tank ammunition. These rounds work best against armored vehicles. The 120-mm cannon also carries an effective high-explosive, antitank multipurpose (HEAT-MP) round.

c. Characteristics. The 120-mm tank cannon has two specific characteristics that affect its employment in built-up areas: limited elevation and depression and short arming ranges. In addition, the M1A1 tank has another characteristic that is not involved with its cannon but that affects Marines working with or around the tank—*extremely hot turbine exhaust.*

(1) The M1A1 tank can elevate its cannon 20 degrees and depress it 10 degrees. The lower depression limit creates a 35-foot (10.8-meter) dead space around a tank. On a street 16 meters wide, this dead space extends to the buildings on each side (Figure B-12). Similarly, there is a zone overhead in which the tank cannot fire (Figure B-13). This dead space offers ideal locations for short-range antiarmor weapons and allows hidden enemy gunners to fire at the tank when the tank cannot fire back. It also exposes the tank's most vulnerable areas: the flanks, rear, and top. Infantry must move ahead, alongside, and to the rear of tanks to provide close protection. The extreme heat produced immediately to the rear of the M1A1 tank prevents dismounted infantry from following closely, but protection from small-arms fire and fragments is still provided by the tank's bulk and armor. The M1A1 tank also has a blind spot caused by the zero-degree depression available over part of the back deck. To engage any target in this area, the tank must pivot to convert the rear target to a flank target.

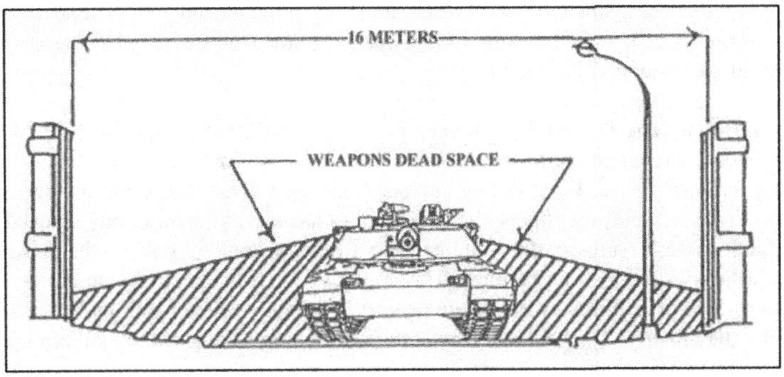

Figure B-12. Tank Cannon Dead Space at Street Level

(2) The 120-mm HEAT-MP round arms at about 36 feet. This arming distance allows the tank to engage targets from short ranges. The armor of the tank protects the crew from both the blowback effects of the round and enemy return fire. The APFSDS round does not need to arm and can, therefore, be fired at almost any range. The discarding portions of the round can be lethal to exposed infantry forward of the tank.

d. Target Effects. HEAT rounds are most effective against masonry walls. The APFSDS round can penetrate deeply into a structure but does not create as large a hole or displace as much spall behind the target. In contrast to lighter HEAT rounds, tank HEAT rounds are large enough to displace enough spall to inflict casualties inside a building. One HEAT round normally creates a breach hole in all but the thickest masonry construction—brick veneer and wood-framed constructions are demolished by a single round. Even the 120-mm HEAT round cannot cut all of the reinforcing rods, which are usually left in place, often hindering entry through the breach hole (Figure B-14). Both HEAT and APFSDS rounds are effective against all field fortifications. Only large earth berms and heavy mass-construction buildings can provide protection against tank fire.

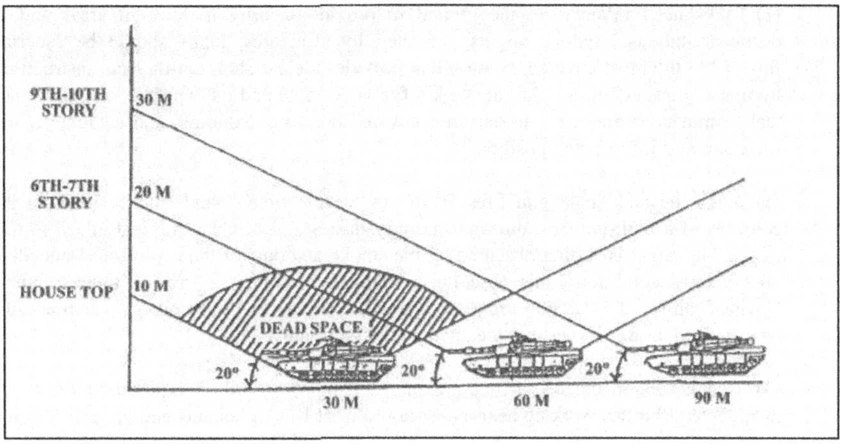

Figure B-13. Tank Cannon Dead Space Above Street Level

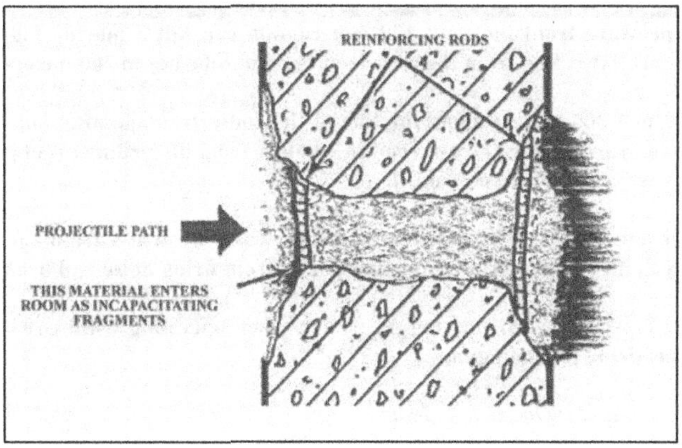

Figure B-14. Tank HEAT Round Effects on Reinforced Concrete Walls

e. Employment. Tank-heavy forces could be at a severe disadvantage during combat in built-up areas, but a few tanks working with the infantry can be most effective, especially if they work well together at the small-unit level. Tank, infantry, and engineer task forces are normally formed to attack a fortified area. The tank platoon or section can work together with rifle squads or platoons.

(1) Tanks need infantry on the ground to provide security in built-up areas and to designate targets. Against targets protected by structures, tanks should be escorted forward to the most covered location that provides a clear shot. On-the-spot instructions by infantry leaders ensure that the tank's fire is accurate and its exposure is limited. The tank commander may have to halt in a covered position, dismount, and reconnoiter his route forward into a firing position.

(2) When the tank main gun fires, it creates a large fireball and smoke cloud. In the confines of a built-up area, dirt and masonry dust are also picked up and added to this cloud. The target is further obscured by the smoke and dust of the explosion. Depending on the local conditions, this obscuration could last as long as two or three minutes. Marines can use this period to reposition or advance unseen by the enemy. Caution must be exercised, however, because the enemy might also move.

(3) A tank cannon creates an overpressure and noise hazard to exposed Marines. All dismounted Marines working near tanks should wear Kevlar helmets and protective vests, as well as ballistic eye protection. If possible, they should also wear earplugs and avoid the tank's frontal 60-degree arc during firing (Figure B-15).

WARNING

- **The overpressure from the tank 120-mm cannon can kill a marine found within a 90-degree arc extending from the muzzle of the gun tube out to 200 meters.**

- **From 200 to 1,000 meters along the line of fire, on a frontage of about 400 meters, dismounted Marines must be aware of the danger from discarding sabot petals, which can kill or seriously injure personnel.**

- **Personnel outside the tank should remain at least 50 meters from the tank in all directions as they may receive damaging effects from firing noise and overpressure.**

- **Personnel must also wear hearing protection when operating within 704 meters of a tank that is firing its main gun.**

TM 9-2350-264-10-2, *Operator's Manual for Operation Under Usual and Unusual Conditions for Tank, Combat, Full-Tracked: 120mm Gun, M1A1 General Abrams*, with Change 4

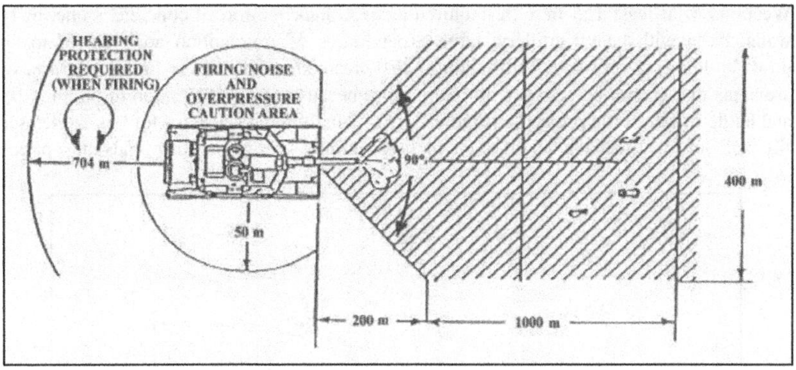

Figure B-15. Danger Areas Around a Tank Firing a 120-mm Gun

(4) Tanks are equipped with powerful thermal sights that can be used to detect enemy personnel and weapons that are hidden in shadows and behind openings. Dust, fires, and thick smoke significantly degrade these sights.

(5) Tanks have turret-mounted grenade launchers that project screening smoke grenades. The grenades use a bursting charge and burning RP particles to create this screen. *Burning particles can easily start uncontrolled fires and are hazardous to dismounted Marines near the tank.* The tank commander and the infantry small-unit leader must coordinate when and under what conditions these launchers can be used. Grenade launchers are a useful feature for protecting the tank but can cause significant problems if unwisely used.

(6) The tank's size and armor can provide dismounted Marines with cover from direct-fire weapons and fragments. With coordination, tanks can provide moving cover for Marines as they advance across small open areas. However, enemy fire that strikes but does not penetrate a tank is a major threat to nearby Marines. Fragmentation generated by antitank rounds and ricochets off of tank armor have historically been a prime cause of casualties while infantry was working with tanks in built-up areas.

11. Artillery. A major source of fire support for infantry forces fighting in built-up areas is field artillery weapons. If the built-up area is near the coast, NGF can be used. Field artillery employment can be in either the indirect- or direct-fire mode.

a. Indirect Fire. Low-angle indirect artillery fire is not effective in attacking targets within walls and masonry structures. It tends to impact on roofs or upper stories rather than on structurally critical wall areas or pillars. High-angle fire can achieve the trajectories necessary to strike targets in urban areas.

Weapons of at least 155 mm are required against thick reinforced concrete, stone, or brick walls. Even with heavy artillery, large expenditures of ammunition are required to knock down buildings of any size. Tall buildings also create areas of indirect-fire dead space, which are areas that cannot be engaged by indirect fire because of a combination of building height and angle of fall of the projectile (Figure B-16). Usually the dead space for low-angle indirect fire is about five times the height of the highest building over which the rounds must pass.

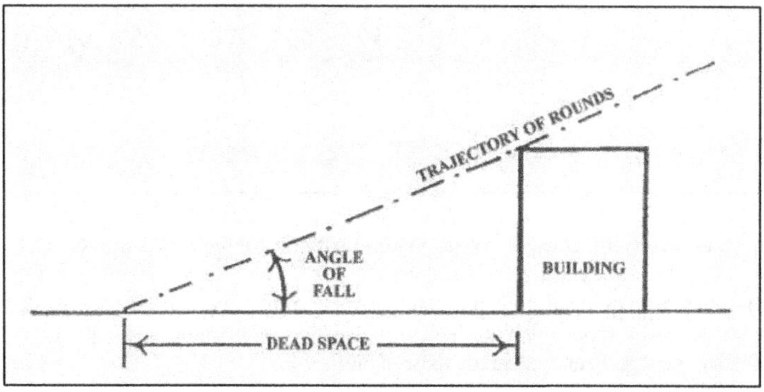

Figure B-16. Indirect-Fire Dead Space (Low Angle)

b. Direct Fire. A likely use of Marine artillery in an urban direct-fire role is to reinforce infantry against tough or important urban targets. Artillery should be used in this role only after an analysis of the need for heavy direct fire and the tradeoff involved in the extreme decentralization of artillery firepower. Artillery has the same need for close security and target designation as tanks.

c. Target Effects. Medium-caliber (155-mm) direct fire has a devastating effect against masonry construction and field fortifications. Smaller artillery (105-mm) pieces in the direct-fire mode are much less destructive than the larger caliber weapons. However, all Marine artillery is towed and, therefore, more difficult to employ in the direct-fire mode.

The 155-mm (M198) towed howitzer is effective because of its rate of fire and penetration. HE rounds can penetrate up to 38 inches of brick and unreinforced concrete. Projectiles can penetrate up to 28 inches of reinforced concrete with considerable damage beyond the wall. HE rounds with concrete-piercing fuzes provide an excellent means of penetrating strong reinforced concrete structures. One round can penetrate up to 46 inches. Five rounds are needed to reliably create a 1.5-meter breach in a wall that is 1 meter thick. About 10 rounds are needed to create the same size breach in a wall that is 1.5 meters thick. Superquick fuzing causes the rubble to be blown into the building, whereas delay fuzing tends to blow the rubble outward into the street. (See Table B-13 for employment considerations for artillery.)

Organization for Combat	Movement/Positioning	Delivery of Fire	Security	Command and Control
Centralized control is required during initial phases; decentralized control is required during later phases to support semi-independent actions of small units.	Movement should occur during night or periods of reduced visibility when possible. There are few displacements, often by platoon or section. Positions should be selected that minimize masking, provide several routes of escape, and afford as much cover and concealment as possible. Use of existing structures (garages, office buildings, highway overpasses) is recommended. Special techniques for emplacing howitzers, such as spades against a curb when the ground is not suitable for emplacement, may be required. Explosives may be required to soften emplacement of howitzers. Reconnaissance, selection, and occupation of position (RSOP) elements should be well armed because they may have to clear areas to be occupied. Extensive route reconnaissance is required. Target acquisition devices are somewhat degraded. Radars should be emplaced to cover likely areas of enemy indirect-fire weapon employment. Radars should not be placed in the midst of an urban area because of masking.	Both direct and indirect fires are delivered for supported units. Destruction of fortifications may require assault fire techniques. High-angle fires may be required. Need for accurate meteorological (MET) and survey data increases because most targets are point targets. Improved conventional munition and variable time (fuze) effects are reduced by structures, although they are effective against personnel on rooftops. HE delay is used for penetration effects. Illumination, chemical incendiary ammunition, and smoke are effective. Ammunition expenditures will be heavy. Lasers and PGMs permit destruction of targets with minimal rubbling of adjacent buildings. Tall buildings may hamper laser use. Batteries must be prepared for hasty survey techniques. Magnetic instruments are impaired.	Positions must be fortified.	Radio communications are impaired by buildings. Wire can usually be run overhead. Make use of civilian communications. A greater use of messengers and prearranged audio and visual signals is required.

Table B-13. Artillery Employment Considerations in Built-Up Areas

12. Mortars. The urban environment greatly restricts low-angle indirect fires because of overhead masking. While all indirect-fire weapons are subject to overhead masking, mortars are less affected than field artillery weapons because of the mortar's higher trajectory. For low-angle artillery fire, dead space is about five times the height of the building behind which the target sits. For mortar fire, dead space is only about one-half the height of the building. Because of these advantages, mortars are even more important than field artillery to the infantry during combat in built-up areas.

a. Employment. Not only can mortars fire into the deep defilade created by tall buildings, but they can also fire out of it. Mortars emplaced behind buildings are difficult for the enemy to accurately locate and even harder to hit with counterfire. Because of their light weight, even heavy mortars can be hand-carried to firing positions that may not be accessible to vehicles.

(1) The 60-mm and 81-mm mortars have limited effects on structural targets. Even with delay fuzes, they seldom penetrate more than the upper stories of light buildings. However, their wide area coverage and multioption fuzes make them useful against an enemy force advancing through streets, through other open areas, or over rubble.

(2) Mortar platoons often operate as separate firing sections during combat in built-up areas. The lack of large open areas can preclude establishing a platoon firing position. Figure B-17 shows how two mortar sections that are separated by only one street can be effective in massing fires and be protected from countermortar fire by employing defilade and dispersion.

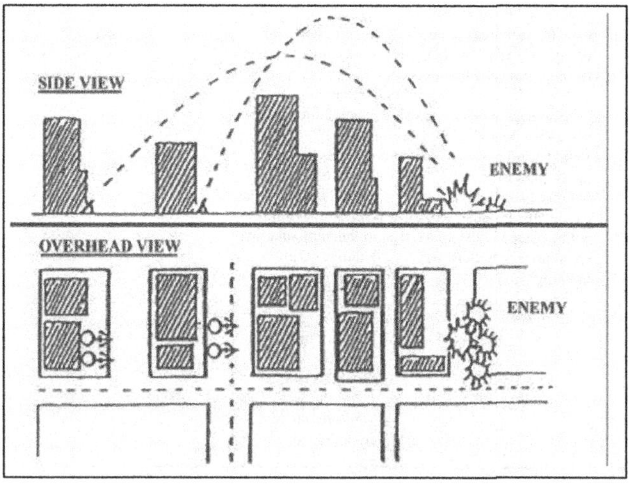

Figure B-17. Split-Section Mortar Operations on Adjacent Streets

(3) All three of the standard mortar projectiles are useful during combat in built-up areas. HE fragmentation rounds are most commonly used. WP is effective for starting fires in buildings and forcing the enemy out of cellars and light-framed buildings, and it is also the most effective mortar round against dug-in enemy tanks. Even near-misses blind and suppress the tank crew, forcing them to button up. Hits are difficult to achieve but are effective.

(4) Because the artificial roughness of urbanized terrain reduces wind speed and increases atmosphere mixing, mortar smoke tends to persist longer and give greater coverage in built-up areas than in open terrain.

(5) Urban masking has an impact on the use of illumination. In built-up areas, it is often necessary to plan illumination behind friendly positions, which places friendly troops in shadows and enemy troops in the light. Illumination rounds are difficult to adjust and are often of limited use because of the deep canyon nature of the urbanized area. Rapidly shifting wind currents in built-up areas also affect mortar illumination, making it less effective.

b. Effects of Mortar Fire. The multioption fuze on newer U.S. mortar rounds makes them effective weapons on urbanized terrain. Delay settings can increase penetration slightly, and proximity bursts can increase the lethal area covered by fragments. Tall buildings can cause proximity-fuzed mortar rounds to detonate prematurely if they pass too closely to the buildings.

(1) 60-mm Mortar. The 60-mm mortar round cannot penetrate most rooftops, even with a delay setting. Small explosive rounds are effective, however, in suppressing snipers on rooftops and preventing roofs from being used by enemy observers. The 60-mm WP round is not normally a good screening round because of its small area of coverage. In combat in built-up areas, however, the tendency of smoke to linger and the small areas to be screened make it more effective. During the battle for Hue in South Vietnam, 60-mm WP rounds were used to create small, short-term smoke screens to conceal movement across open areas such as parks, plazas, and bridges. Fragments from 60-mm HE rounds landing as close as 10 feet cannot penetrate a single sandbag layer or a single-layer brick wall. The effect of a 60-mm mortar HE round that achieves a direct hit on a bunker or fighting position is equivalent to 1 - 2 pounds of TNT. Normally, the blast will not collapse a properly constructed bunker, but it can cause structural damage. The 60-mm mortar will not normally create craters in a hard-surfaced road.

(2) 81-mm Mortar. The 81-mm mortar has much the same effect against urban targets as the 60-mm mortar. It has a slightly greater lethal area, and its smoke rounds (WP and RP) are more effective. A direct hit is equivalent to about 2 pounds of TNT. The 81-mm round cannot cause significant craters in a hard-surfaced road. With a delay setting, the 81-mm round can penetrate the roofs of light buildings.

13. Naval Gunfire. The primary NGF weapon, the 5-inch/54, has a high rate of fire and is roughly equivalent to the 155-mm howitzer in target effect. Target engagement considerations are the same as for artillery; however, the 5-inch/54's flat trajectory is even more affected by terrain masking and a constantly changing gun-target line. NGF can achieve greater target engagement by firing reduced charges at high angles; however, range will be reduced.

14. Aerial Weapons. Both rotary- and fixed-wing aircraft can quickly deliver large volumes of firepower over large built-up areas. Specific targets are hard to distinguish from the air. Good ground-to-air communications are vital in successfully employing aerial firepower. Planners have historically tended to overestimate the effects of HE on defenders. Modern, large buildings are remarkably resistant to damage from bombs and rocket fire.

a. Rotary-Wing Aircraft. Attack helicopters can be used to engage targets in built-up areas. Enemy armored vehicles in small parks, boulevards, or other open areas are good targets for attack helicopters.

(1) The HELLFIRE missile has a larger warhead and greater range than the TOW, but it too is a shaped-charge warhead and is not specifically designed for use against masonry targets. Laser target designation for the HELLFIRE may be difficult as a result of laser reflections off of glass and shiny metal surfaces. The use of attack helicopters to deliver ATGMs against targets in the upper stories of high buildings is sometimes desirable.

(2) The 2.75-inch folding-fin aerial rocket and the 20-mm cannon on Marine attack helicopters are good area weapons to use against enemy forces in the open or under light cover. They are usually ineffective against a large masonry target. The 20-mm cannon produces many ricochets, especially if armor-piercing ammunition is fired into built-up areas.

b. Fixed-Wing Aircraft. Close air support to ground forces fighting in built-up areas is a difficult mission for fixed-wing aircraft. Targets are hard to locate and identify, enemy and friendly forces could be intermingled, and enemy short-range air defense weapons are hard to suppress.

(1) Because enemy and friendly forces may be separated by only one building, accurate delivery of ordnance is critical. Marking panels, lights, electronic beacons, smoke, or some other positive identification of friendly forces is needed.

(2) General-purpose bombs that weigh from 500 to 2,000 pounds are moderately effective in creating casualties among enemy troops located in large buildings. High-dive-angle bomb runs increase accuracy and penetration but also increase the aircraft's exposure to antiaircraft weapons. Low-dive-angle bomb runs using high-drag (retarded) bombs can be used to get bombs into upper stories. Penetration is not good with high-drag bombs. Sometimes aerial bombs pass completely through light-clad buildings and explode on the outside.

(3) Aerial rockets and 20-mm cannons are only moderately effective against enemy soldiers in built-up areas because rockets lack the accuracy to concentrate their effects. The 20-mm cannon rounds penetrate only slightly better than the .50-caliber round, 20-mm armor-piercing rounds can ricochet badly, and tracers can start fires.

(4) The AC-130 aircraft has weapons that can be most effective during combat in built-up areas. This aircraft can deliver accurate fire from a 20-mm Vulcan cannon, 40-mm rapid-fire cannon, and 105-mm howitzer. The 105-mm howitzer round is effective against the roofs and upper floors of buildings. The AC-130 is accurate enough to concentrate its 40-mm cannon and 105-mm howitzer fire onto a single spot to create a rooftop breach that allows fires to be directed deep into the building.

(5) Laser and optically guided munitions can be effective against high-value targets. Specially developed, heavy, laser-guided bombs can penetrate hardened targets. However, problems associated with dense smoke and dust clouds hanging over the built-up area and laser scatter can restrict their use. If the launching aircraft can achieve a successful laser designation and lock-on, these weapons have devastating effects, penetrating deep into reinforced concrete before exploding with great force. If launched without a lock-on, or if the laser spot is lost, these weapons are unpredictable and can travel long distances before they impact.

15. Demolitions. Combat in built-up areas requires the extensive use of demolitions. All Marines, not just engineers, should be trained to employ demolitions. See FM 5-250, *Explosives and Demolitions*, for specific information on the safe use of demolitions.

a. Demolitions. Demolitions come in two types: TNT and C4. Exposed Marines must take cover or move at least 300 meters away from bulk explosives that are being used to breach walls.

(1) TNT comes in 1/4-, 1/2-, and 1-pound blocks. About 5 pounds of TNT are needed to breach a nonreinforced concrete wall that is 12 inches thick if the explosives are laid next to the wall and are not tamped. If the explosives are tamped, about 2 pounds are sufficient.

(2) C4 comes in many different-sized blocks. Normally, it is found in 1 1/4- and 2 1/2-pound blocks within prepackaged satchel charges. About 10 pounds of C4 placed between waist and chest high will blow a hole in the average masonry wall large enough for a man to walk through.

b. Shaped Charges. There are two sizes of shaped charges: a 15-pound M2A3 and a 40-pound M3A3. The M3A3 is the shaped charge that is most likely to be used in built-up areas. It can penetrate 5 feet of reinforced concrete. The hole tapers from 5 inches down to 2 inches. The amount of spall thrown behind the target wall is considerable. There is also a large safety hazard area for friendly forces.

c. Satchel Charges. There are two standard satchel charges: the M183 and the M37. Both come in their own carrying satchel with detonators and blasting cords. Each weighs 20 pounds. The M183 has 16 individual 1 1/4-pound blocks that can be used separately. When used untamped, a satchel breaches a 3-foot-thick concrete wall. Satchel charges are very powerful. Debris is thrown great distances. Friendly forces must move away and take cover before detonation.

d. Cratering Charges. The standard cratering charge is a 43-pound cylinder of ammonium nitrate. This explosive does not have the shattering effect of bulk TNT or C4. It is more useful in deliberate demolitions than in hasty ones.

Appendix C

Attacking and Clearing Buildings

Offensive operations on urbanized terrain will require Marines to attack and clear buildings.

1. Introduction. At the rifle platoon and squad level, the major offensive tasks for combat in built-up areas are attacking and clearing buildings. This involves:

- Suppressing defensive fires
- Advancing Marine assault forces
- Assaulting buildings
- Clearing buildings
- Reorganizing assault forces.

To accomplish these tasks, the assaulting unit task organizes into assault, security, and support forces or elements. (See Appendix A)

2. Requirements. There are five interrelated requirements for attacking a defended building: *fire support, movement, assault, clearing,* and *reorganization.* Proper application and integration of these requirements reduce casualties and hasten accomplishment of the mission. The proper integration of these requirements are determined by METT-T. For example, larger cities may have high-rise apartment buildings and industrial complexes that are separated by parking areas or parks. Increased fire support may be required to suppress and obscure enemy gunners covering the open terrain and spaces between buildings. Conversely, a built-up area with few open areas can provide numerous covered routes that may decrease fire support requirements.

3. Fire Support. Fire support necessary to advance the assault force may include:

- Suppressing and obscuring enemy gunners within the objective building(s) and adjacent structures
- Isolating the objective building(s) with fires to prevent enemy withdrawal, reinforcement, or counterattack
- Breaching walls en route to and in the objective structure
- Destroying enemy positions with direct-fire weapons

- Securing cleared portions of the objective
- Providing replacements for the assault force
- Providing resupply of ammunition and explosives
- Evacuating casualties and prisoners.

a. The size of the support force is determined by the type and size of the objective building(s), whether the adjacent terrain provides open or covered approaches, and the organization and strength of enemy defenses.

b. The support force for a Marine rifle squad may consist of one fire team that could be reinforced with heavy machine guns, antitank weapons, and/or armored vehicles. In situations involving a larger assault force, a rifle platoon or company reinforced with armored vehicles, engineers, and/or artillery may be required to support the movement and assault of an adjacent platoon or company.

c. After seizing objective buildings, the assault force reorganizes and may be required to provide supporting fires for a subsequent assault. Each weapon is assigned a target or area to cover. Individual small-arms weapons place fires on likely enemy weapon positions—loopholes, windows, and roof areas. Snipers are best employed in placing accurate fire through loopholes or engaging long-range targets. M203 fires are directed through windows or loopholes.

d. AT4s and demolitions are employed to breach walls, doors, barricades, and window barriers on the ground level of structures. The tank cannon or LAV-mounted 25-mm gun engages first-floor targets and breaches walls for attacking Marines. Armored vehicle- or wheeled vehicle-mounted machine guns can engage suspected enemy positions on upper floors and in adjacent structures. In addition to destroying or weakening structures, the tank's main gun rounds cause casualties by their explosive effects and by hurling debris throughout the interior of structures.

e. NGF, artillery, and mortars should initially use time fuzes to initially clear exposed enemy personnel, weapons, OPs, and radio sites from rooftops. They should then use delayed fuze action to cause casualties among the defenders inside the structure as a result of shrapnel and falling debris. They may also provide smoke to cover the movement and assault phases of the attack. Artillery can also be used in the direct-fire mode.

f. In support of the movement and assault phases of the attack, buildings can be attacked by fixed-wing aircraft or by attack helicopters.

4. Movement. The assault force minimizes the effects of enemy defensive fires during movement by:

- Using covered routes
- Moving after defensive fires have been suppressed or obscured
- Moving at night or during other periods of reduced visibility
- Selecting routes that will not mask friendly suppressive fires
- Crossing open areas (e.g., streets and spaces between buildings) quickly under the concealment of smoke and suppressive fires
- Moving on rooftops that are not covered by enemy direct fires.

a. The requirement for speed may dictate movement through streets and alleys without clearing all buildings. When moving in this manner, the maneuver force should employ infantry to lead the column, and supported by AAVs, LAVs, wheeled vehicle-mounted machine guns, or tanks following.

b. Marines should move along each side of a street with the leading squad staying abreast of armored vehicles. When not accompanied by armored vehicles, Marines move in a single file along one side of the street under cover of fires from supporting weapons. Individual Marines should be detailed to observe and cover a specific area, such as second-floor windows on the opposite side of the street.

5. Assault. Assault forces quickly and violently execute their assault and subsequent clearing operations. Momentum is maintained to prevent the enemy from organizing a more determined resistance on other floors or in other rooms. Small-unit leaders should keep the assault force moving without allowing the operation to become disorganized.

a. An assault in a built-up area involves the elementary skills of close combat. Leaders must:

- Be trained in the required techniques to defeat the enemy in a face-to-face encounter
- Keep themselves in excellent physical condition
- Have confidence in their abilities and those of their unit.

b. The composition of the assault force will vary depending on METT-T; however, its basic organization at the squad level consists of two assault elements carrying a fighting load of equipment and as much ammunition as possible, especially grenades.

6. Clearing. Entering at the top and fighting downward is the preferred method of clearing a building. Clearing a building is easier from an upper story because gravity and building construction become assets to the assault elements when throwing hand grenades and moving from floor to floor. However, this method is only feasible when access to an upper floor or rooftop can be gained from the windows or roofs of adjoining buildings and assault elements can be transported to the rooftops by helicopter (Figure C-1). Helicopters should land only on those buildings that have special heliports or parking garages on the roofs. When these are unavailable, Marines can rappel or fastrope onto the roof or dismount as the helicopter hovers a few feet above the roof. Assault elements then breach the roof or common walls with explosives and use ropes to enter the lower floors. Stairs are guarded by a covering team or security force when not in use to protect the assault force from a flank attack.

Figure C-1. Helicopters Used To Clear Buildings

a. Although the top-to-bottom method is preferred for clearing a building, assaulting the bottom floor and clearing upward is a common method. When using this technique, the assault force closes on the flank(s) or rear of the building. It then clears each room on the ground floor and then, moving upward, begin a systematic clearance of the remaining floors.

b. Entry can be gained through walls breached by explosives or gunfire. Assault elements should avoid windows and doors because they are usually covered by fire or boobytrapped. If tanks are attached to the company, they can breach the wall by firing their main gun (Figure C-2).

Figure C-2. Main Gun Used To Breach Exterior

c. Shortly before the assault, suppressive fires on the objective should be increased until masked by the advancing assault force. Once masked, fires are shifted to upper windows and continued until the assault force has entered the building. At that time, fires are shifted to adjacent buildings to prevent enemy withdrawal or reinforcement.

d. Hand grenades should be used to clear the space before assault element members enter through a breached wall. Immediately after the explosion, the assault element clearing team enters and uses three round bursts to fire into the interior of the room, concentrating on areas that are possible enemy positions.

e. Once inside the building, the assault force must cover staircases and seize rooms that overlook approaches to the building. These actions are required to isolate enemy forces within the building and to prevent reinforcement. Assault elements should clear each ground floor room and then the basement. Sequence for the assault is that:

(1) The assault force leader determines which room(s) to clear first.

(2) The support force provides suppressive fire while the assault force systematically clears the building. It also provides suppressive fire on adjacent buildings to prevent enemy reinforcements or withdrawal. The support force destroys any enemy trying to exit the building.

(3) The security force isolates the immediate area.

(4) After assault elements establish a foothold in the building, the assault force leader proceeds with the systematic clearing of the building.

f. When clearing a building from the top floor down, assault forces should gain entrance through a common wall or the roof of an adjoining building (Figure C-3). Demolition charges are used to breach the wall and gain entrance to the top floor. Access to lower floors and rooms may be gained by breaching holes in the floor and having the Marines jump or slide down ropes to the lower floors. Stairs can be used if they are first cleared.

g. When using the top-to-bottom method of clearing, requirements remain the same as for other methods. After the floor is breached to gain access to a lower floor, a grenade is allowed to cook-off and is dropped to the lower room. A Marine then sprays the lower room with gunfire using three-round bursts and drops through the mousehole.

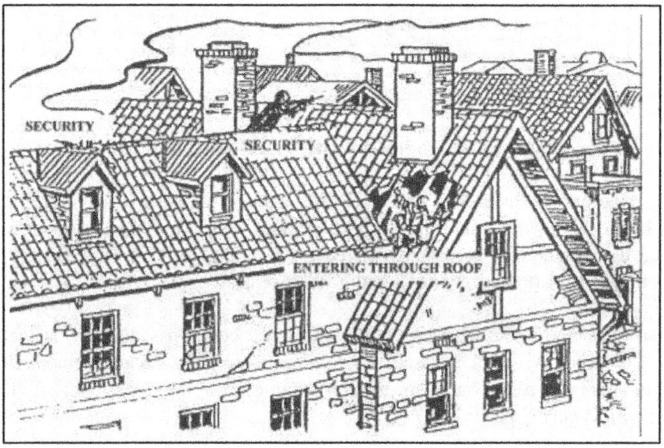

Figure C-3. Upper Floors Secured

h. Marines should avoid predictability when entering and clearing rooms. They should vary entry through doors and mouseholes to keep the enemy off guard (Figure C-4). As rooms are cleared, doors should be left open, and a predetermined mark (cloth, tape, spray paint) should be placed on the doorjamb or over the door, as established by unit SOPs.

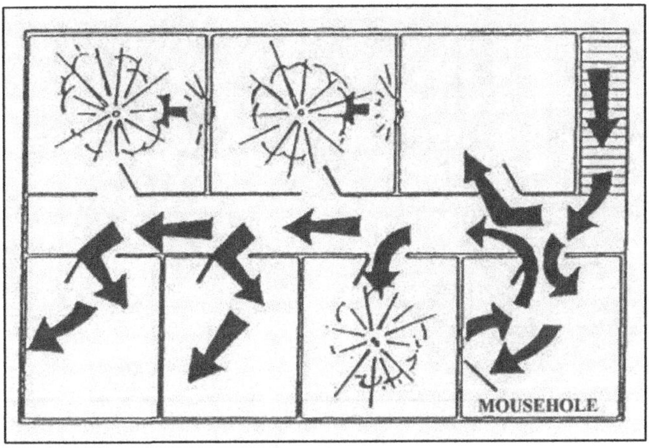

Figure C-4. Varying Techniques for Clearing Rooms

i. If there is a basement, it should be cleared as soon as possible, preferably at the same time as the ground floor. The procedures for clearing a basement are the same as for any room or floor. However, basements often contain entrances to tunnels such as sewers and communications cable tunnels. These should be cleared and secured to prevent the enemy from infiltrating back into cleared areas.

j. Types of buildings that may be cleared include brick buildings and houses, box-wall buildings, heavy-clad framed buildings, and light-clad framed buildings.

(1) Brick Buildings and Houses. The best way to enter a brick building is to blow a breaching hole in the side by using a tank firing HEAT ammunition (Figure C-5). If tanks are not available, a door or window in the rear of the building usually provides better cover and concealment for entry than one in the front. If there is enough cover and concealment, the assault force should enter the rear of the building at an upper level by using a fire escape or grappling hook.

Brick houses usually have similar floor plans on each floor (Figure C-6); therefore, ground floors are cleared the same way as upper floors.

Converely, the floor plans in larger brick buildings typically are different on ground-floor levels than on upper levels (Figure C-7).

Figure C-5. Entering a Brick Building

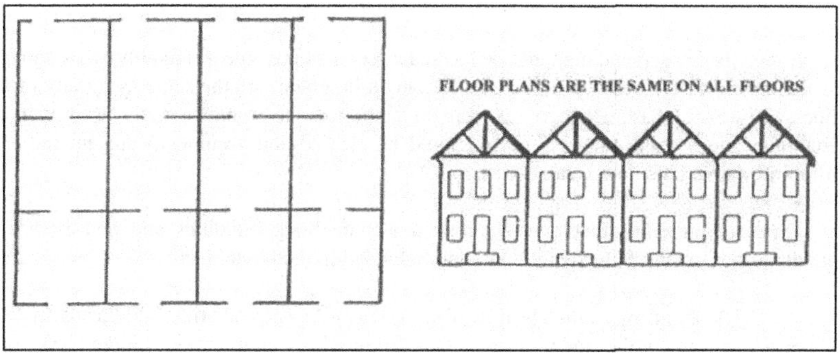

Figure C-6. Brick Houses Have Similar Floor Plans

To clear from building to building, the best method is to move from rooftop to rooftop because the roofs of brick buildings are usually easy to breach. The walls between buildings are at least three bricks thick (total of six bricks between buildings) and require large quantities of demolitions to breach. Walls are normally easier to breach on an upper floor than on a lower floor because the walls are thinner on upper floors. If rooftops are covered by fire and there are not enough demolitions to breach walls between buildings, clearing from rear to rear of buildings is safer than clearing from front to front.

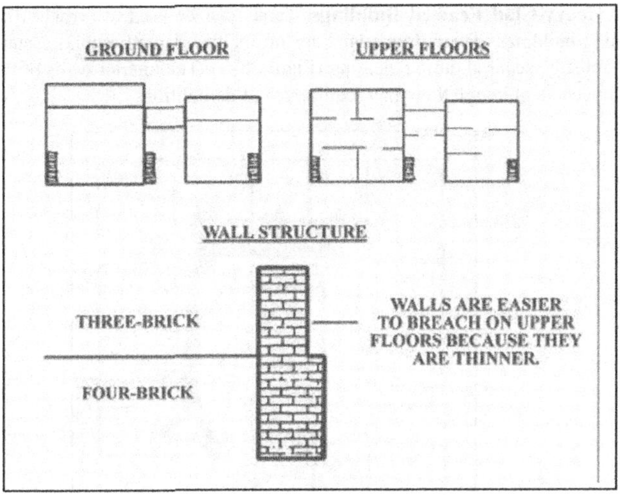

Figure C-7. Floor Plans of Brick Buildings

(2) Box-Wall Buildings. Box-wall buildings often have reinforced concrete walls (Figure C-8), which are difficult to breach because of reinforcing bars. Therefore, the best way to enter is to blow down the door or to blow in one of the side windows. The floor plans of these buildings are predictable; clearing rooms is usually done from one main hallway. Interior walls are also constructed of reinforced concrete and are difficult to breach. The stairways at the ends of the building must also be secured during clearing.

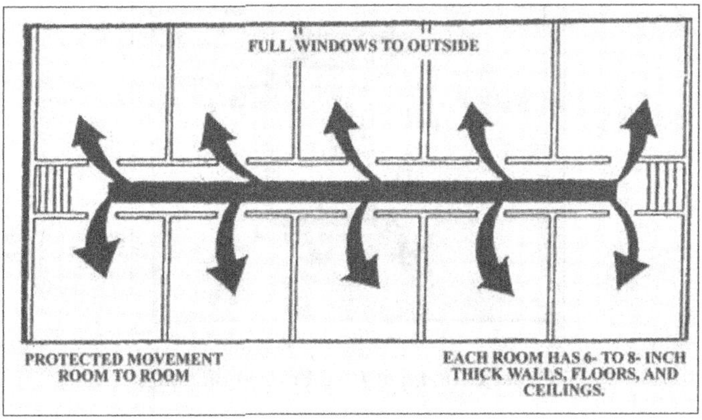

Figure C-8. Box-Wall Buildings

(3) Heavy-Clad Framed Buildings. Tanks can be used to breach a hole in heavy-clad framed buildings. Their floor plans are normally oriented around a stairway or elevator. this must be secured during clearing (Figure C-9). The interior walls of these buildings can be breached, although they may require use of demolitions.

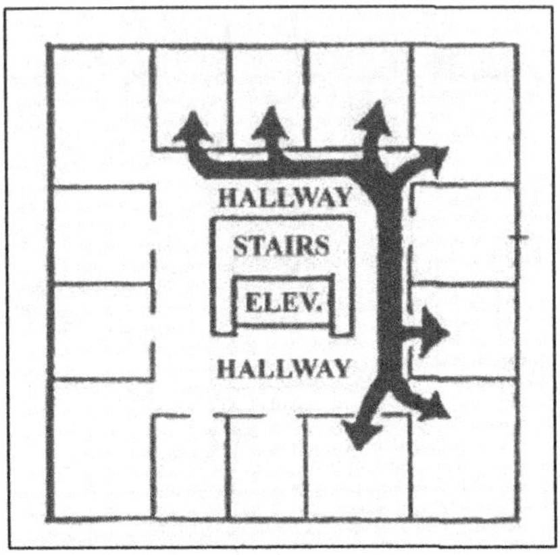

Figure C-9. Heavy-Clad Framed Buildings

(4) Light-Clad Framed Buildings. Light-clad framed buildings (Figure C-10), can be cleared in the same manner as heavy-clad buildings. Secure the central stairway and clear in a circular pattern. Walls are usually thin enough to be breached with an ax.

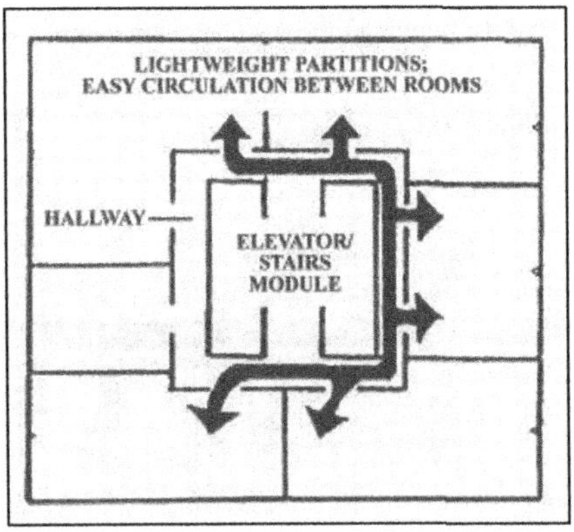

Figure C-10. Light-Clad Framed Buildings

7. Consolidation and Reorganization. Consolidation and reorganization of the assault force in a cleared building must be accomplished quickly to repel enemy counterattacks and to prevent the enemy from infiltrating back into the cleared building. After securing a floor (bottom, middle, or top), selected members of the assault force are assigned to cover potential enemy counterattack routes to the building. Those sentinels alert the assault force and place a heavy volume of fire on enemy forces approaching the building. They should cover:

- Enemy mouseholes between adjacent buildings
- Covered routes to the building
- Underground routes into the basement
- Approaches over adjoining roofs.

As the remainder of the assault force completes search requirements, defensive positions are assigned and tasks are accomplished as established by the priorities of work in the defense. The following actions may then be taken:

- Resupplying and redistributing ammunition
- Marking the building to indicate to friendly forces that the building has been cleared
- Assuming an overwatch mission and supporting an assault on another building
- Treating and evacuating wounded personnel
- Further developing a defensive position if the building is to be occupied for any period.

Appendix D

Fighting Positions

A critical defensive task in combat in built-up areas is the selection and preparation of fighting positions.

1. Considerations. Leaders should consider the following factors when establishing fighting positions:

a. Protection. Leaders should select buildings that provide protection from direct and indirect fires. Reinforced concrete buildings with three or more floors provide suitable protection, while buildings constructed of wood, paneling, or other light materials require reinforcement to gain sufficient protection. One- to two-story buildings without strongly constructed cellars are vulnerable to indirect fires and require construction of overhead protection for each firing position.

b. Dispersion. Normally, a position should not be established in a single building when it is possible to occupy two or more buildings that permit mutually supporting fires. A position in one building without mutual support is vulnerable to bypass, isolation, and subsequent destruction from any direction.

c. Concealment. The fighting position should be concealed. Buildings that are obvious defensive positions should not be selected. The occupation of exposed buildings may be necessary because of requirements for security and fields of fire. Therefore, reinforcement of the structure must be accomplished to provide suitable protection within the building.

d. Fields of Fire. Positions should be mutually supporting and have interlocking fields of fire in all directions. Clearing fields of fire may require the destruction of adjacent buildings by using explosives, engineer equipment, and field expedients. Care should be taken to avoid highlighting the building as a defensive position.

e. Covered Routes. Defensive positions should have at least one covered route that permits resupply, MEDEVAC, reinforcement, or withdrawal from the building. The route can be established by one of the following means:

- Through walls to adjacent buildings
- Through underground systems
- Through communications trenches
- Behind protective buildings.

f. Observation. The building should permit observation of enemy avenues of approach and adjacent defensive sectors.

g. Fire Hazard. Avoid selecting positions in buildings that are a fire hazard. If flammable structures must be occupied, the danger of fire can be reduced by wetting the structure, laying an inch of sand on the floors, and providing fire extinguishers and firefighting equipment. Routes of escape should also be prepared in case of fire.

h. Time. The time available to prepare the defense could be the most critical factor. If enough time is insufficient, buildings that require extensive preparation should not be used. Conversely, buildings located in less desirable areas that require little improvement may be a better choice for a defensive position.

2. Preparation. Preparation of fighting positions depends on proper selection and construction.

a. Selecting Positions. Each weapon should be assigned a primary sector of fire to cover enemy approaches. Alternate positions that overwatch the primary sector should also be selected. These positions are usually located in an adjacent room on the same floor (Figures D-1A, D-1B, and D-1C). Each weapon must be assigned a supplementary position (to engage attacks from other directions) and an FPL.

b. Building Positions. There are many ways to establish a fighting position in a building.

(1) Window Position. Marines should kneel or stand on either side of a window. To fire downward from upper floors, tables or similar objects can be placed against the wall to provide additional elevation, but they must be positioned to prevent the weapon from protruding through the window (Figure D-2). Leaders should inspect positions to determine the width of sector that each position can engage.

(2) Loopholes. To avoid establishing a pattern of always firing from windows, loopholes should be prepared in walls. Marines should avoid firing directly through loopholes to enhance individual protection.

(a) Several loopholes are usually required for each weapon (primary, alternate, and supplementary positions). The number of loopholes should be carefully considered because they can weaken walls and reduce protection. Engineers should be consulted before numerous loopholes are made. Loopholes should be made by punching or drilling holes in walls and should be placed where they are concealed. Blasting loopholes can result in a large hole that can easily be seen by the enemy.

(b) Loopholes should be cone-shaped to obtain a wide arc of fire, to facilitate engagement of high and low targets, and to reduce the size of the exterior aperture (Figure D-3). The edges of a loophole splinter when hit by bullets; therefore, protective linings such as an empty sandbag held in place by wire mesh will reduce spalling effects. When not in use, loopholes should be covered with sandbags to prevent the enemy from firing into or observing through them.

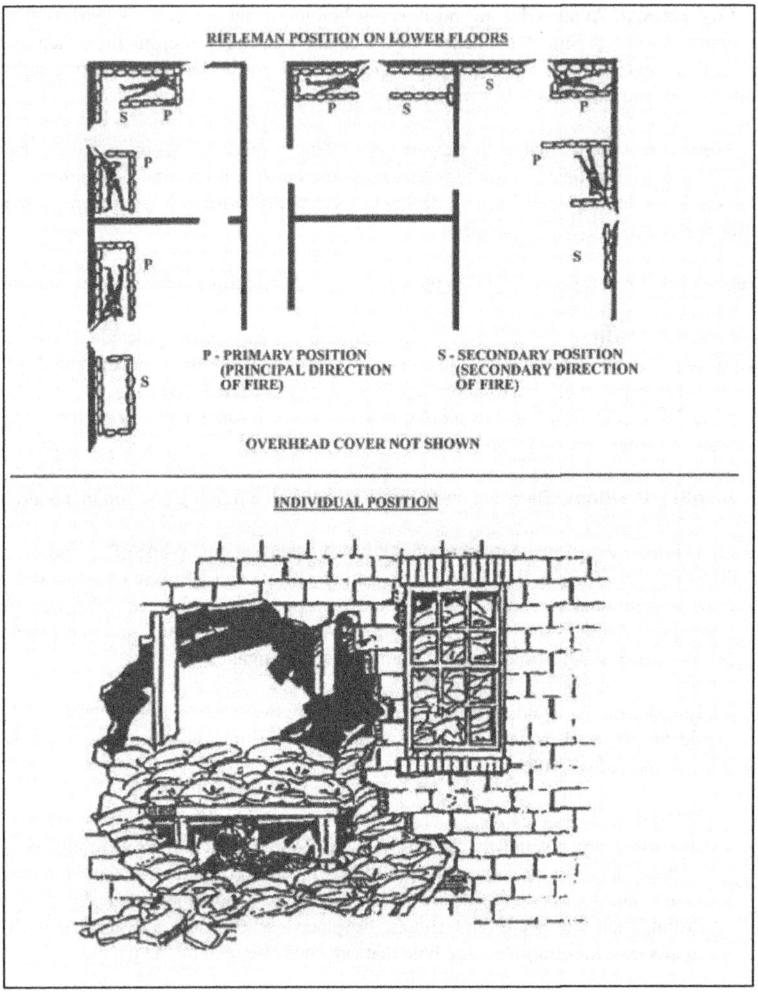

Figure D-1A. Weapon Positions

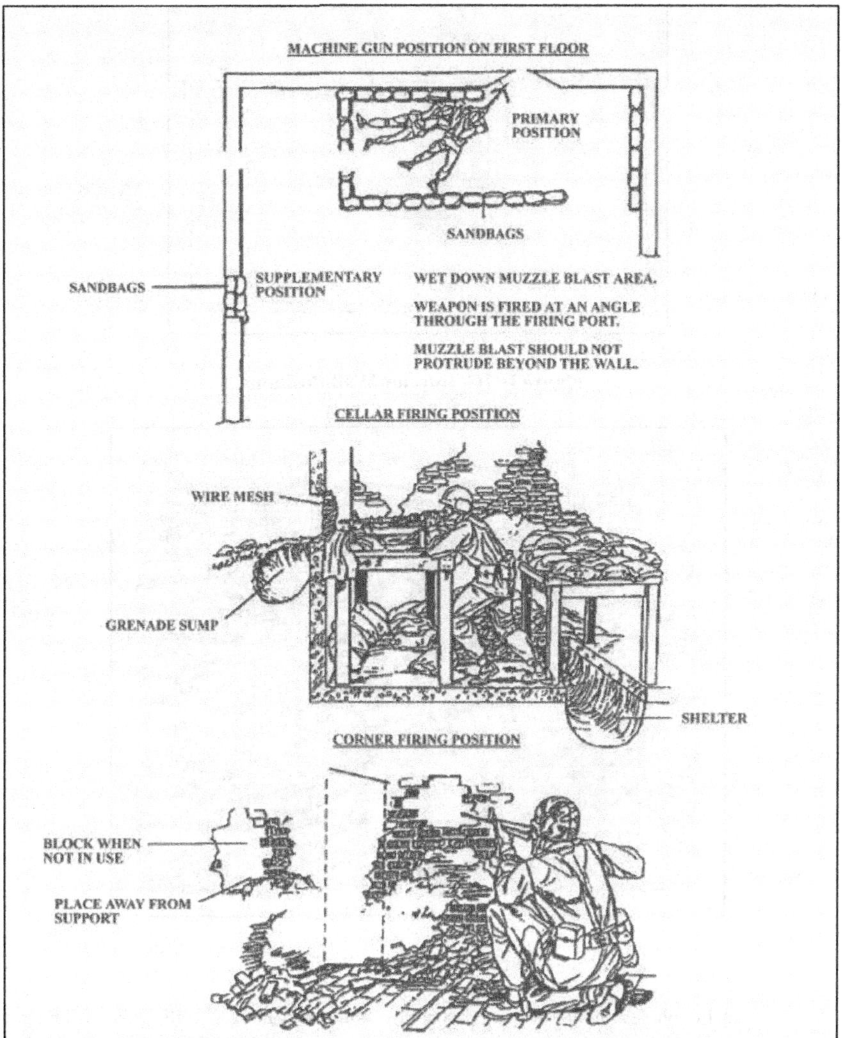

Figure D-1B. Weapon Positions (Continued)

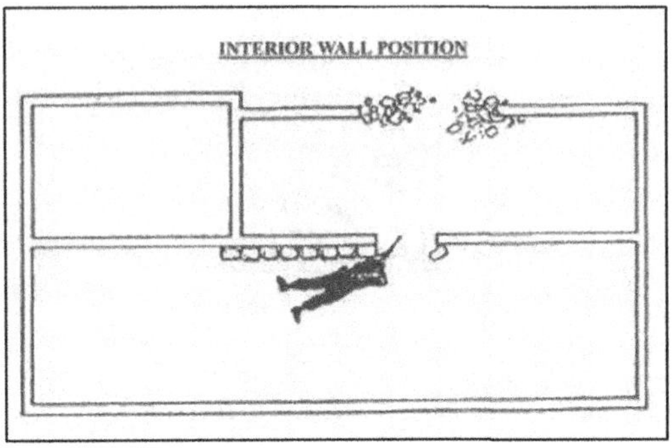

Figure D-1C. Interior Wall Position

Figure D-2. Window Position

(c) Loopholes should also be prepared in interior walls and ceilings of buildings to permit fighting within the position. Interior loopholes should enable overwatch of stairs, halls, and unoccupied rooms and be concealed by pictures, drapes, or furniture. Loopholes in floors permit the defender to engage enemy personnel on lower floors with small-arms fire and grenades.

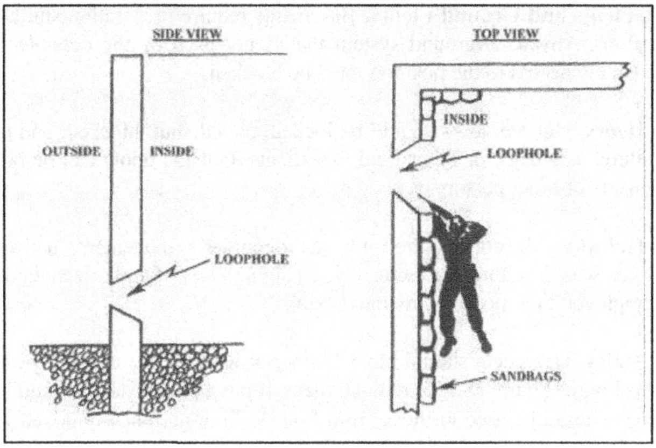

Figure D-3. Cone-Shaped Loopholes

(d) Although walls provide some frontal protection, they should be reinforced with sandbags, furniture filled with dirt, or other expedients. Each position should have overhead and all-around protection (Figure D-4).

Figure D-4. Position With Overhead and All-Around Protection

c. Other Construction Tasks. Other construction tasks may need to be performed. These include work on:

(1) Basements and Ground Floors. Basements require preparation similar to that of the ground floor. Any underground system that is not used by the defender but that could provide enemy access to the position must be blocked.

(a) Doors. Unused doors should be locked, nailed shut, blocked, and reinforced with furniture, sandbags, or other field expedients. Outside doors can be boobytrapped by engineers or other personnel.

(b) Hallways. If not required for the defender's movement, hallways should be blocked with furniture and tactical wire (Figure D-5). If authorized, boobytraps should be employed to impede enemy movement.

(c) Stairs. Defenders should block stairs not used by the defense with furniture and tactical wire (Figure D-5) or remove them. If possible, all stairs should be blocked, and ladders should be used to move from floor to floor and then removed when not being used. Boobytraps should also be employed on stairs.

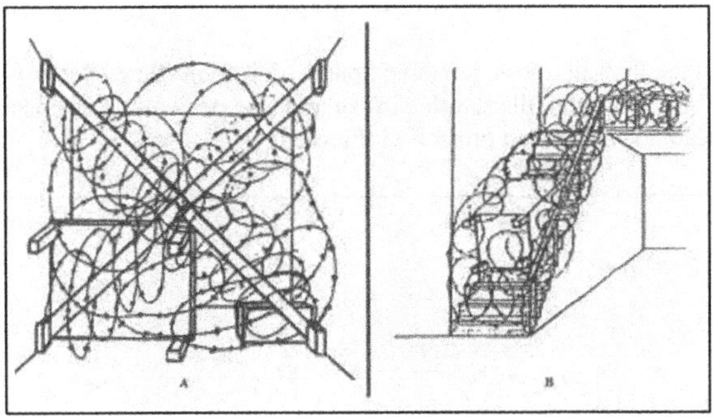

Figure D-5. Blocking Hallways and Stairs

(d) Windows. All glass should be removed. Windows not being used should be blocked with boards or sandbags.

(e) Floors. Fighting positions should be constructed on the ground or basement floors. If there is no basement, fighting positions on the first floor must be constructed to provide additional protection from heavy direct-fire weapons.

(f) Ceilings. Supports that can withstand the weight of rubble from upper floors should be placed under ceilings to provide overhead cover (Figure D-6).

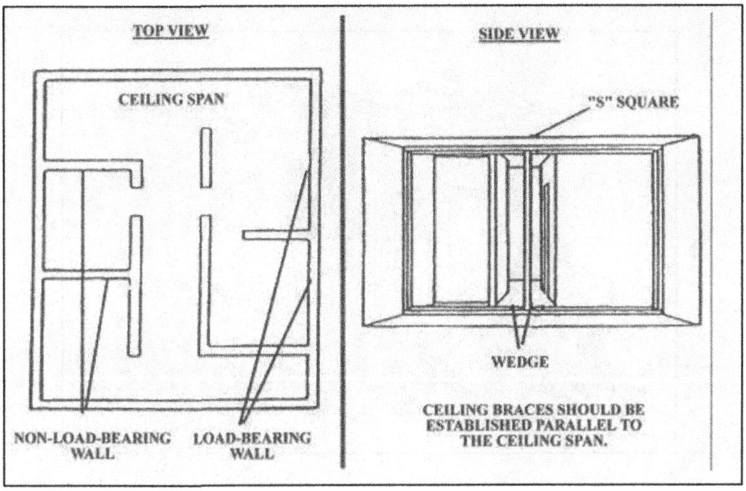

Figure D-6. Ceiling Reinforcement

(g) Unoccupied Rooms. Rooms not required for defense should be blocked with tactical wire, boobytrapped, or both.

(2) Upper Floors. Upper floors require the same preparation as ground floors. Windows need not be blocked, but they should be covered with wire mesh to block grenades thrown from the outside. The wire should be loose at the bottom to permit the defender to drop grenades on enemy forces below.

(3) Interior Routes. Routes are required that permit defending forces to move within the building to engage enemy forces from any direction or reinforce other forces (Figure D-7). Escape routes should also be planned and constructed to permit rapid evacuation of a room or the building. Small holes (called mouseholes) should be made through interior walls to permit movement between rooms. Once the defender has withdrawn to another level, such holes should be clearly marked for both day and night identification. All personnel must be briefed as to where the various routes are located. Rehearsals should be conducted so that everyone becomes familiar with the routes.

(4) Fire Prevention. Buildings that have wooden floors and raftered ceilings require extensive fire prevention measures. The attic and other wooden floors should be covered with about 1 inch of sand or dirt, and buckets of water should be positioned for immediate use. Firefighting materials (dirt, sand, fire extinguishers, and blankets) should be placed on each floor for immediate use. Water basins and bathtubs should be filled as a reserve for firefighting. All electricity and gas should be turned off. Fire breaks can be created by destroying buildings adjacent to the defensive position.

Figure D-7. Movement Between Floors

(5) Communications. Telephone lines should be laid through adjacent buildings or underground systems or buried in shallow trenches. Radio antennas can be concealed by placing them among civilian television antennas, along the sides of chimneys and steeples, or out windows that direct FM communications away from enemy electronic warfare sources and ground observation. Telephone lines within the building should be laid through walls and floors.

(6) Rubbling. Rubbling parts of the building may provide useful cover and concealment for weapons emplacements. Rubbling should be performed only by trained engineers.

(7) Rooftops. Positions in flat-roofed buildings require obstacles that restrict helicopter landings (See Figure F-2). Rooftops that are accessible from adjacent structures should be covered with tactical wire or other expedients and must be guarded. Entrances to buildings from rooftops can be blocked if compatible with the overall defensive plan. Any structure on the outside of a building that could assist in scaling the buildings to gain access to upper floors or to the rooftop should be removed or blocked.

(8) Obstacles. Obstacles should be positioned adjacent to buildings to stop or delay tanks and infantry.

(9) Fields of Fire. Fields of fire should be improved around the defensive position. Selected buildings may be destroyed to enlarge fields of fire. Obstacles to ATGMs, such as telephone or electrical wires, should be cleared. Dead space should be covered with mines and obstacles.

3. Tank and Other Armored Vehicle Positions. Fighting positions for tanks and other armored vehicles are essential to a complete and effective defensive plan in built-up areas.

a. Armored Vehicle Positions. Armored vehicle positions are selected and developed to obtain the best cover, concealment, observation, and fields of fire while retaining the vehicle's ability to move.

(1) If fields of fire are restricted to streets, hull-down positions should be used to gain cover and to fire directly down streets (Figure D-8). From those positions, armored vehicles are protected and can rapidly move to alternate positions. Buildings collapsing from enemy fires are a minimal hazard to armored crews.

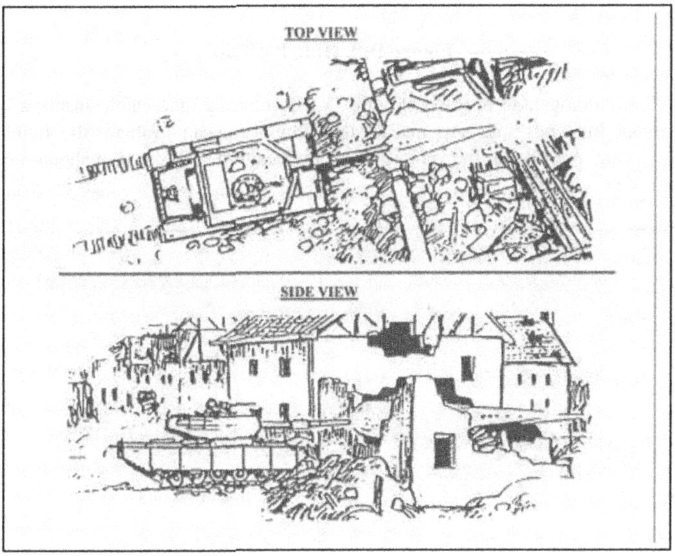

Figure D-8. Hull-Down Position

(2) The hide position (Figure D-9) covers and conceals the vehicle until time to move into position for engagement of targets. Because the crew will not be able to see advancing enemy forces, an observer from the vehicle or a nearby infantry unit must be concealed in an adjacent building to alert the crew. The observer acquires the target and signals the armored vehicle to move to the firing position and fire. After firing, the vehicle moves to an alternate position to avoid compromising one location.

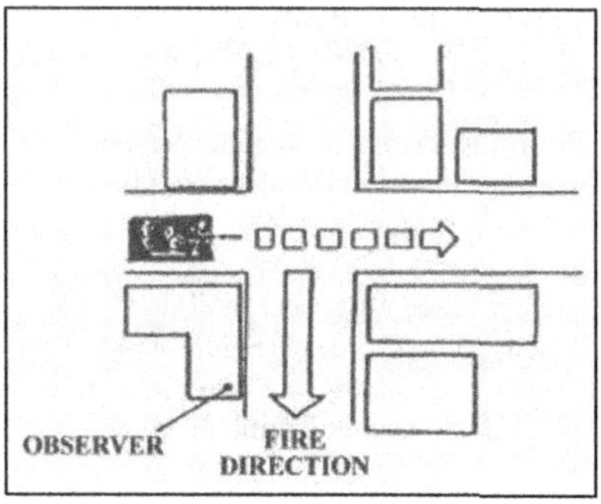

Figure D-9. Hide Position

(3) The building hide position (Figure D-10) conceals the vehicle inside a building. If basement hide positions are inaccessible, engineers can evaluate the building's floor strength and prepare for the vehicle. Once the position is detected, it should be evacuated to avoid enemy fires.

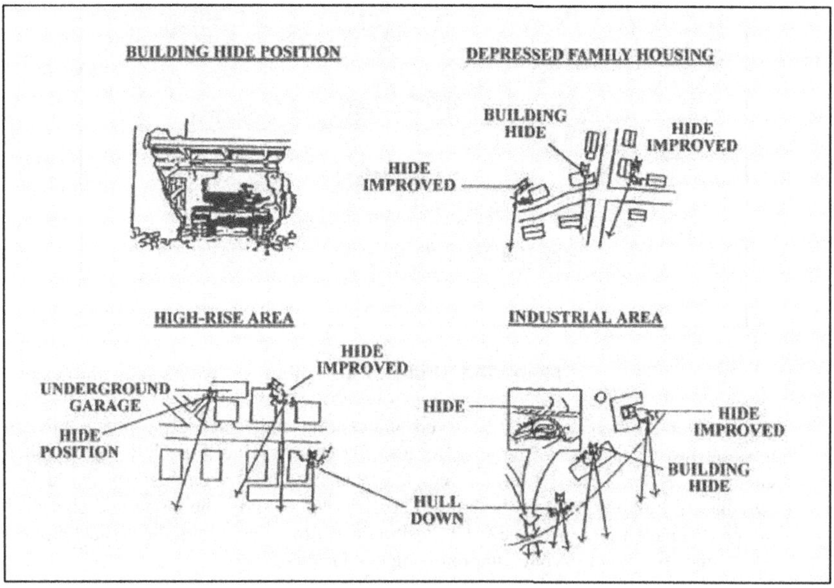

Figure D-10. Building Hide Position

4. Antitank Guided Missile Positions. HMMWVs with ATGMs are employed in areas that maximize their capabilities in the built-up area. Figure D-11 shows mounted and dismounted firing positions.

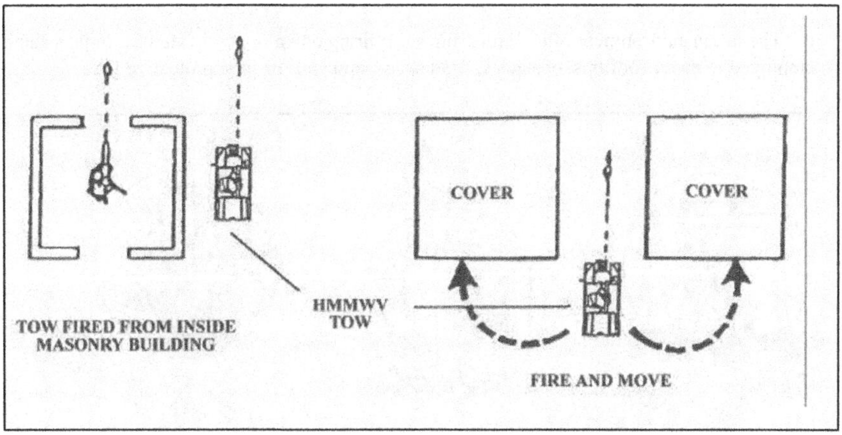

Figure D-11. ATGM Positions

a. Rubble may interfere with missile flight when ATGMs are fired from the HMMWV or from street-level or bottom-floor fighting positions. When firing down streets, missiles must have at least 1 meter of clearance over rubble. Other obstacles to missile flight include trees and brush, vehicles, television antennas, buildings, power lines and wires, walls, and fences (See Figure B-7).

b. An AT4 is best suited for built-up areas because its short arming distance allows employment at close range. AT4s and other light and medium antitank weapons are not effective against the front of modern battle tanks. Because tanks have the least armor protection on the top and rear deck and the tank presents a larger target when engaged from above, AT4s should fire down onto tanks.

5. Sniper Positions. Snipers contribute to combat in built-up areas by firing on selected enemy targets. Effective sniper employment can trouble the enemy far more than its relative personnel numbers reflect.

a. General areas (a building or group of buildings) are designated as sniper positions (Figure D-12), but the sniper selects the best position for engagement. Masonry buildings that offer the best protection, long-range fields of fire, and all-around observation are preferred. The sniper also selects several secondary and supplementary positions to cover his areas of responsibility.

b. Engagement priorities for snipers are determined by the relative importance of the targets to the effective operations of the enemy. Sniper targets may include tank commanders, crewmen of direct fire support weapons, crewmen of crew-served weapons, officers, FOs, and radiotelephone operators.

c. The urban environment often limits snipers to firing down or across streets. Snipers can be employed to cover rooftops, obstacles, dead space, and gaps in final protective fires.

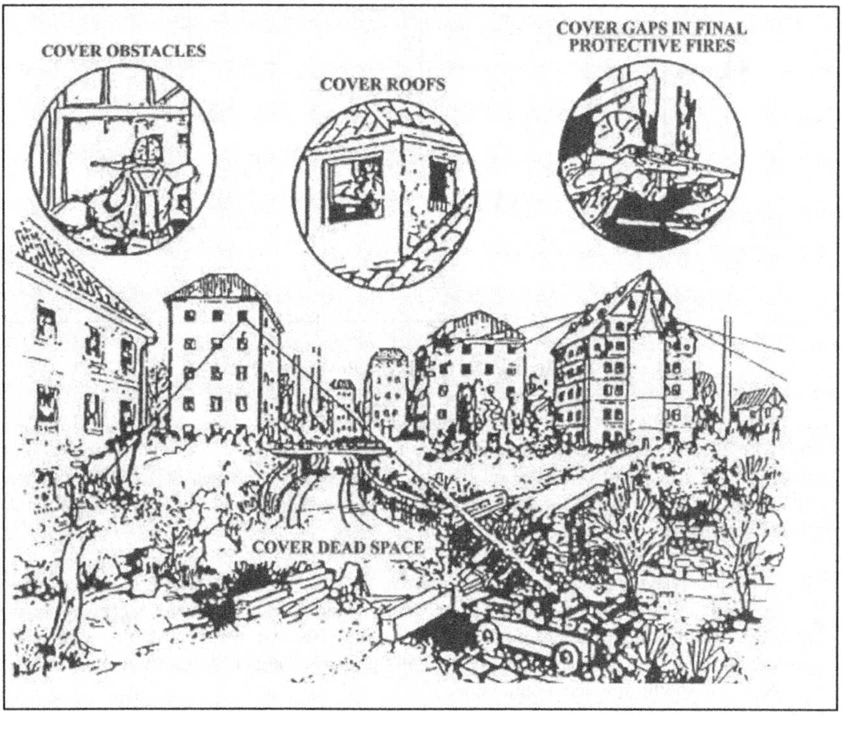

Figure D-12. Sniper Positions

Appendix E

Subterranean Operations

Knowledge of the nature and location of underground facilities is valuable to both the urban attacker and defender.

1. Tactical Value. Fighting in MOUT is multidimensional. Larger cities may have subterranean features that include sunken garages, underground passages, subway lines, utility tunnels, sewers, and storm drains (Figure E-1). Most of these features allow for the movement of units. Even in smaller European towns, sewers and storm drains permit Marines to move beneath the fighting and surface behind the enemy.

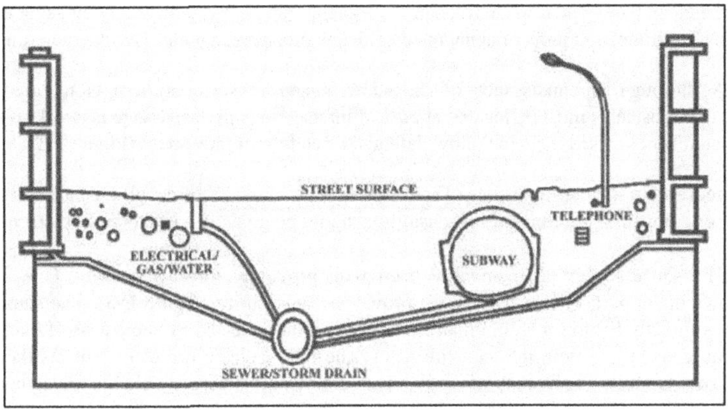

Figure E-1. Potential Underground Passages

a. Subterranean passages may provide the attacker with covered and concealed routes into and through built-up areas. This may allow an enemy to launch his attack along roads that lead into the city while infiltrating a smaller force into the defender's rear. An enemy unit in the defender's rear area may disrupt his defense and obstruct the avenues of withdrawal for his forward defense.

b. An attack along a subterranean avenue of approach could become the main effort. Even if the subterranean effort is not immediately successful, it forces the defender to fight on atleast two levels and to extend resources to more than just street-level fighting.

c. The existence of subterranean passages forces the defender to defend the built-up area above and below ground. While these passages are usually more of a disadvantage to the defender than to the attacker they may offer some advantages in the defense. When thoroughly reconnoitered and controlled by the defender, subterranean passages provide excellent covered and concealed routes to move reinforcements or to launch counterattacks. They also provide ready-made covered lines of communication for the movement of supplies and evacuation of casualties and provide places to cache supplies for forward companies. Subterranean passages also offer the defender a ready-made conduit for communications wire, with protection from armored vehicles and indirect fires.

2. Denial to the Enemy. Subterranean passages are useful to the defender only to the extent that the attacker can be denied their use. The defender has an advantage in that, given the confining, dark environment of these passages, a small group of determined Marines in a prepared position can defeat a numerically superior force.

a. Tunnels afford the attacker little cover and concealment except for the darkness and any manmade barriers. The passageways provide tight fields of fire and amplify the effects of grenades. Obstacles at intersections in tunnels support excellent ambush sites and may turn the subterranean passages into a maze. These obstacles can be created quickly by using chunks of rubble, furniture, and parts of abandoned vehicles interspersed with M18 claymore mines.

b. A thorough reconnaissance of the subterranean or sewer system is required. To be effective, obstacles must be located at critical intersections in the passage network so that they trap attackers in a kill zone but allow defenders freedom of movement (Figure E-2).

3. Subterranean Reconnaissance Techniques. The mission of local reconnaissance for a platoon- or company-sized area of responsibility should be given to a squad-sized element.

a. The squad leader tasks one fire team with providing forward security. One fire team follows in support, and one fire team provides rear security (Figure E-3). The squad leader moves directly behind the pointman and records data collected by the patrol. If available, an engineer should be with the lead fire team. The rear security fire team protects the ground- and subterranean-level point of entry. This fire team is responsible for detecting enemy movement on the patrol's rear and serves as the communications link between the squad leader and higher headquarters. Organization of the patrol may vary based on METT-T.

b. The squad leader should carry a map, compass, street plan, and notebook in which he has detailed the information that he must gather. One Marine is assigned to carry the tools needed to open manhole covers. If the patrol is to move more than 200 to 300 meters, or if the platoon leader directs, the squad may carry a sound-powered telephone (TA-1) and wire dispenser (XM-306A) for communications. Radios are often unreliable in this environment. Marines (especially the pointman, squad leader, engineer, and fire team leaders) should be equipped with night vision devices to maintain surveillance within the sewer. Thermal night devices are also useful as there may not be enough ambient light for accurate visibility with night vision devices.

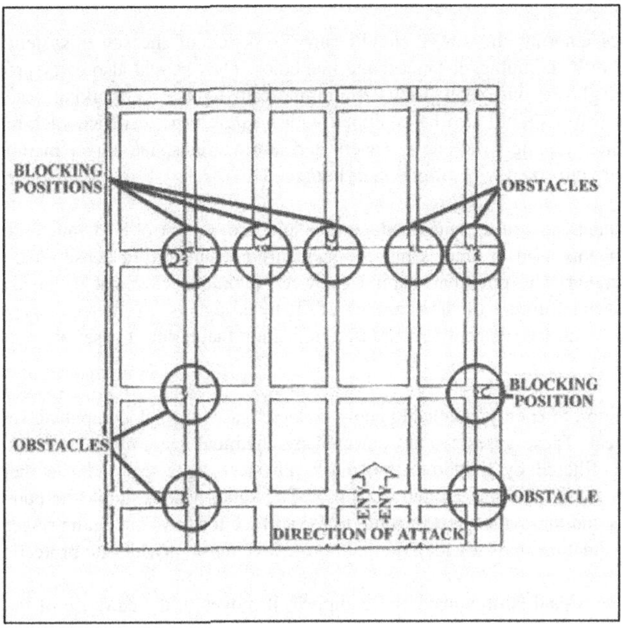

Figure E-2. Defense of a Sewer System

Figure E-3. Organization of the Patrol Unit

c. Marines entering the sewer should carry a sketch of the sewer system that includes magnetic north, azimuths, distances, and manholes. They should also carry protective masks, flashlights, gloves, and chalk (or chemiluminescent lights) for marking features along the route. The patrol should also be equipped with a safety rope to which each man is tied. To improve their footing in slippery sewers and storm drains, the squad members may wrap chicken wire or screen wire around their boots.

d. A constant concern to units conducting a subterranean patrol is chemical defense. Enemy chemical agents used in tunnels may be suddenly encountered in dense concentrations with little dispersion. The pointman should carry an automatic chemical agent alarm system to provide instant warning of the presence of chemical agents. M8 detector paper can also be used to detect chemical agents. At the first indication that harmful gases are present, the squad should mask.

e. In addition to enemy chemical agents, noxious gases from decomposing sewage can also pose a threat. These gases are not detected by chemical agent alarm systems, nor are they completely filtered by the protective mask. Physical signs that indicate their presence in harmful quantities are nausea and dizziness. The squad leader should be constantly alert to these signs and know the shortest route to the surface for fresh air. Some sewer gases deplete oxygen, a situation against which the field protective mask provides no protection.

f. Once the squad is organized and equipped, it moves to the entrance of the tunnel. This may be a manhole. With the manhole cover removed, the squad waits 15 minutes before entry to allow any gases that may be present to dissipate. The pointman then descends into the tunnel to determine whether the air is safe to breathe and whether movement is restricted. The pointman should remain in the tunnel for 10 minutes before the rest of the squad follows. If he becomes ill or is exposed to danger, he can be pulled out by the safety rope.

g. When the squad is moving through the tunnel, the pointman moves about 10 meters in front of the squad leader. Other squad members maintain 5-meter intervals. If water in the tunnel is flowing faster than 2.5 meters per second or the sewer contains slippery obstacles, those intervals should be increased to prevent all patrol members from falling if one man slips. All squad members should stay tied into the safety rope so that they can easily be retrieved from danger. The last Marine in the patrol marks the route with chalk or chemiluminescent lights so other units can find the squad.

h. The squad leader should record the azimuth and pace count of each turn he takes in the tunnel. When he encounters a manhole to the surface, the pointman should open it and determine the location, which the patrol leader then records. The use of recognition signals (Figure E-4) prevents friendly forces from accidentally shooting the pointman as he appears at a manhole.

Figure E-4. Recognition Signals

i. Once the patrol has returned and submitted its report, the platoon leader must decide how, or whether, to use the tunnel. In the offense, a tunnel might provide a covered route to move behind the enemy's defenses. In the defense, a tunnel might provide a covered passage between positions. In any case, the squad members that provided the reconnaissance should act as guides along the route.

j. If the tunnel is to be blocked, the platoon may emplace concertina wire, early warning devices, and antipersonnel mines. A security force should be established at the entrance of the sewer (Figure E-5) to provide warning against any enemy trying to approach the platoon's defense. It should be equipped with command-detonated illumination. The overpressure from grenades, mines, and boobytraps exploding in a sewer or tunnel can have adverse effects on friendly forces, including ruptured eardrums and wounds from flying debris. Also, gases found in sewers can be ignited by the blast effects of these munitions. For these reasons, small-arms weapons should be employed in tunnels and sewers. Friendly personnel should be outside of tunnels or out of range of the effects when mines or demolitions are detonated.

4. Psychological Considerations. Combat operations in subterranean passages are similar to night combat operations. The psychological factors of night operations reduce confidence, cause fear, and increase a feeling of isolation. This feeling of isolation is further magnified by the tight confines of the tunnels. The layout of tunnels could require greater dispersion between positions.

Figure E-5. Security Position Established at the Entrance to a Sewer

a. Leaders must enforce measures to help dispel the feelings of fear and isolation experienced when operating in tunnels. These measures include leadership training, physical and mental fitness, sleep discipline, and stress management.

b. Leaders maintain communication with Marines manning positions in tunnels either through personal visits, messengers, or field telephone. Communication increases situational awareness and provides insight into the mental state of Marines. Marines manning positions below ground should be given as much information as possible on the organization of the tunnels and the importance of the mission. They should be briefed on contingency plans and alternate positions should their primary positions become untenable.

c. Physical and mental fitness can be maintained by periodically rotating Marines out of tunnels so they can stand and walk in fresh air and sunlight. Stress management is also a factor of operations in tunnels. Historically, combat in built-up areas has been one of the most stressful forms of combat. Fighting in continuous darkness and restricted maneuver space causes even more stress to Marines than street fighting.

Appendix F

Obstacles, Mines, and Demolitions

Obstacles are used extensively in combat in built-up areas to allow the defender to canalize the enemy, impede his movement, and disrupt his attack.

Section I

Obstacles

1. Introduction. Obstacles are designed to prevent movement by personnel, to separate infantry from tanks, and to slow or stop vehicles. (See MCRP 3-17A, *MAGTF Breaching Operations.*)

2. Types. Antipersonnel mines, barbed wire, boobytraps, and exploding flame devices are used to construct antipersonnel obstacles (Figure F-1). (See FMFRP 13-5/FM 5-34, *Engineer Field Data*; FM 5-250, *Explosives and Demolitions*; and FM 3-11, *Flame, Riot Control Agent, and Herbicide Operations,* for more detailed information.) These obstacles are used to block the following infantry approaches:

- Streets
- Buildings
- Roofs
- Open spaces
- Dead space
- Underground systems.

a. The approval authority to boobytrap buildings is normally the GCE commander. This authority may be delegated to subordinate commanders.

b. The three types of wire obstacles used in defensive operations are protective, tactical, and supplementary. Wire obstacles should be integrated with other obstacles and covered by fire and observation.

(1) Protective wire is usually located beyond hand-grenade range (40 to 100 meters) from the defensive position.

(2) Tactical wire is positioned to increase the effectiveness of friendly weapons fire. Tactical wire is usually positioned on the friendly side of machine gun FPLs.

(3) Supplementary wire is used to break up the pattern of tactical and protective wire to prevent the enemy from easily locating friendly weapons.

c. Obstacles may be placed in dead spaces to restrict infantry movement in areas that cannot be observed and that are protected from direct fires.

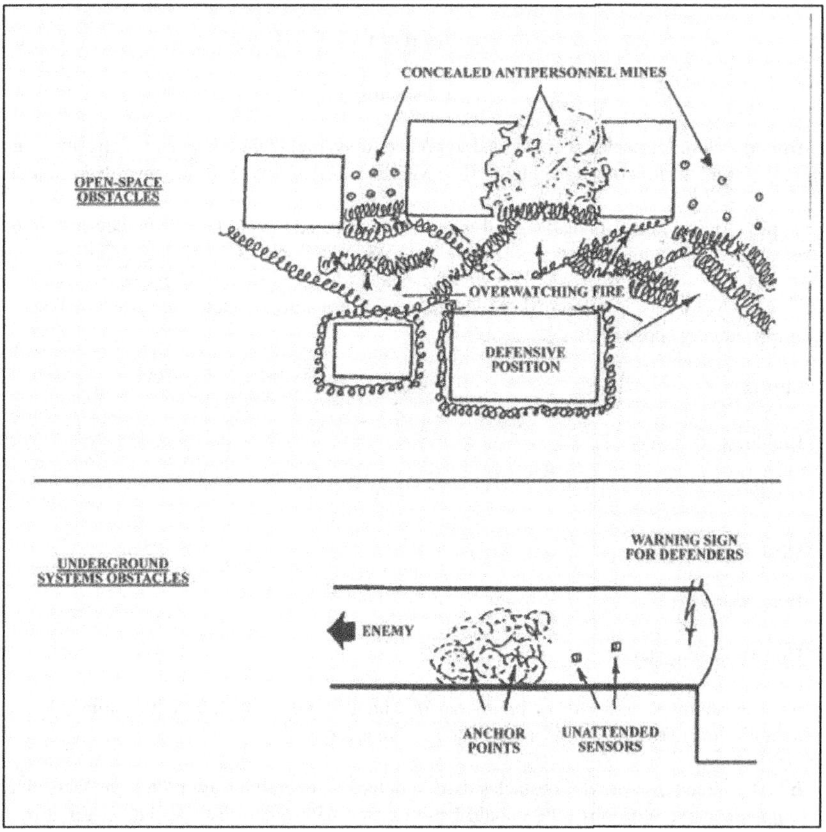

Figure F-1. Antipersonnel Obstacles

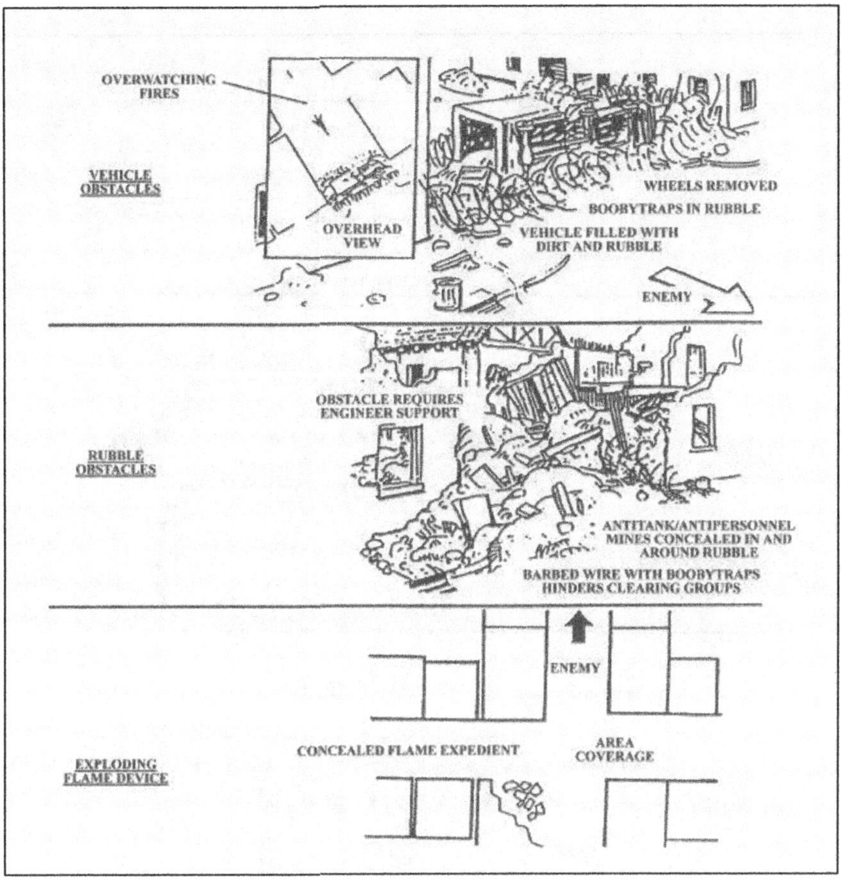

Figure F-1. Antipersonnel Obstacles (Continued)

d. Antiarmor obstacles are mainly restricted to streets and other armored avenues of approach (Figure F-2).

e. Some antitank mines may be sited and employed as antihelicopter mines to cover probable HLZs (Figure F-2).

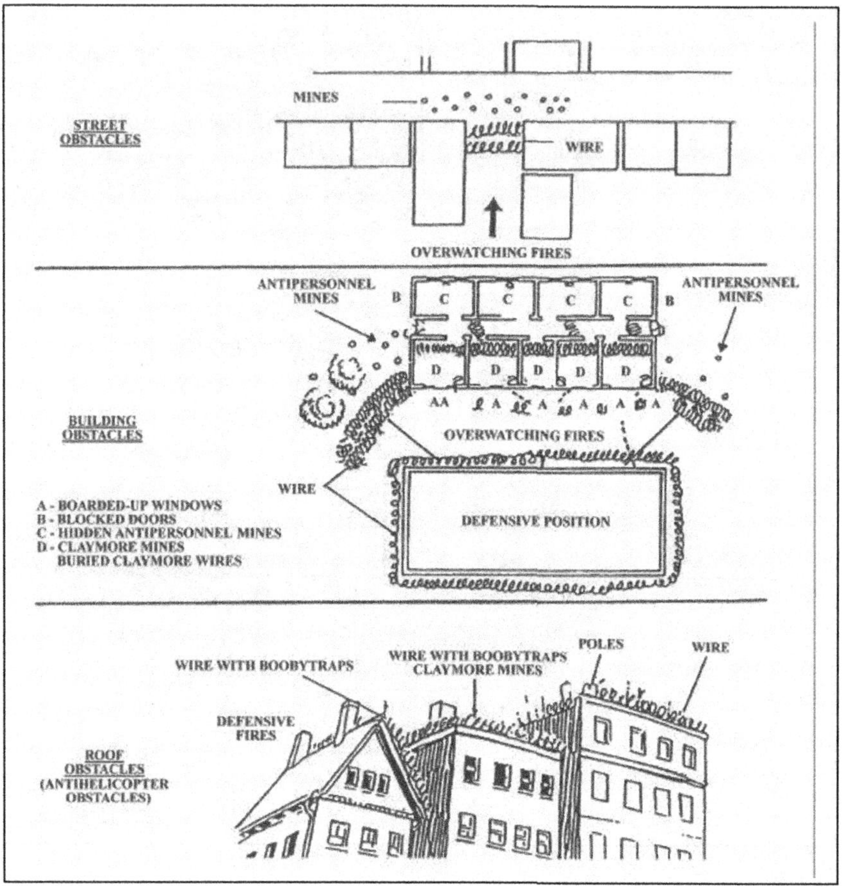

Figure F-2. Antiarmor Obstacles

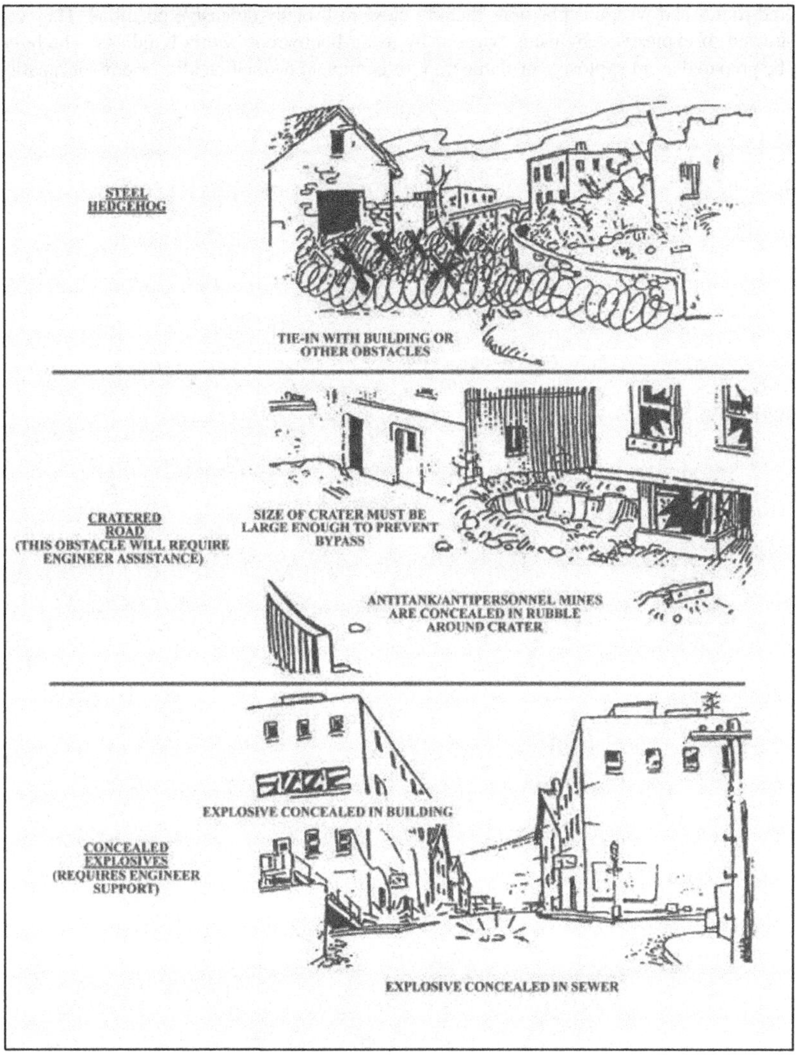

Figure F-2. Antiarmor Obstacles (Continued)

3. Construction of Obstacles. Obstacles are constructed in buildings to deny enemy infantry covered routes and weapons positions that are close to friendly defensive positions. They can be constructed of explosives, by using wire, or by using boobytraps within buildings. The building may be prepared as an explosive or flame trap for command detonation after enemy occupation.

Section II

Mines

4. Introduction. The employment or discovery of mines in built-up areas should be recorded on a building sketch (Figure F-3). (See FM 20-32.) The sketch should include the number of the building (taken from a city map, if available) and all floor plans. It should also include the type of mine and firing device. When possible, mined buildings should be marked on the friendly side (Figure F-4). Clearing areas or buildings that have been mined is extremely difficult and dangerous. Therefore, such areas should be considered "no go" areas for infantry. This factor must be carefully considered when planning and authorizing the placement of mines. (See Table F-1 for notional approving authority for minefields, but with the understanding that information in that table is subject to change with international agreements and ROE.)

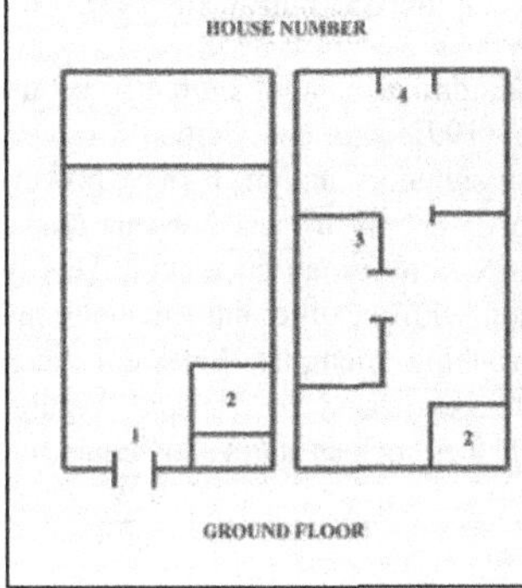

MINE	LOCATION	TYPE	FIRING DEVICE
1	FRONT DOOR	CLAYMORE	COMMAND
2	STAIRS	CLAYMORE	COMMAND
3	CLOSET	CLAYMORE	COMMAND
4	DRESSER	CLAYMORE	COMMAND

Figure F-3. Building Sketch Showing Mines

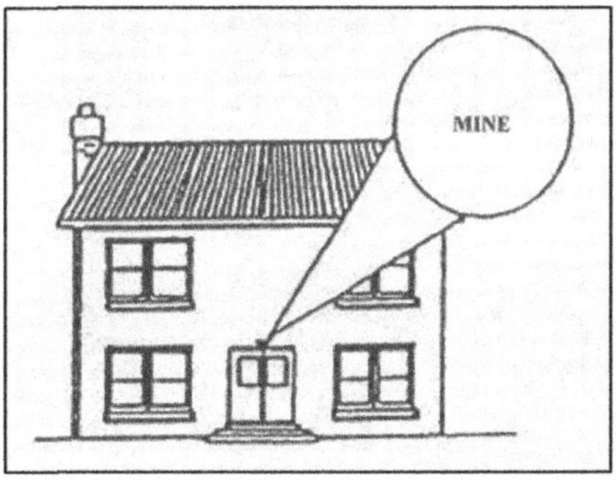

Figure F-4. Marking Mined Buildings

	Minefields (Type)		FASCAM (Duration)		Other Obstacles	
Level	Tactical	Protective	Long	Short	Tactical	Protective
MEF	X	X	X	X	X	X
Division	X	X	X	X	X	X
Regiment	X	X	X	X	X	X
Battalion	X	X		X	X	X
Company		X				X

Table F-1. Notional Minefield Employment Authority Matrix

5. **Types.** Several types of mines may be encountered or employed in built-up areas.

a. The M14 mine (a plastic type of antipersonnel landmine) cannot currently be used by American forces due to the Leahy Amendment of 1995 and the Certain Conventional Weapons Convention, of which the United States is a signatory nation. It may, however, be encountered in MOUT or be employed by coalition or enemy forces. For this reason, the following information is given as a guide for what U.S. forces may encounter. This type of antipersonnel landmine is normally used in conjunction with metallic antipersonnel, antitank, or chemical mines to confuse and hinder enemy breaching attempts. When encountered, it must be carefully handled because its light weight makes it easy to displace (Figure F-5). However, its size makes it ideal for covering obscure places such as stairs and cellars.

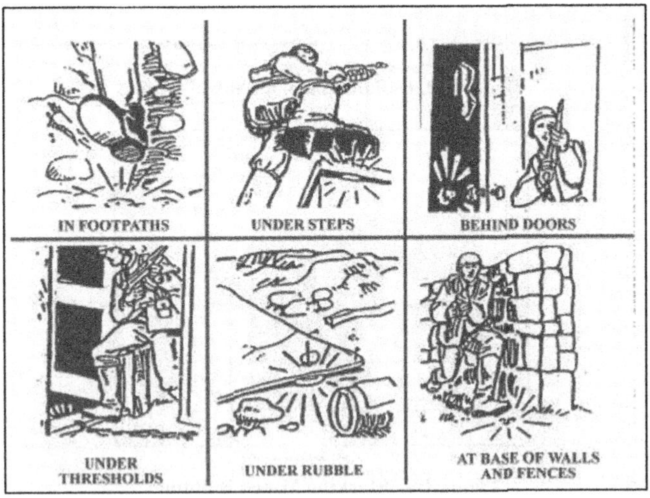

Figure F-5. Typical Employment of the M14 Antipersonnel Mine

b. The M16A2 antipersonnel landmine is subject to the same legal limitations as the M14. It is a boundary type of antipersonnel landmine, which makes it ideal for covering large areas such as rooftops, backyards, parks, and cellars. It can be expediently rigged for command detonation by attaching a rope or piece of communications wire to the pull ring (Figure F-6).

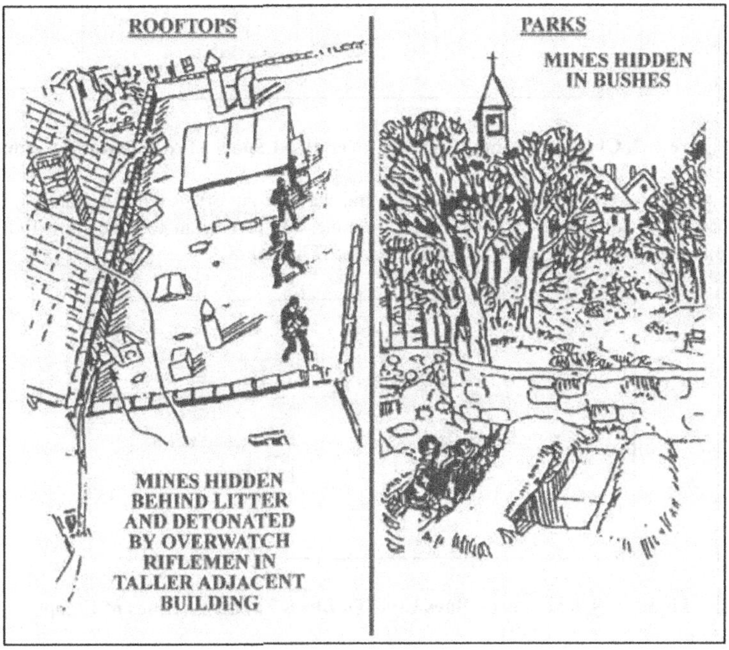

Figure F-6. Emplacements of M16A2 Antipersonnel Mine

c. The M18A1 claymore mine can be employed during the reorganization and consolidation phase on likely enemy avenues of approach. It does not have to be installed in the street, but can be employed on the sides of buildings or any other sturdy structure. The claymore mine, in command-detonated mode, is currently legal for employment by U.S. forces.

(1) Claymore mines can also be used for demolition against thin-skinned buildings and walls.

(2) Claymore mines may be configured for detonation by tripwire. They can help fill the dead space in the final protective fires of automatic weapons (Figure F-7).

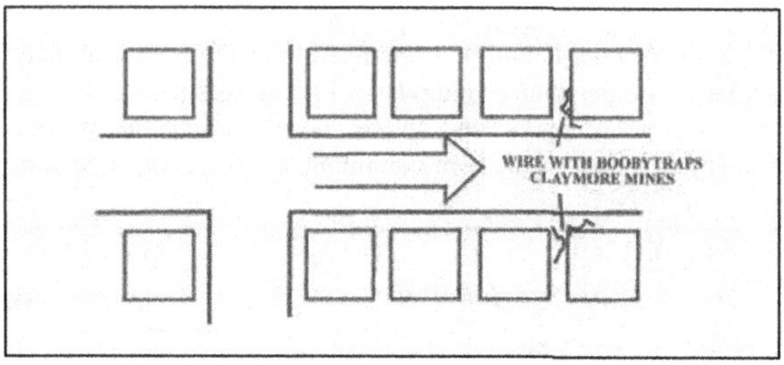

Figure F-7. Claymore Mine Used To Cover Dead Space of Automatic Weapons

(3) Claymore mines can be used in several ways in the offense. For example, if friendly forces are advancing on a city, claymore mines can be used in conjunction with blocking positions to cut off enemy avenues of escape (Figure F-8).

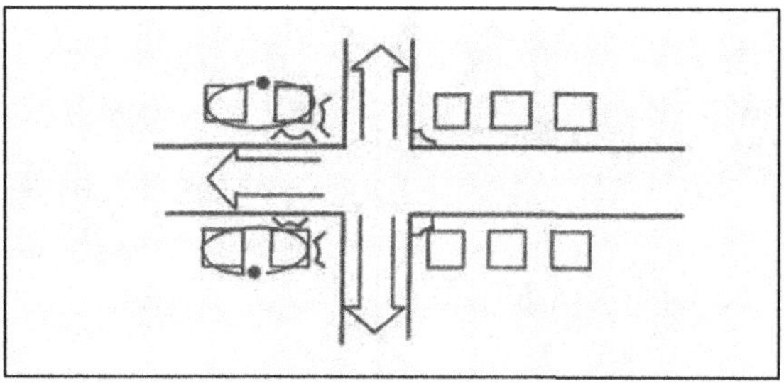

Figure F-8. Claymore Mines Used To Block Enemy Avenues of Escape

d. The M15, M19, and M21 antitank mines are used to stop or turn enemy armor and should be employed (Figure F-9):

- In conjunction with other manmade obstacles and covered by observation and fire
- In streets or alleys to block routes of advance in narrow defiles
- As command-detonated mines with other demolitions.

6. Enemy Mines and Boobytraps. Buildings contain many areas and items that are potential hiding places for boobytraps, for example, doors, windows, telephones, stairs, books, canteens, and so on.

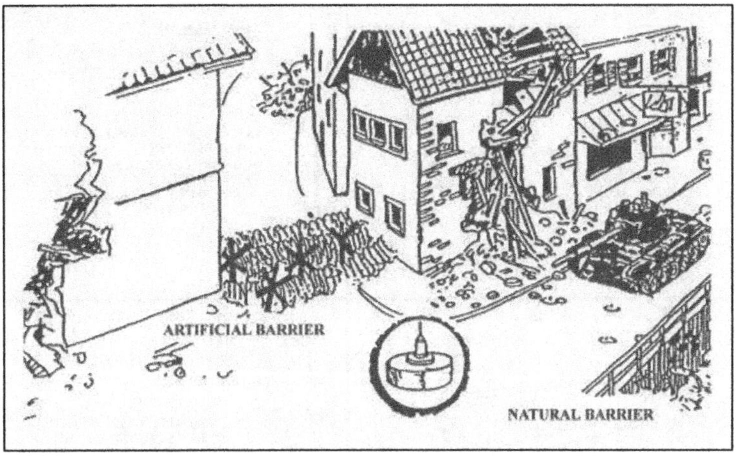

Figure F-9. Emplacement of Antitank Mines

When moving through a building, Marines must not pick up anything—food, valuables, enemy weapons, or other items. Such items could be rigged with pressure devices that explode when they are moved. Marines must be well-dispersed so that if a boobytrap explodes, the number of casualties will be few. Many different types of mines and boobytraps could be encountered during combat in built-up areas (Figure F-10).

a. The equipment commonly used to detect mines in clearing operations includes:

- Mine detectors
- Probes
- Grappling hooks
- Ropes
- Explosives
- Flak vests
- Eye protection
- Engineer tape.

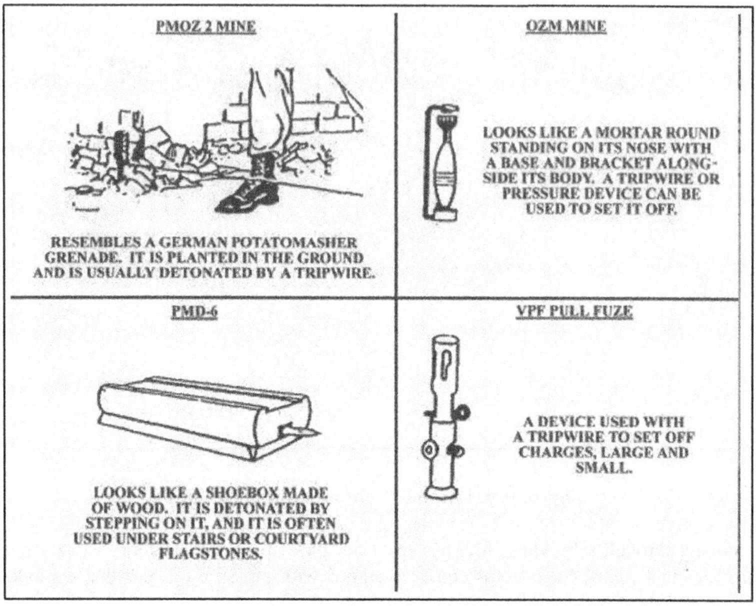

Figure F-10. Common Threat Antipersonnel Mines and Boobytraps

b. To detect tripwires, a Marine can use a 10-foot pole with 5 feet of string tied on one end. A small weight attached to the loose end of the string will snag on the tripwire. This allows the lead Marine to easily detect a tripwire (Figure F-11). Marines can also use a flashlight to cause reflections and shadows to find tripwires. A wand with a dangling thread can be swept over the ground ahead with the thread being watched for any disturbance by a tripwire.

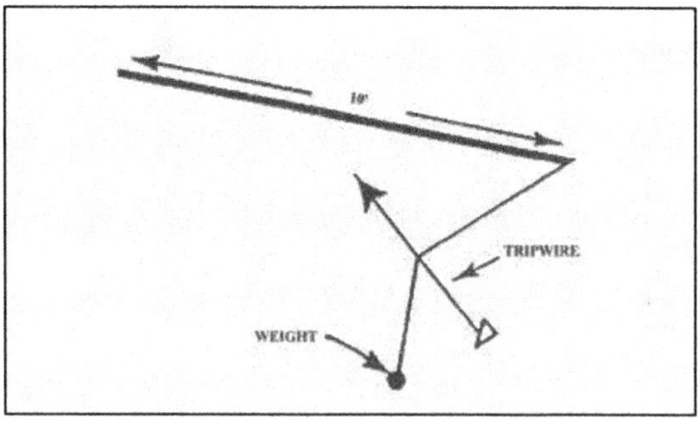

Figure F-11. Tripwire Detection

c. Many standard antipersonnel mines are packed in boxes and crates. If a Marine discovers explosive storage boxes, he should sketch them and turn the sketch over to the platoon commander for forwarding up the chain of command.

d. Most boobytraps should be neutralized by explosive ordnance disposal (EOD) personnel. If EOD teams are not available, boobytraps may be blown in place. Personnel should be protected by adequate cover. If the boobytrap is in a building, all personnel should go outside before the boobytrap is destroyed. Engineer tape placed around the danger area is one method of marking boobytraps. If tape is not available, strips ripped from bedsheets can be used. If possible, a guide should lead personnel through boobytrapped areas. EPWs and civilians may be a good source of information on where and how boobytraps are employed.

Section III

Demolitions

7. Introduction. Demolitions are used more often during combat in built-up areas than during operations in open natural terrain. Demolition operations should be directed by the engineers that support the GCE. However, if engineers are involved in the preparation and execution of the barrier plan, other trained Marines can prepare mouseholes, breach walls, and rubble buildings themselves if assisted and advised by attached engineers.

8. Offensive Use. When assaulting or clearing a built-up area, demolitions may enable the GCE commander to create an avenue of approach through buildings.

a. Every other man in an assault force should carry demolitions, and other selected personnel should carry blasting caps. The same man should *not* carry both the explosives and the blasting caps because of the danger of sympathetic detonation. As the demolitions are expended by the assault force, they should be replaced by explosives carried by the support force.

b. One of the most difficult breaching operations faced by the assault force is the breaching of masonry and reinforced concrete walls. When demolitions must be used, composition C4 is the ideal explosive to use. Normally, building walls are 15 inches thick or less. Assuming that all outer walls are constructed of reinforced concrete, a rule of thumb for breaching is to place 10 pounds of C4 against the target between waist and chest height. When detonated, this normally blows a hole large enough for a man to go through. The amounts of TNT and C4 required to breach concrete are shown in Table F-2.

c. Mouseholes provide the safest method of moving between rooms and floors. They can be created with C4. Because C4 comes packaged with an adhesive backing or can be emplaced using pressure-sensitive tape, it is ideal for this purpose. When using C4 to blow a mousehole in a lath and plaster wall, one block or a strip of blocks should be placed on the wall from neck to knee height. Charges should be primed with detonating cord or electrical blasting caps to obtain simultaneous detonation, which will blow a hole large enough for a man to fit through.

9. Defensive Use. The use of demolitions in defensive operations is similar to their use in offensive operations. When defending a built-up area, demolitions are used to create covered and concealed routes through walls and buildings that can be used for withdrawals, reinforcements, or counterattacks. Demolitions are also used to create obstacles and clear fields of fire.

a. Marines can use demolitions for creating mouseholes and constructing command-detonated mines. Expedient C4 satchel charges can be concealed in firing positions and along movement routes for detonation in the vicinity of the enemy.

	Reinforced Concrete		
Thickness of Material	**TNT**	**C4**	**Size of Opening**
Up to 10 cm (4 inches)	.792 lbs	.591 lbs	10 to 15 cm (4 to 6 inches)
10 to 20 cm (4 to 8 inches)	3.4 lbs	2.54 lbs	15 to 30 cm (6 to 12 inches)
	Nonreinforced Concrete Masonry		
Thickness of Material	**TNT**	**C4**	**Size of Opening**
Up to 35 cm (14 inches)	3.66 lbs	2.73 lbs	35 cm (14 inches)
35 to 45 cm (14 to 18 inches)	7.5 lbs	5.6 lbs	45 cm (18 inches)
45 to 50 cm (18 to 20 inches)	10.2 lbs	7.61 lbs	50 cm (20 inches)

Table F-2. Amounts of TNT and C4 Required To Breach Concrete

b. The engineers should furnish technical assistance for selective rubbling. Normally, buildings can be rubbled by using shaped charges or C4 on the supports and major beams.

c. Charges should be placed directly against the surface to be breached unless a shaped charge is used. Whenever possible, demolitions should be tamped to increase their effectiveness. Tamping materials could be sandbags, rubble, or desks and chairs (Figure F-12).

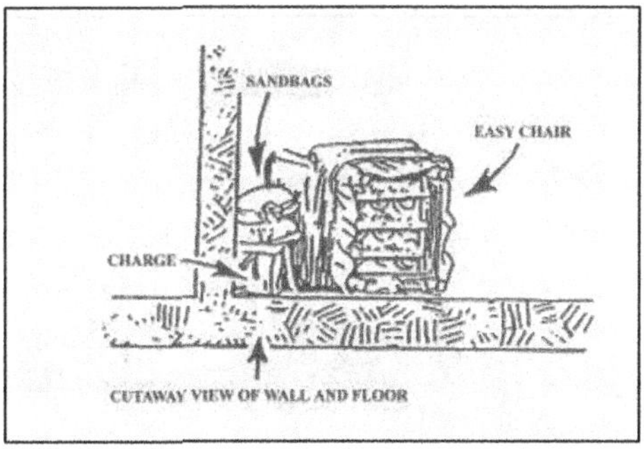

Figure F-12. Chair Used To Tamp Breaching Charge

d. For exterior walls, tamping of breaching charges may be impossible because of enemy fire. Thus, the untamped charge normally requires twice the explosive charge to produce the same effect as a tamped charge (Figure F-13).

METHODS OF PLACEMENT				
THICKNESS OF CONCRETE				
FEET	POUNDS OF TNT	POUNDS OF C4	POUNDS OF TNT	POUNDS OF C4
2	14	11	28	21
2 1/2	27	21	54	41
3	39	30	78	59
3 1/2	62	47	124	93
4	93	70	185	138
4 1/2	132	99	263	196
5	147	106	284	211
5 1/2	189	141	376	282
6	245	186	490	366

Figure F-13. Breaching Reinforced Concrete

e. When enemy fire prevents an approach to the wall, the breaching charge can be attached to a pole and slid into position for untamped detonation at the base of the wall (Figure F-14).

Figure F-14. Charge Placement When Small-Arms Fire Cannot Be Suppressed

f. The internal walls of most buildings function as partitions rather than as load-bearing walls. Therefore, smaller explosive charges can be used to breach them. In the absence of C4 or other military explosives, internal walls can be breached by using a claymore mine (Figure F-15). The claymore should be tamped to increase its effectiveness and to reduce the amount of explosive force directed to the rear.

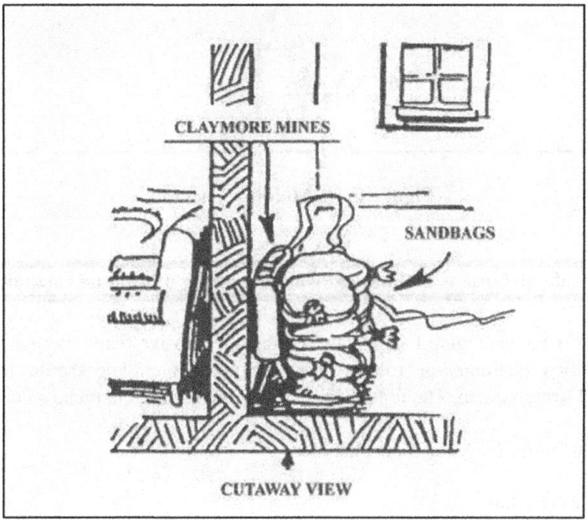

Figure F-15. Tamping a Claymore Mine To Breach Internal Walls

g. The Molotov cocktail (Figure F-16) is an expedient device for disabling both wheeled and tracked vehicles. It is easy to make because the materials are readily available. The results are most effective because of the close engagement in built-up areas. The objective is to ignite a flammable portion of the vehicle such as the fuel or ammunition that it is transporting. The following materials are needed to make a Molotov cocktail:

- Bottle or other glass container
- Gas (60 percent)
- Oil (40 percent)
- Rag for use as a wick.

The gas and oil are mixed thoroughly (60 percent gas to 40 percent oil). The rag is soaked with the mixture, then the mixture is placed into the bottle. The rag is then inserted in the opening of the bottle as a wick. When a target is sighted, the wick is lit and the bottle is thrown hard enough to break.

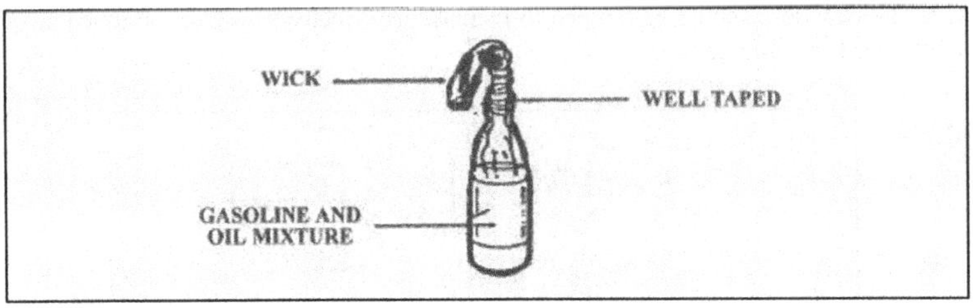

Figure F-16. Molotov Cocktail

WARNING

Ensure that a safe distance is maintained when throwing a Molotov cocktail.

h. The bunker bomb (Figure F-17) is an expedient, explosive flame weapon that is best used against fortified positions or rooms. This expedient munition should be used with a nonelectrical firing system. The following materials are required to make a bunker bomb:

- 1 metal small-arms ammunition container
- 1 gallon of gasoline
- 50 feet of detonating cord
- 1 nonelectrical blasting cap
- 1 M60 fuze igniter
- 7 1/2 feet of M700 time fuze
- 1 M49 trip flare or M34 WP grenade.

(1) Step 1. Fill the ammunition can 3/4 full with fuel and secure the lid.

(2) Step 2. "Hasty whip" the device with 15 turns around the center of the container using 44 feet of detonating cord. Leave 2-foot "pigtails" for attaching the igniter and fuze igniter.

(3) Step 3. Tape the igniter (M49 trip flare or M34 WP grenade) to the container handle.

(4) Step 4. Place one detonating cord pigtail end under the igniter spoon handle. Tape it in place.

(5) Step 5. Attach the M60 fuze igniter and the nonelectrical blasting cap to the M700 time fuze.

(6) Step 6. Attach the nonelectrical firing system to the other pigtail by making a loop in the detonating cord and attaching the blasting cap to it.

(7) Step 7. Remove the safety pin from the igniter (M49 trip flare or M34 WP grenade). The device is ready to be fired.

WARNING

Never carry the device by the handle or igniter. Remove the igniter safety pin only when it is time to use the device. Use extreme care when handling or carrying nonelectrical firing systems. Protect blasting caps from shock and extreme heat. Do not allow the time fuze to kink or become knotted. Doing so may disrupt the powder train and may cause a misfire. Prime detonating cord and remove the time fuze igniter safety pin only when it is time to use the device.

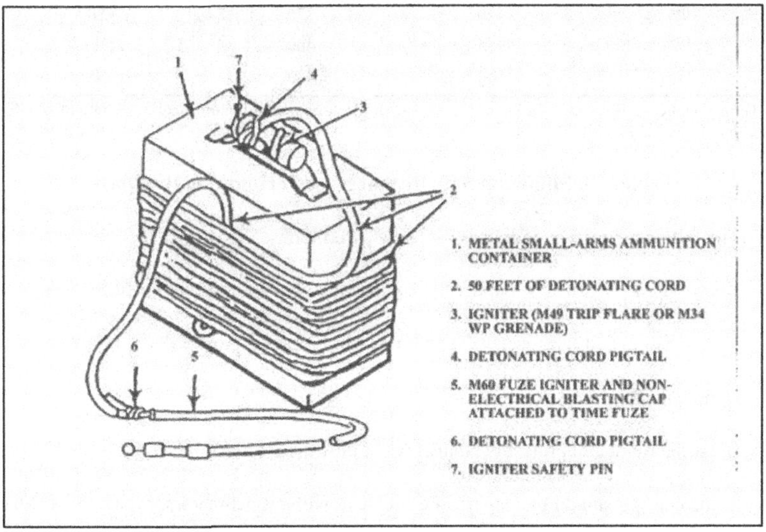

Figure F-17. Bunker Bomb Made From an Ammunition Can

10. Safety. The greatest danger to friendly personnel from demolitions is the debris thrown by the explosion. Leaders must ensure that protective measures are enforced. The safe distance listed in Table F-3 indicates the danger of demolition effects.

Pounds of Explosive	Safe Distance in Meters[1]	Pounds of Explosive	Safe Distance in Meters[1]
1 to 29	300	150	534
30	311	175	560
35	327	200	585
40	342	225	609
45	356	250	630
50	369	275	651
60	392	300	670
70	413	325	688
80	431	350	705
90	449	375	722
100	465	400	737
125	500	425	750

[1]These distances will be modified in combat when troops are in other buildings, around corners, or behind intervening walls.

Table F-3. Minimum Safe Distances for Personnel in the Open

a. The following are the guidelines for using demolitions:

- Keep the blasting machine under the control of a noncommissioned officer.
- Wear helmets and flack jackets at all times while firing explosives.
- Handle misfires with extreme care.
- Clear the room and protect personnel when blowing interior walls.

b. Some charges should be prepared beforehand, without detonators, to save time. Examples include 10- or 20-pound breaching charges of C4 and expedient-shaped charges in No. 10 cans.

- Use C4 to breach hard targets (masonry construction).

- Do not take chances.
- Do not divide responsibility for demolition work.
- Do not mix demolitions and detonators until ready to use.
- Do not carry demolitions and caps together.

Appendix G

Nuclear, Biological, and Chemical Considerations

Current U.S. policy regarding lethal or incapacitating agents is that their use against an armed enemy requires approval at the NCA level. Potential enemies may not operate under the same restrictions. Field commanders must be prepared to assume an adequate NBC defensive posture when engaged in urban fighting.

1. Protection From Nuclear, Biological, and Chemical Weapons. Generally, the lowest floor or basement of a reinforced concrete or steel-formed building offers good protection from nuclear hazards and initial liquid chemical contamination although some chemical agents tend to collect in lower areas. Tanks and LAVs/AAVs also provide protection. (See FMFM 11/MCWP 3-37 series manuals for greater detail.)

a. Biological attacks are difficult to detect and recognize. Biological agents can be disseminated by using aerosols, vectors, and covert methods. Because biological agents can be sprayed or dropped in bomblets, personnel who observe such indicators should promptly report them. Prompt reporting and treatment of the sick speed the employment of medical countermeasures. Although buildings and shelters provide some protection against spraying, they provide little protection against biological agents.

b. Chemical agents cause casualties by being inhaled or absorbed through the skin. They may afford Marines a few seconds to mask. Buildings have a channeling effect and tend to contain the effects of an agent, causing great variations in chemical concentration from room to room or from building to building. Chemical agents usually settle in low places, making sewers and subways hazardous hiding places. A prepared defender should include some collective protective measures in the defensive network. Personnel using fans may be able to put enough overpressure into tunnels to keep some chemical agents from entering. The protective mask and battle dress overgarment provide the best protection against chemical agents.

c. Personal hygiene is a critical defensive measure against infection and disease. Unfortunately, built-up areas are characterized by sophisticated sanitation systems. When those systems are destroyed, the resulting sanitary conditions become much worse than those in areas where sanitation facilities do not exist. Commanders must ensure that personnel employ sanitation measures and that their immunizations are current.

d. Commanders should plan their mission-oriented protective posture (MOPP) with the realization that increased logistics demands in built-up areas also apply to NBC equipment. Protective clothing, detection and decontamination equipment, and sealed containers of food and water must be stockpiled the same as other critical supplies. When operating in protective clothing, commanders must make allowances for the strenuous activities normally associated with combat in built-up areas.

(1) Detection. After an NBC attack, battalions should dispatch their detection and survey teams. Detection in built-up areas is complicated by the numerous rooms within buildings. Teams should conduct tests and surveys of major streets, intersections, and buildings in their area for inclusion in initial NBC reports. A systematic survey of all buildings, rooms, and underground facilities must be accomplished before occupation by unmasked personnel. All data should be forwarded using the appropriate NBC report.

(2) Decontamination. Personnel begin decontamination operations as soon after an NBC attack as the mission allows. Personnel should conduct individual decontamination of themselves and their personal equipment. Unit commanders determine the need for MOPP gear exchange and the requirements for a hasty or deliberate decontamination operation.

(a) Nuclear. Personnel should wear wet-weather gear for certain decontamination operations (hosing down buildings) to prevent radioactive material from touching the skin.

(b) Chemical and Biological. Roads, sidewalks, and other hard surfaces are best decontaminated by weathering, if time permits. Agents can also be covered with several inches of dirt or sand to provide protection. Fragment testing should be conducted periodically to ensure that the agent has not seeped through the covering. For critical sections of roads, a truck-mounted M12A1 (power-driven decontamination apparatus (PDDA)) can be used to spray supertropical bleach (STB) slurry; this aids rapid decontamination. Buildings are difficult to decontaminate, especially wooden ones. Some techniques for their decontamination are scrubbing with STB slurry; washing with hot, soapy water; washing or spraying with a soda solution; and airing.

2. Smoke Operations. The use of smoke is an integral part of either offensive or defensive operations in built-up areas. In the offense, smoke can support the maneuver of combat units and deception operations. Smoke employed in the defense obscures enemy air and ground observation, limiting the accuracy of enemy fires and intelligence collection.

a. Smoke should not be used when it degrades the effectiveness of friendly forces. Likewise, an extremely dense concentration of smoke in a closed area displaces the oxygen, suffocating Marines even when they are wearing protective masks.

b. Smoke pots, generators, or artillery smoke munitions should be used to cover the withdrawal of defending forces or the movement of attacking forces. Artillery-delivered WP can be effective against enemy forces by screening, in addition to causing casualties and fires. Consideration should be given to the incendiary effects of WP, smoke (HC and M825), and illumination munitions on the litter and debris of built-up areas.

c. Smoke grenades can be massed to provide a hasty screen for concealing personnel moving across streets and alleys. Smoke grenades can also be used for signaling; those launched by an M203 can be used to mark targets for fixed and rotary-wing air support.

d. The use of smoke in built-up areas is affected by the complex wind patterns caused by buildings. When covering a built-up area with a smoke haze or blanket, personnel must include all buildings. Failure to obscure tall buildings, towers, and steeples provides enemy observers with reference points for fire placement within the built-up area.

3. Riot Control Agents. When authorized, riot control agents such as CS and CN can be used to drive enemy forces from positions or to deny them areas for occupation. Riot control agents are incapacitating but generally have no lasting effects. They may be appropriate when preventing civilian casualties is a planning consideration. However, riot control agents may not be effective against an enemy that is well trained in chemical defense.

Appendix H

MOUT Under Limited-Visibility Conditions

To be successful, leaders must use limited-visibility conditions to their advantage.

1. Advantages. When fighting in built-up areas during night or periods of limited visibility, attacking or defending forces have several advantages.

a. In many cases, Marines may have a technological advantage in thermal imagery and light intensification over their opponents. This enables Marines to identify, engage, and destroy enemy targets before being detected by the enemy.

b. Possessing the capability to conduct operations as well at night or during periods of reduced visibility as during daylight gives Marines a tremendous advantage over an enemy with reduced capability. It also aids the generation of tempo and allows Marines to gain or retain the initiative.

c. Generally, ranges of direct-fire engagements are greatly reduced in the MOUT environment. During periods of limited visibility, effective unaided target acquisition ranges are further reduced. This enables attacking forces to close to shorter ranges, thus increasing the lethality and accuracy of weapons. Attacking forces can also take advantage of the enemy's reduced ability to see and can more effectively engage the enemy before being detected with thermal imagery or light intensification devices.

d. Helicopterborne assaults are best conducted during periods of limited visibility because the enemy's air defenses are degraded.

e. Attacking during periods of limited visibility gives the attacker a greater chance of achieving surprise.

2. Disadvantages. When fighting in built-up areas during periods of limited visibility, attacking and defending forces also face some disadvantages.

a. Command and control is difficult in any operation in a built-up area; periods of limited visibility increase this difficulty.

b. Marines have an instinctive tendency to move close together during periods of limited visibility. Constant attention must be given to prevent Marines from "bunching up."

c. Marines may become disoriented easily because of the combined effects of low visibility and the characteristics of built-up areas.

d. Target identification becomes more difficult in limited-visibility conditions. Depending on the training of the individual, the Marine may fire at anything seen or may hesitate too long before firing. Improper target identification is one of the leading causes of fratricide, so leaders must pay close attention to individual target engagement.

3. Fratricide Avoidance. The risk of fratricide is much greater during periods of limited visibility. The key to avoiding fratricide is increased situational awareness by leaders and individuals, coupled with realistic training in target identification. (See *Fratricide: Reducing Self-Inflicted Losses*, Newsletter, No. 92-4, April 92, Center for Army Lessons Learned, U.S. Army Combined Arms Command, Fort Leavenworth, Kansas.) Other considerations include:

a. Graphic control measures should be clearly defined and obvious. Examples include distinctive buildings, large boulevards, rivers, and so forth.

b. Leaders must exercise firm control when engaging targets. Movements should also be coordinated and controlled.

c. Cleared rooms and buildings should be distinctly marked to identify cleared areas and friendly forces to base-of-fire elements supporting the maneuver.

d. Visible markers (for example, glint tape or thermal strips) should be attached to individual Marines for rapid identification.

e. Far and near recognition symbols should be coordinated and used properly.

f. Units employing close air support must exercise firm control. Failure to do so may lead to the pilot becoming disoriented and engaging friend and foe alike.

4. Urban Environmental Effects on Night Vision Devices. Built-up areas affect standard night vision devices and sights differently than do open areas. This may cause some confusion because the images Marines receive through their night vision devices are unusual compared to those they may be used to.

a. Most built-up areas have electric power; therefore, street lights and/or building lights may "white out" some light intensification devices unless the power is disrupted or lights are turned off.

b. Fires may be burning in the urban area. Open flames cause problems not only for light intensification devices, but also for thermal devices.

c. Subterranean areas and the interiors of buildings will not have ambient light if the power is off. Passive night vision devices must then use an artificial light source, such as infrared radiation, to provide enough ambient light for the devices to work effectively.

d. The many reflective surfaces found in built-up areas may cause false images, particularly for laser range finders and laser target designators.

e. Thermal imaging devices may not be able to see through large amounts of dust particles suspended in the air.

f. Smoke and dust degrade the effectiveness of night vision devices.

g. Fog degrades long-range target acquisition from thermal sights.

h. Weapons flashes within enclosed areas appear to be much brighter. This causes Marines to lose their night vision and washes out light intensification devices.

5. Considerations. The environment of built-up areas presents special challenges and considerations during periods of limited visibility.

a. The use of glint tape, thermal tape, or chemiluminescent lights is an important consideration. These can be used to mark the FLOT, casualties, cleared buildings and rooms, weapons positions, and individual Marines. Their use must be clearly addressed in the unit's SOP. When markers are used for extended periods, their meanings should change because the enemy may be able to capture or manufacture and use these marking devices to their advantage.

b. The use of tracer and incendiary ammunition may be restricted to prevent fires. The light of fires whites out some night vision devices and may interfere with or confuse thermal devices.

c. The control of power stations may be essential to operations during periods of limited visibility. Control of power stations enables friendly forces to control, to a degree, background illumination. Shutting off the power to the street lights is much easier than shooting out the lights. Commanders must balance the tradeoff between force protection and maintaining law and order after the battle is over. During cold weather, the control of power stations may be critical to the welfare of the civilian population.

d. Identifying friendly units, noncombatant civilians, and enemy troops becomes more difficult during limited-visibility operations.

e. Locating the source of sounds becomes more difficult because of the natural echoing in built-up areas and the tendency of sounds to carry farther at night.

f. Locating boobytraps and obstacles also becomes more difficult at night. Movement rates are normally slower than during periods of normal visibility.

6. Special Equipment. Fighting during periods of limited visibility requires some specialized equipment to maximize advantages.

a. As a rule, thermal imaging devices such as the AN/PAS-7 infrared viewer and the AN/TAS-5 Dragon infrared sight are better for limited-visibility operations than are light intensification devices such as the AN/PVS-7B. Light intensification devices are more easily washed out by background light, weapons flashes in enclosed areas, and fires. Thermal devices, while also affected by fires, are not as easily washed out.

b. The AN/PAQ-4A infrared aiming device is similar to its off-the-shelf laser aiming sight counterparts, except it is not visible to the naked eye. Infrared pen lights can also be attached to weapons and, when used in conjunction with night vision devices, can provide a quick sight picture, illuminate rooms and hallways, identify obstacles and boobytraps, and identify friendly forces.

c. Other night sights for weapons include the AN/TVS-5 crew-served weapon night vision sight, the AN/PVS-4 individual weapons night vision sight, and the AN/UAS-12 night vision sight, which mounts to the modular universal laser equipment (MULE) and the TOW missile system.

d. Trip flares, flares, illumination from mortars, and artillery and spotlights (visible light or infrared radiation) can be used to blind enemy night vision devices or to artificially illuminate the battlefield (Figure H-1). (See FMFM 6-8, *Supporting Arms Observer, Spotter, and Controller*, for more information on illumination from mortars and artillery.)

e. Spare batteries for the night vision devices should be carried to keep the devices operational. Soft, clean rags should be used to clean the lenses.

7. Combat Support. Coordinating employment of forces and fire support in limited-visibility conditions is a major concern to leaders during MOUT. The concentration of forces and fires at the point of decision is facilitated by the technological edge that night vision devices provide.

a. Any degradation of accuracy in artillery fire will likely be a result of the limitations of target acquisition assets. While FOs and FACs may have thermal sights and laser range finders, most Marines on the battlefield do not yet have devices that will enable them to acquire targets accurately. The following are some devices and techniques to improve target acquisition for indirect fires:

(1) If the target is within LOS, tanks and LAVs can rapidly identify the target and provide an accurate range.

(2) Preregistered targets are effective if the target reference point can be observed and the observer has clear communications with the firing unit.

(3) Fixed-wing and rotary-wing aircraft can be used to identify targets and adjust supporting arms.

Figure H-1. Use of Indirect-Fire Illumination During MOUT

8. Combat Service Support. Maneuver unit commanders and their Marines are not the only individuals who must adjust to combat under limited-visibility conditions in MOUT. Logisticians at every level must anticipate requirements for these conditions.

a. Units conducting resupply operations during periods of limited visibility should consider:

- Issuing drivers and vehicle commanders night vision devices so that the vehicles going to and from logistic release points do not require illumination (This also prevents the enemy from acquiring resupply locations by following vehicles with blackout lights on.)

- Maintaining strict noise and light discipline

- Following a clearly marked route to avoid obstacles and prevent the resupply vehicle(s) from becoming disoriented

- Providing radios to resupply vehicles

- Issuing each vehicle a map of the AO (preferably a city map with the street names).

b. Combat units operating for an extended time under conditions of limited visibility should have enough batteries to keep the night vision devices functioning at optimum power and sensitivity.

c. A large operational readiness float of night vision devices should be maintained.

d. Casualty collection during periods of limited visibility is more difficult. Clear methods for marking any casualties must be established before the operation begins.

e. CSS operations located in existing structures should not be visible from a distance. This includes limiting vehicle traffic to an absolute minimum, sealing doors and windows to prevent light leakage, and dispersing assets as much as possible.

9. Operational Considerations. Marine units conduct attacks during periods of limited visibility to gain or sustain the momentum of the attack. Before conducting a limited-visibility attack, the commander must balance the risks and ensure that every Marine understands the mission, intent, and control measures. Rehearsals and strict command and control reduce casualties and greatly enhance the chances for mission accomplishment.

a. To reduce confusion, Marines should clear buildings and rooms using the same techniques they use during the day.

b. Movement rates are slower in the dark. Each Marine must remain alert for mines, boobytraps, and enemy positions.

c. Rifle squads and fire teams should be equipped with a mixture of both thermal imaging and light intensification devices whenever possible. This enables the squads and fire teams to obtain a better picture of the night environment and enables Marines to balance the strengths and weaknesses of each type of night vision device for maximum results.

d. When moving through buildings, assault forces mark cleared rooms and buildings and communicate with the support forces. Marking cleared rooms is especially critical if more than one assault force or element is in the same building. Communications with supporting forces are imperative to avoid fratricide.

e. If flashlights or chemiluminescent lights are used, they should be held away from the head or chest area. This will make it harder for enemy soldiers firing at the light to kill the Marine holding the flashlight or chemiluminescent light.

f. Units must know where everyone is during offensive operations. This not only reduces the risk of fratricide, but also decreases the time spent identifying, locating, and treating casualties. Also, it greatly reduces the chance of Marines becoming disoriented and separated from the unit.

g. Assault units should be aware of adjacent fires that may diminish the effectiveness of night vision devices. Weapons flashes within small rooms cause Marines to lose their night vision and can wash out light intensification devices. Also, enemy soldiers may use flares inside and outside of buildings to deliberately render night vision devices ineffective.

h. Leaders must ensure that all Marines follow the ROE and the laws of land warfare. This is critical if the enemy is intermixed with the local civilian population. Also, leaders must follow all control measures.

i. Enemy forces can be expected to use periods of limited visibility to the same advantages as Marine forces do.

j. Enemy forces may have access to sophisticated night vision devices manufactured in Europe, the United States, Japan, Korea, and the former Soviet Union.

Appendix I

Urban Building Analysis

As in other types of operations, success in urban combat depends largely on the ability to analyze the military aspects of terrain. Marines must be able to recognize certain terrain features when evaluating urbanized terrain.

1. Types of Mass-Construction Buildings. Mass-construction buildings are those in which the outside walls support the weight of the building and its contents. Additional support, especially in wide buildings, comes from using load-bearing interior walls, strongpoints (called pilasters) on the exterior walls, cast-iron interior columns, and arches or braces over the windows and doors (Figure I-1). Modern types of mass-construction buildings are wall and slab structures, such as many modern apartments and hotels, and "tilt-up" structures commonly used for industry or storage. Mass-construction buildings are built in many ways:

- The walls can be built in place by using brick, block, or poured-in-place concrete.
- The walls can be prefabricated and then tilted up, or they can be reinforced-concrete panels.
- The walls can be prefabricated and assembled like boxes.

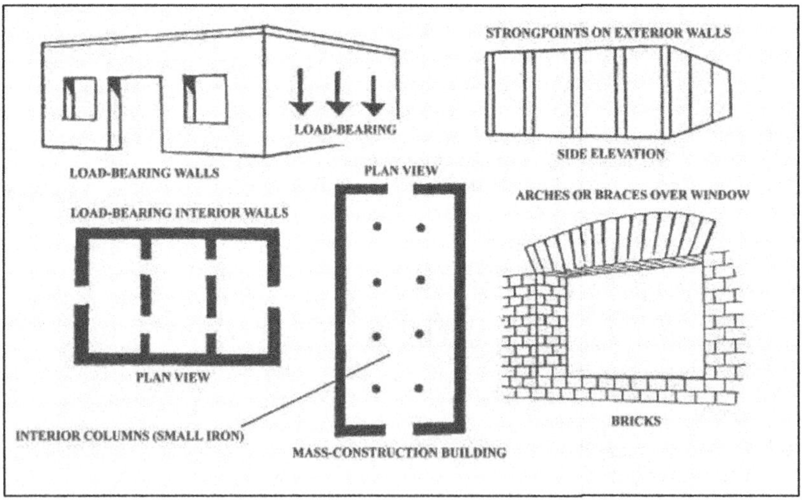

Figure I-1. Mass-Construction Buildings

a. Brick buildings are, in some regions, the most common and most important of the mass-construction buildings. In Europe, brick buildings are commonly covered with a stucco veneer so that bricks do not show (Figure I-2). One of the most common uses of brick buildings is the small store. These buildings are found in all built-up areas but are most common in the core periphery (Figure I-3).

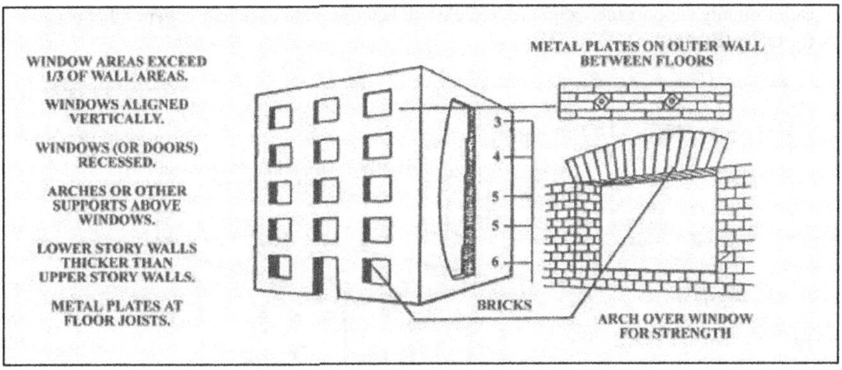

Figure I-2. Brick Buildings

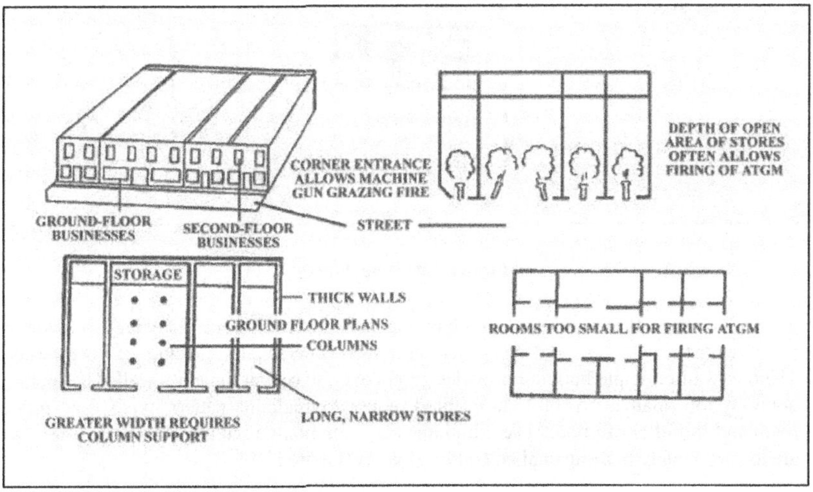

Figure I-3. Brick Store

b. Another common mass-construction building found mainly in industrial areas and along commercial ribbons is the warehouse. It is built of poured-in-place concrete reinforced with steel bars or of prefabricated tilt-up walls. The walls of warehouses provide good cover, although the roofs are generally light and vulnerable. The warehouses' large open bays permit firing of ATGMs or surface-to-air missiles and, because they are normally found in outlying areas, often afford adequate fields of fire for ATGMs. These buildings are built on slabs that can normally support the weight of vehicles and can provide excellent cover and concealment for tanks (Figure I-4).

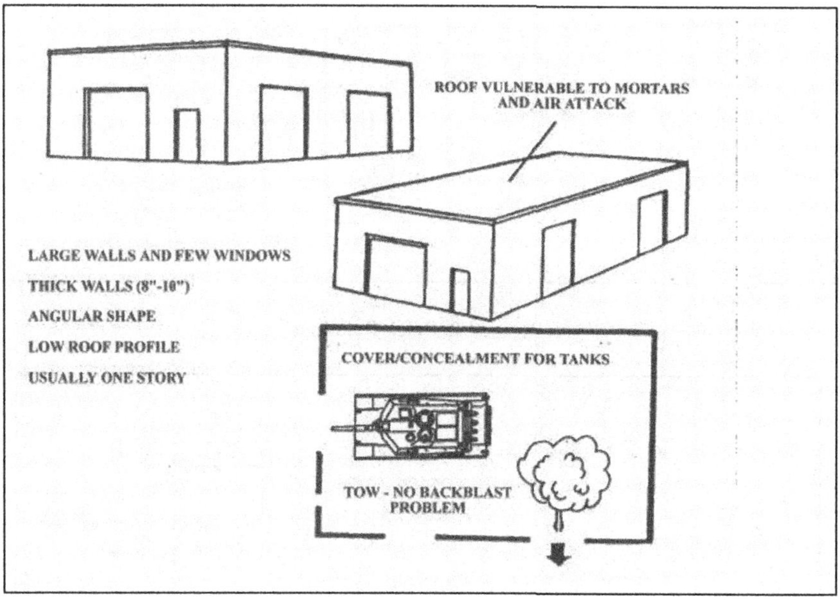

Figure I-4. Warehouse

c. Another type of mass-construction building is the box-wall-principle type. It is made from prefabricated panels of 6- to 8-inch-thick reinforced concrete. The outside wall is often glass. The box-wall-principle building provides good cover, except at the glass wall. The rooms are normally too small for ATGMs to be fired. A good circulation pattern exists from room to room and from floor to floor. These buildings are commonly used as hotels or apartments and are located mainly in residential and outlying areas (Figure I-5).

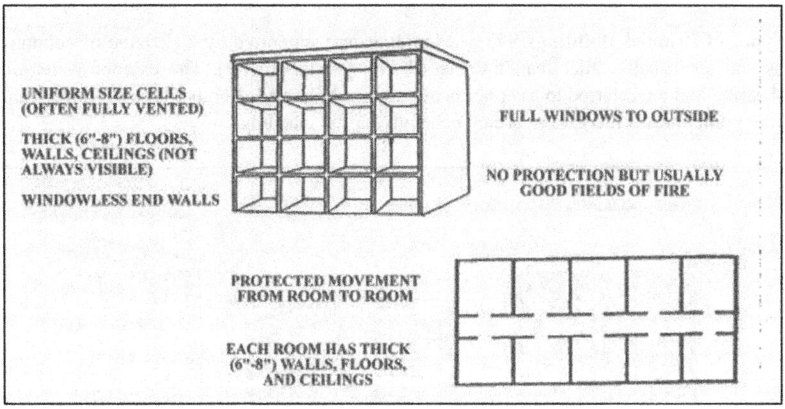

Figure I-5. Box-Wall-Principle Building

d. Public gathering places (churches, theaters) are mass-construction buildings with large, open interiors. The walls provide good cover, but the roof does not. The interior walls are not load-bearing and are normally easy to breach or remove. These buildings have adequate interior space for firing ATGMs or surface-to-air missiles. They are often located next to parks or other open areas and, therefore, have long fields of fire. Public gathering places are most common in core, core periphery, residential, and outlying high-rise areas (Figure I-6).

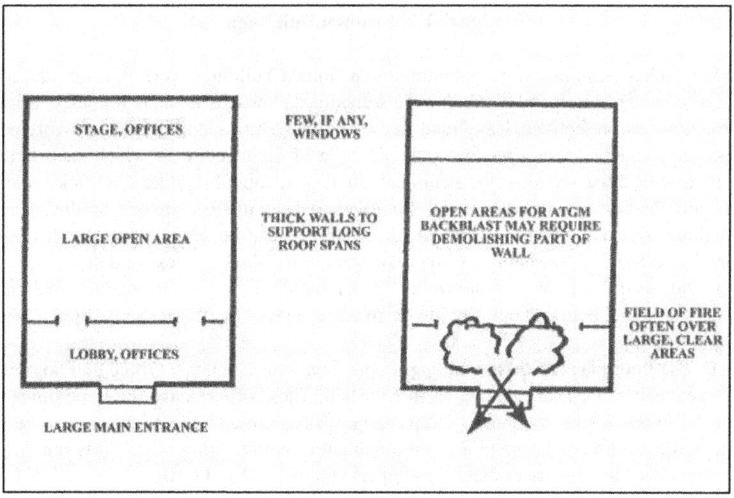

Figure I-6. Public Gathering Places

2. Types of Framed Buildings. Framed buildings are supported by a skeleton of columns and beams and are usually taller than frameless buildings (Figure I-7). The exterior walls are not load-bearing and are referred to as either heavy clad or light clad. Another type of framed building often found in cities is the garage, which generally has no cladding.

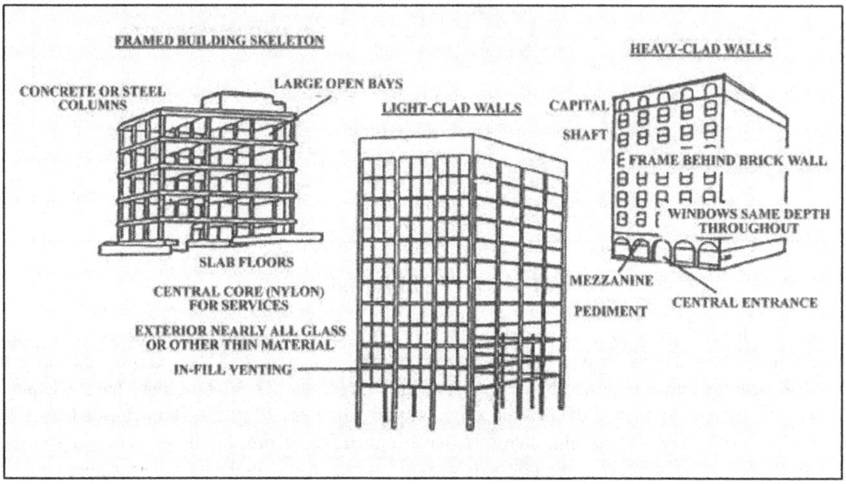

Figure I-7. Framed Buildings

a. Heavy-clad buildings were common when framed buildings were first introduced. Their walls are made of brick and block that are sometimes almost as thick as frameless brick walls, although not as protective. Heavy-clad framed buildings are found in core and core periphery areas. They can be recognized by a classic style or architecture in which each building is designed with three sections: the pediment, shaft, and capital. Unlike the brick building, the walls are the same thickness on all floors, and the windows are set at the same depth throughout. Often the frame members (the columns) can be seen, especially at the ground floor. The cladding, consisting of layers of terra cotta blocks, brick, and stone veneer, does not provide as good a cover as the walls of brick buildings. It protects against small-arms fire and light shrapnel but does not provide much cover against heavy weapons (Figure I-8).

(1) The floor plans of these buildings depend on their functions. Office buildings normally have small offices surrounding an interior hall. These offices have the same dimensions as the distance between columns (some large offices are as large as two times the distance between columns). These rooms are too small to permit firing of ATGMs but do provide some cover for snipers or machine gunners (Figures I-9 and I-10).

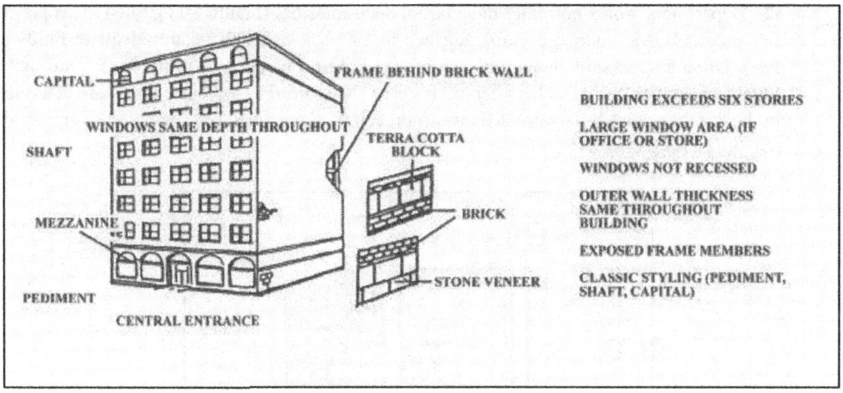

Figure I-8. Heavy-Clad Framed Building

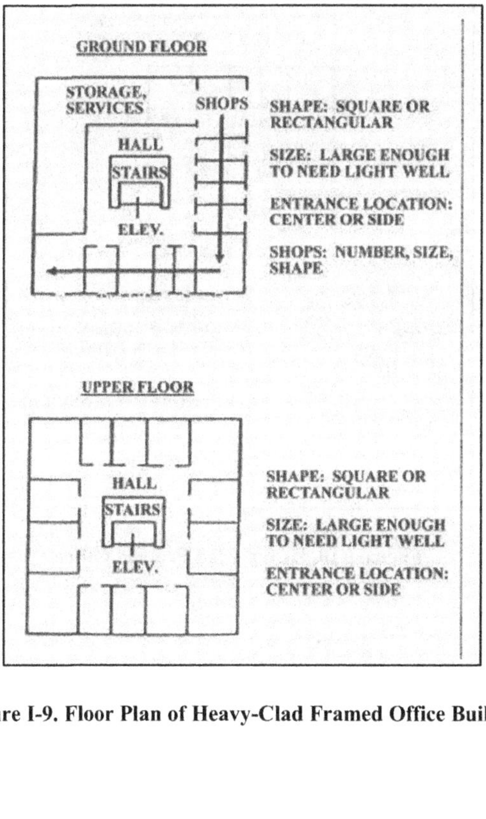

Figure I-9. Floor Plan of Heavy-Clad Framed Office Building

(2) Department stores normally have large, open interiors (Figure I-11). Such areas permit firing of ATGMs or man-portable surface-to-air missiles (if there are adequate fields of fire). Often a mezzanine level with a large backblast area permits firing down onto tanks. Steel fire doors often exist between sections of the store. The steel fire doors are activated by heat. Once closed, they are difficult to breach or force open, but they effectively divide the store into sections (Figure I-12).

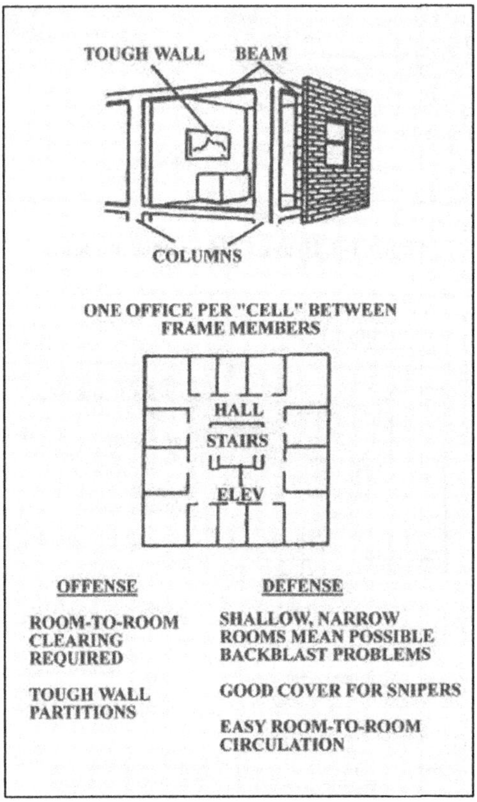

Figure I-10. Heavy-Clad Framed Office

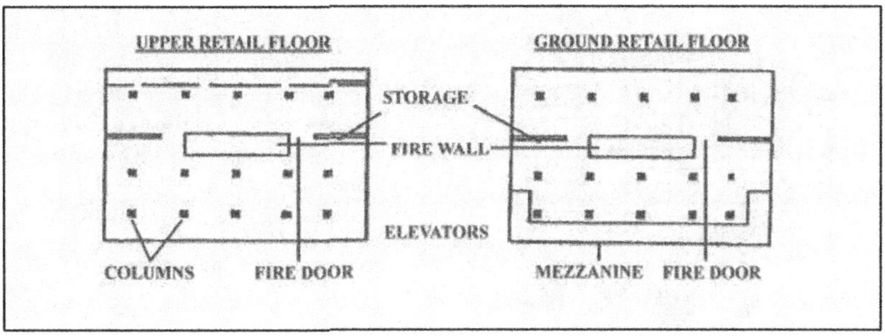

Figure I-11. Heavy-Clad Framed Department Store

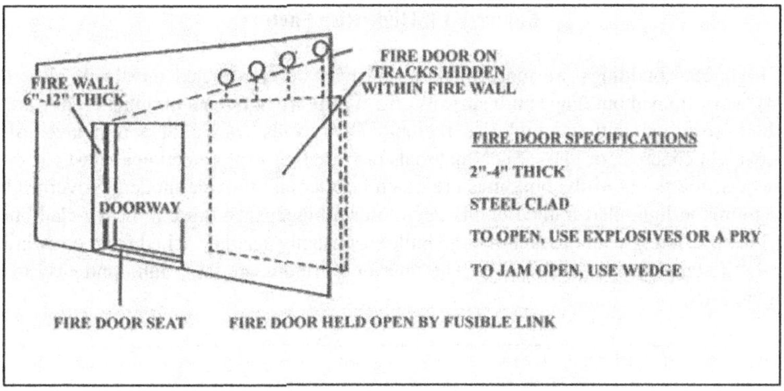

Figure I-12. Fire Wall and Fire Door

(3) Another type of heavy-clad framed building is used as a high-rise factory (Figure I-13). Such buildings are normally easily recognized because the concrete beams and columns are visible from the outside. They are usually located in older industrial areas. The large windows and open interior favor the use of ATGMs. Because the floors are often made to support heavy machinery, these buildings provide good overhead cover.

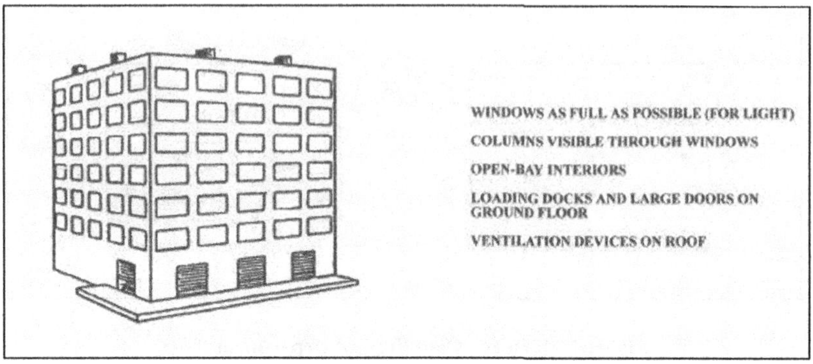

Figure I-13. High-Rise Factory

b. Light-clad buildings are more modern and may be constructed mostly of glass (Figure I-14). Most framed buildings built since World War II are light-clad buildings. They are found in both core and outlying high-rise regions. Their walls consist of a thin layer of brick, lightweight concrete, or glass. Such materials provide minimal protection against any weapon. However, the floors of the buildings are much heavier and provide moderate overhead cover. The rooms in light-clad framed buildings are much bigger than those in heavy-clad buildings. This feature, along with the fact that the buildings usually stand detached from other buildings, favors the employment of ATGMs. The interior partitions are thin, light, and easy to breach (Figure I-15).

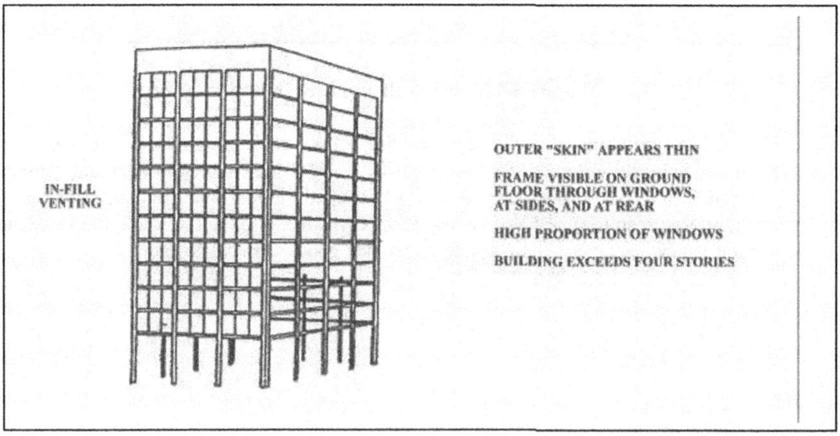

Figure I-14. Light-Clad Framed Building

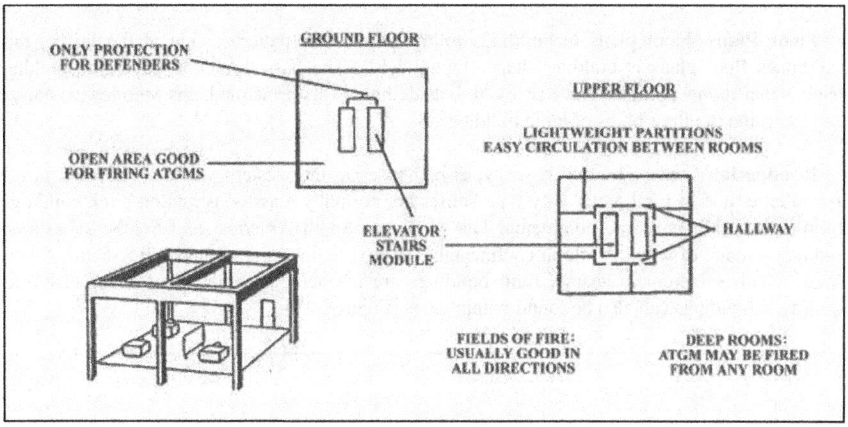

Figure I-15. Light-Clad Framed Room

c. The garage is one of the few buildings in an urban area in which all floors support vehicles. It provides a means to elevate vehicle-mounted TOWs, and the open interiors permit firing of ATGMs. Garages are normally high enough to provide a 360-degree field of fire for antiaircraft weapons. For example, a Stinger could hide under the top floor of the garage, come out to engage an aircraft, and then take cover again (Figure I-16).

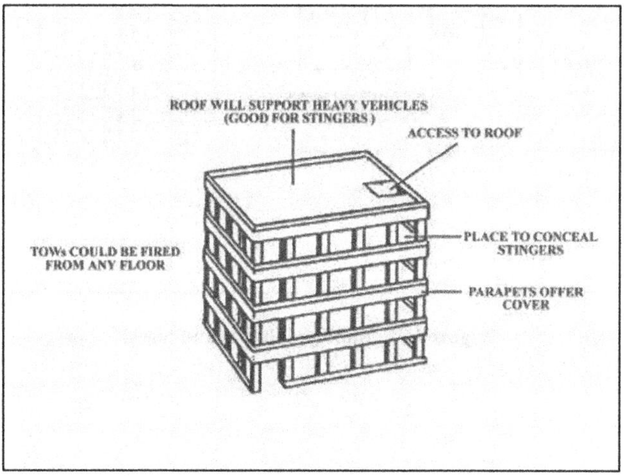

Figure I-16. Garage

3. Floor Plans. Floor plans in buildings follow predictable patterns. One of the factors that determines floor plans is building shape (Figure I-17). The basic principle governing building shape is that rooms normally have access to outside light. This principle helps Marines to analyze and determine the floor plans of large buildings.

4. Residential Areas. The two basic types of houses in the western world are located in and around cities and in rural areas. City-type houses are normally mass-construction brick buildings. Rural-type buildings in the continental United States, South America, and Southeast Asia are commonly made of wood, while in continental Europe, Southwest Asia, and sub-Saharan Africa, where wood is extremely scarce, rural buildings are normally constructed of concrete blocks. Rural-type buildings can also be found within cities (Figure I-18).

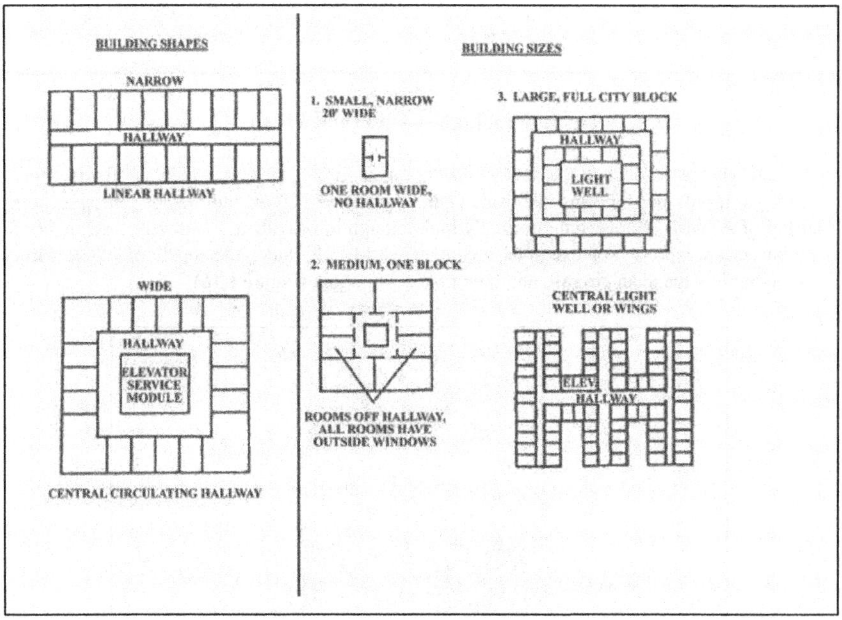

Figure I-17. Building Shapes and Sizes

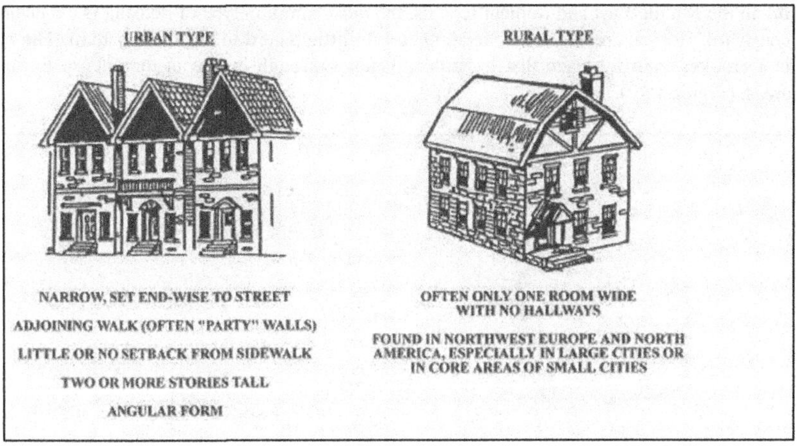

Figure I-18. Types of Housing

a. Another common type of building structure in cities with European influence is called the Hof-style apartment building (Figure I-19).

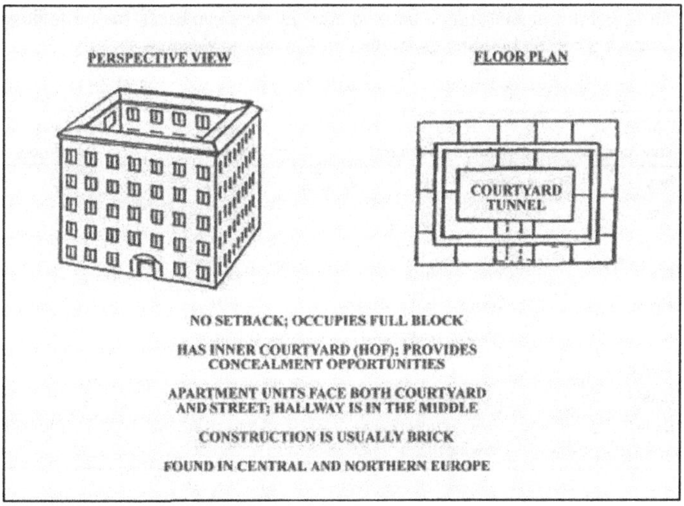

Figure I-19. Hof-Style Apartment Building

b. In the Middle East and tropical regions, the most common type of housing is the enclosed courtyard. Houses are added to one another with little regard to the street pattern. The result is a crooked, narrow maze that is harder to move through or fire in than dense European areas. (Figure I-20).

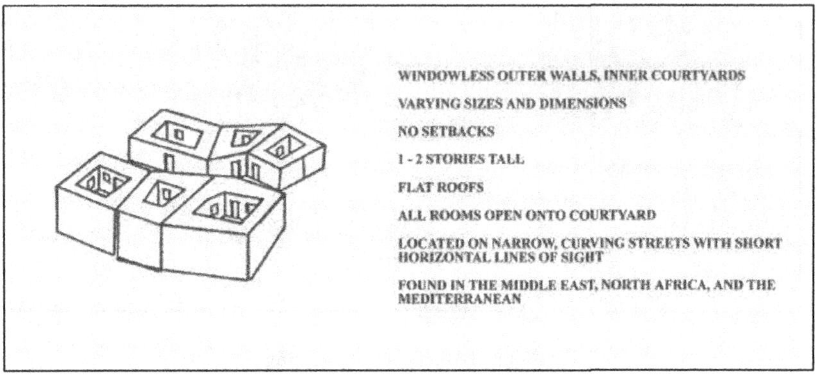

Figure I-20. Enclosed Courtyard

5. Characteristics of Buildings. Certain characteristics of both mass-construction and framed buildings can be helpful in analyzing a built-up area. Leaders can use Table I-1 to determine how to defend or attack a certain building given the unit's available weapons systems.

Type of Construction	Building Material	Height (Stories)	Average Wall Thickness (cm)
Mass	Stone	1 to 10	75
Mass	Brick	1 to 3	22
Mass	Brick	3 to 6	38
Mass	Concrete block	1 to 5	20
Mass	Concrete wall and slab	1 to 10	22 to 38
Mass	Concrete tilt-ups	1 to 3	18
Framed	Wood	1 to 5	3
Framed	Steel (heavy cladding)	3 to 50	30
Framed	Concrete/steel (light cladding)	3 to 100	2 to 8

Table I-1. Characteristics of Buildings

6. Distribution of Building Types. Certain types of buildings dominate certain parts of a city; this establishes patterns within a city. Analysis of the distribution and nature of these patterns has a direct bearing on military planning and weapon selection (Figure I-21).

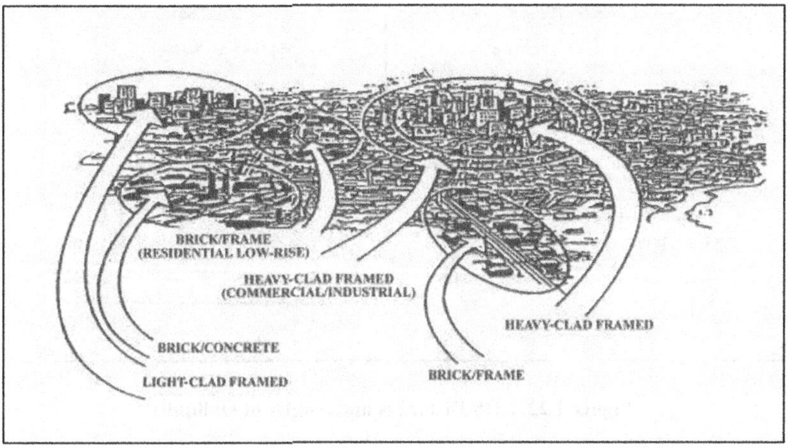

Figure I-21. Distribution of Building Types

a. Mass-construction buildings are the most common structures in built-up areas, forming about two-thirds of all building types. Brick structures account for nearly 60 percent of all buildings, especially in Europe.

b. Steel- and concrete-framed multistory buildings have an importance far beyond their one-third contribution to total ground floor area. They occupy core areas—a city's most valuable land—and as centers of economic and political power, they have a high potential military significance.

c. Open space accounts for about 15 percent of an average city's area. Many open spaces are grass-covered and are used for parks, athletic fields, and golf courses; some are broad, paved areas. The largest open spaces are associated with suburban housing developments where large tracts of land are recreation areas.

d. Streets serving areas consisting of mostly one type of building normally have a common pattern. In downtown areas, for example, high land values result in narrow streets. Street widths are grouped into three major classes: 7 to 15 meters, located in medieval sections of European cities; 15 to 25 meters, located in newer, planned sections of most cities; and 25 to 50 meters, located along broad boulevards or set far apart on large parcels of land. When a street is narrow, observing or firing into windows of a building across the street can be difficult because an observer is forced to look along the building rather than into windows.

When the street is wider, the observer has a better chance to look and fire into the window openings (Figure I-22).

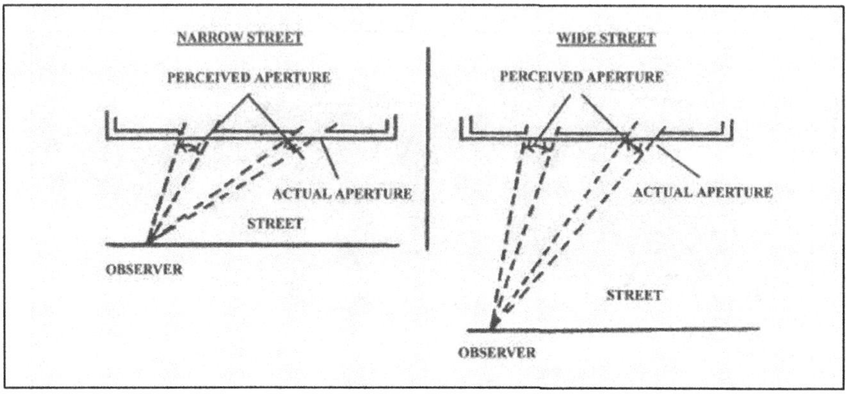

Figure I-22. LOS Distances and Angles of Obliquity

Appendix J

Lessons Learned from Russian Military Operations in Chechnya 1994-1996

1. Introduction. Many historical studies of urban combat have become dated. This is primarily the result of the impact that technology has had on military capabilities and evolving tactics. However, certain tenets of urban combat remain constant. The battles for the city of Grozny during the Russian intervention in the Republic of Chechnya represent a recent and critically important example of large scale operations in urban combat. Combat in Grozny was characterized by a large, technologically sophisticated military force (Russian) engaging and ultimately being defeated by a small, relatively primitive irregular force (Chechen). Grozny provides a number of fresh insights, and reinforcement of time honored tenets of urban warfare, across the scope of activities germane to modern urban combat.

2. Operations in Chechnya. Russia's war in Chechnya essentially began with the collapse of the Soviet Union, immediately after which pro-independence factions in Chechnya began agitating to free the Republic from Russian domination. This eventually led to civil war with pro-Russian factions within the Republic. Russia committed troops to Chechnya on 11 December 1994, for what many in the Russian government and military thought would be a quick campaign to restore Russian primacy. What followed was a bitter, costly, and protracted struggle that ended with the complete withdrawal of Russian military forces from Chechnya.

A primary focus of Russian operations in Chechnya was the capture of the capital city, Grozny. The initial attack on Grozny was conducted by a mechanized force, consisting of 6,000 troops mounted in tanks, BMP infantry fighting vehicles, and BTR armored personnel carriers. The Russians anticipated light opposition, but instead met a determined and heavy resistance from Chechens armed with a large number of antitank weapons. The attack was repulsed with significant Russian losses, both in personnel and armored vehicles (105 of 120 vehicles were lost). A second mechanized attack conducted on New Years Eve, 1994, was also repulsed with the loss of 140 of 200 tanks employed. The Russians captured Grozny some two months later, primarily through the use of excessive, overwhelming firepower, and at great cost to themselves and the local Chechen population.

The capture of Grozny marked the end of the first phase of Russian combat operations in Chechnya. Beginning in March 1995, the Russians became heavily engaged in anti-partisan operations as they tried to gain control of the country. Though the Russians controlled several major Chechen cities and a portion of the countryside by May 1995, they were never able to fully isolate the Republic. Chechen separatist fighters continued to receive a steady flow of arms and supplies from neighboring countries. Consequently, the Chechen fighters were able to maintain the initiative throughout this period, engaging Russian forces whenever and wherever they chose.

Even late in the campaign, Russian intelligence failed to detect Chechen fighters infiltrating Grozny, which enabled the Chechens to launch a major attack against the Russian-controlled city in early August 1996. The Chechens succeeded in capturing the city less than two weeks later.

.3

The recapture of Grozny was a significant loss for the Russians, precipitating a general cease-fire and the withdrawal of Russian forces from the Republic.

Besides the military aspects of combat in Chechnya, other issues such as ethnic, cultural, and religious divides impacted upon the nature of the conflict. The Russians failed to culturally orient their troops, which resulted in many serious cultural mistakes when dealing with the Chechen civilians. Once insulted or mistreated, the Chechen civilians became active participants in the struggle against Russia. The war eventually degenerated into one of abject ferocity and brutality on both sides. The Chechens are reported to have executed Russian prisoners, decapitated bodies, and boobytrapped the dead. Russian wounded and bodies were hung upside down outside of Chechen defensive positions in buildings, causing the Russians to have to shoot through their own people to engage the Chechens. The Russians, initially hesitant to cause civilian casualties, increasingly engaged Chechen units with a growing disregard for the safety of the local populace. As Russian discipline degraded, the incidences of Russian troops engaging in looting, arson, torture, even the summary execution of civilians, rose significantly. The result was the deaths of tens of thousands of civilians.

3. Lessons Learned. The intervention in Chechnya and the battles for Grozny generated many important lessons across the warfighting functions of combat in urbanized terrain. The following information is primarily from the Russian point of view, though some Chechen capabilities are included. The lessons learned are arrayed as warfighting functional considerations.

a. Command and Control.

(1) Russian military commanders lacked clear policy guidance. The Russian military's ability to attain and maintain steady, logical operational progress was hampered, if not negated, by a lack of clear governmental policy. The Russian government was split at the highest levels over the wisdom of committing forces to Chechnya, and the conduct of the intervention.

(2) Poor Russian command and control structure throughout all levels. There was no direct, unified chain of command for the operations in Chechnya. Operational directives often flowed from a number of governmental agencies directly to local unit commanders, often bypassing intermediate military commanders. This resulted in a lack of overall situational awareness, and exceptionally poor coordination among tactical forces.

(3) Russians were unable to contain/prohibit Chechen operations against Russia. When the Russian forces finally began to achieve progress in late May 1995, the Chechens initiated attacks against targets inside Russia. The result was a major propaganda victory for the Chechens.

(4) Excessive senior officer (flag level) interference contributed to operational confusion. There were eight major changes at the senior command level during the course of the intervention. Additionally, there were approximately one hundred general officers on the operational scene, about one general officer per 3,000 troops, which resulted in confusion as they got in each others way.

(5) Poor small unit leadership, particularly at the NCO level, was a primary cause of Russian tactical failures in Grozny. While the Russians recognize that urban combat is small unit intensive, they do not possess a professional NCO corps. Additionally, Russian forces lacked some 12,000 platoon leaders when the intervention began. This had a crippling effect on small unit operations.

(6) Russian inability to isolate Chechnya/Grozny allowed a constant flow of reinforcements and supplies into the country/city. Russian control of any given area was at best, porous. The Chechens continued to receive outside support throughout the campaign, as well as, a large supply of arms and equipment from poorly disciplined Russian troops inside Chechnya.

(7) Both sides used commercial off-the-shelf communication assets. The information war was as vitally important as any other aspect of the campaign. The Chechens had to make do with commercially available communication equipment. They fully exploited the use of cellular telephones, Motorola radios, improvised television stations, video cameras, and the internet. The Russians lacked sufficient military communication equipment, and had to supplement with commercial assets.

(8) Russian tactical communication was very difficult in Grozny. Urban structures in Grozny interfered with Russian military radios, severely hindering tactical communications. The Russians had to establish ground-based and aircraft-based relay stations to overcome the interference.

(9) Boundaries between units were tactical weak points. Units had to think of boundaries in both a horizontal and vertical context. There were many instances when the Chechens held the third floor and above, while the Russians held the first two floors and maybe the roof. This caused a lack of coordination, and mutual support. Additionally, due to poor command and control and coordination between units, boundaries became seams through which the Chechens could pass unopposed.

b. Intelligence.

(1) Lack of high quality intelligence made operations much more difficult and dangerous. Russian intelligence collection and dissemination capabilities were poor throughout the intervention. In March 1996, the Russian Minister of the Interior complained that poor reconnaissance and intelligence allowed Chechen fighters to move in and out of Grozny without opposition, which contributed to the heavy Russian losses.

(2) Directives to minimize civilian casualties ultimately put the Russians at a disadvantage. An advantage accrues to the side with less concern for the safety of the civilian population. Initially, the Russians attempted to limit civilian casualties. The Chechen

.3

fighters took advantage of this reticence, often using civilians, schools, and apartment buildings as a shield. While this may seem to be cold-blooded of the Chechen fighters, what must be understood is that the majority of the populace in Grozny were actually ethnic Russians, and that the ethnic Chechens had no remorse in using them in this manner.

(3) Russian concerns for the safety of the Chechen populace and property declined. As the Russian troops became increasingly frustrated in their attempts to separate Chechen fighters from the local populace, and their casualties began to mount, concerns for civilian safety waned, resulting in greater civilian casualties and further alienating the Chechen people.

(4) Rigorous communications security is essential, even against less sophisticated enemies.Much of the Russian tactical radio traffic was broadcast in the clear, which allowed the Chechens to intercept it. The Chechens were able to redirect Russian airstrikes onto Russian targets, and engage Russian artillery forward observers who broadcast their own position coordinates.

(5) The Russians made extensive use of RPVs for reconnaissance. The Chechen campaign saw the first Russian use of unmanned aerial vehicles for reconnaissance and intelligence gathering. There were primarily used by Russian airborne forces.

c. Maneuver.

(1) Inadequate training in the most basic maneuver and combat skills inhibited Russian operations. Overall, poor Russian combat performance could be traced to a lack of training in fundamental military skills, a situation that was then exacerbated by an additional lack of training in specific urban combat skills. Individual skills training was so inadequate that some units were entirely incapable of undertaking combat operations. The Russian's ability to conduct large scale operations had been seriously eroded by a complete cessation of exercises at the division level in the 2 years prior to the Chechen campaign. Even regimental and battalion level exercises had been curtailed by 75 percent or more. Additionally, The air forces were not any better trained or prepared for their operational role in Chechnya.

(2) Lack of cohesiveness among the Russian forces led to confusion and antagonism. The Russian ground forces consisted primarily of regular army units and Ministry of Interior Internal Troops, which were essentially a police force. With no real unified command and control, the two groups were usually conducting uncoordinated missions, which led to confusion and fratricide.

(3) The nature of cities channels combat operations along narrow lanes of activity. Combat conditions in Grozny were characterized by narrow fields of view, limited fields of fire, and constricted avenues of approach. For the Russians, this meant movement along approaches well known and heavily defended by the Chechens.

(4) Night fighting was the most difficult activity for Russian infantry. The Russian infantry, at all levels, was inadequately trained for night operations, and lacked night vision equipment.

(5) Armored vehicles cannot operate in cities without extensive dismounted infantry support. The Chechens formed antitank hunter-killer teams equipped with RPG-7 and RPG-18 antitank missiles. They engaged Russian tanks with volley fire from above, behind, and the sides. Armor columns not supported by dismounted infantry suffered great losses.

(6) Armored engineer vehicles are indispensable for removing obstacles. The Russians found that armored engineer vehicles were critical to restoring or improving mobility in Grozny. Eventually, the Russians formed special assault groups that included armored engineer vehicles in each group.

(7) The Chechens preferred small unit, hit-and-run ambush tactics. The Chechens relied on small units (15 to 20 troops) to conduct quick, sharp ambushes, from which they would disengage before the Russians could bring significant firepower to bear. Ambushes were often conducted in three tiers, with Chechens in the underground, on the ground floor, and on the roof, with each element having a different task in the ambush. They also used ambush tactics against Russian helicopters.

(8) Urban combat is extremely manpower intensive, and results in significant attrition of personnel and materiel for the attacker. The Russians discovered that a 5 to 1 manpower ratio was often not enough, due to the high casualty rates consistent with urban combat and the requirement to guard virtually every building taken.

(10) Tracked vehicles are preferable to wheeled vehicles in the urban environment. Tracked vehicles provided better mobility than wheeled vehicles due to their significantly greater ability to negotiate obstacles and rubble.

(11) Armored vehicles required additional protection (i.e., reactive armor, sandbags, wire mesh, etc.) to negate attacks from the sides, rear, and top. Russian tank design emphasized greater armor protection on the frontal arc, and less on other areas of the vehicle. This proved inadequate for urban combat, where attacks can come from any direction. The majority of lethal hits came from above, easily penetrating turrets and engine decks, and from the rear.

(12) The spatial qualities and perspectives of urban combat are more "vertical" than non-urban combat. The Chechens regularly fought from basements, rooftops, and upper-floor rooms, making them difficult to locate and engage. Russians soldiers trained in open areas, had to reorient to a confined, highly vertical environment.

.3

(13) Standard Russian military unit configurations were inappropriate for urban combat. Typical Russian army unit organizations, oriented to open spaces and mechanized operations, proved to be inadequate for urban combat. They eventually had to adopt specially configured assault teams that included, among other assets, significantly greater firepower and engineer assets. Essentially, they moved towards a better combined arms team.

(14) The Russians found their helicopters to be far too vulnerable to rooftop snipers and ambushes in the urban setting. While Russian doctrine had long specified the capture of a building from the top-down, it became apparent in Grozny that this tactic was too expensive in terms of helicopter losses. The Chechens made effective use of snipers and RPG-equipped ambush teams to damage or destroy helicopters supporting Russian ground forces.

d. Fires.

(1) Artillery was invaluable in the direct fire mode. The Russians found artillery, to include rocket launchers and mortars, to be very effective for direct fire against point targets , usually at ranges of 150 to 200 meters.

(2) Obscurants (smoke, white phosphorous, tear gas) are especially useful in urban combat. The Russians used significant amounts of smoke and white phosphorus to mask the movement of forces. Every fourth or fifth artillery or mortar round was either smoke or WP. WP also had a toxic effect, easily penetrating Chechen protective masks.

(3) The Chechens effectively used RPGs against helicopters. The Chechens made effect use of ambush teams equipped with RPGs to engage Russian helicopters, either by damaging or destroying the helicopter or simply prohibiting helicopterborne operations.

(4) Air defense guns (i.e., ZSU-23-4) were valuable in suppressing ground targets. The Russians found air defense guns effective against multi-story buildings because they had sufficient elevation to hit targets in the upper stories. Air defense guns became a primary asset in the Russian special assault groups.

(5) The Russians determined that overwhelming firepower can make-up for organizational and tactical deficiencies in the short run if collateral damage is disregarded. Whenever Russian military units proved incapable of effective urban operations, the Russians simply fell back on massive, overwhelming firepower to reduce a strong point. It was through the use of excessive, unrestrained firepower, and a complete lack of regard for collateral damage, that finally enabled the Russians to gain control of Grozny.

(6) The Chechens negated Russian supporting fires by "hugging" Russian units. Whenever possible, Chechen fighters adopted the tactic of staying as physically close to Russian units as possible, so that the Russians could not employ supporting arms without the risk of injuring their own troops.

(7) The Russians found "bunker busting" weapons to be invaluable. The Russians came to rely heavily on "bunker busting" weapons for engaging Chechens, primarily snipers, entrenched in buildings.

(8) Russian helicopters needed stand-off weapons, that would allow them to engage targets while outside the range of Chechen weapons. The Chechens employed 23mm cannons and heavy machine guns against Russian helicopters, causing the Russian's helicopter gunships to fire from ranges of 3,000 meters or more. This required the use of wire-guided munitions and large caliber (240mm) rockets.

(9) The Russians made extensive use of precision-guided weapons. When the weather allowed, the Russians were able to use precision guided weapons, such as laser-guided bombs and missiles. They had great effect against priority targets such as bridges, major road intersections, and buildings.

e. Logistics.

(1) Deficiencies in the Russian logistics system degraded combat effectiveness. Significant inadequacies in the Russian logistics apparatus effected virtually all combat forces and operations. Some soldiers actually entered combat in Grozny without weapons or ammunition. The Russian transportation system failed. Supply officers were unprepared for the high volume demands for ammunition, hand grenades, smoke grenades, demolition charges, and one-shot antitank weapons.

(2) Russian logistics personnel were ill-prepared in basic soldier skills. Russian infantry often had to be used to conduct logistics operations into Grozny, because the logistics personnel lack the basic skills to defend themselves against the Chechens.

(3) The Russian ground forces required special equipment not normally found in unit T/Es. The T/E for the typical Russian infantry unit did not include such items as ropes, grappling hooks, and ladders, all of which proved essential for urban combat.

(4) Armored vehicle recovery is extremely difficult and dangerous. The Russians found the recovery of vehicles to be extremely difficult due to the terrain (i.e., narrow streets, rubble and debris), and Chechen opposition.

(5) Some Russian equipment was modified in the field to counter Chechen tactics and equipment. The Russians had to adopt add-on armor for their vehicles. They fabricated metal screens and wire mesh guards, which helped to defeat RPG and Molotov cocktail attacks.

(6) Medical support was of great importance. Russian casualty evacuation was difficult due to the lack of vehicular mobility, and the vulnerability of helicopters to Chechen fire. Many casualties were psychological, due in great part to the intense, violent nature of close quarter combat. The Russians were ill-prepared for the level of mental health problems they encountered. Additionally, logistics units were often unable to provide fresh drinking water, which caused Russian troops to consume contaminated water that resulted in other health problems.

f. Force Protection.

(1) Operations took much longer than expected, and the cost in terms of casualties was much greater than anticipated. The Russians committed unprepared, poorly trained and poorly equipped forces to Chechnya, based on an unrealistic operational time frame. Ultimately, this cost the Russians significant casualties and the war.

(2) Fratricide was a serious and continuing problem throughout the campaign. The situation facing Russian combat forces was ripe for fratricide. Poorly trained units, operating in a confused and uncertain urban environment, often unable to tell friend from foe, and lacking quality leadership and inter-unit coordination, were often as dangerous to themselves as they were to the Chechens .

(3) Chechen use of adhoc air defense systems eliminated Russian air support, particularly helicopters, in Grozny. The Chechens were able to improvise effective anti-air defenses using a mixture of anti-air and ground (i.e., heavy machine guns, RPGs) weapons, and ambush tactics specifically adapted to the urban environment.

(4) The Chechens made extensive and effective use of snipers. Snipers fired from well inside rooms versus near window openings, as well as, from rooftops and basements. The Russians lacked an effective sniper and counter-sniper capability of their own.

(5) The Chechens improvised crude chemical weapons. The Chechens, lacking access to military chemical weapons, Used off-the-shelf materials and equipment to make chemical mines (i.e., chlorine gas), detonated remotely by radio.

(6) The Chechens made extensive and effective use of boobytraps. It seemed that the Chechens mined and boobytrapped 7-8everything. They had a good understanding of the average Russian soldier's actions and reactions, and boobytrapped accordingly. Additionally, the Russians found it difficult to maintain boobytrap awareness among their troops.

Appendix K

Glossary

I. Acronyms

AAA .... antiaircraft artillery
AAAV .... advanced assault amphibious vehicle
AAV .... assault amphibious vehicle
ACE .... aviation combat element
AO .... area of operations
APC .... armored personnel carrier
APDS-T .... armor-piercing, discarding sabot with tracer
APFSDS .... armor-piercing, fin-stabilized, discarding sabot
API .... armor-piercing incendiary
ARVN .... Army of the Republic of Vietnam
ATF .... amphibious task force
ATGM .... antitank guided missile
BDA .... battle damage assessment
BMNT .... beginning of morning nautical twilight
CA .... civil affairs
CAG .... civil affairs group
CE .... command element
CI .... counterintelligence
COC .... combat operations center
CONAD .... consolidated administration center
CP .... command post
CSS .... combat service support
CSSE .... combat service support element
DOD .... Department of Defense
DS .... direct support
EOD .... explosive ordnance disposal
EPW .... enemy prisoner of war
EW .... electronic warfare
FAC(A) .... forward air controller (airborne)
FASCAM .... family of scatterable mines
FDC .... fire direction center
FEBA .... forward edge of the battle area
FLOT .... forward line of own troops
FM .... U.S. Army field manual
FM .... Frequency modulated
FMFM .... Fleet Marine Force manual
FMFRP .... Fleet Marine Force reference publication
FO .... forward observer
FPL .... final protective line
FRAGO .... fragmentary order
FSC .... fire support coordinator
GCE .... ground combat element
GPS .... global positioning system
GS .... general support
HE .... high explosive
HEAT .... high-explosive antitank
HEAT-MP .... HEAT-multipurpose
HEDP .... high-explosive dual-purpose
HEI-T .... high-explosive incendiary with tracer
HLZ .... helicopter landing zone
HMMWV .... high-mobility, multipurpose wheeled vehicle
HSS .... health services support
HUMINT .... human intelligence
IDF .... Israeli Defense Forces
IFV .... infantry fighting vehicle
IPB .. intelligence preparation of the battlefield/battlespace
ITOW .... improved TOW
JTF .... joint task force
KOCOA .... key terrain, observation and fields of fire, cover and concealment, obstacles, and avenues of approach
LAV .... light armored vehicle
LCAC .... landing craft air cushion
LD .... line of departure
LOS .... line of sight
LZ .... landing zone
MAGTF .... Marine air-ground task force
MCDP .... Marine Corps doctrinal publication
MCRP .... Marine Corps reference publication
MCWP .... Marine Corps warfighting publication
MEDEVAC .... medical evacuation
MEF .... Marine expeditionary force
MET .... meteorological
METT-T ... mission, enemy, terrain and weather, troops and support available-time available
MEU .... Marine expeditionary unit
MOOTW .... military operations other than war
MOPP .... mission-oriented protective posture
MOUT .... military operations on urbanized terrain
MP .... military police
MPF .... maritime prepositioning force
MULE .... modular universal laser equipment
NBC .... nuclear, biological, and chemical
NCA .... National Command Authorities
NCF .... naval construction force
NEF .... naval expeditionary force
NEO .... noncombatant evacuation operation
NGF .... naval gunfire
NSFS .... naval surface fire support
NVA .... North Vietnamese Army
NWP .... naval warfare publication

Appendix J

Glossary

I. Acronyms

OP .... observation post
OPLAN .... operation plan

OPORD
operation order
OPSEC
operations security
PDDA
power-driven decontamination apparatus
PGM
precision-guided munition
PLO
Palestine Liberation Organization
PLRS
position location reporting system
POL
petroleum, oils, and lubricants
PSYOP
psychological operations
ROE
rules of engagement
ROK
Republic of Korea
RP
red phosphorus
RSOP . . . . reconnaissance, selection, and occupation

of

position
SASR
special application sniper rifle
SMAW
shoulder-launched multipurpose assault

weapon
SOP
standing operating procedure
STANAG
NATO standardization agreement
STB
supertropical bleach
TOW
tube-launched, optically tracked, wire-guided

missile
UAV
unmanned aerial vehicle
UCMJ
Uniform Code of Military Justice
U.N.
United Nations
VC
Viet Cong
VHF
very high frequency

WP . . . . white phosphorus

II. Definitions

A

assault—1. The climax of an attack, closing with the enemy in hand-to-hand fighting. 2. To make a short, violent, but well-ordered attack against a local objective, such as a gun emplacement, a fort, or a machine gun nest. (Joint Pub 1-02)

assault—The culmination of an attack which closes with the enemy. (MCRP 5-2A)

attack—An offensive action characterized by movement supported by fire with the objective of defeating or destroying the enemy. (FMFRP 0-14)

B

battlespace—All aspects of air, surface, subsurface, land, space, and electromagnetic spectrum which encompass the area of influence and area of interest. (FMFRP 0-14)

block—1. A tactical task assigned to a unit that requires it to deny the enemy access to a given area or to prevent enemy advance in a given direction or an avenue of approach. It may be for a specified time. Units assigned this mission may have to retain terrain and accept decisive engagement. 2. An obstacle effect that integrates fire planning and obstacle effort to stop an attacker on a specific avenue of approach or to prevent an enemy from exiting an engagement area. (MCRP 5-2A)

breach—A tactical task where any means available are employed to break through or secure a passage through an enemy defense, obstacle, minefield, or fortification. (MCRP 5-2A)

built-up area—A concentration of structures, facilities, and populations, such as villages, cities, and towns, that forms the economic and cultural focus for the surrounding area. (MCRP 5-2A, modified)

bypass—A tactical task that involves maneuvering around an obstacle, position, or enemy force to maintain the momentum of advance. Bypassed obstacles and enemy forces are reported to higher headquarters. (MCRP 5-2A)

C

clear—A tactical task to remove all enemy forces and eliminate organized resistance in an assigned zone, area, or location by destroying, capturing, or forcing the withdrawal of enemy forces such that they cannot interfere with the friendly unit's ability to accomplish its mission. (MCRP 5-2A)

combat arms—Units who close with and destroy enemy forces or provide firepower and destructive capabilities on the battlefield. (MCRP 5-2A, modified)

combat operations center (COC)—The primary operational agency required to control the tactical operations of a command that employs ground and aviation combat, combat support, and combat service support elements or portions thereof. The combat operations center continually monitors, records, and supervises operations in the name of the commander and includes the necessary personnel and communications to do the same. (FMFRP 0-14)

combat service support (CSS)—The essential capabilities, functions, activities, and tasks necessary to sustain all elements of operating forces in theater at all levels of war. Within the national and theater logistic systems, it includes but is not limited to that support rendered by service forces in ensuring the aspects of supply, maintenance, transportation, health services, and other services required by aviation and ground combat troops to permit those units to accomplish their missions in combat. Combat service support encompasses those activities at all levels of war that produce sustainment to all operating forces on the battlefield. (Joint Pub 1-02)

combat support (CS)—Fire support and operational assistance provided to combat elements. (Joint Pub 1-02)

combat support (CS)—Units that provide critical combat functions in conjunction with combat arms units to secure victory (MCRP 5-2A, modified).

command post (CP)—A unit's or subunit's headquarters where the commander and the staff perform their activities. In combat, a unit's or subunit's headquarters is often divided into echelons; the echelon in which the unit or subunit commander is located or from which such commander operates is called a command post. (Joint Pub 1-02)

contain—A tactical task to restrict enemy movement. (MCRP 5-2A)

cover—1. Shelter or protection from enemy observation that reduces the effects of enemy direct and indirect fire. 2. A type of security operation that protects the force from surprise, develops the situation, and gives commanders time and space in which to respond to the enemy's actions. (MCRP 5-2A)

D

defend—A combat operation designed to defeat an attacker and prevent him from achieving his objectives. It employs all means and methods available to prevent, resist, or destroy an enemy attack. Forms of defensive operations are area and mobile. Choices of defensive operations are in-depth and forward. The defensive techniques are defend in sector, defend a battle position, and defend a strong point. (MCRP 5-2A)

destroy—1. A tactical task to physically render an enemy force combat-ineffective unless it is reconstituted. 2. To render a target so damaged that it cannot function as intended nor be restored to a usable condition without being entirely rebuilt. (MCRP 5-2A)

disrupt—A tactical task or obstacle effect that integrates fire planning and obstacle effort to break apart an enemy's formation and tempo, interrupt the enemy's timetable, or cause premature commitment of enemy forces, or the piecemealing of his attack. (MCRP 5-2A)

II. Definitions

F

fire support—Assistance to elements of the Marine air-ground task force engaged with the enemy rendered by other firing units, including (but not limited to) artillery, mortars, naval surface fire support, and offensive air support. (FMFRP 0-14)

fix—1. A tactical task in which actions are taken to prevent the enemy from moving any part of his forces either from a specific location or for a specific period of time by holding or surrounding them to prevent their withdrawal for use elsewhere. 2. A tactical obstacle effect that integrates fire planning and obstacle effort to slow an attacker within a specified area—normally an engagement area. (MCRP 5-2A)

H

helicopterborne operation—A military action in which combat forces and their equipment maneuver about the battlefield by helicopters or vertical-landed aircraft. Aviation activities are under the control of the aviation combat element commander who is assigned in direct or general support of one or more combat element(s). (FMFRP 0-14)

hold—1. To maintain or retain possession of by force, as a position or an area. 2. In an attack, to exert sufficient pressure to prevent movement or redisposition of enemy forces. (Joint Pub 1-02)

I

infiltration—The movement through or into an area or territory occupied by either friendly or enemy troops or organizations. The movement is made, either by small groups or by individuals, at extended or irregular intervals. When used in connection with the enemy, it infers that contact is avoided. (Joint Pub 1-02)

isolate—A tactical task given to a unit to seal off (both physically and psychologically) an enemy from his sources of support, to deny an enemy freedom of movement, and prevent an enemy unit from having contact with other enemy forces. An enemy must not be allowed sanctuary within his present position. (MCRP 5-2A)

L

linkup—An operation wherein two friendly ground forces join together in a hostile area. (FMFRP 0-14)

M

Marine air-ground task force (MAGTF)—A task organization of Marine forces (division, aircraft wing, and service support groups) under a single command and structured to accomplish a specific mission. (Joint Pub 1-02)

military operations on urbanized terrain (MOUT)—All military actions that are planned and conducted on a topographical complex and its adjacent natural terrain where manmade construction is the dominant feature. It includes combat in cities, which is that portion of MOUT involving house-to-house and street-by-street fighting in towns and cities (MCRP 5-2A, modified)

military operations other than war (MOOTW)—Operations that encompass the use of military capabilities across the range of military operations short of war. These military actions can be applied to complement any combination of the other instruments of national power and occur before, during, and after war. (Joint Pub 1-02)

N

neutralize—1. To render enemy personnel or material incapable of interfering with a particular operation. 2. To render safe mines, bombs, missiles, and boobytraps. (MCRP 5-2A)

O

occupy—A tactical task in which a force moves onto an objective, key terrain, or other man-made or natural terrain area without opposition, and controls that entire area. (MCRP 5-2A)

P

peace enforcement—Application of military force, or the threat of its use, normally pursuant to international authorization, to compel compliance with resolutions or sanctions designed to maintain or restore peace and order. (Joint Pub 1-02)

peacekeeping—Military operations undertaken with the consent of all major parties to a dispute, designed to monitor and facilitate implementation of an agreement (ceasefire, truce, or other such agreement) and support diplomatic efforts to reach a long-term political settlement. (Joint Pub 1-02)

protect—A tactical task to prevent observation of or engagement or interference with, a force or location. (MCRP 5-2A)

pursuit—An offensive operation designed to catch or cut off a hostile force attempting to escape, with the aim of destroying it. (Joint Pub 1-02)

R

reduce—1. A tactical task to gain control over an enemy position or objective. 2. A task to create lanes through or over an obstacle sufficient to allow the attacking force to accomplish its mission. (MCRP 5-2A)

retain—A tactical task to occupy and hold a terrain feature to ensure it is free of enemy occupation or use. (MCRP 5-2A)

rules of engagement (ROE)—Directives issued by competent military authority which delineate the circumstances and limitations under which United States forces will initiate and/or continue combat engagement with other forces encountered. (Joint Pub 1-02)

S

screen—A task to maintain surveillance, provide early warning to the main body, or

II. Definitions

impede, destroy, and harass enemy reconnaissance within its capability without becoming decisively engaged. (MCRP 5-2A)

search—A systematic reconnaissance of a defined area, so that all parts of the area have passed within visibility. (Joint Pub 1-02)

secure—A tactical task to gain possession of a position or terrain feature, with or without force, and to deploy in a manner which prevents its destruction or loss to enemy action. The attacking force may or may not have to physically occupy the area. (MCRP 5-2A)

seize—To clear a designated area and obtain control of it. (FMFRP 0-14)

suppression—A tactical task to employ direct or indirect fires, electronic attack, or smoke on enemy personnel, weapons, or equipment to prevent or degrade enemy fires and observation of the friendly forces. (FMFRP 0-14)

T

terrorism—The calculated use of violence or threat of violence to inculcate fear; intended to coerce or to intimidate governments or societies in the pursuit of goals that are generally political, religious, or ideological. (Joint Pub 1-02)

Appendix L

References and Related Publications

Field Manuals (FMs)

3-11	Flame, Riot Control Agent, and Herbicide Operations
5-34	Engineer Field Data
5-102	Countermobility
5-250	Explosives and Demolitions
7-98	Operations in a Low-Intensity Conflict
9-15	Explosive Ordnance Disposal Service and Unit Operations
20-32	Mine/Countermine Operations
21-75	Combat Skills of the Soldier
3-30	Grenades and Pyrotechnic Signals
23-34	TOW Weapon System
27-10	Law of Land Warfare
33-1	Psychological Operations
34-130	Intelligence Preparation of the Battlefield
41-5	Joint Manual for Civil Affairs
41-10	Civil Affairs Operations
90-8	Counterguerrilla Operations
100-20	Military Operations In Low Intensity Conflict

Technical Manual (TM)

9-2350-264-10-2	Operator's Manual for Operation Under Usual and Unusual Conditions for Tank, Combat, Full-Tracked: 120mm Gun, M1A1 General Abrams, with Change 4

Marine Corps Concept Paper

A Concept for Future Military Operations on Urbanized Terrain

Marine Corps Doctrinal Publications (MCDPs)

1	Warfighting
1-3	Tactics

Marine Corps Reference Publications (MCRPs)

5-2A	Operational Terms and Graphics

Marine Corps Warfighting Publications (MCWPs)

3-15.3	Scout Sniping (under development)
3-37.1	Chemical Operations
3-37.4	NBC Reconnaissance

Fleet Marine Force Manuals (FMFMs)

0-9	Field Firing for the M16A2
0-25	The Law of Land Warfare
1-3	Tactics
2-6	MAGTF Rear Area Security
2-11	Antiarmor Operations
3-5	Employment of Military Police in Combat
3-30	Communications
6	Ground Combat Operations
6-7	Scouting and Patrolling for Infantry Units
6-8	Supporting Arms Observer, Spotter, and Controller
6-9	Marine Artillery Support
6-21	Tactical Fundamentals of Helicopterborne Operations
7-14	Combating Terrorism
7-34	MAGTF Civil Affairs
7-34-1	Civil Affairs Guidance for the Small-Unit Leader
7-40	Helicopter Insertion/Extraction
8-2	Counterinsurgency Operations
11-9	NBC Protection
11-10	NBC Decontamination
11-11	Treatment of Chemical Agent Casualties and Conventional Military Chemical Injuries
11-12	Chemical Hazard Assessment Guide
11-13	NBC Handbook
11-17	Chemical and Biological Contamination Avoidance
11-18	Nuclear Contamination Avoidance
11-20	NBC Reconnaissance
13-7	MAGTF Breaching Operations

Fleet Marine Force Reference Publications (FMFRPs)

0-6	Marine Troop Leader's Guide
0-14	Marine Corps Supplement to DOD Dictionary of Military and Associated Terms
4-26	Enemy Prisoners of War and Civilian Internees
7-14A	The Individual's Guide for Understanding and Surviving Terrorism
7-37	Vehicle Bomb Search
12-15	Small Wars Manual

12-18	Mao Tse-Tung on Guerrilla Warfare
12-25	The Guerrilla and How to Fight Him
13-5	Engineer Field Data

Joint Publications

1-02	DOD Dictionary of Military and Associated Terms
3-07	Joint Doctrine for Military Operations Other Than War
3-11	Joint Doctrine for Nuclear, Biological, and Chemical (NBC) Defense
3-15	Joint Doctrine for Barriers, Obstacles, and Mine Warfare
3-53	Doctrine for Joint Psychological Operations
4-06	JTTP for Mortuary Affairs in Joint Operations

Standard NATO Agreements (STANAGs)

2003	Patrol Reports by Army Forces
2014	Operations Orders
2029	Method of Describing Ground Locations, Areas, and Boundaries
2070	Emergency War Burial Procedures
2088	Battlefield Illumination
2101	Establishing Liaison
2868	Land Force Tactical Doctrine (ATP-35(B))

Miscellaneous

Fratricide: Reducing Self-Inflicted Losses, Newsletter No. 92-4, Center for Army Lessons Learned (CALL), U.S. Army Combined Arms Command, Fort Leavenworth, Kansas, April 1992.

HR, 25, Annex to Hague Convention No. IV, *Embodying Regulations Respecting the Laws and Customs of War on Land*, 18 October 1907.

Modern Experience in City Combat, U.S. Army Human Engineering Laboratory, Aberdeen Proving Ground, Maryland, 1987.

Made in United States
North Haven, CT
04 November 2026

21775163R00320